Plant Science: Concepts, Tools and Methods

Plant Science: Concepts, Tools and Methods

Edited by **Nancy Cahoy**

SYRAWOOD
PUBLISHING HOUSE

New York

Published by Syrawood Publishing House,
750 Third Avenue, 9th Floor,
New York, NY 10017, USA
www.syrawoodpublishinghouse.com

Plant Science: Concepts, Tools and Methods
Edited by Nancy Cahoy

International Standard Book Number: 978-1-68286-202-5 (Hardback)

Printed in the United States of America.

Contents

Preface

This book has been a concerted effort by a group of academicians, researchers and scientists, who have contributed their research works for the realization of the book. This book has materialized in the wake of emerging advancements and innovations in this field. Therefore, the need of the hour was to compile all the required researches and disseminate the knowledge to a broad spectrum of people comprising of students, researchers and specialists of the field.

Plant science has progressed significantly in the last few decades. A multitude of researches have been conducted across diverse branches of plant sciences like genetics, evolution, etc. This book is well equipped to familiarize the reader with various significant topics like plant cell biology, genomics, functional plant breeding, interaction of plants with their environments, etc. Replete with details to enhance the knowledge of readers about the traditional and modern tools available in the field, the students of biology and botany in particular will find this book a suitable guide for their respective fields of study.

At the end of the preface, I would like to thank the authors for their brilliant chapters and the publisher for guiding us all-through the making of the book till its final stage. Also, I would like to thank my family for providing the support and encouragement throughout my academic career and research projects.

Editor

Fluorescent labelling of the actin cytoskeleton in plants using a cameloid antibody

Alessandra Rocchetti, Chris Hawes* and Verena Kriechbaumer

Abstract

Background: Certain members of the *Camelidae* family produce a special type of antibody with only one heavy chain. The antigen binding domains are the smallest functional fragments of these heavy-chain only antibodies and as a consequence have been termed nanobodies. Discovery of these nanobodies has allowed the development of a number of therapeutic proteins and tools.

In this study a class of nanobodies fused to fluorescent proteins (chromobodies), and therefore allowing antigen-binding and visualisation by fluorescence, have been used. Such chromobodies can be expressed in living cells and used as genetically encoded immunocytochemical markers.

Results: Here a modified version of the commercially available Actin-Chromobody® as a novel tool for visualising actin dynamics in tobacco leaf cells was tested. The actin-chromobody binds to actin in a specific manner. Treatment with latrunculin B, a drug which disrupts the actin cytoskeleton through inhibition of polymerisation results in loss of fluorescence after less than 30 min but this can be rapidly restored by washing out latrunculin B and thereby allowing the actin filaments to repolymerise.

To test the effect of the actin-chromobody on actin dynamics and compare it to one of the conventional labelling probes, Lifeact, the effect of both probes on Golgi movement was studied as the motility of Golgi bodies is largely dependent on the actin cytoskeleton. With the actin-chromobody expressed in cells, Golgi body movement was slowed down but the manner of movement rather than speed was affected less than with Lifeact.

Conclusions: The actin-chromobody technique presented in this study provides a novel option for *in vivo* labelling of the actin cytoskeleton in comparison to conventionally used probes that are based on actin binding proteins.

The actin-chromobody is particularly beneficial to study actin dynamics in plant cells as it does label actin without impairing dynamic movement and polymerisation of the actin filaments.

Keywords: Actin, Nanobody, Chromobody, Golgi body, Actin dynamics

Background

Expression and applications of antibody constructs

In 1989, a novel type of antibody was identified first in the sera of dromedaries and later on in various members of the *Camelidae* family [1]. These antibodies differ from the typical antibody composition of two heavy and two light chains in that they are composed of just one heavy chain. Camelids produce both conventional and heavy-chain only antibodies (HcAbs) in ratios differing by species; 45% of llama serum antibodies are HcAbs and 75% in camels [1]. Isolation of the antigen binding domain (V_HH, variable heavy chain of a heavy-chain antibody), the smallest functional fragment of these heavy-chain only antibodies, called nanobodies, lead to the development of various therapeutic proteins and tools.

Antibodies have the potential to bind to and therefore detect any molecule and cell structure making them a powerful research tool. Nanobodies only have a molecular mass of around 13 kDa and a size of 2 nm × 4 nm [2,3]. This small size offers several advantages over conventional antibodies or even antibody fragments such as monovalent antibody fragments (Fab) and single-chain variable fragments (scFv). For instance, for expression studies, only one protein domain has to be cloned and expressed. Nanobodies also show high stability and solubility even at high temperatures and under denaturing conditions [4,5]. Due to their stable and soluble nature,

* Correspondence: chawes@brookes.ac.uk
Biological and Medical Sciences, Oxford Brookes University, Oxford OX3 0BP, UK

plus small size, high levels of expression are possible in heterologous systems in a reproducible manner and such features also allow for fusions to fluorescent proteins or protein tags [6]. Specific nanobodies can be screened for in a phage display system [7]. Nanobodies have been shown to be produced and functional in cellular compartments and environments that do not allow formation of disulphide bonds and are therefore functional in living cells [8]. In contrast to the flat or concave antigen binding site of conventional antibodies nanobodies display a convex conformation [9,3], allowing binding into otherwise inaccessible clefts and pockets which has proven a useful tool for inhibiting specific molecules such as lysozyme enzymes [9]. Furthermore, nanobodies still show binding affinities, like scFvs, in the nanomolar or even picomolar range [5].

Nanobodies have been used and tested in various applications. For instance they are considered for inhibitory therapeutic applications against viruses such as Influenza A, Respiratory Syncytial virus and Rabies virus [10] or even HIV-1 [11,12] to name a few [reviewed in 13].

A growing tool for manipulating animal and plant systems is the use of antibodies not only for inhibiting but altering the function of molecules. Nanobodies are the system of choice for such due to their ability to function intracellularly. In potatoes it was shown that they can target to the correct organelle and inhibit the function of the potato starch branching enzyme A more efficiently than an antisense construct [14]. A recent application of nanobodies has been the detection of the castor bean plant toxin ricin, a notorious bioterrorism agent. The nanobodies not only show high sensitivity towards ricin but also high specificity in distinguishing ricin from the non-toxic castor bean protein RCA120 [15].

The class of biomarkers used in this study have been termed "chromobodies" as they consist of nanobodies fused to fluorescent proteins generating fluorescent antigen-binding nanobodies that can be expressed in living cells [16]. Chromobodies have been shown to be useful tools in the real-time detection of dynamic changes in chromatin, nuclear lamina and the cytoskeleton in animal cells [16]. Such fusions have been shown to label and visualise endogenous cellular structures without disturbing cellular functions allowing real time studies of live cells processes [16].

Actin cytoskeleton

The actin-cytoskeleton in animal cells is central to cell shaping, polarity and motility [17]. Most, but not all, plant cells contain a vacuole occupying up to 90% of the intercellular volume and are caged into a rigid cell wall limiting the cell expansion [18]. The cytoplasm is therefore constrained to a thin layer at the cell cortex and the actin-cytoskeleton sustains both the organisation of the cortical endomembrane system and cytoplasmic streaming [19,20].

The actin cytoskeleton is a network composed of fine 7 nm diameter filaments that can form bundles. It is continuously rearranging and actin dynamics have been described according to a stochastic model: filaments rapidly elongate at the barbed end, change shape, slide one along the other to bundle and finally break down [21]. Actin bundles and fine filaments have different fluorescence intensity when labelled as well as differences in resistance to depolymerising agents and dynamics. Bundles are brighter, more stationary over time and depolymerise more slowly; the latter have faint fluorescence, are more dynamic and can depolymerise rapidly [22].

Different labelling strategies have been developed to study the organisation and dynamics of actin filaments in plants. The expression of fluorescent actin has not proved useful in plants because most of it stays in monomeric form and diffuse in the cytoplasm resulting in a strong fluorescent background [23]. Phalloidin, a toxin extracted from death cup *Amanita phalloides*, binds and stabilizes F-actin and when conjugated to the fluorescent dye rhodamine selectively stains actin filaments in permeabilised and fixed plant cells. Rhodamine-phalloidin staining is also effective in unfixed cells but favours the formation of bundles [24]. As such it is not useful for any study of actin dynamics.

Actin binding proteins (ABPs) are involved in regulating the assembly of actin filaments and therefore are good marker candidates [25]. The actin binding domain of different ABPs have been fused to fluorescent proteins and expressed in plants. Lifeact, the most recently developed probe, is a 17 amino acid peptide from the yeast protein Abp140 that decorates F-actin [26]. In *Arabidopsis thaliana* Lifeact fused to the fluorescent protein Venus affects the reorganisation rate of bundles and cytoplasmic strands of the actin cytoskeleton at higher expression levels but has proven to be most valuable at optimised lower expression levels as it is currently the best probe to labels dynamic populations of actin filaments [27]. The actin binding domain of mouse talin fused to fluorescent proteins has been used to label plant actin filaments but has severe effects on the actin cytoskeleton and its depolymerisation [28]. One of the two actin-binding domains of the *A. thaliana* fimbrin1 protein (AtFIM1) fused to GFP (GFP-fABD2) labels the fine actin dynamic scaffold in different species and cell types. Stable expression in *A. thaliana* did not show adverse effects on general morphology or development [29].

All of the fluorescent reporters available so far depict varying organisations of the actin network. This may be due to a preferential binding to fine actin filaments rather than bundles or because the marker is derived from an actin-bundling protein therefore causing the aggregation of actin filaments. Considering that the actin cytoskeleton is a continuously re-arranging scaffold that provides tracks for movement and positioning of organelles such as Golgi bodies [30], a more reliable and less interfering fluorescent marker is needed for *in vivo* imaging.

In this study we used a modified version of the commercially available Actin-Chromobody® (ChromoTek, Martinsried, Germany) as a novel tool for visualising actin dynamics in tobacco leaf cells. The originally supplied plasmid contains the 13 kDa actin-binding alpaca $V_H H$ fused to a C-terminal GFP protein. This chromobody was previously used to transfect HeLa cells to show the recovery of the actin filaments after Cytochalasin D treatment (ChromoTek homepage) where it was shown that the transient binding does not influence cell viability or motility.

Results and discussion

In planta expression of the actin-chromobody

Constructs fusing the antibody sequence with both N- and C-terminal fluorescent protein tags, respectively, were prepared. *Agrobacterium tumefaciens* was transformed with these constructs and *Nicotiana tabacum* leaves were infiltrated with the transformed agrobacteria, either singly or with the Golgi marker consisting of the signal anchor sequence of a rat sialyl transferase fused to GFP [ST-GFP, 31] as described in [32]. In mammalian cells the C-terminal fusion expressed and actin targeting of the chromobody was reported (http://www.chromotek.com/products/chromobodies/actin-chromobody), whereas in plant cells the antibody C-terminal fusion remained cytosolic (Figure 1, lane 1A) with no fluorescence in other organelles such as the endoplasmic reticulum (Figure 1, lane 1B) but was found in the nucleoplasm (Figure 1, lane 1C), which is common for cytosolic proteins [33]. The N-terminal YFP-fusion, however, clearly labelled actin filaments in a specific manner (Figure 1, lane 2A-C).

To determine optimal expression conditions that would allow investigation of actin dynamics as well as provide sufficient expression levels for visualisation, tobacco leaves

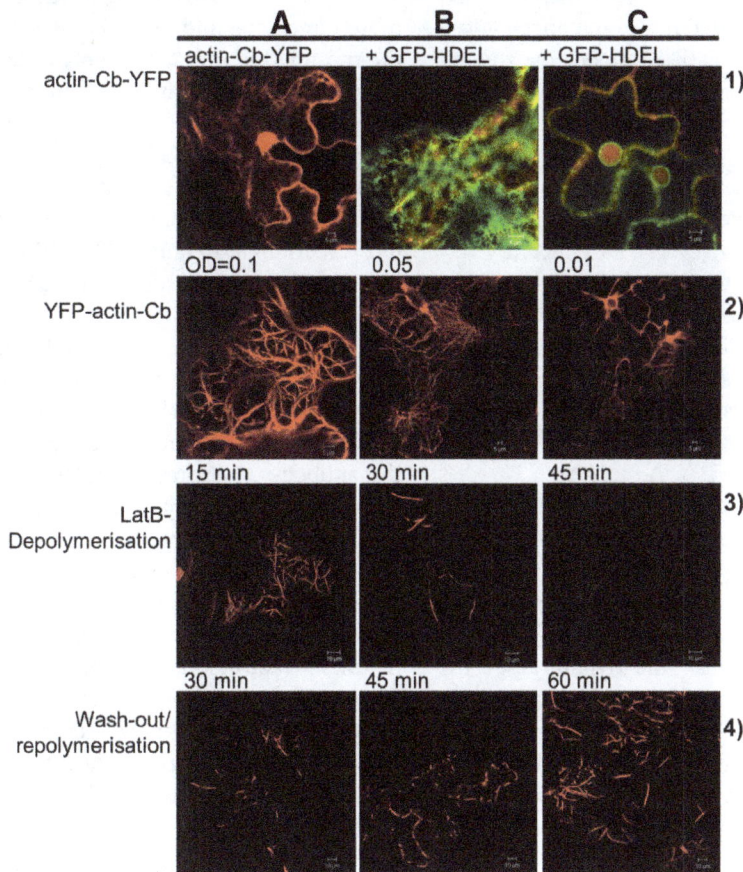

Figure 1 Transient expression of the actin-chromobody. Lane 1) Transient expression of the actin-chromobody (Actin-Cb) construct with a C-terminal YFP-fusion in *Nicotiana tabacum* leaves (**A**); co-expression with the endoplasmic reticulum marker HDEL fused to GFP is shown (**B, C**). Lane 2) Transient expression of the actin-chromobody (actin-Cb) construct with an N-terminal YFP-fusion in *Nicotiana tabacum* leaves at different Agrobacterium concentrations: $OD_{600} = 0.1$ (**A**), $OD_{600} = 0.05$ (**B**), $OD_{600} = 0.01$ (**C**). Lane 3) Depolymerisation of actin cytoskeleton (labelled by the actin-chromobody) after 15 in (**A**), 30 min (**B**) and (45 min (**C**) treatment with 25 μM latrunculin B (LatB). Lane 4) Repolymerisation of the actin cytoskeleton by washing out the LatB after 30 min (**A**), 45 min (**B**) and 60 min (**C**), respectively.

were infiltrated with three different concentrations of *Agrobacterium tumefaciens*: OD_{600} of 0.1, 0.05 and 0.01 with 0.1 being the conventional infiltration OD. The highest OD of 0.1 resulted in major bundling of actin filaments (Figure 1, lane 2A) and at the lowest OD of 0.01 the construct mainly bound to thicker actin bundles (Figure 1, lane 2C). The OD of 0.05 labelled both thicker filaments as well as finer ones (Figure 1, lane 2B) and was therefore chosen for follow-up experimentation. In general at OD 0.05 what appeared to be a more complete overview of the actin cytoskeleton with thick bundles and thinner filaments was obtained compared to that with Lifeact expression (Figure 2B). On coexpression with the ST-GFP Golgi marker, Golgi bodies could clearly be seen to be associated with the actin filament bundles as previously reported [31].

To investigate if actin dynamics was impaired by the chromobody binding leaf segments were treated with 25 µM latrunculin B (LatB), an agent isolated from a Red Sea sponge known to disrupt the actin cytoskeleton of cells. LatB binds monomeric actin with 1:1 stoichiometry and thereby blocks F-actin polymerisation without affecting cell viability [34]. After 15 min of LatB treatment the thinner strands were completely absent (Figure 1 lane 3A) and after 30 min and 45 min only the bundled actin strands were visible (Figure 1 lane 3B, C); these bundles remain even with overnight treatment in the drug (data not shown).

With LatB being a relatively small molecule of less than 0.4 kDa it is possible to reverse its effects by immersing the leaf cuttings in water and thereby washing out the drug and allowing the actin filaments to repolymerise. A rapid recovery of filamentous actin within less than one hour of washing was observed (Figure 1, lane 4B) with a visible increase in strands in 30 min (Figure 1, lane 4A).

Impact of actin-chromobody labelling on actin function

The motility of Golgi bodies is mainly dependent on the actin cytoskeleton and treatment with the actin depolymerising agent cytochalasin D caused the organelles to stop moving [35]. However, *in vivo* labelling of the actin cytoskeleton might compromise the dynamics of the organelle changing the organisation of the actin network [27].

In order to compare the effect of Lifeact-GFP and the YFP-actin-Cb on the movement of Golgi bodies in *N. tabacum*, the cytoskeleton markers were transiently coexpressed with the Golgi marker ST-GFP (Figure 2A, B) and compared to the transient expression of ST-GFP only. For this, the infiltration OD and therefore the expression levels for Lifeact were chosen for optimal Golgi movement and as little bundling as possible. Movies were collected for each combination and analysed with Volocity software to track Golgi bodies and calculate parameters such as velocity, displacement rate and meandering index. The values are represented as Cumulative Distribution Frequency (CDF) and means are normalised against data from ST-GFP expression alone.

The expression of either the cytoskeleton markers significantly slowed the motility of Golgi bodies as described by their velocity which is the length of the track over time (Figure 2C). The displacement rate, which is the linear distance between the initial and final position of the organelle, was not significantly affected by the expression of Lifeact-GFP but was significantly reduced in the presence of YFP-actin-Cb (Figure 2D). Both velocity and displacement rate in the combinations of ST-GFP and YFP-actin-Cb or Lifeact-GFP, respectively, were normalised to the datasets for expression of ST-GFP alone and presented as a percentage of the latter (Figure 2E). The final parameter assessed was the Meandering Index which is the ratio between the displacement rate and velocity (the ratio between the linear distance and the overall path of the Golgi body), describing the type of Golgi movement. The closer the meandering index is to a value of 1, the more directional and linear is the movement. The smaller the meandering index, the more saltatory is the organelle movement. The meandering index therefore gives an indication of the complexity of the dynamics. Upon the expression of Lifeact-GFP, the relative mean of the meandering index was significantly increased by 19% compared to ST-GFP alone indicating that the organelles moved more uni-directionally (Figure 2E). YFP-actin-Cb significantly reduced the meandering index by 11% with respect to ST-GFP (Figure 2E) indicating that the saltatory movement was slightly favoured. The emerging model (Figure 2F) is that given a fixed time span of 1 second Golgi bodies move more slowly, with a shorter linear distance but conserve the complexity of the movement pattern in the presence of the antibody. Coexpression with Lifeact-GFP results in a reduction of the velocity, with the same linear distance but a much less saltatory movement.

These effects of the markers on actin dynamics might be explained by the different effects the two markers have on cytoskeleton rearrangement and thus its dynamic. Lifeact derives from a cross bundling factor and favours the formation of actin cables that might lead Golgi bodies along more directional pathways whilst binding of the actin-Cb might interfere less with the filament organisation therefore having a less of an effect on the movement type.

Conclusion

Mainly due to their small size and stability in combination with production advantages, nanobodies have been shown to be valuable tools for inhibiting or manipulating cell processes with a great potential for genetically encoded *in vivo* immunocytochemical labelling.

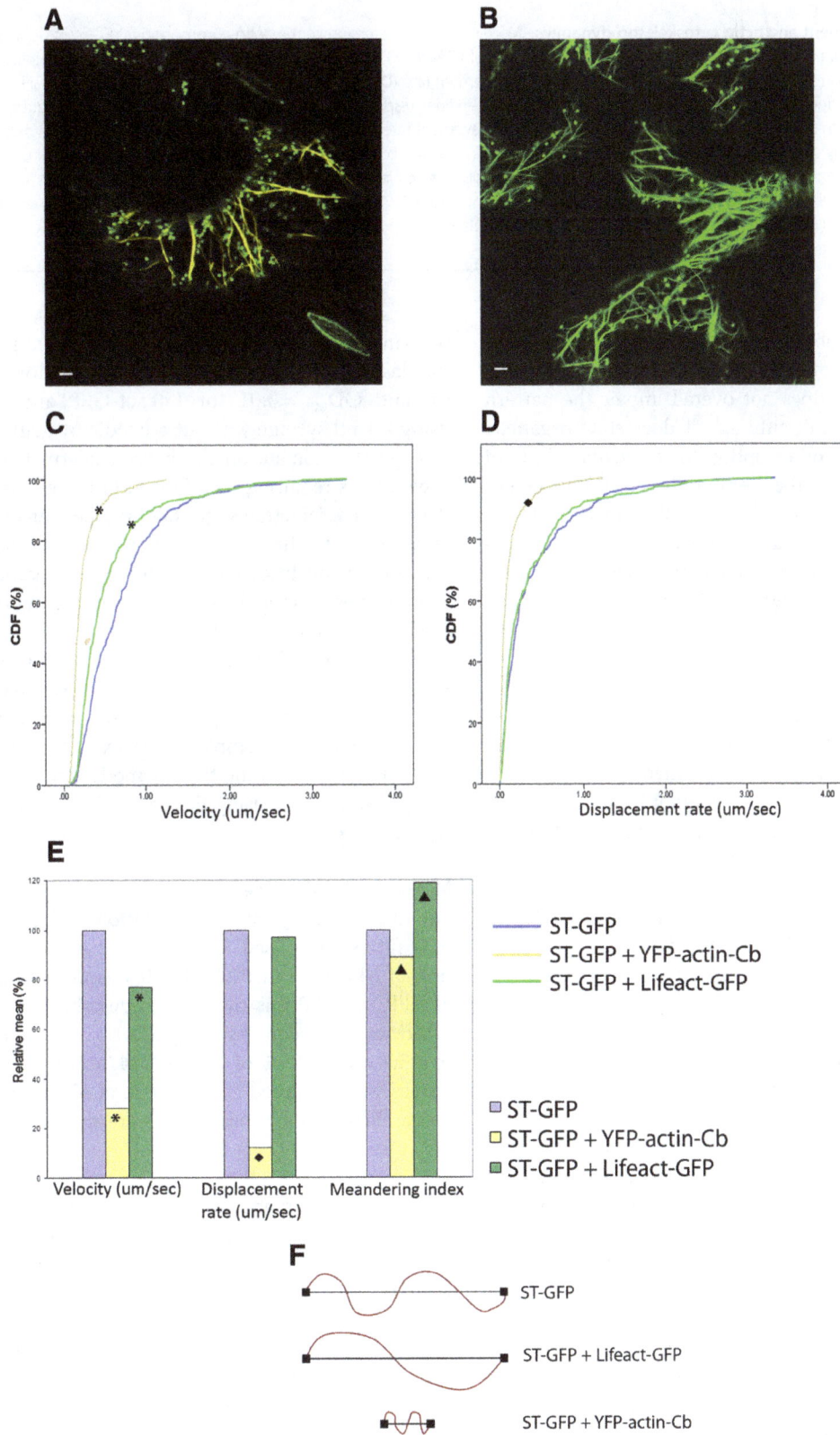

Figure 2 (See legend on next page.)

(See figure on previous page.)
Figure 2 Golgi movement and actin cytoskeleton dynamics. A) Transient co-expression of the actin-chromobody (YFP-actin-Cb in yellow) construct with the Golgi marker ST-GFP (green dots); **B)** co-expression of Lifeact GFP (green lines) with ST-GFP (green dots). Scale bars = 5 μm. Cumulative Distribution Frequency (CDF) plots of velocity **(C)** and displacement rate **(D)** of *N. tabacum* transiently expressing only ST-GFP (blue line), both ST-GFP and YFP-actin-Cb (yellow line) or ST-GFP and Lifeact-GFP (green line), respectively. Curves marked with shapes (*, ♦) indicate a statistically significant difference from the control ST-GFP of $p < 0.05$. **E)** Velocity, displacement rate and meandering index values calculated with Volocity software for *N. tabacum* transiently expressing ST-GFP (blue line), both ST-GFP and YFP-actin-Cb (yellow line) or ST-GFP and Lifeact-GFP (green line), respectively. Mean values are expressed as a percentage of the control (ST-GFP). Symbols (*, ♦, ▲) indicate that the means are significantly different from the control at $p < 0.05$. **F)** Schematic representation of the path and movement pattern of Golgi bodies. When ST-GFP is coexpressed with Lifeact-GFP, Golgi bodies move same linear distance as the control but have a less salutatory path. The coexpression of the antibody determines Golgi bodies moving shorter linear distance and slightly more salutatory.

The actin-chromobody described here is especially useful for the study of actin dynamics in plant cells as it labels the actin, but does not overall impair the pattern of organelle movement, although it does slow organelle velocity. It is thus another option for *in vivo* labelling of actin compared with the commonly used fluorescent protein probes based around actin binding proteins or their active domains. It may be possible to exploit the reduction in organelle movement and by implication myosin activity, for the study of organelle dynamics and their relationship with the actin cytoskeleton.

Methods

Cloning of expression plasmids

The Actin-Chromobody® plasmid containing the alpaca actin-antibody gene was obtained from Chromo-Tek (Martinsried, Germany). Primers were ordered from Eurofins MWG Operon (Ebersberg, Germany). Q5 high-fidelity DNA polymerase (New England Biolabs, Herts, UK) was used for all polymerase chain reaction (PCR) reactions. The actin-Ab-PCR product was cloned into the binary vectors PB7WGY2 and PB7YWG2 providing an N- or C-terminal YFP-tag, respectively, using Gateway® technology (Invitrogen life sciences).

Plant material and transient expression system

For Agrobacterium-mediated transient expression, 4-week-old tobacco (*Nicotiana tabacum* SR1 cv Petit Havana) plants grown in the greenhouse were used. Briefly, each expression vector was introduced into Agrobacterium strain GV3101 (pMP90) by heat shock. A single colony from the transformants was inoculated into 5 ml of YEB medium (per litre: 5 g of beef extract, 1 g of yeast extract, 5 g of sucrose and 0.5 g of $MgSO_4 \cdot 7H_2O$) supplemented with 50 μg/ml spectinomycin and rifampicin. After overnight shaking at 25°C, 1 ml of the bacterial culture was pelleted in a 1.5-ml tube by centrifugation at 2200 × g for 5 min at room temperature. The pellet was washed twice with 1 ml of infiltration medium (50 mM MES, 2 mM $Na_3PO4 \cdot 12H_2O$, 0.1 mM acetosyringone and 5 mg/ml glucose) and then resuspended in 1 ml of infiltration buffer. The bacterial suspension was diluted with the same buffer to adjust the inoculum concentration to the desired final OD_{600} (0.1, 0.05 or 0.01) for YFP-actin-Cb and $OD_{600} = 0.01$ for Lifeact-GFP and inoculated using a 1 ml syringe without a needle by gentle pressure through the stomata on the lower epidermal surface. For experiments requiring co-infection of more than one construct, bacterial strains containing the constructs were mixed prior to the leaf infiltration, with the inoculum of each mixed construct adjusted to the required final OD_{600}. Transformed plants then were incubated under normal growth conditions for 48 h.

Images were taken using a Zeiss LSM510 Meta laser scanning confocal microscope (http://www.zeiss.com/) with 40× and 63× oil immersion objectives. For imaging of GFP/YFP combinations, samples were excited using 458 and 514 nm laser lines in multi-track mode with line switching. Images were edited using the LSM510 image browser and Adobe Photoshop.

Microscopy and movies

For dual imaging 488 nm excitation and 505-530 band pass filters were used for eGFP and for YFP an excitation of 514 nm and BP 470-500 was used; dual imaging of GFP and YFP was captured as described above.

Movies were acquired using 63X objective lens, zoomed to ×3.7 and a ROI of 244 × 244 pixels. Movies of 50 frames were acquired at scan time of 470 msec. Example movie files used for this analysis are shown in Additional file 1.

Organelle tracking and statistical analysis

Organelle tracking was done using the Volocity 6.3 (Improvision - PerkinElmer). Intensity and size parameters were set and the software identified and tracked Golgi bodies according to shortest path model. The velocity, displacement rate and meandering index of 100-344 Golgi bodies per condition were calculated by the software. Statistical analysis and graphs were done with SPSS 21.0 and the Kolmogorov-Smirnov test was used to assess the statistical difference in the distribution of velocity, displacement rate and meandering index values for $p < 0.05$.

Additional file

Additional file 1: Example movies of Golgi body movement.
Movement of Golgi bodies (ST-GFP labelled, green dots) in co-expression
with YFP-actin-Cb (A) or Lifeact-GFP (B), respectively. Green fluorescent
(Golgi) dots in movies were analysed with the Volocity software to calculate
velocity, displacement rate and meandering index of Golgi body movement.

Abbreviations
HcAbs: Heavy-chain only antibodies; V_HH: Variable heavy chain of a
heavy-chain antibody; scFv: Single-chain variable fragments; ABP: Actin
binding protein; GFP: Green fluorescent protein; YFP: Yellow fluorescent
protein; ST: Rat sialyl transferase; LatB: Latrunculin B; actin-Cb: Actin
chromobody; CDF: Cumulative Distribution Frequency.

Competing interests
The authors declare that they have no competing interests.

Authors' contributions
AR participated in the study design, carried out expression studies, analysed
Golgi dynamics and helped draft the manuscript. CH conceived of the study,
and participated in its design and coordination and helped to draft the
manuscript. VK participated in the study design, made the constructs for
expression, carried out expression studies and helped draft the manuscript.
All authors read and approved the final manuscript.

Acknowledgments
AR is supported by an Oxford Brookes University Nigel Groome studentship.
VK is funded by the BBSRC grant no. BB/J005959/1.

References
1. Hamers-Casterman C, Atarhouch T, Muyldermans S, Robinson G, Hamers C,
   Songa EB, Bendahman N, Hamers R: **Naturally occurring antibodies devoid
   of light chains.** *Nature* 1993, **363**:446–448.
2. Muyldermans S, Atarhouch T, Saldanha J, Barbosa JA, Hamers R: **Sequence
   and structure of VH domain from naturally occurring camel heavy chain
   immunoglobulins lacking light chains.** *Protein Eng* 1994, **7**:1129–1135.
3. Muyldermans S, Cambillau C, Wyns L: **Recognition of antigens by
   singledomain antibody fragments: the superfluous luxury of paired
   domains.** *Trends Biochem Sci* 2001, **26**:230–235.
4. Ladenson RC, Crimmins DL, Landt Y, Ladenson JH: **Isolation and characterization
   of a thermally stable recombinant anti-caffeine heavy-chain antibody
   fragment.** *Anal Chem* 2006, **78**:4501–4508.
5. van der Linden RH, Frenken LG, de Geus B, Harmsen MM, Ruuls RC, Stok W,
   de Ron L, Wilson S, Davis P, Verrips CT: **Comparison of physical chemical
   properties of llama VHH antibody fragments and mouse monoclonal
   antibodies.** *Biochim Biophys Acta* 1999, **1431**:37–46.
6. Olichon A, Surrey T: **Selection of genetically encoded fluorescent single
   domain antibodies engineered for efficient expression in Escherichia
   coli.** *J Biol Chem* 2007, **282**:36314–36320.
7. Schmidthals K, Helma J, Zolghadr K, Rothbauer U, Leonhardt H: **Novel
   antibody derivatives for proteomeand high-content analysis.** *Anal Bioanal
   Chem* 2010, **397**:3203–3208.
8. Klooster R, Eman MR, le Duc Q, Verheesen P, Verrips CT, Roovers RC, Post JA:
   **Selection and characterization of KDEL-specific VHH antibody fragments
   and their application in the study of ER resident protein expression.**
   J Immunol Methods 2009, **342**:1–12.
9. Desmyter A, Transue TR, Ghahroudi MA, Thi MH, Poortmans F, Hamers R,
   Muyldermans S, Wyns L: **Crystal structure of a camel single-domain VH
   antibody fragment in complex with lysozyme.** *Nat Struct Biol* 1996,
   3:803–811.
10. Hultberg A, Temperton NJ, Rosseels V, Koenders M, Gonzalez-Pajuelo M,
    Schepens B, Ibañez LI, Vanlandschoot P, Schillemans J, Saunders M, Weiss RA,
    Saelens X, Melero JA, Verrips CT, Van Gucht S, de Haard HJ: **Llama-derived
    single domain antibodies to build multivalent, superpotent and broadened
    neutralizing anti-viral molecules.** *PLoS One* 2011, **6**:e17665.
11. Forsman A, Beirnaert E, Aasa-Chapman MMI, Hoorelbeke B, Hijazi K, Koh W,
    Tack V, Szynol A, Kelly C, McKnight A, Verrips T, de Haard H, Weiss RA: **Llama
    antibody fragments with cross-subtype human immunodeficiency virus
    type 1 (HIV-1)-neutralizing properties and high affinity for HIV-1 gp120.**
    J Virol 2008, **82**:12069–12081.
12. Koh WWL, Steffensen S, Gonzalez-Pajuelo M, Hoorelbeke B, Gorlani A, Szynol A,
    Forsman A, Aasa-Chapman MMI, de Haard H, Verrips T, Weiss RA: **Generation
    of a family-specific phage library of llama single chain antibody fragments
    that neutralize HIV-1.** *J Biol Chem* 2010, **285**:19116–19124.
13. Vanlandschoot P, Stortelers C, Beirnaert E, Ibañez LI, Schepens B, Depla E,
    Saelens X: **Nanobodies: New ammunition to battle viruses.** *Antivir Res*
    2011, **92**:389–407.
14. Jobling SA, Jarman C, Teh MM, Holmberg N, Blake C, Verhoeyen ME:
    **Immunomodulation of enzyme function in plants by single-domain
    antibody fragments.** *Nat Biotechnol* 2003, **21**:77–80.
15. Shia WW, Bailey RC: **Single domain antibodies for the detection of ricin
    using silicon photonic microring resonator arrays.** *Anal Chem* 2013,
    85:805–810.
16. Rothbauer U, Zolghadr K, Tillib S, Nowak D, Schermelleh L, Gahl A,
    Backmann N, Conrath K, Muyldermans S, Cardoso MC, Leonhardt H:
    Targeting and tracing antigens in live cells with fluorescent nanobodies.
    Nat Methods 2006, **3**:887–889.
17. Blanchoin L, Boujemaa-Paterski R, Sykes C, Plastino J: **Actin dynamics,
    architecture, and mechanics in cell motility.** *Physiol Rev* 2014, **94**:235–263.
18. Kost B, Chua N-H: **The Plant Cytoskeleton: Vacuoles and Cell Walls Make
    the Difference.** *Cell* 2002, **108**:9–12.
19. Avisar D, Abu-Abied M, Belausov E, Sadot E, Hawes C, Sparkes IA: **A
    comparative study of the involvement of 17 Arabidopsis myosin family
    members on the motility of Golgi and other organelles.** *Plant Physiol*
    2009, **150**:700–709.
20. Shimmen T, Yokota E: **Cytoplasmic streaming in plants.** *Curr Opin Cell Biol*
    2004, **16**:68–72.
21. Staiger CJ, Sheahan MB, Khurana P, Wang X, McCurdy DW, Blanchoin L:
    **Actin filament dynamics are dominated by rapid growth and
    severing activity in the Arabidopsis cortical array.** *J Cell Biol* 2009,
    184:269–280.
22. Henty-Ridilla JL, Li J, Blanchoin L, Staiger CJ: **Actin dynamics in the cortical
    array of plant cells.** *Curr Opin Plant Biol* 2013, **16**:678–687.
23. Lemieux MG, Janzen D, Hwang R, Roldan J, Jarchum I, Knecht DA:
    **Visualization of the actin cytoskeleton: different F-actin-binding probes
    tell different stories.** *Cytoskeleton* 2013, **71**:157–169.
24. Sonobe S, Shibaoka H: **Cortical fine actin filaments in higher plant cells
    visualized by rhodamine-phalloidin after pretreatment with
    m-maleimidobenzoyl N-hydroxysuccinimide ester.** *Protoplasma* 1989,
    148:80–86.
25. Higaki T, Sano T, Hasezawa S: **Actin microfilament dynamics and actin
    side-binding proteins in plants.** *Curr Opin Plant Biol* 2007, **10**:549–556.
26. Riedl J, Crevenna AH, Kessenbrock K, Yu JH, Neukirchen D, Bista M, Bradke F,
    Jenne D, Holak TA, Werb Z, Sixt M, Wedlich-Soldner R: **Lifeact: a versatile
    marker to visualize F-actin.** *Nat Met* 2008, **5**:605–607.
27. Van der Honing HS, van Bezouwen LS, Emons AMC, Ketelaar T: **High
    expression of Lifeact in Arabidopsis thaliana reduces dynamic
    reorganization of actin filaments but does not affect plant development.**
    Cytoskeleton 2010, **68**:578–587.
28. Ketelaar T, Anthony RG, Hussey PJ: **Green fluorescent protein-mTalin
    causes defects in actin organization and cell expansion in Arabidopsis
    and inhibits actin depolymerizing factor's actin depolymerizing activity
    in vitro.** *Plant Physiol* 2004, **136**:3990–3998.
29. Sheahan MB, Staiger CJ, Rose RJ, McCurdy DW: **A green fluorescent
    protein fusion to actin-binding domain 2 of Arabidopsis fimbrin
    highlights new features of a dynamic actin cytoskeleton in live plant
    cells.** *Plant Physiol* 2004, **136**:3968–3978.
30. Akkerman M, Overdijk EJR, Schel JHN, Emons AMC, Ketelaar T: **Golgi Body
    motility in the plant cell cortex correlates with actin cytoskeleton
    organization.** *Plant Cell Physiol* 2011, **52**:1844–1855.
31. Boevink P, Santa Cruz S, Hawes C, Harris N, Oparka KJ: **Virus-mediated
    delivery of the green fluorescent protein to the endoplasmic reticulum
    of plant cells.** *Plant J* 1996, **10**:935–941.
32. Sparkes IA, Runions J, Kearns A, Hawes C: **Rapid, transient expression of
    fluorescent fusion proteins in tobacco plants and generation of stably
    transformed plants.** *Nat Protoc* 2006, **1**:2019–2025.

33.   Brandizzi F, Fricker M, Hawes C: **A greener world: the revolution in plant bioimaging.** *Nat Rev Mol Cell Biol* 2002, **3:**520–530.

34.   Brandizzi F, Snapp EL, Roberts AG, Lippincott-Schwartz J, Hawes C: **Membrane protein transport between the endoplasmic reticulum and the Golgi in tobacco leaves is energy dependent but cytoskeleton independent: Evidence from selective photobleaching.** *Plant Cell* 2002, **14:**1293–1309.

35.   Boevink P, Oparka K, Santa Cruz S, Martin B, Betteridge A, Hawes C: **Stacks on tracks: the plant Golgi apparatus traffics on an actin/ER network.** *Plant J* 1998, **15:**441–447.

2

Rapid fluorescent reporter quantification by leaf disc analysis and its application in plant-virus studies

Fabio Pasin[*], Satish Kulasekaran, Paolo Natale, Carmen Simón-Mateo and Juan Antonio García

Abstract

Background: Fluorescent proteins are extraordinary tools for biology studies due to their versatility; they are used extensively to improve comprehension of plant-microbe interactions. The viral infection process can easily be tracked and imaged in a plant with fluorescent protein-tagged viruses. In plants, fluorescent protein genes are among the most commonly used reporters in transient RNA silencing and heterologous protein expression assays. Fluorescence intensity is used to quantify fluorescent protein accumulation by image analysis or spectroscopy of protein extracts; however, these methods might not be suitable for medium- to large-scale comparisons.

Results: We report that laser scanners, used routinely in proteomic studies, are suitable for quantitative imaging of plant leaves that express different fluorescent protein pairs. We developed a microtiter plate fluorescence spectroscopy method for direct quantitative comparison of fluorescent protein accumulation in intact leaf discs. We used this technique to measure a fluorescent reporter in a transient RNA silencing suppression assay, and also to monitor early amplification dynamics of a fluorescent protein-labeled potyvirus.

Conclusions: Laser scanners allow dual-color fluorescence imaging of leaf samples, which might not be acquired in standard stereomicroscope devices. Fluorescence microtiter plate analysis of intact leaf discs can be used for rapid, accurate quantitative comparison of fluorescent protein accumulation.

Keywords: Fluorescent protein, Fluorescence spectroscopy, Microtiter plate, RNA silencing, Plant virus

Background

Reporter genes and their products are valuable tools for plant studies, due to the ease of imaging and quantification of the proteins encoded [1]. Fluorescent proteins are widely employed as reporters, since they have no requirements for exogenous substrate/co-factors and do not interfere with cell growth or function [2]. These proteins can be detected and imaged in live tissue without cell lysis or biochemical analysis, and they allow optical exploration of cell structures and molecule dynamics as well as pathogen monitoring with minimal sample preparation [3].

Use of fluorescent protein as a quantitative reporter includes evaluation of new vectors for heterologous protein expression and of promoter activity, translational regulation and transient RNA silencing [4-8]. In plant pathology and symbiosis studies, fluorescent proteins are an important aid for monitoring infection/colonization onset and spreading, and thus facilitate comprehension of host-microbe interactions. Since the first demonstrations that plant viruses are useful vectors for foreign sequence transfer to their hosts [9-12], several genes were shown to be suitable RNA virus reporters; they include those that encode chloramphenicol acetyltransferase, firefly and *Renilla* luciferases, β-glucuronidase, anthocyanin biosynthesis transcription factors, and *Aequorea victoria* green fluorescent protein (GFP) [12-18].

Compared to other markers, fluorescent protein genes inserted into viral genomes offer good reporter stability [19], viral localization to individual cells, and monitoring of co-infection with differently-labeled viruses [20,21]. A further advantage of these proteins is that their fluorescence intensity is directly proportional to protein amount and can be used for quantification [22,23]. Although GFP fluorescence can be quantified by image analysis [24,25], this involves time-consuming steps that can be overcome

* Correspondence: fpasin@cnb.csic.es
Centro Nacional de Biotecnología (CNB-CSIC), Darwin 3, Madrid 28049, Spain

by spectrofluorometric measurement of intact plant organs or protein extracts from GFP-expressing samples [23,26,27].

A microplate assay was recently described that measures luciferase activity in intact leaf discs [28]. In a similar approach, here we evaluated the use of 96-well plate readers for rapid quantification of two *A. victoria* GFP variants, the ultraviolet (UV)-excitable mGFP5 [29] and a mutant with enhanced brightness sGFP(S65T) [30]. The method was applied in viral RNA silencing suppressor studies and in accumulation monitoring of GFP-labeled *Plum pox virus* (PPV) clones. A palette of engineered monomeric fluorescent proteins was expressed transiently in plants (Table 1) and shown to be easily quantifiable by direct leaf disc analysis.

Results and discussion

Laser scanner imaging of *Nicotiana benthamiana* leaves

GFP variants such as mGFP5 [29], which can be excited by long-wavelength ultraviolet (UV) light, are used frequently in plant studies of species other than the small-sized *Arabidopsis*, since fluorescence imaging of whole specimens is constrained by objective lens size of fluorescence (stereo)microscopes. The need for fluorescence microscopes is overcome by use of UV lamps as excitation sources, although this restricts fluorophore choice and limits multi-fluorescence imaging. Scanners with excitation lasers at 457, 488, 532, and 633 nm are used for fluorescence imaging in two-dimensional difference gel analysis systems [36] and have a relatively large glass platen (for example, 35 cm × 43 cm, in the Typhoon 9400). As a 633 nm laser might be unsuited to leaf tissue imaging due to interference from chlorophyll autofluorescence [37], we tested whether 457, 488 and 532 nm lasers can be used for imaging *N. benthamiana* leaves that transiently express fluorescent proteins. Plant expression vectors bearing coding sequences for mGFP5 or a monomeric red fluorescent protein TagRFP-T [35] were delivered to plants by *Agrobacterium* infiltration. *Tomato bushy stunt virus* p19 RNA silencing suppressor was co-expressed to increase yield of

the heterologous proteins delivered [38]. At 6 days post-agro-infiltration (dpa), *N. benthamiana* leaf fluorescence was acquired after excitation with 488 nm and 532 nm lasers. A strong signal was detected in leaf patches expressing the fluorescent proteins. Only background signal was detected in non-infiltrated leaf areas and when non-optimal excitation/emission conditions were used, i.e., mGFP5-expressing patches imaged with TagRFP-T settings (Ex532/Em580) and TagRFP-T-expressing patches imaged with mGFP5 settings (Ex488/Em526) (Figure 1A, C). To expand fluorophore choice, we tested a cyan (mTFP1; [32]) and a yellow (mPapaya1; [34]) fluorescent protein, and found them to be easily imaged in agro-infiltrated leaves (Figure 1B, D). These results support the suitability of mGFP5/TagRFP-T and mTFP1/mPapaya1 pairs for laser scanner bicolor imaging in plants.

Spectral properties and quantification of plant-expressed fluorescent proteins

A fluorescence signal acquired by laser scanner imaging is suitable for quantitative comparisons (Figure 1C, D), as is done routinely in proteomic studies [36]. Image analysis can be a lengthy process, however, and signal quantification can be affected if leaf lamina occupy different focal planes during the acquisition step. As microtiter plate readers are available for medium-high throughput analysis, we used a monochromator-based plate reader to analyze the fluorescence signal from intact leaf discs collected from agro-infiltrated patches (Figure 2A). We found that fluorescence properties of mGFP5 could be measured without extract preparation (Figure 2B), and excitation and emission spectra closely resembled those reported [29]. Five-fold dilutions of the mGFP5-*Agrobacterium* strain were used in a transient expression assay. Fluorescence intensity values were consistent with the amount of bacteria delivered (Pearson $R^2 = 0.9855$; $n = 4$; Infinite M200 values were considered) and independent of the fluorescent plate reader used (Figure 2C).

Table 1 Reporter proteins and fluorescence analysis conditions evaluated

Reporter	Laser scanner imaging		Plate reader FI quantification		Species	Structure	Ref.
	Laser (nm)	Em (nm)	Ex (nm)	Em (nm)			
mTagBFP2	n.a.[1]	n.a.	400/9	455/20	*Entacmaea quadricolor*	Monomer	[31]
mTFP1	457	526SP	450/9	480/20	*Clavularia sp.*	Monomer	[32]
mGFP5	488	526SP	485/9	535/20	*Aequorea victoria*	Weak dimer	[29]
sGFP(S65T)	488	526SP	485/9	535/20	*Aequorea victoria*	Weak dimer	[30]
mNeonGreen	n.t.[2]	n.t.	500/9	530/20	*Branchiostoma lanceolatum*	Monomer	[33]
mPapaya1	532	555/20	520/9	550/20	*Zoanthus sp.*	Monomer	[34]
TagRFP-T	532	580/30	560/9	595/20	*Entacmaea quadricolor*	Monomer	[35]

[1]n.a., not applicable.
[2]n.t., not tested.

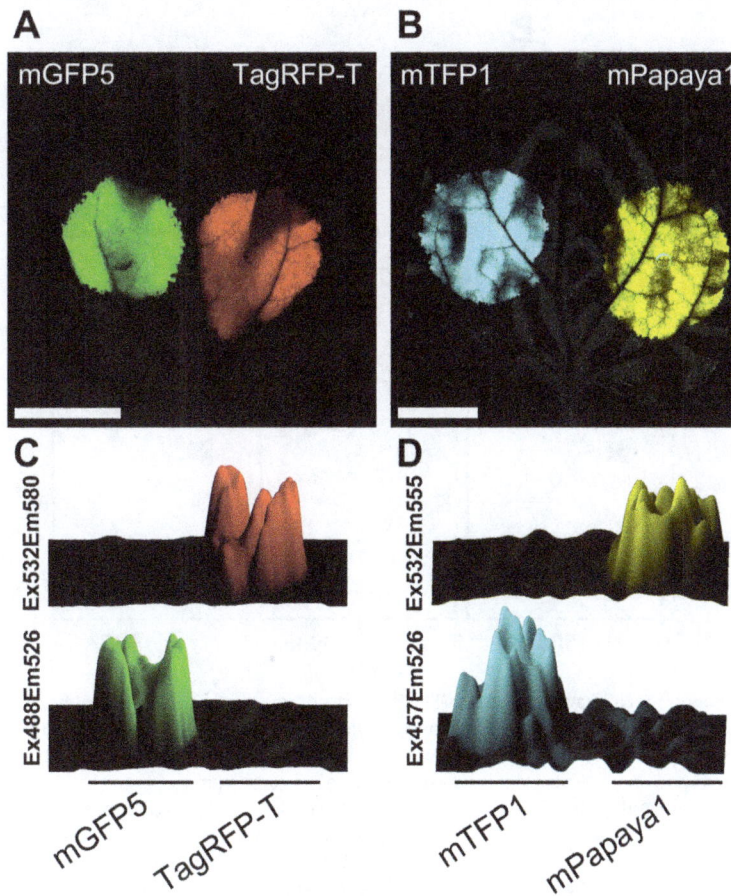

Figure 1 Laser scanner imaging of fluorescent protein-expressing leaves. Fluorescent proteins were transiently expressed by co-infiltrating *N. benthamiana* leaf tissue with an *Agrobacterium* pSN.5 p19 culture plus cultures of *Agrobacterium* containing pBin-35S-mGFP5 (mGFP5), pSN.5 TagRFP-T (TagRFP-T), pSN.5 mTFP1 (mTFP1) or pSN.5 mPapaya1 (mPapaya1). Fluorescence was imaged by leaf laser scanning. **(A)** Signal acquired at 6 dpa for TagRFP-T (red) and mGFP5 (green); green and red channel images were merged. Scale bar, 2 cm. **(B)** Signal at 3 dpa for mTFP1 (cyan) and mPapaya1 (yellow); cyan and yellow channel images were merged. Scale bar, 2 cm. **(C,D)** Surface plots of infiltrated patches from above images.

Rapid fluorometer GFP quantification in transient RNA silencing assays

To determine whether leaf disc fluorescence intensity can be used for quantitative analysis of GFP accumulation in leaf tissue, we co-expressed mGFP5 with PPV silencing suppressor constructs. These included HCPro with the parent sequence (WT), with the L134H substitution (LH; which abolishes RNA silencing suppression activity [40,41]), and HCPro into which amino acids REN-239, 240, 241 were replaced by alanines (AS9). The AS9 construct was tested since the corresponding HCPro mutants in *Tobacco etch virus* (TEV) and *Turnip mosaic virus* (TuMV) are silencing suppression-defective [42-44], but no data are available for PPV. The red TagRFP-T was also included to test for interference with mGFP5 fluorescence analysis (Figure 3A). At 6 dpa, laser scanner imaging detected bright fluorescence in patches in which mGFP5 was co-delivered with wild-type HCPro (WT, Figure 3A). Analysis on a 96-well plate reader showed a significantly higher fluorescence signal in WT samples than in those of

the other constructs tested, i.e., LH, AS9 and red fluorescent protein samples (Figure 3B). Fluorescence intensity in AS9 samples was equivalent to that in silencing suppression mutant L134H samples. These results suggest that the PPV HCPro AS9 (REN-239, 240, 241 replacement) construct behaves like the TEV and TuMV HCPro AS9 mutants. In immunoblot analysis, mGFP5 protein accumulation correlated positively with fluorescence signal quantification values (Pearson R^2 = 0.9989; Figure 3C). In a parallel experiment, transient delivery of HCPro proteins was confirmed by anti-PPV HCPro immunoblot analysis of samples co-infiltrated with p19 (Figure 3D). We detected no TagRFP-T interference in mGFP5 quantification assays (Figure 3A, B).

Monitoring of plant viral amplification dynamics by fluorometer analysis

We used sGFP(S65T), a synthetic GFP version with enhanced brightness [30], as a sensitive reporter to follow PPV early amplification in plant tissue. The pSN-PPV

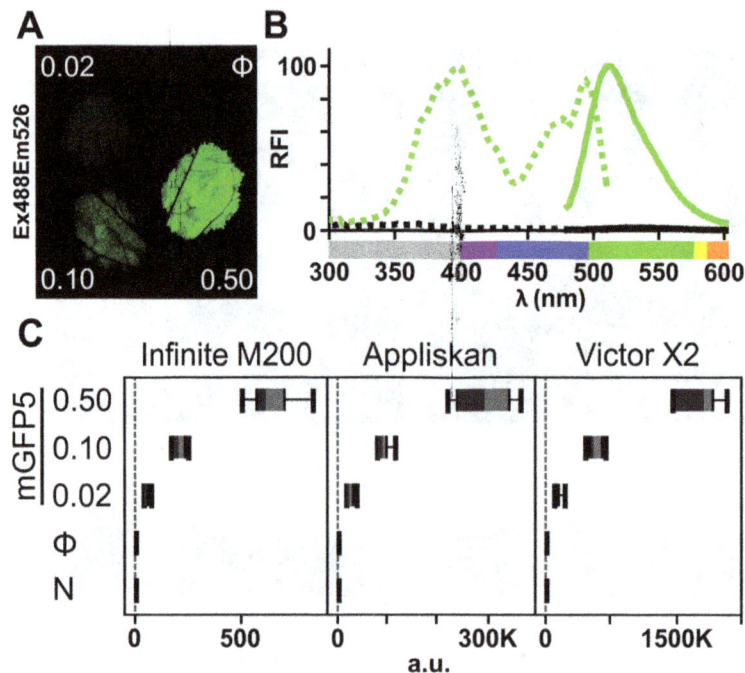

Figure 2 Spectral properties and fluorescence quantification of GFP. Fluorescent protein was transiently expressed by co-infiltrating *N. benthamiana* leaf tissue with an *Agrobacterium* pSN.5 p19 culture plus a strain with no expression vector (Φ), or 5-fold dilutions of *Agrobacterium* containing pBin-35S-mGFP5 (mGFP5 at OD_{600} 0.50, 0.10 and 0.02). **(A)** At 3 dpa, mGFP5 (green) fluorescence was imaged by leaf laser scanning. **(B)** In a plate reader, excitation (dotted lines) and emission spectra (solid lines) were measured from leaf discs of tissue agro-infiltrated with mGFP5 (green) or no expression vector (black). Relative fluorescence intensity (RFI) was plotted using mGFP5 peaks equal to 100. Ultraviolet (UV) wavelengths are in gray, visible spectrum colors were assigned as described [39]. **(C)** Box-plot graphs show quantification values from $n = 8$ samples/condition. Fluorescence intensity of leaf discs agro-infiltrated with mGFP5 strain dilutions, no expression vector (Φ) or non-treated samples (N) was acquired in monochromator-based (Infinite M200) and two filter-based (Appliskan and Victor X2) plate readers. Fluorescence intensity is expressed in arbitrary units (a.u.).

binary vector [45] was used to deliver sGFP(S65T)-tagged PPV by agro-inoculation (Figure 4A). As anticipated, sGFP(S65T) fluorescence was readily detected in infected leaves (Figure 4B). Fluorophore spectra were confirmed by analysis of leaf discs from inoculated leaves. Compared to mGFP5, sGFP(S65T) retained the blue light excitation peak but lacked the UV peak (Figure 4C).

We further compared GFP fluorescence intensity (FI) signal dynamics of leaves agro-inoculated with pSN-PPV (wtPPV) or with pSN-PPV P1-S (S259A), a cDNA clone of a non-infectious PPV mutant with silencing suppression defects [45]. Whereas the FI of the PPV S259A clone peaked at 2 dpa, FI of wtPPV continued to increase over the 6-day time course (Figure 4D). In agro-inoculated leaves, fluorescence quantification results were corroborated by immunoblot analysis of GFP and PPV coat protein (CP, Figure 4E). We developed a strand-specific quantification of PPV RNA by RT-qPCR assay (Additional file 1), and viral RNA amounts at 6 dpa were consistent with protein determinations (Figure 4F). β-glucuronidase and luciferase genes can be used to analyze potyviral accumulation, genome amplification rates and cell-to-cell movement [14,46-48]; here we show that detection of a

GFP-tagged virus is quite straightforward, since no substrates/co-factors are needed and sample preparation requirements are minimal.

Direct leaf disc analysis of engineered monomeric fluorescent proteins

There is a wide variety of engineered fluorescent proteins with improved optical and stability properties and many spectral variants were obtained by evolution of the *A. victoria* GFP sequence. For multicolor experiments, however, fluorescent proteins with minimal sequence similarity are desirable, to reduce post-transcriptional gene silencing events and assure immunodetection specificity. We evaluated the novel bright fluorescent proteins blue mTagBFP2 [31], cyan mTFP1 [32], green mNeonGreen [33], yellow mPapaya1 [34] and red TagRFP-T [35], all derived from species other than *A. victoria* (Table 1), for transient expression in plants. Fluorophore spectral properties and fluorescence intensity were easily determined using intact leaf discs collected from tissue agro-infiltrated with the corresponding constructs (Figure 5). We also show that the FI of different fluorophores can be measured simultaneously and, in multicolor experiments, the choice of

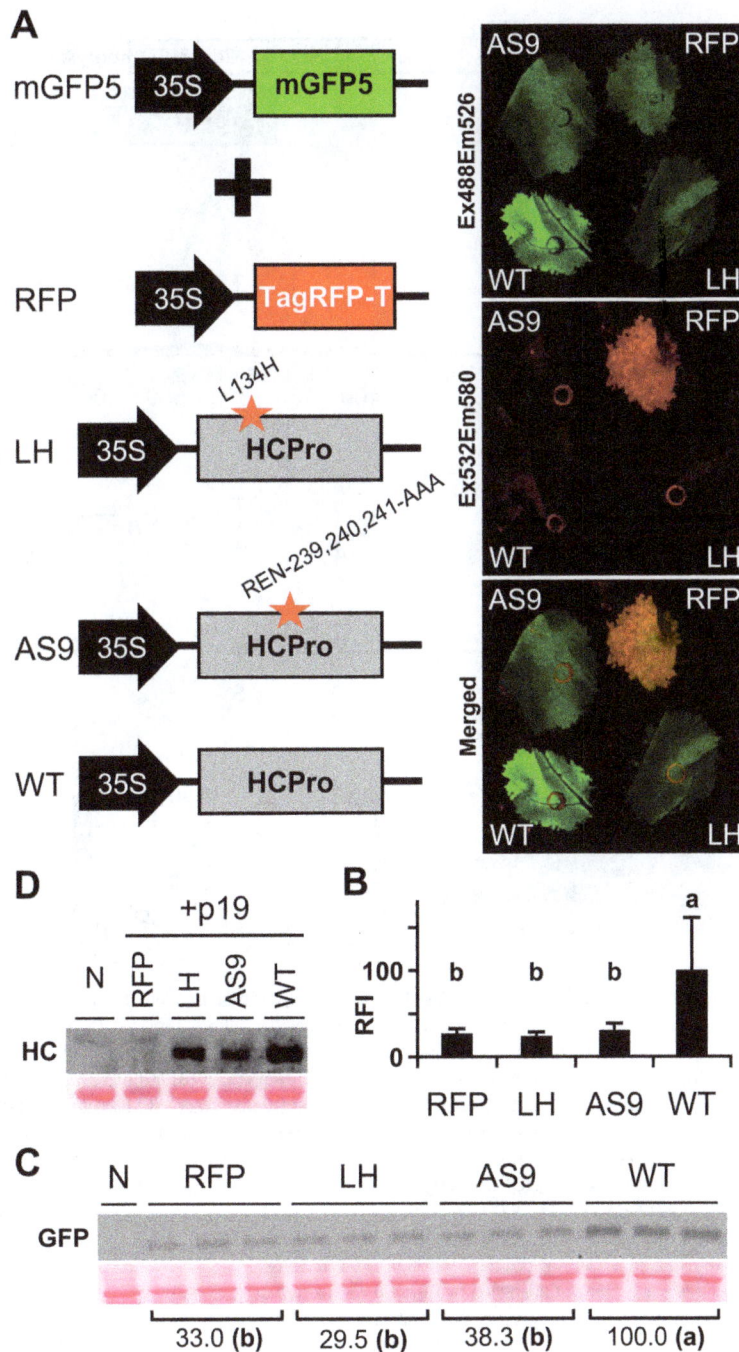

Figure 3 Quantification of GFP accumulation in transient RNA silencing assay. (A) GFP was transiently expressed by co-infiltrating *N. benthamiana* leaf tissue with an *Agrobacterium* pBin-35S-mGFP5 culture plus cultures of *Agrobacterium* containing pSN.5 TagRFP-T (RFP), pSN.5 HC-L134H (LH, producing PPV HCPro L134H mutant), pSN.5 HC-AS9 (AS9, producing a PPV HCPro mutant in which amino acids REN-239, 240, 241 were replaced by alanines) or pSN.5 wtHC (WT, producing wild-type PPV HCPro). At 6 dpa, leaf fluorescence was acquired by laser scanning using Ex488/Em526 (green) and Ex532/Em580 (red); the image overlay is shown (Merged). **(B)** GFP fluorescence intensity of the agro-infiltrated leaf patches was quantified in a 96-well plate reader. RFI was plotted using WT mean value equal to 100. Bar graph shows mean ± SD ($n = 14$ biological replicates from two independent *Agrobacterium* cultures); the difference between the results marked with different letters is statistically significant, $p < 0.01$, one-way Anova and Tukey's HSD test. **(C)** GFP protein accumulation in infiltrated leaves at 6 dpa was assessed by immunoblot analysis. Relative GFP signal intensities are indicated using average WT equal to 100; the difference between the values marked with different letters is statistically significant, $p < 0.01$, one-way Anova and Tukey's HSD test. Each lane represents a pool of 3 or 4 leaf samples infiltrated with two independent *Agrobacterium* cultures. N, non-treated leaf sample. Ponceau red-stained blot as loading control. **(D)** HCPro expression by the binary vectors tested was assessed by HCPro immunoblot analysis of leaf co-infiltrated with an *Agrobacterium* pSN.5 p19 culture (6 dpa). Each lane represents a pool of infiltrated leaf samples. N, non-treated leaf sample. Ponceau red-stained blot as loading control.

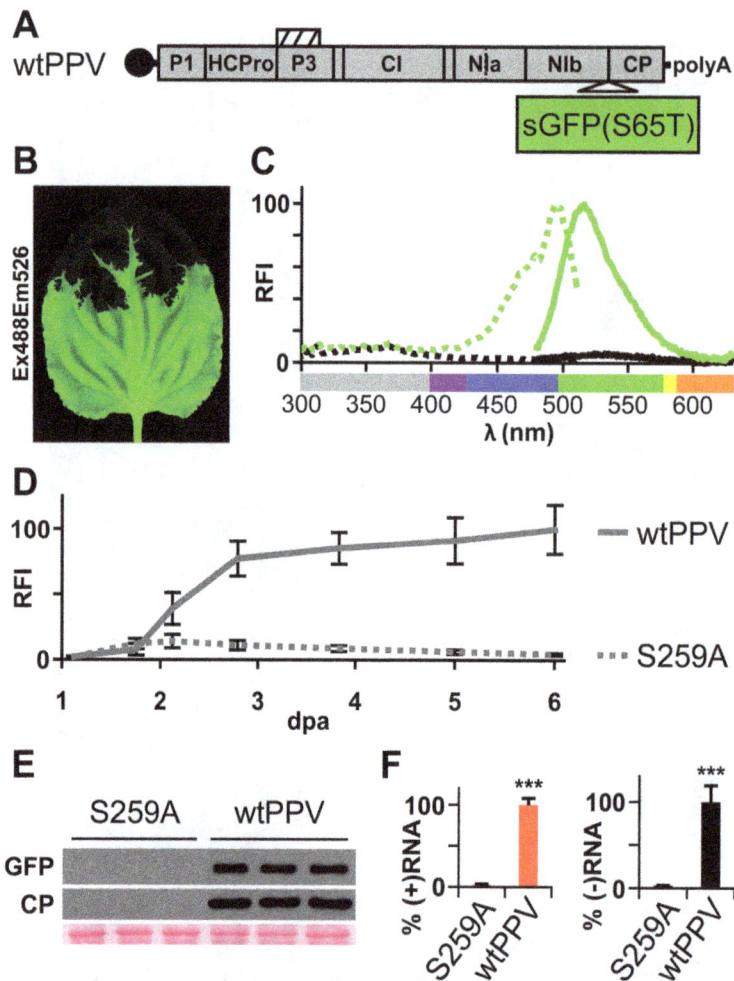

Figure 4 Monitoring of GFP-tagged virus amplification dynamics by fluorescence spectroscopy. GFP-tagged viral cDNA clones pSN-PPV (wtPPV, wild-type PPV) and pSN-PPV P1-S (S259A, in which P1 protease catalytic amino acid S259 was replaced by alanine) were delivered to plants by agro-infiltration. **(A)** Diagram of wild-type PPV (wtPPV) genome originated following pSN-PPV agro-infiltration. Hatched box indicates P3N-PIPO protein. The reporter sGFP(S65T) gene is inserted between NIb and CP coding sequences. **(B)** *N. benthamiana* plants were challenged with pSN-PPV, and fluorescence of systemically infected leaves was detected by laser scanning (10 dpa; green). **(C)** Excitation (dotted lines) and emission spectra (solid lines) of sGFP(S65T) were measured from pSN-PPV agro-inoculated leaf discs (green); leaves infiltrated with an *Agrobacterium* culture without expression vectors were used as control (black). Relative fluorescence intensity (RFI) was plotted using sGFP(S65T) peaks equal to 100. UV wavelengths are in gray, visible spectrum colors were assigned as described [39]. **(D)** GFP fluorescent intensity (RFI) from infiltrated leaves was quantified in a 96-well plate reader and plotted using average wtPPV value at 6 dpa equal to 100. Line graph shows mean ± SD (*n* = 16 samples/condition, from two independent *Agrobacterium* cultures). **(E)** Amount of GFP protein and PPV CP in infiltrated leaves at 6 dpa was assessed by immunoblot analysis. Each lane represents a pool of 3 or 4 leaf samples infiltrated with two independent *Agrobacterium* cultures. Ponceau red-stained blot is shown as loading control. **(F)** Amount of viral (+)RNA and (−)RNA from inoculated leaves at 6 dpa was quantified by RT-qPCR and plotted using average wtPPV value equal to 100. Bar graph shows mean ± SD (*n* = 4 biological replicates, from two independent *Agrobacterium* cultures); ***$p < 0.001$, Student's *t*-test.

reporters with minimal spectral overlap assures signal specificity (Additional file 2).

Conclusions

We present laser scanning as an alternative method for fluorescence imaging of plant samples that, due to their size, cannot be acquired in their entirety in standard fluorescence stereomicroscopes. Dual-color fluorescence imaging of leaf samples is achieved using fluorophore combinations with minimal spectral overlap, such

as mGFP5/TagRFP-T and mTFP1/mPapaya1, and image analysis can be used for raw quantitative comparisons.

We show that fluorescence plate readers are extremely powerful tools for medium-high throughput analysis of fluorescent proteins expressed in plant tissue, making it feasible to collect data from a 96-well plate in a few minutes. Fluorescence intensity is readily quantified in leaf discs, with no need to prepare protein extracts. A large number of improved fluorescent proteins have been developed, and proteins with reduced biological half-life,

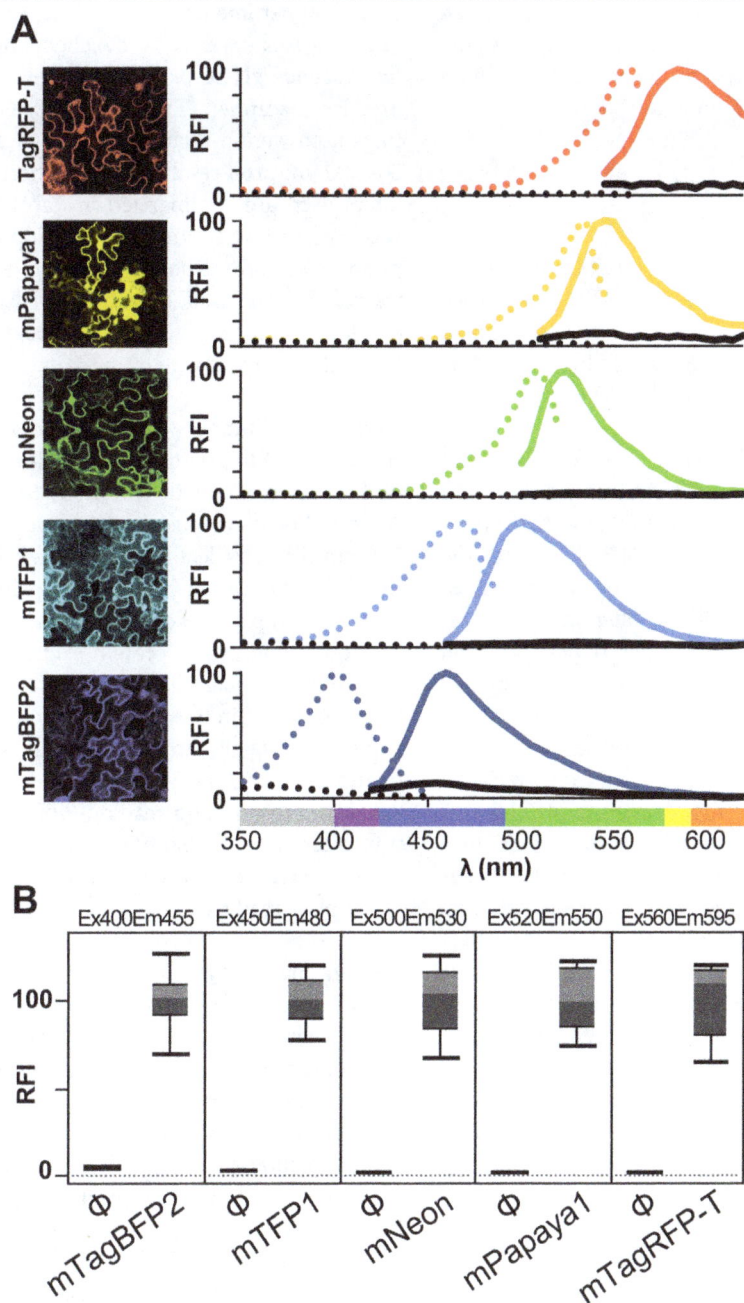

Figure 5 Direct leaf disc analysis of engineered monomeric fluorescent proteins. Fluorescent proteins were transiently expressed by co-infiltrating *N. benthamiana* leaves with an *Agrobacterium* pSN.5 p19 culture plus cultures of *Agrobacterium* containing pSN.5 mTagBFP2 (mTagBFP2), pSN.5 mTFP1 (mTFP1), pSN.5 mNeon (mNeon), pSN.5 mPapaya1 (mPapaya1), pSN.5 TagRFP-T (TagRFP-T) or a strain with no expression vector (Φ). **(A)** At 6 dpa, cell fluorescence was imaged by confocal microscopy. Fluorophore excitation (dotted lines) and emission spectra (solid lines) from agro-infiltrated leaf discs were measured in a 96-well plate reader. Leaves infiltrated with an *Agrobacterium* culture without expression vectors were used as control (black lines). Relative fluorescence intensity (RFI) was plotted using fluorophore peaks equal to 100. UV wavelengths are in gray, visible spectrum colors were assigned as described [39]. **(B)** Box-plot graphs show quantification values from *n* = 8 samples/condition. Fluorescence intensity of the leaf discs agro-infiltrated with the indicated fluorescent protein-expressing plasmid or without expression vector (Φ) was measured in a monochromator-based plate reader. RFI was plotted using each fluorophore mean value equal to 100.

rapid choromophore maturation and photoactivable variants [3,49-51] might be used to increase assay sensitivity and temporal resolution for kinetic studies. We show that co-expression of TagRFP-T has no appreciable effect on fluorescence intensity quantification of mGFP5. A battery of fluorescent proteins that have minimal sequence identity with the widely used *A. victoria* GFP sequence was quantified easily in a monochromator-type plate reader. We anticipate that the method presented will aid in the design of fluorescence-based experiments with single and multiple reporter genes and facilitate comparisons of fluorophore amounts.

Methods
DNA plasmids and constructs
The binary vector pSN-PPV bearing a full-length cDNA copy of a PPV isolate and its variant pSN-PPV P1-S were reported [45]. An *Agrobacterium* strain GV3101 containing the binary vector pBin-35S-mGFP5 was kindly provided by D. Baulcombe (University of Cambridge, Cambridge, UK). For the remaining transient expression vectors, genes of interest were inserted into XbaI/PmlI-digested pSN2-ccdB [45] by Gibson assembly [52]. Briefly, to obtain pSN.5 TagRFP-T (encoding a mutant red TagRFP), the TagRFP sequence was amplified from pSITEII-6C1 [53] and the S158T mutation [35] was inserted by the overlap extension method [54]. For pSN.5 mTagBFP2, blue mTagBFP sequence was amplified from pGGC024 [55] (kindly provided by J. Forner, Universität Heidelberg, Heidelberg, Germany), and the I174A mutation [31] was inserted. For pSN.5 mTFP1, the cyan mTFP1 sequence was synthesized *de novo* (GeneArt, Life Technologies). For pSN.5 mNeon, green the mNeonGreen sequence was amplified from pICSL80019 [56], kindly provided by M. Youles (The Sainsbury Laboratory, Norwich, UK). For pSN.5 mPapaya1, the yellow mPapaya1 sequence was synthesized *de novo* (GeneArt, Life Technologies). For pSN.5 wtHC, PPV HCPro was amplified from pSN-PPV ΔP1 [45]; for pSN.5 HC-L134H, PPV HCPro was amplified from pSN-PPV ΔP1 and the L134H mutation inserted, whereas for pSN.5 HC-AS9, PPV HCPro was amplified from pSN-PPV ΔP1 and amino acids REN-239,240,241 were replaced by alanines. For pSN.5 p19, tomato bushy stunt virus p19 was amplified from pBIN61-P19 [38]. In all the newly-generated constructs, coding sequences are driven by a double enhancer *Cauliflower mosaic virus* 35S promoter, flanked by PPV 5'UTR and 3'UTR, followed by a nopaline synthase terminator.

Plant agro-infiltration
Nicotiana benthamiana and *N. clevelandii* were grown in a greenhouse maintained at a 16 h light/8 h dark photoperiod, temperature range 19-23°C. Agro-infiltration of *N. benthamiana* and *N. clevelandii* plants was as described [6]; whenever possible, tested constructs were delivered in individual patches of the same leaf. The viral replication assay was conducted in three-week-old *N. clevelandii* plants following agroinfiltration and sampling guidelines [14], with the exception that a saturating concentration of *Agrobacterium* (OD$_{600}$ 1.0) was used.

Laser scanner imaging
Plant leaves were sandwiched between two low-fluorescence glasse plates and fluorescence was acquired in a laser scanner (Typhoon 9400, GE Healthcare). Settings used were normal sensitivity, focal plane +3 mm and 50–100 μm pixel resolution; excitation lasers and emission filters used are summarized in Table 1. Signal saturation was avoided by adjusting photomultiplier tube voltage. Typhoon data were exported to 16-bit .tiff files. ImageJ software [57] was used to produce false-color images and overlays, and to generate 3D-projections through the Interactive 3D Surface Plot plug-in.

Fluorescence intensity measurements
Black 96-well flat-bottom plates (Nunc) with 50 μL water/well (to limit sample dehydration) were used for the assay. A cork borer was used for tissue sampling; individual 5.0 mm-diameter leaf discs, collected at the same distance from the infiltration point, were placed upside down in the prepared plates. Top reading measurements were used to acquire fluorescent protein excitation, emission spectra and intensity quantification in a monochromator-based plate reader (Infinite M200, Tecan Group). Gain value was adjusted manually to avoid signal saturation. RFI was quantified using the excitation and emission bands indicated in Table 1. Top reading GFP fluorescence intensity was alternatively quantified in an Appliskan (Thermo Fisher Scientific) and/or Victor X2 (PerkinElmer) filter-based plate readers.

Western blot assays
Liquid nitrogen-frozen plant tissue was homogenized in a TissueLyzer bead mill (Qiagen). Total proteins were extracted, separated by glycine-SDS-PAGE and electroblotted onto a nitrocellulose membrane, as reported [45]. Proteins were detected using rabbit anti-PPV CP and -PPV HCPro sera, and mouse anti-GFP monoclonal antibody (clones 7.1 and 13.1, Roche) as primary antibodies; horseradish peroxidase-conjugated goat anti-rabbit IgG (Jackson) or sheep anti-mouse IgG (GE Healthcare) were used as secondary antibody. For signal quantification, chemiluminescence was acquired in a ChemiDoc XRS imager (BioRad) and analyzed with ImageJ.

RT-qPCR
Total RNA was extracted with the FavorPrep Plant Total RNA Mini kit (Favorgen), including on-column DNAseI treatment. Purified RNA was quantified spectrophotometrically by NanoDrop (Thermo Fisher Scientific) and concentration adjusted to 50 ng/μL. Strand-specific cDNA for PPV RNA was synthesized for at least three biological replicates per condition using tagged cDNA primers in the RT step [58]. The 10-μL RT reactions contained 100 ng of total RNA and (at final concentrations) 1x Superscript III

first-strand buffer, 0.5 mM of each dNTP, 5.0 mM dithio-threitol, 1.0 U/μL RiboLock (Fermentas), 5.0 U/μL Super-script III (Invitrogen) and 50 nm primer Q26_R or Q29_F (Additional file 1) to transcribe cDNA from positive and negative PPV genomes, respectively. Mixtures were incubated (35 min at 56°C, 10 min at 95°C), cooled to room temperature and diluted 1/10 - 1/25 with nuclease-free water. Technical triplicate 8 μL qPCR reactions were prepared in 384-well optical plates using 4 μL diluted cDNA sample, 1x Hot FIREPol EvaGreen qPCR Mix Plus (Solis BioDyne), 195 nM each of primer pair Q27_F/Q28_R, or 300 nM each of primer pair Q30_F/Q31_R (Additional file 1) for quantification of positive and negative PPV genomes, respectively. In a 7900HT Fast Real-Time PCR System (Applied Biosystems), reactions were subjected to 10 min at 95°C activation step, 40 cycles of 95°C, 30 s and 60°C, 60 s, followed by a final dissociation curve analysis step. Absolute quantification was done using external DNA standard curves [59]. Briefly, *Nicotiana* plants were agro-inoculated with pSN-PPV, total RNA was purified from systemically infected tissue and reverse-transcribed using the High-Capacity cDNA Archive Kit (Applied Biosystems). cDNA was used as template for PCR reactions which contained primer pair Q25_F/Q26_R or primer pair Q23_R/Q29_F for positive and negative PPV genomes, respectively. Amplicons were gel-purified and serially diluted to generate qPCR standard templates. Strand specificity of RT-qPCR assays was evaluated using synthetic positive and negative strand PPV RNA fragments. The T7 Φ2.5 promoter sequence was incorporated into PCR fragments amplified using primers Q22_F/Q23_R and Q24_R/Q25_F for positive and negative RNA strand templates, respectively. *In vitro* transcription and RNA purification were as described [45]. Healthy *Nicotiana* total RNA was used as carrier for 10-fold dilutions of target RNA alone or with a fixed amount of complementary RNA. RNA samples were reverse-transcribed in triplicate and used as template in qPCR reactions, as above.

Additional files

Additional file 1: Primers and PPV target region used in RT-qPCR viral RNA quantification. (A) Sequence and use of the RT-qPCR primers. Nucleotides identical to pSN-PPV-derived viral RNA sequence are shown in uppercase letters. Non-viral tag sequences are in bold, 5′ clamps to increase annealing stability are underlined and the T7 Φ2.5 promoter sequence is double-underlined. Application as follows: T7(+), *in vitro* transcription of positive strand RNA with T7 RNA polymerase; T7(−), *in vitro* transcription of negative strand RNA with T7 RNA polymerase; S(+), generation of template for positive strand standard curve; RT(+), positive strand-specific cDNA synthesis; Q(+), qPCR amplification of positive strand; S(−), generation of template for negative strand standard curve; RT(−), negative strand-specific cDNA synthesis; Q(−), qPCR amplification of negative strand. (B) Detailed scheme of the pSN-PPV binary vector used for PPV delivery to plants. P3N-PIPO protein was omitted for clarity. A 189 bp intron from the potato ST-LS-1 gene, inserted in the P3 sequence [GenBank: EF569215.1] to increase cDNA vector stability [60], is shown as a hatched

box. Region flanking the P3 splicing site of pSN-PPV-derived viral RNA is shown. Positive (+) and negative (−) PPV sequences are represented with the primers used for RT-qPCR quantifications. Reverse transcription primers were designed to span the P3 intron junction of spliced viral RNAs. Brackets indicate qPCR amplicon regions. Diagram is not to scale. (C) Strand specificity of RT-qPCR assays. Standard curves were generated from cDNA synthesis reactions into which target RNA was mixed with 100 ng of *Nicotiana* total RNA alone (circles) or in the presence of a competing strand (squares). Cycle threshold numbers were plotted against the logarithm of target RNA.

Additional file 2: Quantification of engineered monomeric fluorescent proteins in multicolor experiments. Fluorescent proteins were transiently expressed by co-infiltrating *N. benthamiana* leaves with an *Agrobacterium* pSN.5 p19 culture plus cultures of *Agrobacterium* containing pSN.5 mTagBFP2 (mTagBFP2), pSN.5 mTFP1 (mTFP1), pSN.5 mNeon (mNeon), pSN.5 mPapaya1 (mPapaya1) or pSN.5 TagRFP-T (TagRFP-T). At 6 dpa, fluorescence intensity of the leaf discs agro-infiltrated with the indicated fluorescent protein-expressing plasmid was measured in a monochromator-based plate reader. Evaluated excitation and emission wavelengths are shown on the left, and summarized in Table 1. Box-plot graphs show quantification values from $n = 8$ samples/condition. FI is expressed in arbitrary units.

Abbreviations
GFP: Green fluorescent protein; PPV: *Plum pox virus*; UV: Ultraviolet; (R)FI: (Relative) Fluorescence intensity; dpa: Days post-agro-infiltration; CP: Coat protein.

Competing interests
The authors declare that they have no financial or other competing interests.

Authors' contributions
Conceived and designed the experiments: FP, SK, PN, CS-M, JAG. Performed the experiments: FP. Analyzed the data: FP, JAG. Contributed reagents/materials/analysis tools: FP, CS-M, JAG. Wrote the paper: FP, JAG. All authors read and approved the final manuscript.

Acknowledgements
We thank S. Landeras-Bueno for fluorometer analysis assistance, J.L. Martínez for use of the Infinite M200 fluorometer, M. Fernández for help with laser scanner imaging, J.A. Abelenda and L. Almonacid for help with qPCR, and G. Castrillo and S. Prat for discussion. We are grateful to D. Baulcombe, J. Forner and M. Youles for supply of material. Special thanks to C. Mark for editorial assistance. FP is financed by a La Caixa PhD fellowship and acknowledges support from S. Pasin and L. Lievore. This work was funded by grants BIO2010-18541 and BIO2013-49053-R from the Spanish government. The publication fee was covered partially by the CSIC Open Access Publication Support Initiative through the Unit of Information Resources for Research (URICI).

References
1. Rosellini D: **Selectable markers and reporter genes: a well furnished toolbox for plant science and genetic engineering.** *Crit Rev Plant Sci* 2012, 31:401–453.
2. Chalfie M, Tu Y, Euskirchen G, Ward WW, Prasher DC: **Green fluorescent protein as a marker for gene expression.** *Science* 1994, 263:802–805.
3. Shaw SL, Ehrhardt DW: **Smaller, faster, brighter: advances in optical imaging of living plant cells.** *Annu Rev Plant Biol* 2013, 64:351–375.
4. Johansen LK, Carrington JC: **Silencing on the spot. Induction and suppression of RNA silencing in the *Agrobacterium*-mediated transient expression system.** *Plant Physiol* 2001, 126:930–938.
5. Giner A, Lakatos L, García-Chapa M, López-Moya JJ, Burgyán J: **Viral protein inhibits RISC activity by argonaute binding through conserved WG/GW motifs.** *PLoS Pathog* 2010, 6:e1000996.
6. Maliogka VI, Calvo M, Carbonell A, García JA, Valli A: **Heterologous RNA-silencing suppressors from both plant- and animal-infecting viruses support plum pox virus infection.** *J Gen Virol* 2012, 93:1601–1611.
7. Haikonen T, Rajamäki M-L, Valkonen JPT: **Improved silencing suppression and enhanced heterologous protein expression are achieved using an**

engineered viral helper component proteinase. *J Virol Methods* 2013, 193:687–692.

8. Sainsbury F, Thuenemann EC, Lomonossoff GP: **pEAQ: versatile expression vectors for easy and quick transient expression of heterologous proteins in plants.** *Plant Biotechnol J* 2009, 7:682–693.

9. Gronenborn B, Gardner RC, Schaefer S, Shepherd RJ: **Propagation of foreign DNA in plants using cauliflower mosaic virus as vector.** *Nature* 1981, 294:773–776.

10. Brisson N, Paszkowski J, Penswick JR, Gronenborn B, Potrykus I, Hohn T: **Expression of a bacterial gene in plants by using a viral vector.** *Nature* 1984, 310:511–514.

11. Takamatsu N, Ishikawa M, Meshi T, Okada Y: **Expression of bacterial chloramphenicol acetyltransferase gene in tobacco plants mediated by TMV-RNA.** *EMBO J* 1987, 6:307–311.

12. French R, Janda M, Ahlquist P: **Bacterial gene inserted in an engineered RNA virus: efficient expression in monocotyledonous plant cells.** *Science* 1986, 231:1294–1297.

13. Joshi RL, Joshi V, Ow DW: **BSMV genome mediated expression of a foreign gene in dicot and monocot plant cells.** *EMBO J* 1990, 9:2663–2669.

14. Eskelin K, Suntio T, Hyvärinen S, Hafren A, Mäkinen K: *Renilla* **luciferase-based quantitation of *Potato virus A* infection initiated with *Agrobacterium* infiltration of *N. benthamiana* leaves.** *J Virol Methods* 2010, 164:101–110.

15. Dolja VV, McBride HJ, Carrington JC: **Tagging of plant potyvirus replication and movement by insertion of beta-glucuronidase into the viral polyprotein.** *Proc Natl Acad Sci U S A* 1992, 89:10208–10212.

16. Chapman S, Kavanagh T, Baulcombe D: **Potato virus X as a vector for gene expression in plants.** *Plant J* 1992, 2:549–557.

17. Bedoya LC, Martínez F, Orzáez D, Daròs J-A: **Visual tracking of plant virus infection and movement using a reporter MYB transcription factor that activates anthocyanin biosynthesis.** *Plant Physiol* 2012, 158:1130–1138.

18. Baulcombe DC, Chapman S, Santa Cruz S: **Jellyfish green fluorescent protein as a reporter for virus infections.** *Plant J* 1995, 7:1045–1053.

19. Fernández-Fernández MR, Mouriño M, Rivera J, Rodríguez F, Plana-Durán J, García JA: **Protection of rabbits against rabbit hemorrhagic disease virus by immunization with the VP60 protein expressed in plants with a potyvirus-based vector.** *Virology* 2001, 280:283–291.

20. Dietrich C, Maiss E: **Fluorescent labelling reveals spatial separation of potyvirus populations in mixed infected *Nicotiana benthamiana* plants.** *J Gen Virol* 2003, 84:2871–2876.

21. Tromas N, Zwart MP, Lafforgue G, Elena SF: **Within-host spatiotemporal dynamics of plant virus infection at the cellular level.** *PLoS Genet* 2014, 10:e1004186.

22. Remans T, Schenk PM, Manners JM, Grof CP, Elliott AR: **A protocol for the fluorometric quantification of mGFP5-ER and sGFP (S65T) in transgenic plants.** *Plant Mol Biol Report* 1999, 17:385–395.

23. Richards HA, Halfhill MD, Millwood RJ, Stewart CN Jr: **Quantitative GFP fluorescence as an indicator of recombinant protein synthesis in transgenic plants.** *Plant Cell Rep* 2003, 22:117–121.

24. Dhillon T, Chiera JM, Lindbo JA, Finer JJ: **Quantitative evaluation of six different viral suppressors of silencing using image analysis of transient GFP expression.** *Plant Cell Rep* 2009, 28:639–647.

25. Stephan D, Slabber C, George G, Ninov V, Francis KP, Burger JT: **Visualization of plant viral suppressor silencing activity in intact leaf lamina by quantitative fluorescent imaging.** *Plant Methods* 2011, 7:25.

26. Millwood RJ, Halfhill MD, Harkins D, Russotti R, Stewart CN: **Instrumentation and methodology for quantifying GFP fluorescence in intact plant organs.** *Biotechniques* 2003, 34:638–643.

27. Lukhovitskaya NI, Solovieva AD, Boddeti SK, Thaduri S, Solovyev AG, Savenkov EI: **An RNA virus-encoded zinc-finger protein acts as a plant transcription factor and induces a regulator of cell size and proliferation in two tobacco species.** *Plant Cell* 2013, 25:960–973.

28. Castrillo G, Sánchez-Bermejo E, De Lorenzo L, Crevillén P, Fraile-Escanciano A, TC M, Mouriz A, Catarecha P, Sobrino-Plata J, Olsson S, Leo Del Puerto Y, Mateos I, Rojo E, Hernández LE, Jarillo JA, Piñeiro M, Paz-Ares J, Leyva A: **WRKY6 transcription factor restricts arsenate uptake and transposon activation in *Arabidopsis*.** *Plant Cell* 2013, 25:2944–2957.

29. Siemering KR, Golbik R, Sever R, Haseloff J: **Mutations that suppress the thermosensitivity of green fluorescent protein.** *Curr Biol* 1996, 6:1653–1663.

30. Chiu W, Niwa Y, Zeng W, Hirano T, Kobayashi H, Sheen J: **Engineered GFP as a vital reporter in plants.** *Curr Biol* 1996, 6:325–330.

31. Subach OM, Cranfill PJ, Davidson MW, Verkhusha VV: **An enhanced monomeric blue fluorescent protein with the high chemical stability of the chromophore.** *PLoS One* 2011, 6:e28674.

32. Ai H, Henderson JN, Remington SJ, Campbell RE: **Directed evolution of a monomeric, bright and photostable version of *Clavularia* cyan fluorescent protein: structural characterization and applications in fluorescence imaging.** *Biochem J* 2006, 400:531–540.

33. Shaner NC, Lambert GG, Chammas A, Ni Y, Cranfill PJ, Baird MA, Sell BR, Allen JR, Day RN, Israelsson M, Davidson MW, Wang J: **A bright monomeric green fluorescent protein derived from *Branchiostoma lanceolatum*.** *Nat Methods* 2013, 10:407–409.

34. Hoi H, Howe ES, Ding Y, Zhang W, Baird MA, Sell BR, Allen JR, Davidson MW, Campbell RE: **An engineered monomeric *Zoanthus sp.* yellow fluorescent protein.** *Chem Biol* 2013, 20:1296–1304.

35. Shaner NC, Lin MZ, McKeown MR, Steinbach PA, Hazelwood KL, Davidson MW, Tsien RY: **Improving the photostability of bright monomeric orange and red fluorescent proteins.** *Nat Methods* 2008, 5:545–551.

36. Marouga R, David S, Hawkins E: **The development of the DIGE system: 2D fluorescence difference gel analysis technology.** *Anal Bioanal Chem* 2005, 382:669–678.

37. Chapman S, Oparka KJ, Roberts AG: **New tools for in vivo fluorescence tagging.** *Curr Opin Plant Biol* 2005, 8:565–573.

38. Voinnet O, Rivas S, Mestre P, Baulcombe D: **An enhanced transient expression system in plants based on suppression of gene silencing by the p19 protein of tomato bushy stunt virus.** *Plant J* 2003, 33:949–956.

39. Orna MV: **Discovery of the physics of color.** In *Chem Hist Color.* Springer; 2013:11–28.

40. González-Jara P, Atencio FA, Martínez-García B, Barajas D, Tenllado F, Díaz-Ruíz JR: **A single amino acid mutation in the *Plum pox virus* helper component-proteinase gene abolishes both synergistic and RNA silencing suppression activities.** *Phytopathology* 2005, 95:894–901.

41. Valli A, Gallo A, Calvo M, Pérez JJ, García JA: **A novel role of the potyviral helper component proteinase contributes to enhance the yield of viral particles.** *J Virol* 2014, doi:10.1128/JVI.01010-14.

42. Kasschau KD, Carrington JC: **Long-distance movement and replication maintenance functions correlate with silencing suppression activity of potyviral HC-Pro.** *Virology* 2001, 285:71–81.

43. Torres-Barceló C, Martín S, Daròs J-A, Elena SF: **From hypo- to hypersuppression: effect of amino acid substitutions on the RNA-silencing suppressor activity of the *Tobacco etch potyvirus* HC-Pro.** *Genetics* 2008, 180:1039–1049.

44. Garcia-Ruiz H, Takeda A, Chapman EJ, Sullivan CM, Fahlgren N, Brempelis KJ, Carrington JC: *Arabidopsis* **RNA-dependent RNA polymerases and Dicer-like proteins in antiviral defense and small interfering RNA biogenesis during *Turnip mosaic virus* infection.** *Plant Cell* 2010, 22:481–496.

45. Pasin F, Simón-Mateo C, García JA: **The hypervariable amino-terminus of P1 protease modulates potyviral replication and host defense responses.** *PLoS Pathog* 2014, 10:e1003985.

46. Carrington JC, Haldeman R, Dolja VV, Restrepo-Hartwig MA: **Internal cleavage and *trans*-proteolytic activities of the VPg-proteinase (NIa) of tobacco etch potyvirus in vivo.** *J Virol* 1993, 67:6995–7000.

47. Dolja VV, Haldeman R, Robertson NL, Dougherty WG, Carrington JC: **Distinct functions of capsid protein in assembly and movement of tobacco etch potyvirus in plants.** *EMBO J* 1994, 13:1482–1491.

48. Hafrén A, Hofius D, Rönnholm G, Sonnewald U, Mäkinen K: **HSP70 and its cochaperone CPIP promote potyvirus infection in *Nicotiana benthamiana* by regulating viral coat protein functions.** *Plant Cell* 2010, 22:523–535.

49. Li X, Zhao X, Fang Y, Jiang X, Duong T, Fan C, Huang C-C, Kain SR: **Generation of destabilized green fluorescent protein as a transcription reporter.** *J Biol Chem* 1998, 273:34970–34975.

50. Chapman S, Faulkner C, Kaiserli E, Garcia-Mata C, Savenkov EI, Roberts AG, Oparka KJ, Christie JM: **The photoreversible fluorescent protein iLOV outperforms GFP as a reporter of plant virus infection.** *Proc Natl Acad Sci U S A* 2008, 105:20038–20043.

51. Nienhaus K, Nienhaus GU: **Fluorescent proteins for live-cell imaging with super-resolution.** *Chem Soc Rev* 2014, 43:1088–1106.

52. Gibson DG, Young L, Chuang R-Y, Venter JC, Hutchison CA III, Smith HO: **Enzymatic assembly of DNA molecules up to several hundred kilobases.** *Nat Methods* 2009, 6:343–345.

53. Martin K, Kopperud K, Chakrabarty R, Banerjee R, Brooks R, Goodin MM: **Transient expression in *Nicotiana benthamiana* fluorescent marker lines**

provides enhanced definition of protein localization, movement and interactions *in planta*. *Plant J* 2009, **59**:150–162.

54. Horton RM, Hunt HD, Ho SN, Pullen JK, Pease LR: **Engineering hybrid genes without the use of restriction enzymes: gene splicing by overlap extension.** *Gene* 1989, **77**:61–68.

55. Lampropoulos A, Sutikovic Z, Wenzl C, Maegele I, Lohmann JU, Forner J: **GreenGate - A novel, versatile, and efficient cloning system for plant transgenesis.** *PLoS One* 2013, **8**:e83043.

56. Engler C, Youles M, Gruetzner R, Ehnert T-M, Werner S, Jones JD, Patron NJ, Marillonnet S: **A Golden Gate modular cloning toolbox for plants.** *ACS Synth Biol* 2014, doi:10.1021/sb4001504.

57. Schneider CA, Rasband WS, Eliceiri KW: **NIH Image to ImageJ: 25 years of image analysis.** *Nat Methods* 2012, **9**:671–675.

58. Komurian-Pradel F, Perret M, Deiman B, Sodoyer M, Lotteau V, Paranhos-Baccalà G, André P: **Strand specific quantitative real-time PCR to study replication of hepatitis C virus genome.** *J Virol Methods* 2004, **116**:103–106.

59. Pfaffl MW, Hageleit M: **Validities of mRNA quantification using recombinant RNA and recombinant DNA external calibration curves in real-time RT-PCR.** *Biotechnol Lett* 2001, **23**:275–282.

60. López-Moya JJ, García JA: **Construction of a stable and highly infectious intron-containing cDNA clone of plum pox potyvirus and its use to infect plants by particle bombardment.** *Virus Res* 2000, **68**:99–107.

3

A novel cost effective and high-throughput isolation and identification method for marine microalgae

Martin T Jahn[1,2], Katrin Schmidt[1] and Thomas Mock[1*]

Abstract

Background: Marine microalgae are of major ecologic and emerging economic importance. Biotechnological screening schemes of microalgae for specific traits and laboratory experiments to advance our knowledge on algal biology and evolution strongly benefit from culture collections reflecting a maximum of the natural inter- and intraspecific diversity. However, standard procedures for strain isolation and identification, namely DNA extraction, purification, amplification, sequencing and taxonomic identification still include considerable constraints increasing the time required to establish new cultures.

Results: In this study, we report a cost effective and high-throughput isolation and identification method for marine microalgae. The throughput was increased by applying strain isolation on plates and taxonomic identification by direct PCR (dPCR) of phylogenetic marker genes in combination with a novel sequencing electropherogram based screening method to assess the taxonomic diversity and identity of the isolated cultures. For validation of the effectiveness of this approach, we isolated and identified a range of unialgal cultures from natural phytoplankton communities sampled in the Arctic Ocean. These cultures include the isolate of a novel marine Chlorophyceae strain among several different diatoms.

Conclusions: We provide an efficient and effective approach leading from natural phytoplankton communities to isolated and taxonomically identified algal strains in only a few weeks. Validated with sensitive Arctic phytoplankton, this approach overcomes the constraints of standard molecular characterisation and establishment of unialgal cultures.

Keywords: Marine microalgae, Direct PCR, Isolation, Cultivation, Taxonomy

Background

Marine microalgae are unicellular photosynthetic eukaryotes of major ecological and economic importance worldwide. Ecologically, they are the base of the marine food web and contribute to at least 30% of annual CO_2 fixation worldwide and therefore massively impact global biogeochemical cycles [1,2]. Economically, diverse marine microalgae are used or have the potential to be used as nutraceuticals, for the production of pharmaceuticals [3,4], cosmetics [5], for bioremediation [6-8], and biofuels [9].

In recent years, the emerging application of culture-independent omics approaches like metagenomics and metatranscriptomics delivered comprehensive insights into the gene repertoire and activity of marine microalgal communities [10-13]. However, results from high-throughput omics approaches ideally need to be scrutinized by experiments with isolated strains from the same communities if the scientific endeavour goes beyond purely describing the diversity and abundance of genes and transcripts in relation to environmental conditions. Similarly, in the field of microalgae biotechnology, novel isolation and identification protocols are essential for identifying specific traits like lipid content [14,15] or any other bioactive compounds [16]. Thus, there is a high demand to develop novel isolation and identification protocols. However, laborious standard procedures such as single-cell isolation of

* Correspondence: t.mock@uea.ac.uk
[1]School of Environmental Sciences, University of East Anglia, Norwich Research Park, Norwich NR4 7TJ, UK
Full list of author information is available at the end of the article

strains, DNA extraction, purification, amplification, sequencing and taxonomic identification include several time, cost and space consuming constraints.

To overcome these constraints, we developed a new cost effective and high-throughput isolation and identification method for marine microalgae. We combined high throughput isolation by streaking cells from enrichment cultures on agar plates with subsequent cultivation in multi-well plates. To assess as to whether a culture was unialgal or not, we applied direct PCR (dPCR) by only using boiling MiliQ water to lyse the cells in combination with a novel sequencing electropherogram based assessment method. While using the V4 as the most variable small subunit (SSU) [17], the underlying idea was that molecular marker sequences of different species possess different bases at the same position. This concept is similar to the detection of intraspecific point mutations exploiting sequencing electropherogram tracefiles [18]. The ambiguous base-calls detected as a biased uncalled/called peak ratio increase the position specific error probability (Pe) [19], which decreases the per-base Quality Values (QV = -10 $\log_{10}$(Pe)) as a standard quality metric [20]. The per-base quality values were used in our approach to evaluate the presence or absence of an unialgal culture.

This new approach is relatively cost effective, time saving and high throughput to overcome the constraints of standard molecular characterisation (e.g. by DGGE or RFLP) and establishment of unialgal cultures without the need of DNA extraction and cloning. To validate the efficiency of this approach, we isolated and identified algal strains from natural phytoplankton communities of the Arctic Ocean.

Results

The objective of this study was to establish a cost and time effective method for microalgal isolation and identification. Using the methods described below, we were able to obtain 24 unialgal cultures consisting of 7 unique ribotypes based on the V4 18S rDNA region.

Efficiency of growing algae on plates and dPCR

Using the high-throughput isolation technique of streaking enriched natural microalgal communities across agar plates, on 59.3% (35 of 59) of the incubated plates algal growth was detected. From about three quarters (77.1%; 27 of 35) of these plates, it was possible to pick single colonies. Moreover, all (158 of 158) of the picked colonies transferred to 12-well plates showed visible growth under the microscope after 1.5 weeks of cultivation. In a preliminary study, primers amplifying the whole 18S rDNA (~1750 bp) region were used for unialgal assessment and taxonomic identification. However, dPCR amplicon sequencing from the 5′ end of the whole 18S rDNA region

lacked sufficient variability compared to the V4 sub-region on the 18S rRNA gene. By combining the dPCRs of the whole 18S rDNA and of the V4 region of 18S rDNA, the dPCR approach succeeded in 70.25% (85 of 121) of the reactions. Furthermore, the amplicons obtained by dPCR, as shown in Figure 1, had identical size compared to the control PCR conducted with extracted DNA. Also, no additional bands were visible for dPCRs.

About 65% of the screened cultures (24 of 37) were identified as unialgal based on our new electropherogram-based assessment. Figure 2 illustrates the discrimination principle between sequences from unialgal cultures and mixed populations.

Taxonomy and geographic origin

In total, 6 different taxonomic groups were identified based on V4-18S rDNA sequences. NCBI nucleotide BLAST searches (Table 1) revealed that all groups comprised microalgae including an array of 4 different classes with Bacillariophyceae, Fragilariophyceae, Coscinodiscophyceae and Chlorophyceae (Table 1; Figure 3). Noticeable morphological features of the novel Chlorophyceae strain are its contractile vacuoles, two isokont flagella, stigmata, pyrenoids and the size of 10 μm (Figure 4). With the exception of this novel Chlorophyceae strain (Figure 3b), diatoms made up the vast majority of isolated species (Table 1). Amongst diatoms pennate species were twice as often isolated and identified by BLAST searches as centric species, which is in agreement with our microscopic observations (10 of 16 isolates). However, it was found that the V4 region failed to resolve differences within the family Fragilariaceae between the genera *Syndrea*, *Fragilaria* and *Synedropsis* (Table 1, Figure 3a), despite equal sequence quality and length. A similar situation was found in two cultures between the best hits *Nitzschia thermalis* and *Amphora* sp. (1-80-1-M and 2-80-27-M). However, taxonomic groups clustered with high bootstrap support (Figure 3a).

Using our approach, we were most successful in isolating the pennate diatom *Cylindrotheca closterium* 9 times from a variety of 5 different sampling locations along most of the latitudinal transect (latitude: 65.246- 78.839) of this study (Figure 5). The Fragilariaceae-cluster (Figure 3a) in contrast was only recovered as an isolate from samples originating from the northernmost sites (Figure 5). On the west side of the transect, a novel Chlorophyceae was isolated from the chlorophyll maximum in a depth of 10 m. *Chaetoceros* cf. *neogracile* and *Skeletonema marinoi* were collected from location 2-80-51-M and 2-80-8-M, respectively (Figure 5).

Discussion

In recent years, huge efforts were made to establish culture collections holding thousands of marine algae strains

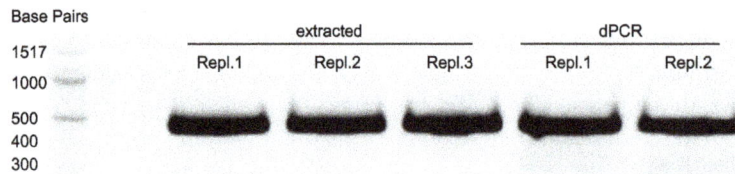

Figure 1 Comparison of PCR- Products utilizing extracted DNA and direct culture as template. Kit extracted DNA is amplified in three replicates (Repl.) using the primers TAReuk454FWD1 and TAReukREV3 [36]. Amplification from direct culture (dPCR) in two replicates using same primers as described in the methods section. – represents negative control. Whole PCR products (50 µl) are separated and visualised by ethidium bromide staining on a 1% TAE agarose gel.

like in the National Center for Marine Algae and Micro-biota (NCMA). Novel approaches of cryopreservation [21,22] reduced culture maintenance efforts considerably [23]. This study reports an approach that enabled us to establish a range of unialgal cultures from Arctic Ocean samples (1) under cost effective conditions due to the omission of DNA-extraction and cloning (2) with low space requirements due to the use of 12-well format (3) within processing times of three weeks.

In accordance with previous studies, the isolated species *Cylindrotheca closterium* [24], *Skeletonama marinoi* [25] and *Chaetoceros* cf. *neogracile* [26] were already identified in the Arctic Ocean and are available in culture collections. The given morphological features of the novel Chlorophyceae strain together with the clustering of its V4 18S rDNA ribotype into *Chlamydomonas* may indicate closer affiliation towards this genus. Even though several different *Chlamydomonas* species were identified in Antarctic saline lakes [27,28], on Arctic glaciers [29], or in sea ice of the brackish Baltic Sea [30], this would be, to our knowledge, the first record of a marine *Chlamydomonas* strain from the deep Chlorophyll maximum layer in the open ocean. However, further characterisation of this

strain is needed what is beyond the scope of this methodical paper. It remains to be seen how significant marine Chlorophyceae species are in terms of diversity, abundance and activity in relation to members of the class Prasinophyta.

Every isolation method has biases towards specific groups to be successfully isolated. Plating, as our method of choice, was reported to exclude some flagellates and coccoids and most dinoflagellates [23]. Alternatively, combining dPCR with other isolation techniques like single-cell sorting [31,32] may increase the spectrum of isolated strains and especially those that won't grow well on agar plates. However, the costs of single-cell isolation and its biases (e.g. selection against filamentous and larger algae) seem to object to our approach.

The success rates of our dPCR approach clearly emphasise the advantages of using microalgae cultures as they grow without the need of DNA extraction as described previously [33-35]. However, a limitation of dPCR might be the use of the V4 region. Nevertheless, the V4 region used as a molecular marker in this study represents the most variable SSU region [17]. However, dinoflagellates possess less variability in this region [36] making it more

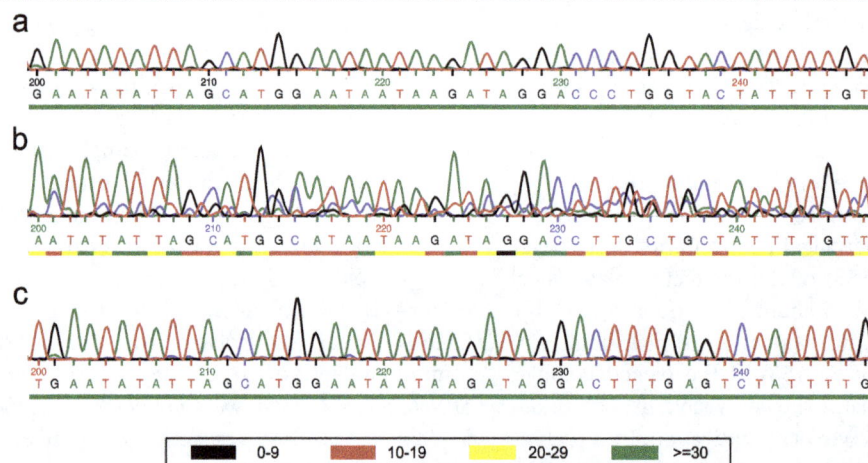

Figure 2 Representative sequencing electropherogram sections. Compared are the base calling signal noise of **(a)** unialgal *Thalassiosira pseudonana* laboratory culture **(b)** non-unialgal culture 1-80-15-M with 2 morphospecies **(c)** unialgal classified culture of *Skeletonema marinoi* (2-80-8-M). The color code refers to the per base Quality Values (QV) as the −10 log10(Pe), with Pe as the base call error probability.

Table 1 Closest BLAST matches against NCBI- database of Sequences recovered from isolated Arctic Ocean samples

SampleID	% < QV20	Closest species BLAST search hits	Last certain common taxonomic assignment	Times isolated	NCBI Sequence ID	Score	Expect	Identities	Gaps
1-80-1-M	0	Uncultured marine eukaryote	Class: Bacillariophyceae		GU385607.1	599 (324)	3.00E-167	324/324(100%)	0/324(0%)
		Bacillariophyta sp.			KF177731.1	593 (321)	2.00E-165	323/324(99%)	0/324(0%)
		Nitzschia thermalis			AY485458.1	588 (318)	7.00E-164	322/324(99%)	0/324(0%)
		Amphora sp.			AY485451.1	588 (318)	7.00E-164	322/324(99%)	0/324(0%)
1-80-3-S	0	Cylindrotheca closterium	Species: closterium		HM070405.1	595 (322)	4.00E-166	322/322(100%)	0/322(0%)
1-80-5-M	0- 0.30	Cylindrotheca closterium	Species: closterium	3	HM070405.1	608 (329)	6.00E-170	329/329(100%)	0/329(0%)
	0.93	Cylindrotheca closterium	Species: closterium		HM070405.1	597 (323)	1.00E-166	323/323(100%)	0/323(0%)
2-80-8-M	0	Skeletonema marinoi (5)	Species: marinoi		HM805045.1	665 (360)	0	360/360(100%)	0/360(0%)
1-80-15-M	0.61	Fragilariaceae sp.	Family: Fragilariaceae		JF794051.1	608 (329)	6.00E-170	329/329(100%)	0/329(0%)
		Synedra hyperborea Grunow			HQ912621.1	608 (329)	6.00E-170	329/329(100%)	0/329(0%)
		Synedra minuscula			EF423415.1	608 (329)	6.00E-170	329/329(100%)	0/329(0%)
		Fragilaria sp.			EU090021.1	608 (329)	6.00E-170	329/329(100%)	0/329(0%)
		Fragilaria cf. striatula			AJ971377.1	608 (329)	6.00E-170	329/329(100%)	0/329(0%)
	0	Cylindrotheca closterium	Species: closterium		HM070405.1	595 (322)	4.00E-166	322/322(100%)	0/322(0%)
2-80-27-M	0.30	Uncultured marine eukaryote	Class: Bacillariophyceae		GU385607.1	610 (330)	2.00E-170	330/330(100%)	0/330(0%)
		Bacillariophyta sp.			KF177731.1	604 (327)	7.00E-169	329/330(99%)	0/330(0%)
		Nitzschia thermalis			AY485458.1	599 (324)	3.00E-167	328/330(99%)	0/330(0%)
		Amphora sp.			AY485451.1	599 (324)	3.00E-167	328/330(99%)	0/330(0%)
1-80-30-M	0- 0.31	Fragilariaceae sp.	Family: Fragilariaceae	3	JF794051.1	595 (322)	4.00E-166	322/322(100%)	0/322(0%)
		Synedra hyperborea			HQ912621.1	595 (322)	4.00E-166	322/322(100%)	0/322(0%)
		Synedra minuscula			EF423415.1	595(322)	4.00E-166	322/322(100%)	0/322(0%)
		Fragilaria sp.			EU090021.1	595 (322)	4.00E-166	322/322(100%)	0/322(0%)
		Fragilaria cf. striatula			AJ971377.1	595 (322)	4.00E-166	322/322(100%)	0/322(0%)
1-80-37-M	0.31	Synedropsis cf. recta	Family: Fragilariaceae		HQ912616.1	584 (316)	1.00E-162	318/319(99%)	0/319(0%)
		Fragilaria striatula			EU090018.1	584 (316)	1.00E-162	318/319(99%)	0/319(0%)
		Fragilaria barbararum			AJ971376.1	584 (316)	1.00E-162	318/319(99%)	0/319(0%)
2-80-51-M	0- 0.62	Chaetoceros cf. neogracile	cf. species: neogracile	2	JN934684.1	595 (322)	4.00E-166	322/322(100%)	0/322(0%)
	0.55	Cylindrotheca closterium	Species: closterium		HM070405.1	667 (361)	0	361/361(100%)	0/361(0%)
2-80-61-M	0.94	Uncultured Chlorophyta	Class: Chlorophyceae		FN690710.1	582 (315)	3.00E-162	317/318(99%)	0/318(0%)
		Chlamydomonas raudensis			AJ781313.1	555 (300)	7.00E-154	312/318(98%)	0/318(0%)
NA	0	Cylindrotheca closterium	Species: closterium		HM070405.1	606 (328)	2.00E-169	328/328(100%)	0/328(0%)
	0	Synedropsis cf. recta	Family: Fragilariaceae		HQ912616.1	636 (344)	3.00E-178	346/347(99%)	0/347(0%)

Table 1 Closest BLAST matches against NCBI- database of Sequences recovered from isolated Arctic Ocean samples *(Continued)*

Fragilaria striatula		EU090018.1	636 (344)	3.00E-178	346/347(99%)	0/347(0%)
Fragilaria barbararum		AJ971376.1	636 (344)	3.00E-178	346/347(99%)	0/347(0%)
Species: *closterium*	0	*Cylindrotheca closterium*	HM070405.1	665 (360)	0	360/360(100%)	0/360(0%)
Species: *marinoi*	0.62	*Skeletonema marinoi*	HM805045.1	597 (323)	1.00E-166	323/323(100%)	0/323(0%)
Class: Chlorophyceae	2	Uncultured Chlorophyta	FN690710.1	588 (318)	7.00E-164	320/321(99%)	0/321(0%)
	0	*Chlamydomonas raudensis*	AJ781313.1	560 (303)	2.00E-155	315/321(98%)	0/321(0%)

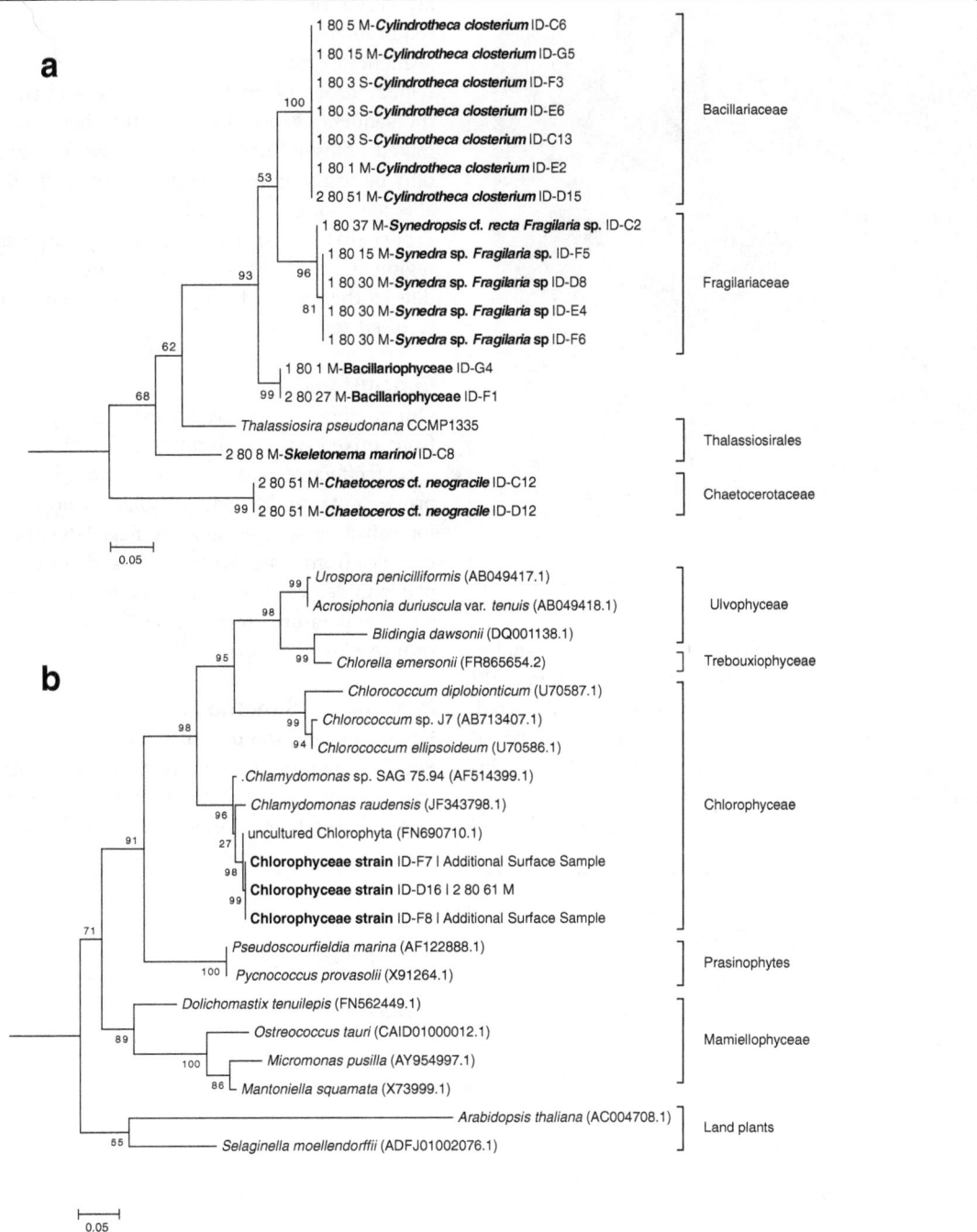

Figure 3 Maximum-likelihood (ML) trees built from the alignments of V4 18S rDNA sequences. Molecular phylogeny of **(a)** isolated diatom groups and **(b)** Chlorophyceae with related clades. Nucleotide sequences obtained in the underlying study indicated by species names in bold. Further sequences were obtained from the SILVA database (www.arb-silva.de) given with accession numbers. The trees with the highest log likelihood (**(a)** -1355.5135; **(b)** -2321.7603) were inferred using the Maximum Likelihood method based on the Kimura 2-parameter model with MEGA6. The fraction of replicate trees in which the associated taxa clustered together is shown next to the branches (1000 bootstraps). The outgroups were **(a)** *Arabidopsis thaliana* and **(b)** *Mus musculus*. All positions with less than 80% site coverage were excluded for tree construction. The scale bar represents number of substitutions per site.

difficult to taxonomically characterise isolates without ambiguity. Despite the fact that we had longer reads (average 361 bp) available for BLAST searches against NCBI compared with Stoeck et al. [36] (average 270 bp), it was still not sufficient to resolve taxonomies within Fragilariaceae and between *Nitzschia* sp. and *Amphora* sp.. In fact, the V4- region as a molecular marker was found to be too conserved to allow taxonomic resolution in these cases.

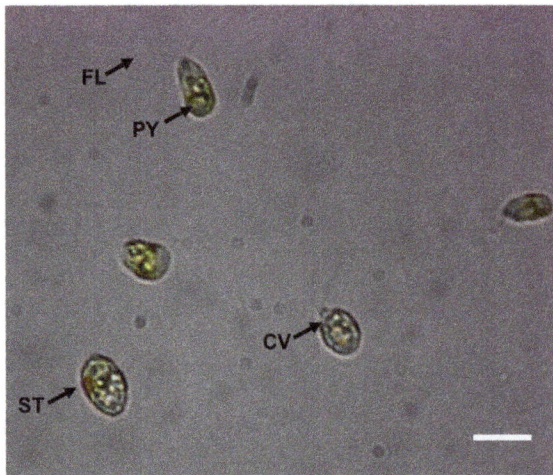

Figure 4 Phase contrast micrograph of novel Chlorophyceae isolate with isokont flagella (FL), pyrenoid (PY), stigma (ST) and contractile vacuole (CV). The cells, about 10 μm in size, were grown at 4°C, 24 h day light, 150 μmol photon m^{-2} s^{-1}. Magnification 100×, scale bar = 10 μm.

algorithm for an objective trimming of the sequences. The requirements in this context are twofold. On one hand, sequences from unialgal cultures have to be trimmed at regular drops of quality at the end and the beginning of the sequence reads. On the other hand, sequences from mixed communities containing low quality reads should only be trimmed to a distinct lower length limit for a reliable assessment as described in the methods section. We expect that interspecific length polymorphisms of the V4 region increase the sensitivity of our culture assessment due to the fact that only one base shift would lead to a screwed sequence.

Conclusions

Our method is suitable for establishing unialgal cultures from mixed natural communities within a few weeks on a cost effective and high-throughput basis. Further improvements could include isolation on low-meting agar for sensitive species such as flagellates, picking of algal colonies from plates with robots and cultivation in 96-well plates under various conditions (e.g. different media, light and temperature) to increase the likelihood of isolating rare species or strains.

Materials and methods

Study sites and sample collection

For the low cost and high throughput isolation and identification of marine arctic microalgae a total of 27 water samples was taken along a latitudinal gradient (65.25°N

The use of sequencing electropherograms for analytical purposes like the detection of point mutations [18] or multiple clone sequences [37] is frequently reported. In our case, using the novel electropherogram based analysis allowed distinction between sequences from a single strain/species and sequences from multiple strains/species. A crucial step is the formulation of a well-defined

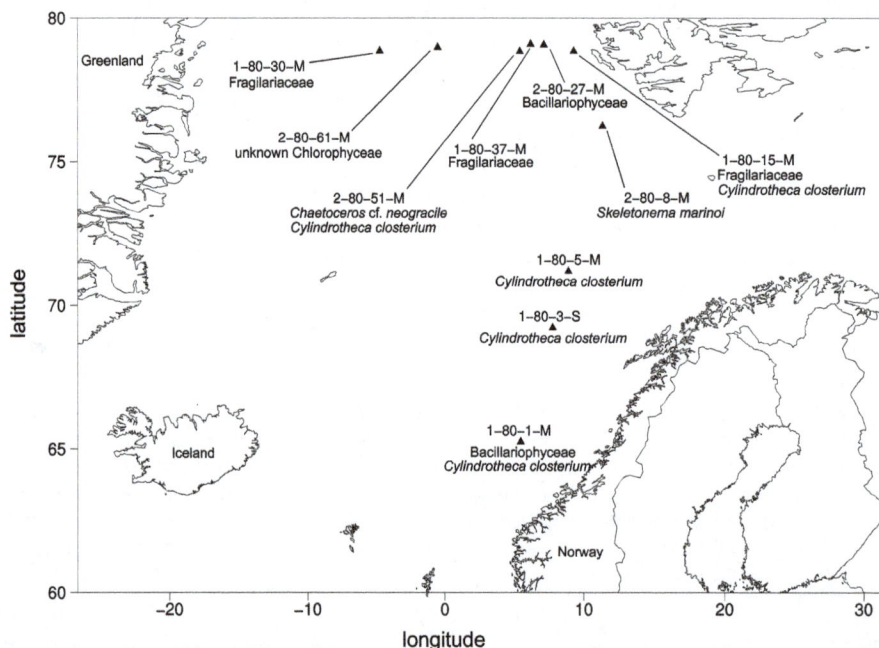

Figure 5 Arctic Ocean Map representing distribution of isolated algae. Each ▲ denotes a sampling point with hyphen separated unique identifier (trial-Polarstern cruise number- Cruise stop number-depth [S = Surface/M = Chlorophyll max]) and closest certain taxonomy according to a BLAST search against GeneBank.

to 79.37°N) from the Arctic Ocean during June and July 2012. Briefly, 12 L seawater was sampled either at the chlorophyll maximum (23 samples; depths 7-110 m) or at the surface (4 samples; depth 5 m) using a Niskin bottle rosette sampler. Additionally, at each sampling depth, temperature, salinity, surface irradiance as well as chlorophyll and nutrient concentration (NO_3, NH, PO_4, Silicate) were measured (see Additional file 1). Sea water was pre-filtered through a 100 μm mesh and the flow-through fraction (<100 μm) was transferred into f/2-medium [38] for enrichment of natural microalgal communities. Whilst transferred regularly into fresh medium, the samples were enriched cultured 425 days at 4°C and about 150 μmol photon m^{-2} s^{-1} for ca. 50 generations before unialgal cultures were isolated. However, the time for enrichment is variable depending on the temperature-dependent growth rates of the algal communities.

High throughput microalgae isolation

Isolation of microalgae into unialgal cultures was done by streaking the enriched microalgal communities across agar plates as described previously [23]. In short, environmental sample cultures were plated on chilled petri dishes containing f/2-medium solidified with 1% (w/v) agar. Subsequently, the agar plates were incubated at 4°C, 24 h day cycle, 150 μmol photon m^{-2} s^{-1} in a light thermostat (Rumed, Rubarth Apparate GmbH, Laatzen, Germany) for 1-2 weeks. Clearly separated colonies were picked from the plates at the end of the striping and transferred each to 3 ml fresh f/2-medium provided in space efficient 12- well plates. Plates without clearly separated colonies where discarded. Inoculated 12-well plates were incubated for 1.5 weeks at 4°C, 24 h day light, 150 μmol photon m^{-2} s^{-1} to increase cell density. These cultures were screened for a) the presence of algae cells (fluorescence emission from chlorophyll a) and b) for visual inspection of having unialgal cultures based on uniform morphology of at least 200 individual algal cells using a phase contrast microscope at 400× maginfication (Olympus BX40, Olympus Optical Co., Ltd., Japan) equipped with Olympus Camedia C-7070 wide-zoom digital camera. Cultures that met both criteria were kept for further molecular analysis.

Direct polymerase chain reaction

For the direct PCR (dPCR)- amplification of ribosomal DNA, a volume of 500 μl suspended culture from each of the positive wells according to the visual inspection criteria (see above) was transferred to 1.5 ml centrifuge tubes and incubated for 5 min at 100°C (Dry Bath Heating System, Starlab, Milton Keynes, United Kingdom) to inactivate protease activity. Then algal cells were harvested by centrifugation at 16,000 rpm for 10 min at room temperature (Eppendorf centrifuge 5418 R, Germany) and

the supernatant was discarded. In order to disrupt the algal cell integrity the pellet was re-suspended properly with 100 μl boiling MiliQ-water. The 4°C chilled suspension was either used directly for PCR or stored at -20°C until further use.

Primers TAReuk454FWD1 (5'-CCAGCA(G/C)C(C/T)G CGGTAATTCC-3') and TAReukREV3 (5'-ACTTTCGT TCTTGAT(C/T)(A/G)A-3') [36] were used to amplify the V4- region of the 18S rDNA using TC-512 PCR System (Techne Co. Staffs, UK). The dPCR was carried out in 50 μL reaction tubes with 10 μl prepared suspension as template, 2.5 U/μl Taq DNA polymerase (GoTaq® Flexi DNA polymerase, Promega, Madison, WI, USA), 1× Taq reaction Buffer, 2 mM $MgCl2$, 0.2 mM each dNTP, and 0.4 μM of each primer. The parameters of thermal cycling of Stoeck et al. (2010) [36] were slightly modified to 30 s initial denaturation at 98°C, 10 × (98°C, 10 s; 53°C, 30 s; 72°C, 30 s), 20 × (98°C, 10 s; 48°C, 30 s; 72°C, 30 s) and 10 min final extension at 72°C.

Gel purification and sequencing

The dPCR-products were visualised on 1% (w/v) TAE-agarose gels stained with ethidium bromide. Amplicon bands of the expected size of 421 bp (*Fragilariopsis cylindrus*) were cut and gel purified using the NucleoSpin® Gel and PCR Clean-up kit (Macherey-Nagel GmbH & Co. KG, Düren, Germany) according to the manufacturer's instructions. The DNA yield and purity of the purified dPCR-products were evaluated using the NanoDrop ND-1000 spectrophotometer (NanoDrop Technologies, Wilmington, USA). Finally, utilising the TA-Reuk454FWD1 forward primer, the amplicons were Sanger-sequenced on a ABI 3730XL sequencer by Eurofins MWG Operon (Ebersberg, Germany).

Nucleotide sequence analysis

The sequencing chromatogram trace (.ab1- format) was analysed and trimmed using the ABI Sequence Scanner v1.0 (Applied Biosystems™). Sequence trimming as well as evaluation of the unialgal status was based on implemented per-base Quality Values (QV) as $-10 \log_{10}(Pe)$, with Pe as the base call error probability [19]. These QV consider chromatogram features like peak spacing, uncalled/called peak ratio and peak resolution. The sequences were trimmed: a) at the 5'end after the first 25-35 bp when the QV consecutive was >20 in a 20 bp window and b) at the 3'end starting after 350 bases, before the first 20 consecutive basecalls contained more than 1 bases with < QV20. Whilst taking the sequencing machine basecalling accuracy of 98.5% [39] into account, the trimmed sequences were classified as unialgal, when the fraction of <20QV basecalls was smaller than one percent. For taxonomic identification BLAST sequence similarity searches [40] of as unialgal classified cultures against the NCBI database (http://www.ncbi.nlm.nih.gov; release 199) were performed using the megablast algorithm. Multiple sequence alignments of the obtained V4 18S rDNA-sequences were done using ClustalX v1.6 [41] and curated using Gblocks v0.91b [42]. A rooted phylogenetic tree was produced by MEGA v6.0 [43] using the maximum likelihood method based on the Kimura 2-parameter model [44] excluding positions with less than 80% site coverage. The robustness of the inferred tree was estimated using a bootstrap analysis consisting of 1000 resampling's of the data.

The nucleotide sequences have been deposited in GenBank and a representative set of cultures was deposited in the Culture Collection of Algae and Protozoa (CCAP) under accession numbers given in Additional file 2.

Additional files

Additional file 1: Metadata of study sites.
Additional file 2: Culture accession numbers of this study.

Competing interests
The authors declare that they have no competing interests.

Authors' contributions
The experiments were conceived by MTJ, KS and TM and performed by MTJ. The data was analysed by MTJ. KS performed microscopy of the Chlorophyceae strain and collected the samples. MTJ and TM co-wrote the paper. All authors read and approved the final manuscript.

Acknowledgements
TM acknowledges the Natural Environment Research Council (NERC) for funding (NE/K004530/1). We would like to thank Klaus Valentin from the Alfred-Wegener-Institute for Polar and Marine Research for supporting the field campaigns. Furthermore, we would like to thank Captain Schwarze and the Polarstern crews of the ARK27-1 and ANT29-1 expeditions for their vital help during sampling.

Author details
[1]School of Environmental Sciences, University of East Anglia, Norwich Research Park, Norwich NR4 7TJ, UK. [2]Current address: Department of Botany II, Julius-Maximilians University Würzburg, Julius-von-Sachs-Platz 3, 97082 Würzburg, Germany.

References
1. Platt T, Fuentes-Yaco C, Frank KT: Marine ecology: spring algal bloom and larval fish survival. Nature 2003, 423:398–399.
2. Gosselin M, Levasseur M, Wheeler PA, Horner RA, Booth BC: New measurements of phytoplankton and ice algal production in the Arctic Ocean. Deep Sea Res Part II Top Stud Oceanography 1997, 44:1623–1644.
3. Schwartz RE, Hirsch CF, Sesin DF, Flor JE, Chartrain M, Fromtling RE, Harris GH, Salvatore MJ, Liesch JM, Yudin K: Pharmaceuticals from cultured algae. J Ind Microbiol 1990, 5:113–123.
4. Borowitzka MA: Microalgae as sources of pharmaceuticals and other biologically active compounds. J Appl Phycol 1995, 7:3–15.
5. Kim SK, Ravichandran YD, Khan SB, Kim YT: Prospective of the cosmeceuticals derived from marine organisms. Biotechnol Bioproc E 2008, 13:511–523.
6. Lee RF, Valkirs AO, Seligman PF: Importance of microalgae in the biodegradation of tributyltin in estuarine waters. Environ Sci Technol 1989, 23:1515–1518.

7. Cardinale BJ: Biodiversity improves water quality through niche partitioning. Nature 2011, 472:86–89.
8. El-Sheekh M, Ghareib M, EL-Souod GA: Biodegradation of phenolic and polycyclic aromatic compounds by some algae and Cyanobacteria. J Bioremediation Biodegradation 2011, 3:133.
9. Waltz E: Biotech's green gold? Nat Biotech 2009, 27:15–18.
10. Toseland A, Daines SJ, Clark JR, Kirkham A, Strauss J, Uhlig C, Lenton TM, Valentin K, Pearson GA, Moulton V, Mock T: The impact of temperature on marine phytoplankton resource allocation and metabolism. Nat Clim Change 2013, 3:979–984.
11. Cuvelier ML, Allen AE, Monier A, McCrow JP, Messié M, Tringe SG, Woyke T, Welsh RM, Ishoey T, Lee J-H, Binder BJ, DuPont CL, Latasa M, Guigand C, Buck KR, Hilton J, Thiagarajan M, Caler E, Read B, Lasken RS, Chavez FP, Worden AZ: Targeted metagenomics and ecology of globally important uncultured eukaryotic phytoplankton. Proc Natl Acad Sci 2010, 107:14679–14684.
12. Worden AZ, Lee J-H, Mock T, Rouzé P, Simmons MP, Aerts AL, Allen AE, Cuvelier ML, Derelle E, Everett MV, Foulon E, Grimwood J, Gundlach H, Henrissat B, Napoli C, McDonald SM, Parker MS, Rombauts S, Salamov A, Von Dassow P, Badger JH, Coutinho PM, Demir E, Dubchak I, Gentemann C, Eikrem W, Gready JE, John U, Lanier W, Lindquist EA, et al: Green evolution and dynamic adaptations revealed by genomes of the marine picoeukaryotes Micromonas. Science 2009, 324:268–272.
13. Marchetti A, Schruth DM, Durkin CA, Parker MS, Kodner RB, Berthiaume CT, Morales R, Allen AE, Armbrust EV: Comparative metatranscriptomics identifies molecular bases for the physiological responses of phytoplankton to varying iron availability. Proc Natl Acad Sci 2012, 109:E317–E325.
14. Chen W, Zhang C, Song LR, Sommerfeld M, Hu Q: A high throughput Nile red method for quantitative measurement of neutral lipids in microalgae. J Microbiol Meth 2009, 77:41–47.
15. Slocombe SP, Zhang QY, Black KD, Day JG, Stanley MS: Comparison of screening methods for high-throughput determination of oil yields in micro-algal biofuel strains. J Appl Phycol 2013, 25:961–972.
16. Plaza M, Santoyo S, Jaime L, García-Blairsy Reina G, Herrero M, Señoráns FJ, Ibáñez E: Screening for bioactive compounds from algae. J Pharm Biomed Anal 2010, 51:450–455.
17. Wuyts J, De Rijk P, Van de Peer Y, Pison G, Rousseeuw P, De Wachter R: Comparative analysis of more than 3000 sequences reveals the existence of two pseudoknots in area V4 of eukaryotic small subunit ribosomal RNA. Nucleic Acids Res 2000, 28:4698–4708.
18. Davies H, Bignell GR, Cox C, Stephens P, Edkins S, Clegg S, Teague J, Woffendin H, Garnett MJ, Bottomley W, Davis N, Dicks E, Ewing R, Floyd Y, Gray K, Hall S, Hawes R, Hughes J, Kosmidou V, Menzies A, Mould C, Parker A, Stevens C, Watt S, Hooper S, Wilson R, Jayatilake H, Gusterson BA, Cooper C, Shipley J, Hargrave D, et al: Mutations of the BRAF gene in human cancer. Nature 2002, 417:949–954.
19. Ewing B, Green P: Base-calling of automated sequencer traces using phred. II. error probabilities. Genome Res 1998, 8:186–194.
20. Applied Biosystems™: User bulletin- KB™ basecaller software v1.4.1. [http://tools.lifetechnologies.com/content/sfs/manuals/cms_079032.pdf]
21. Day JG, Fleck RA, Benson EE: Cryopreservation-recalcitrance in microalgae: novel approaches to identify and avoid cryo-injury. J Appl Phycol 2000, 12:369–377.
22. Bui TVL, Ross IL, Jakob G, Hankamer B: Impact of procedural steps and cryopreservation agents in the cryopreservation of Chlorophyte microalgae. PLoS One 2013, 8:e78668.
23. Andersen RA, Kawachi M: Traditional microalgae isolation techniques. In Algal culturing techniques. Edited by Andersen RA. Burlington, USA: Elsevier Academic Press; 2005:83–100.
24. Booth BC, Horner RA: Microalgae on the arctic ocean section, 1994: species abundance and biomass. Deep Sea Res Part II Top Stud Oceanography 1997, 44:1607–1622.
25. Zhang F, He JF, Xia LH, Cai MH, Lin L, Guang YZ: Applying and comparing two chemometric methods in absorption spectral analysis of photopigments from Arctic microalgae. J Microbiol Meth 2010, 83:120–126.
26. Vaulot D, Le Gall F, Marie D, Guillou L, Partensky F: The Roscoff Culture Collection (RCC): a collection dedicated to marine picoplankton. Nova Hedwigia 2004, 79:49–70.
27. Lizotte MP, Priscu JC: Photosynthesis- irradiance relationships in phytoplankton from the physically stable water column of a perennially ice- covered lake (Lake Bonney, Antarctica). J Phycol 1992, 28:179–185.

28. Pocock T, Lachance M-A, Pröschold T, Priscu JC, Kim SS, Huner NPA: **Identification of a psychrophilic green alga from Lake Bonney Antarctica: Chlamydomonas raudensis Ettl. (UWO 241) Chlorophyceae.** *J Phycol* 2004, **40:**1138–1148.

29. Säwström C, Mumford P, Marshall W, Hodson A, Laybourn-Parry J: **The microbial communities and primary productivity of cryoconite holes in an Arctic glacier (Svalbard 79°N).** *Polar Biol* 2002, **25:**591–596.

30. Majaneva M, Rintala J-M, Piisilä M, Fewer DP, Blomster J: **Comparison of wintertime eukaryotic community from sea ice and open water in the Baltic Sea, based on sequencing of the 18S rRNA gene.** *Polar Biol* 2012, **35:**875–889.

31. Doan TTY, Sivaloganathan B, Obbard JP: **Screening of marine microalgae for biodiesel feedstock.** *Biomass Bioenergy* 2011, **35:**2534–2544.

32. Montero MF, Aristizábal M, García Reina G: **Isolation of high-lipid content strains of the marine microalga Tetraselmis suecica for biodiesel production by flow cytometry and single-cell sorting.** *J Appl Phycol* 2011, **23:**1053–1057.

33. Wan MX, Rosenberg JN, Faruq J, Betenbaugh MJ, Xia JL: **An improved colony PCR procedure for genetic screening of Chlorella and related microalgae.** *Biotechnol Lett* 2011, **33:**1615–1619.

34. Radha S, Fathima AA, Iyappan S, Ramya M: **Direct colony PCR for rapid identification of varied microalgae from freshwater environment.** *J Appl Phycol* 2013, **25:**609–613.

35. Zamora I, Feldman J, Marshall W: **PCR-based assay for mating type and diploidy in Chlamydomonas.** *Biotechniques* 2004, **37:**534–536.

36. Stoeck T, Bass D, Nebel M, Christen R, Jones MD, Breiner HW, Richards TA: **Multiple marker parallel tag environmental DNA sequencing reveals a highly complex eukaryotic community in marine anoxic water.** *Mol Ecol* 2010, **19**(Suppl 1):21–31.

37. Eurofins MWG Operon: **Sequencing result guide.** 2011, 2014. http://www.eurofinsgenomics.eu.

38. Guillard RRL: **Culture of Phytoplankton for feeding marine invertebrates.** In *Culture of Marine Invertebrate Animals.* Edited by Smith WL, Chanley MH. New York, USA: Plenum Press; 1975:29–60.

39. Applied Biosystems™: **System profile applied biosystems 3130 and 3130xl genetic analyzers.** [www3.appliedbiosystems.com/cms/groups/mcb_marketing/documents/generaldocuments/cms_041990.pdf]

40. Altschul SF, Gish W, Miller W, Myers EW, Lipman DJ: **Basic local alignment search tool.** *J Mol Biol* 1990, **215:**403–410.

41. Thompson JD, Gibson TJ, Plewniak F, Jeanmougin F, Higgins DG: **The CLUSTAL X windows interface: flexible strategies for multiple sequence alignment aided by quality analysis tools.** *Nucleic Acids Res* 1997, **25:**4876–4882.

42. Talavera G, Castresana J: **Improvement of phylogenies after removing divergent and ambiguously aligned blocks from protein sequence alignments.** *Syst Biol* 2007, **56:**564–577.

43. Tamura K, Stecher G, Peterson D, Filipski A, Kumar S: **MEGA6: Molecular Evolutionary Genetics Analysis version 6.0.** *Mol Biol Evol* 2013, **30:**2725–2729.

44. Kimura M: **A simple method for estimating evolutionary rates of base substitutions through comparative studies of nucleotide sequences.** *J Mol Evol* 1980, **16:**111–120.

4

Isolation and kinetic characterisation of hydrophobically distinct populations of form I Rubisco

Kerry O'Donnelly[1], Guangyuan Zhao[2], Priya Patel[2], M Salman Butt[1], Lok Hang Mak[2], Simon Kretschmer[2], Rudiger Woscholski[1] and Laura M C Barter[1*]

Abstract

Background: Rubisco (Ribulose-1,5-bisphosphate carboxylase/oxygenase) is a Calvin Cycle enzyme involved in CO_2 assimilation. It is thought to be a major cause of photosynthetic inefficiency, suffering from both a slow catalytic rate and lack of specificity due to a competing reaction with oxygen. Revealing and understanding the engineering rules that dictate Rubisco's activity could have a significant impact on photosynthetic efficiency and crop yield.

Results: This paper describes the purification and characterisation of a number of hydrophobically distinct populations of Rubisco from both *Spinacia oleracea* and *Brassica oleracea* extracts. The populations were obtained using a novel and rapid purification protocol that employs hydrophobic interaction chromatography (HIC) as a form I Rubisco enrichment procedure, resulting in distinct Rubisco populations of expected enzymatic activities, high purities and integrity.

Conclusions: We demonstrate here that HIC can be employed to isolate form I Rubisco with purities and activities comparable to those obtained via ion exchange chromatography (IEC). Interestingly, and in contrast to other published purification methods, HIC resulted in the isolation of a number of hydrophobically distinct Rubisco populations. Our findings reveal a so far unaccounted diversity in the hydrophobic properties within form 1 Rubisco. By employing HIC to isolate and characterise *Spinacia oleracea* and *Brassica oleracea*, we show that the presence of these distinct Rubisco populations is not species specific, and we report for the first time the kinetic properties of Rubisco from *Brassica oleracea* extracts. These observations may aid future studies concerning Rubisco's structural and functional properties.

Keyword: *Brassica oleracea*, Hydrophobic interaction chromatography, Purification, Rubisco, *Spinacia oleracea*

Background

Rubisco (Ribulose-1,5-bisphosphate carboxylase/oxygenase) is the CO_2-fixing enzyme in the Calvin cycle, the primary pathway of carbon assimilation in photosynthetic organisms. It catalyses the reaction between Ribulose-1,5-bisphosphate (RuBP) and CO_2 to produce two molecules of 3-phosphoglycerate [1,2]. It is reported that Rubisco is responsible for a net fixation of 10^{11} tons of CO_2 from the atmosphere to the biosphere per year [3]. Despite its importance, Rubisco is a remarkably inefficient enzyme, suffering from both poor specificity and a slow catalytic rate. It is not surprising that this enzyme has generated much interest, as it is suggested to be one of the major bottlenecks limiting maximum photosynthetic efficiency [4]. The efficiency of carbon fixation is reduced by side reactions, the most notable being photorespiration, where Rubisco fixates O_2 instead of CO_2. Photorespiration imposes a significant metabolic constraint, lowering the efficiency of carbon fixation by up to 25-50% and constantly draining the pool of available RuBP [5,6]. Rubisco is the most abundant protein found on the planet, making up approximately 50% of the total leaf protein [7]. This however comes at a cost in terms of the plant's nitrogen requirements [1].

* Correspondence: l.barter@imperial.ac.uk
[1]Institute of Chemical Biology, Department of Chemistry, Imperial College, Flowers Building, South Kensington Campus, Exhibition Road, London SW7 2AZ, UK
Full list of author information is available at the end of the article

Rubisco is a multimeric enzyme comprising varying numbers of the large (50–55 kDa) and small subunits (12–18 kDa), which in higher plants and green algae are encoded by the chloroplast genome and the nuclear genome respectively [2,8]. There are a number of different forms of Rubisco (forms I, II and III, as well as the more diverse Rubisco-like proteins (RLPs) called form IV), reviewed by Tabita et al. [9]. Form I Rubisco is the most common, found in higher plants, cyanobacteria and eukaryotic algae [2], and its structure has been solved in many species, but first reports were for tobacco [10] and spinach [11]. It consists of eight large and eight small subunits in a hexadecameric structure, forming a barrel of four large subunit dimers arranged around a four-fold axis of symmetry, capped by four small subunits at each end [8,12]. The large subunits play a catalytic role, however the precise role of the small subunit is not clear, as it is not essential for catalysis [13,14]. Interestingly hybrid enzymes which contain large subunits and small subunits from different species have been reported to show differences in their stability and specificity [15-17].

There is a lack of understanding of the interactions and engineering rules that control and regulate Rubisco's activity, although this is critical if we are to increase photosynthetic efficiency. A prerequisite for the characterisation of the structure and function of Rubisco, along with investigations into its interactome, is the need for rapid methods that can isolate Rubisco with high purity. Purification methods in the past have taken advantage of the enzyme's negative net charge under physiological pH, which make it suitable for anion exchange chromatography [18-21], as well as the relatively high molecular weight of the hexadecameric holoenzyme, which made it an ideal candidate for size exclusion chromatography, or sucrose gradient centrifugation [22-25]. In addition, differential ammonium sulphate (($NH_4)_2SO_4$) precipitation has been employed prior to the ion exchange chromatography (IEC) [19,24]. A number of the reported high purity Rubisco isolation protocols are lengthy, due to the need for dialysis or ultra-centrifugation to remove sucrose prior to assaying the sample, which leads to an increased risk of endoproteolytic activity on the homogenized leaf tissue [26]. Given the wide use and effectiveness of $(NH_4)_2SO_4$ precipitation it is surprising that hydrophobic interaction chromatography (HIC), which is based on similar separation principles, has only been explored once as a purification technique for form III Rubisco from recombinant A. fulgidu, but has yet to be employed for form I Rubisco [27].

The potential of HIC as an alternative to IEC to obtain form I Rubisco with high purity was explored, since it is ideally suited as the proceeding step to the commonly employed $(NH_4)_2SO_4$ precipitation in Rubisco purification protocols [28]. Employing $(NH_4)_2SO_4$ in purification

protocols may have the added benefit of removing endogenous inhibitors, since the presence of sulphate ions in Rubisco extraction buffers has been shown to remove and prevent the binding of inhibitors such as 2 carboxy-D-arabinitol 1-phosphate (CA1P) [29,30]. $(NH_4)_2SO_4$ precipitation results in samples with high salt content, which require dialysis, desalting or dilution prior to the subsequent IEC purification step. In contrast, HIC can exploit the high salt content of the $(NH_4)_2SO_4$ precipitate, therefore avoiding any further dilution or desalting steps. HIC, like IEC, offers the added advantage of being able to bind proteins from large sample volumes, which is in contrast to gel filtration chromatography and sucrose gradient centrifugation, often employed in addition to IEC [19,24].

We demonstrate here a rapid purification method that, for the first time, is able to reveal hydrophobically distinct Rubisco populations. Our results confirm that the HIC protocol, detailed here, can be employed to obtain highly purified Rubisco from Spinacia oleracea (S. oleracea, Spinach), with kinetic properties in agreement with literature values from samples subjected to commonly employed purification protocols [20,31].

Once verified by this comparison, the HIC method was utilised to purify Rubisco from another species, the so far uncharacterised enzyme from Brassica oleracea (B. oleracea, Cabbage), revealing for the first time the kinetic properties of Rubisco from this species.

Results and discussion

In order to test the suitability of HIC, lysed samples from both S. oleracea and B. oleracea were subjected to $(NH_4)_2SO_4$ precipitation, with the majority of Rubisco protein precipitating between 35 and 60% saturation of $(NH_4)_2SO_4$ (Figure 1). As expected, this observation is in agreement with previous reports where $(NH_4)_2SO_4$ precipitation has been employed to extract Rubisco from higher plant species, including S. oleracea and Arabidopsis [24,32].

HIC reveals the presence of hydrophobically distinct Rubisco fractions

As the HIC column binds proteins at high salt concentration, the $(NH_4)_2SO_4$ fraction containing Rubisco was directly subjected to our HIC protocol (without the need to desalt the sample). Our method is summarised in Figure 2, which also shows a typical Rubisco isolation and purification protocol taken from Salvucci et al. for comparison [24]. Bound proteins were step-eluted in order to gain reproducibility, while conserving simplicity in handling (facilitating the potential use of syringe operated columns). A typical UV trace of the HIC elution profile demonstrates that a substantial proportion of the loaded proteins from S. oleracea (Figure 3A) and B.

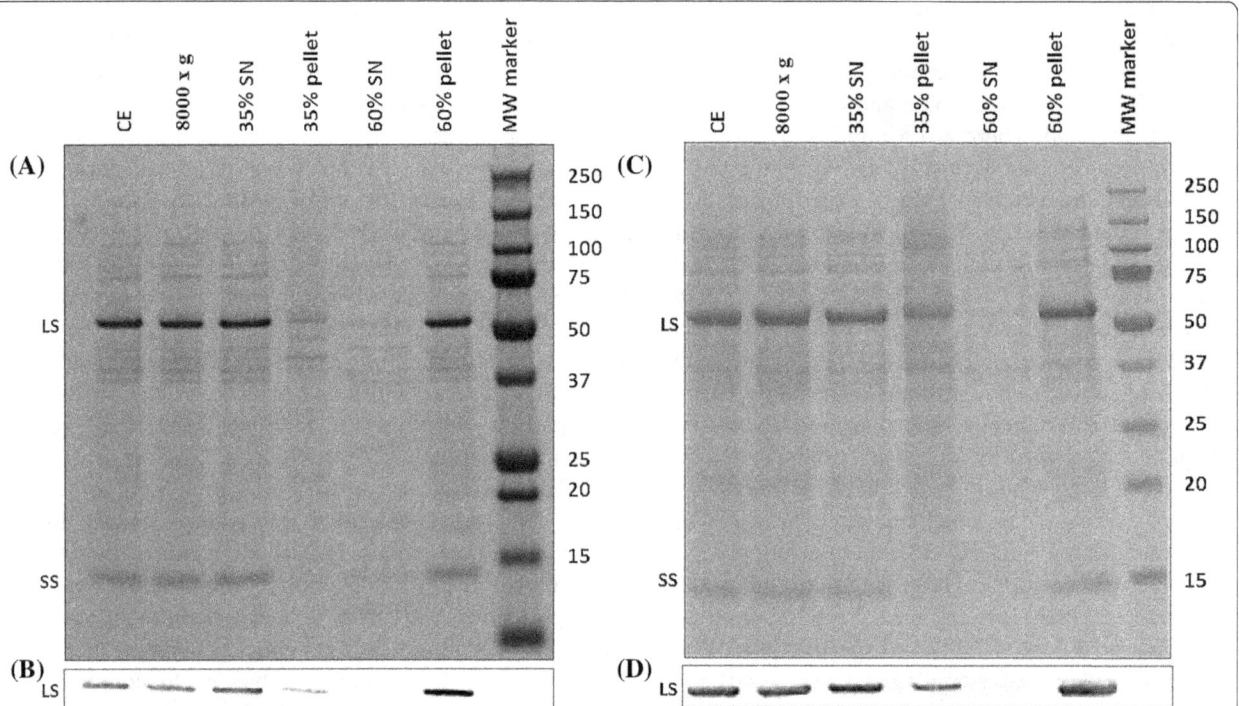

Figure 1 SDS-PAGE gels and western blots showing protein content in ammonium sulphate fractions from *S. oleracea* and *B. oleracea*.
(A) SDS-PAGE gel showing protein content in ammonium sulphate fractions from *S. oleracea*; Extracts were subjected to ammonium sulphate precipitation as described in Methods. Samples were separated by SDS-PAGE and subsequently stained for protein. The position of Rubisco's large (LS) and small (SS) subunits are indicated in the figure. 0.4 µg of sample was added per lane. **(B)** Western blot showing Rubisco content in ammonium sulphate fractions from *S. oleracea*; The relative intensities of the LS were calculated to be 0.85 ± 0.04 (CE), 0.89 ± 0.01 ($8000 \times$ g), 0.92 ± 0.02 (35% SN), 0.55 ± 0.15 (35% pell) and 1 (60% pell). **(C)** SDS-PAGE gel showing protein content in ammonium sulphate fractions from *B. oleracea*; Extracts were subjected to ammonium sulphate precipitation as described in Methods. Samples were separated by SDS-PAGE and subsequently stained for protein. The position of Rubisco's large (LS) and small (SS) subunits are indicated in the figure. 0.6 µg of sample was added per lane. **(D)** Western blot showing Rubisco content in ammonium sulphate fractions from *B. oleracea*; The relative intensities of the LS were calculated to be 0.83 ± 0.06 (CE), 0.85 ± 0.03 ($8000 \times$ g), 0.89 ± 0.06 (35% SN), 0.61 ± 0.12 (35% pell) and 1 (60% pell).

oleracea (Figure 4A), either did not bind to the column, or were eluted in the 1 M $(NH_4)_2SO_4$ wash step or in the 500 mM step elution (see Additional file 1: Tables S1 and S2 for HIC purification tables). Interestingly there were five prominent peaks, observed at 500 mM, 400 mM, 300 mM, 200 mM and 0 M salt concentrations, all of which contained Rubisco (shown in the SDS-PAGE gel, Figures 3B and 4B, and confirmed by the presence of the large subunit in the western blot, Figures 3C and 4C), although it is of note that the 500 mM fractions had relatively low purities of 65% (*S. oleracea*) and 59% (*B. oleracea*) (calculated by analysis of SDS-PAGE gels, Figures 3B and 4B). The UV trace of these step elution fractions revealed the presence of relatively broad peaks and tails, supporting the use of step elution. For each concentration of ammonium sulphate used in the step elution of protein in our HIC protocol, the column was allowed to equilibrate, such that the UV absorbance returned to the baseline. This ensured that all protein at the corresponding salt concentration had been eluted prior to proceeding with the next step elution.

The 0 M Rubisco fraction isolated from *S. oleracea* had a purity of 77%, with the main impurities (as judged by densitometric scans of Coomassie stained gels, Figure 3B) being proteins with molecular weights of ~45 kDa and ~26 kDa. In contrast, the 400 mM, 300 mM and 200 mM Rubisco fractions (isolated from *S. oleracea*) showed relatively high purities (between 90-93%) with the main impurity present in all three fractions being a ~41 kDa protein. The 200 mM fraction contained an additional ~26 kDa protein.

Interestingly, the HIC fractions obtained from *B. oleracea* extracts revealed some species-specific hydrophobic differences. While extracts from both species could be separated into fractions at 400 mM, 300 mM, 200 mM and 0 M salt concentrations, differences in the purity of these fractions were observed when compared with the results from *S. oleracea*. For example, the results from *B. oleracea* showed that the 300 mM, 200 mM and 0 mM HIC fractions had high purity (93-96%), with the 300 mM fraction also containing a ~39 kDa protein impurity. Interestingly the fraction that eluted at 400 mM salt showed a

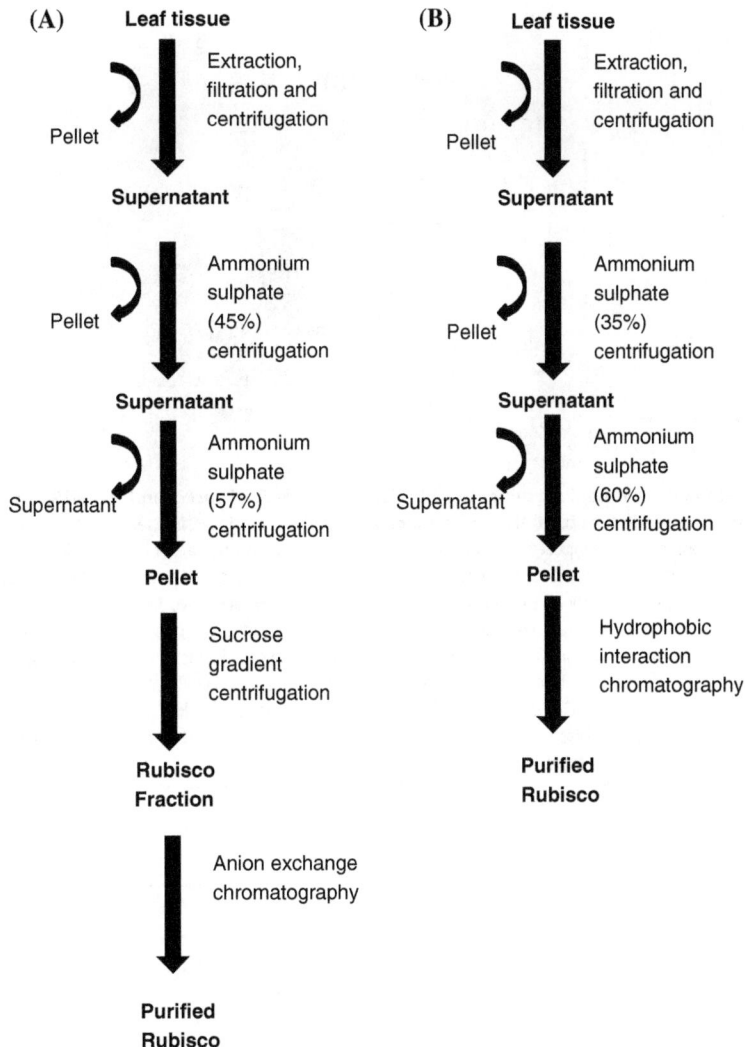

Figure 2 Flow-chart outlining a comparison of HIC and IEC Rubisco purification protocols. A comparison is made between **(A)** a commonly employed IEC protocol reported by Sulvucci *et al.* [24] and **(B)** the combined ammonium sulphate and HIC protocol reported in this paper.

lower purity of 82% (Figure 4B), with the main impurities being due to ~43 kDa and ~39 kDa proteins. This is in contrast to the results from *S. oleracea* where we observed the 0 M eluate to have a lower purity.

The protein impurities (41 kDa and 45 kDa for *S. oleracea*) in these HIC fractions are of comparable molecular weight to the smaller and larger isoforms of Rubisco activase [33,34], (a protein that has been indicated to associate with Rubisco [35]). Although there is no available data on the molecular weights of Rubisco activase isoforms from *B. oleracea*, it was speculated that the protein impurities (39 kDa and 43 kDa) in these HIC fractions could be due to the presence of the smaller and larger isoforms of Rubisco Activase. However, probing these fractions with antibodies against Rubisco activase did not support this notion in either species (data not shown).

Taken together, our data demonstrates the suitability of HIC as a method for obtaining high purity Rubisco. Rubisco extracted from the two species tested here (*S. oleracea* and *B. oleracea*) could be purified to similar levels to those reported using a rapid FPLC method, which obtained Rubisco from *S. oleracea* with 93% purity [20]. It is worth noting that no significant degradation products of the large Rubisco subunit could be observed in the HIC fractions (Figures 3C and 4C), indicating that Rubisco was not subject to proteolytic damage throughout the purification described here. This corroborates the protein integrity preserving character of the HIC method.

To ensure that the hydrophobically distinct populations of Rubisco were not a result of the use of $(NH_4)_2SO_4$ during the precipitation step or during HIC, crude extracts from *S. oleracea* were loaded straight onto the HIC column

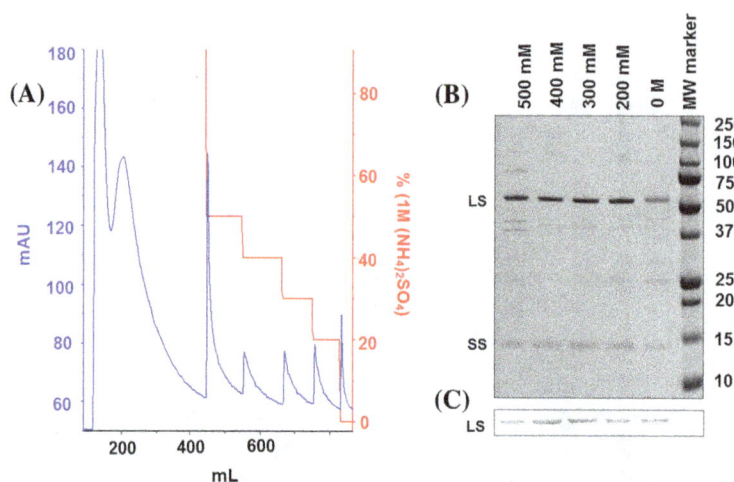

Figure 3 HIC elution profile from the 60% ammonium sulphate pellet isolated from *S. oleracea*, and the SDS-PAGE gel and western blot showing protein content from the elution steps of the HIC purification. **(A)** HIC elution profile generated by loading the 60% $(NH_4)_2SO_4$ pellet isolated from *S. oleracea*; The ammonium sulphate pellet (60% pell; see Figure 1) was loaded onto a 5 mL HiTrap HIC column, washed and eluted as described in Methods. The gradient used for the separation (red line) and the positions of the major absorption peaks obtained by measuring OD at 280 nm (blue line) following the HIC elution profile is shown. Elutions were performed at 500 mM, 400 mM, 300 mM, 200 mM, 0 M ammonium sulphate. **(B)** SDS-PAGE gel showing protein content from the different steps involved in the HIC purification protocol of Rubisco isolated from *S. oleracea*; Samples were separated by SDS-PAGE and subsequently stained for protein. The position of Rubisco's large (LS) and small (SS) subunits are indicated in the figure. 0.6 µg of sample was added per lane. The purity of the Rubisco fractions were calculated as 65.2 ± 4.7% (500 mM), 91.9 ± 3.1%, (400 mM), 92.6 ± 2.2%, (300 mM), 90.2 ± 2.4% (200 mM) and 76.6 ± 3.1% (0 M). **(C)** Western blot showing Rubisco content from HIC purified *S. oleracea* Rubisco fractions; The relative intensities of the LS for the four high purity Rubisco fractions were 0.97 ± 0.03, (400 mM), 0.98 ± 0.02, (300 mM), 0.94 ± 0.04 (200 mM) and 0.84 ± 0.06 (0 M). The gels and western blot figures are representative of 3 sample sets.

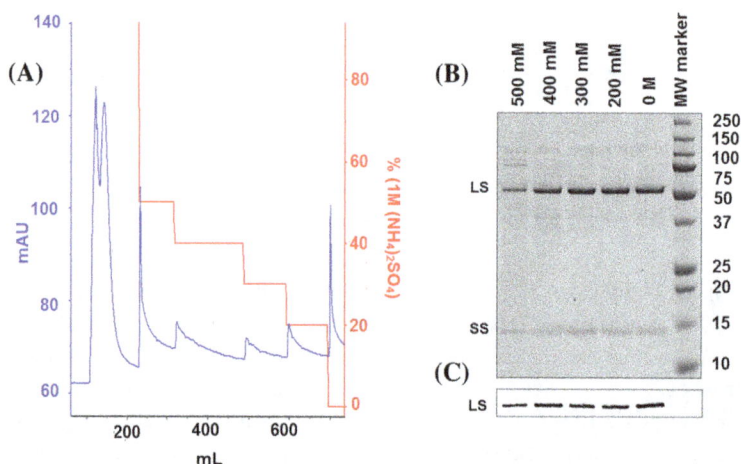

Figure 4 HIC elution profile from the 60% ammonium sulphate pellet isolated from *B. oleracea*, and the SDS-PAGE gel and western blot showing protein content from the elution steps of the HIC purification. **(A)** HIC elution profile generated by loading the 60% $(NH_4)_2SO_4$ pellet isolated from *B. oleracea*; The ammonium sulphate pellet (60% pellet; see Figure 1) was loaded onto a 5 mL HiTrap HIC column, washed and eluted as described in Methods. The gradient used for the separation (red line) and the positions of the major absorption peaks obtained by measuring OD at 280 nm (blue line) following the HIC elution profile is shown. Elutions were performed at 500 mM, 400 mM, 300 mM, 200 mM, 0 M ammonium sulphate. **(B)** SDS-PAGE gel showing protein content from the different steps involved in the HIC purification protocol of Rubisco isolated from *B. oleracea*; Samples were separated by SDS-PAGE and subsequently stained for protein. The position of Rubisco's large (LS) and small (SS) subunits are indicated in the figure. 0.6 µg of sample was added per lane. The purity of the Rubisco fractions were calculated as 59.2 ± 4.2% (500 mM), 82.3 ± 6.0% (400 mM), 92.8 ± 1.3% (300 mM), 96.1 ± 0.2% (200 mM) and 95.2 ± 2.7% (0 M). **(C)** Western blot showing Rubisco content from HIC purified *B. oleracea* Rubisco fractions; The relative intensities of the LS for the four high purity Rubisco fractions were calculated as 0.76 ± 0.11 (400 mM), 0.92 ± 0.07 (300 mM), 0.98 ± 0.02 (200 mM) and 0.97 ± 0.06 (0 M). The gels and western blot figures are representative of 3 sample sets.

(omitting the $(NH_4)_2SO_4$ precipitation step), followed by elution with potassium chloride (KCl) instead of $(NH_4)_2SO_4$. We again observed hydrophobically distinct fractions of Rubisco under these conditions, using this HIC protocol, suggesting that the different fractions of Rubisco are not a result of the use of $(NH_4)_2SO_4$ (Additional file 2).

At this stage, we can only speculate as to why we observe hydrophobically distinct populations of Rubisco in both species. Separation by HIC clearly implies that the Rubisco populations differ in their hydrophobicity, at least on the surface of the protein. As Rubisco is a hugely complex enzyme, HIC could potentially be separating Rubisco populations where one or more of the enzyme's eight subunits might have undergone a conformational change, with the HIC fractions capturing the resulting populations generated through subunit heterogeneity. Alternatively, the difference in hydrophobicity could be due to protein – protein interactions (e.g. potentially with one of the impurities found in these fractions) or post-translational modifications. Furthermore, it is possible that the distinct Rubisco fractions are capturing diverse holoenzyme populations of Rubisco, with differing combinations of small subunits. As the small subunits provide structural stability [8], any difference in small subunit combinations could affect Rubisco's hydrophobic character. The distinct populations, and the protocol to obtain them, could therefore lead to exciting developments in understanding structural and interactome studies of this vital enzyme that will benefit current and future photosynthetic researchers.

Kinetic analysis of the HIC fractions from *S. oleracea* and *B. oleracea* extracts

The observed elution profiles from *S. oleracea* and *B. oleracea* extracts imply that there are differences in the hydrophobic character of Rubisco eluted in these different fractions. To probe whether the difference in the hydrophobic character could influence Rubisco's enzymological properties, we determined the K_M and K_{CAT} values for all Rubisco containing fractions. The Michaelis-Menten plot reveals that all the different fractions containing Rubisco, isolated using HIC, from *S. oleracea* (Figure 5A) have similar enzymological properties. The Michaelis-Menten plot reveals that the enzymological properties of the 300 mM, 200 mM and 0 M HIC Rubisco containing fractions isolated from *B. oleracea* are also similar (Figure 6A). In contrast, the Rubisco activity in the 400 mM HIC fraction from *B. oleracea* extracts has a significant lower K_{CAT} and a higher K_M as compared to the 200 mM and 0 M HIC fractions.

The K_{CAT} and K_M values for Rubisco from *S. oleracea* (see Table 1) are in reasonable agreement with previously published data [20,31]. As enzymological studies on Rubisco extracted from *B. oleracea* have not been previously published, we report here for the first time the K_{CAT} for *B. oleracea* Rubisco (for the 300, 200 and 0 mM HIC fractions), and note that although a factor of 2 lower than *S. oleracea*, it is in reasonable agreement with values obtained from other C3 species (Table 1) [31]. The K_M (RuBP) values (for the 300, 200 and 0 mM HIC fractions) for *B. oleracea* (Table 1) are 2–3 times higher than that for *S. oleracea*, but are again in

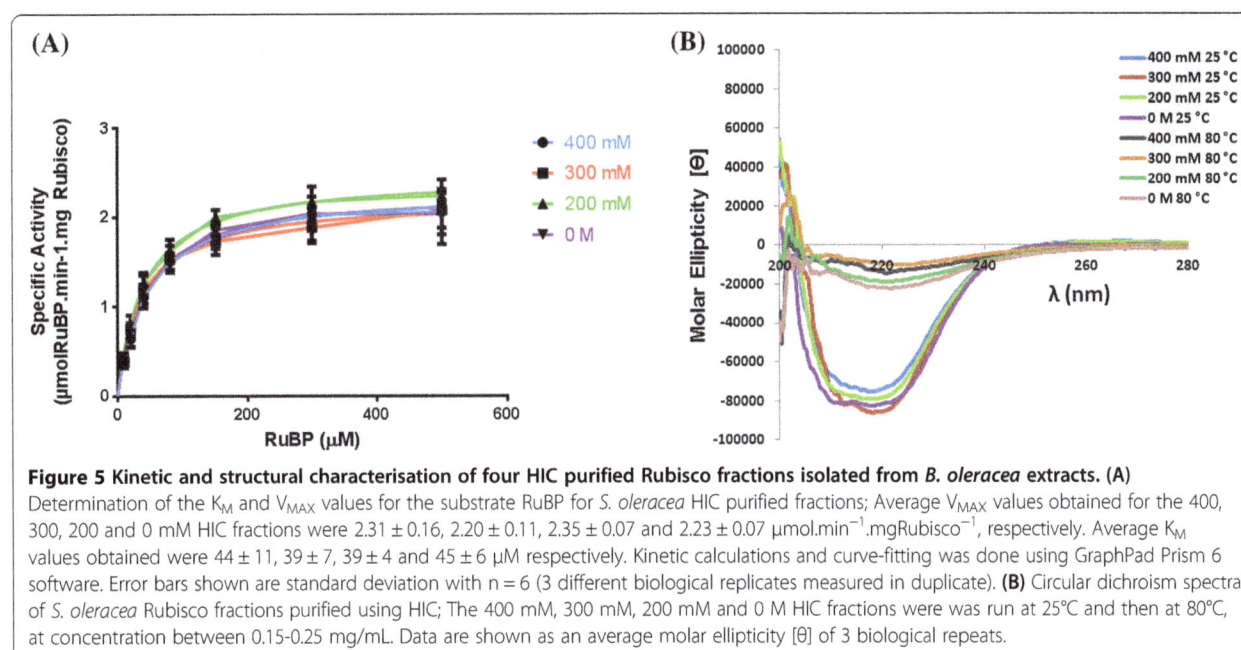

Figure 5 Kinetic and structural characterisation of four HIC purified Rubisco fractions isolated from *B. oleracea* extracts. (A) Determination of the K_M and V_{MAX} values for the substrate RuBP for *S. oleracea* HIC purified fractions; Average V_{MAX} values obtained for the 400, 300, 200 and 0 mM HIC fractions were 2.31 ± 0.16, 2.20 ± 0.11, 2.35 ± 0.07 and 2.23 ± 0.07 $\mu mol.min^{-1}.mgRubisco^{-1}$, respectively. Average K_M values obtained were 44 ± 11, 39 ± 7, 39 ± 4 and 45 ± 6 μM respectively. Kinetic calculations and curve-fitting was done using GraphPad Prism 6 software. Error bars shown are standard deviation with n = 6 (3 different biological replicates measured in duplicate). **(B)** Circular dichroism spectra of *S. oleracea* Rubisco fractions purified using HIC; The 400 mM, 300 mM, 200 mM and 0 M HIC fractions were was run at 25°C and then at 80°C, at concentration between 0.15-0.25 mg/mL. Data are shown as an average molar ellipticity [θ] of 3 biological repeats.

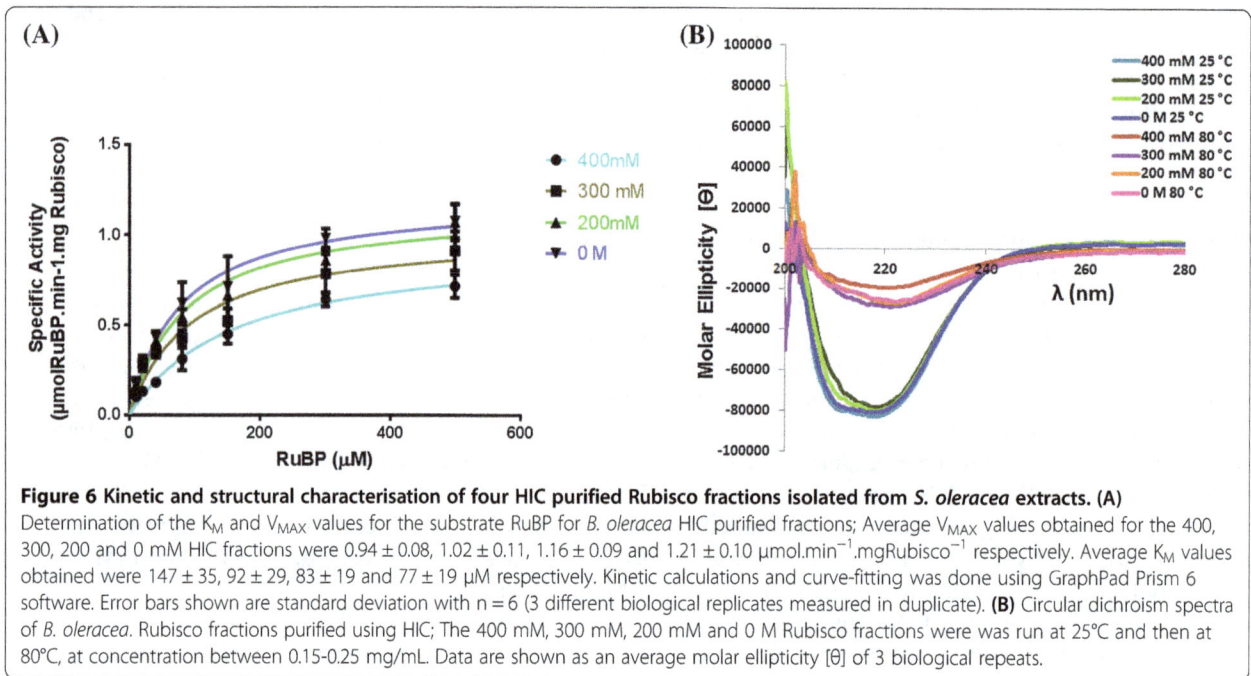

Figure 6 Kinetic and structural characterisation of four HIC purified Rubisco fractions isolated from *S. oleracea* extracts. (A) Determination of the K_M and V_{MAX} values for the substrate RuBP for *B. oleracea* HIC purified fractions; Average V_{MAX} values obtained for the 400, 300, 200 and 0 mM HIC fractions were 0.94 ± 0.08, 1.02 ± 0.11, 1.16 ± 0.09 and 1.21 ± 0.10 $\mu mol.min^{-1}.mgRubisco^{-1}$ respectively. Average K_M values obtained were 147 ± 35, 92 ± 29, 83 ± 19 and 77 ± 19 μM respectively. Kinetic calculations and curve-fitting was done using GraphPad Prism 6 software. Error bars shown are standard deviation with n = 6 (3 different biological replicates measured in duplicate). **(B)** Circular dichroism spectra of *B. oleracea*. Rubisco fractions purified using HIC; The 400 mM, 300 mM, 200 mM and 0 M Rubisco fractions were was run at 25°C and then at 80°C, at concentration between 0.15-0.25 mg/mL. Data are shown as an average molar ellipticity [θ] of 3 biological repeats.

reasonable agreement to C3 species reported by Yeoh *et al.* [36].

Since there is a significant difference in sequence homology between Rubisco from *S. oleracea* and *B. oleracea* (71.7% for the small subunits of *S. oleracea* (RBCS2), and *B. oleracea* (BRA034024) and 91.4% for the large subunits), it is not unexpected that we should observe species specific differences in the HIC elution profiles and enzymological kinetics of Rubisco. However, the presence of distinct hydrophobic populations of Rubisco in both *S. oleracea* and *B. oleracea* suggests this phenomenon is not species specific.

An additional experiment was designed to investigate whether HIC could separate distinct subpopulations of Rubisco from Rubisco previously purified using IEC. Extracts from *S. oleracea* were purified by IEC and the Rubisco containing IEC fraction was then loaded onto

an HIC column. Hydrophobically distinct fractions of Rubisco were again observed when eluted from the HIC column (Additional file 3). This further highlights the significance and potential use of this newly developed HIC protocol for Rubisco purification, and demonstrates the potential use of HIC in conjunction with the well-established IEC protocols.

Circular dichroism studies reveal no significant structural differences in the Rubisco populations

Circular dichroism (CD) can be a useful spectroscopic method to study conformational changes of proteins [37-39]. CD studies were carried out on the four HIC fractions derived from both *S. oleracea* and *B. oleracea* extracts, to probe the cause of the differences in the hydrophobicity of the samples (Figures 5B and 6B respectively). For comparison, an IEC fraction of the

Table 1 Summary of V_{MAX}, K_{CAT} and K_M values of the HIC purified Rubisco fraction from *S. oleracea* and *B. oleracea* extracts

Species	$[(NH_4)_2SO_4]$ mM	VMAX ($\mu mol.min^{-1}.mgRubisco^{-1}$)	K^*_{CAT} (s^{-1})	KM (RuBP) (μM)
S. oleracea	400	2.31 ± 0.16	2.63 ± 0.20	44 ± 11
	300	2.20 ± 0.11	2.52 ± 0.14	39 ± 7
	200	2.35 ± 0.07	2.76 ± 0.10	39 ± 4
	0	2.23 ± 0.06	2.59 ± 0.08	45 ± 6
B. oleracea	400	0.94 ± 0.08	1.09 ± 0.09	147 ± 35
	300	1.02 ± 0.11	1.17 ± 0.13	92 ± 29
	200	1.16 ± 0.09	1.34 ± 0.10	83 ± 19
	0	1.21 ± 0.10	1.39 ± 0.11	77 ± 19

*K_{CAT} was calculated based on 8 active sites and a molecular weight of 550 K g/mol.

corresponding extracts was also analysed (Additional file 4). Interestingly, each of the hydrophobically distinct fractions of Rubisco had similar values of molar ellipticites [θ] (~ −8000 deg cm^2 dmol^{-1}) and structures of the CD spectra when compared to each other, to the IEC fraction, and with CD spectra of Rubisco previously reported in the literature [40,41]. For comparison, maximum denaturation of the fractions was achieved by heating at 80°C (shown in Figures 5B and 6B), which caused a large drop in the molar ellipticities of the samples (≤2000 deg cm^2 dmol^{-1}), as well as a loss of structure in the spectra. This analysis demonstrates that there are no significant changes in the secondary structures of the Rubisco populations. Fractions purified using HIC or IEC showed similar CD spectra, suggesting that the Rubisco in the HIC fractions has the same structural integrity as the IEC purified Rubisco. With the latter being currently the standard method for obtaining pure Rubisco [20], the HIC method investigated here seems to produce pure Rubisco of comparable nature and quality. Furthermore, the CD experiments confirm that the heterogeneity of the Rubisco populations is not due to denaturation.

Conclusions

Taking these findings together, we can conclude that HIC is not only a valuable alternative purification method for form I Rubisco, but it also has the unique capability of being able to resolve distinct Rubisco populations. Since Rubisco isolated using HIC has similar values of purity and catalytic activity to that obtained using existing IEC purification methods, HIC has the potential to replace IEC, providing the additional benefit of avoiding high dilutions, dialysis and gel filtration chromatography. In addition, the ability to separate distinct Rubisco populations makes HIC a valuable method for current and future Rubisco researchers. This is of particular importance for investigators seeking to determine Rubisco's structure and interactome, since conformational alterations could significantly impact upon these results.

Methods
Chemicals and equipment
Chemicals for SDS-PAGE were obtained from Invitrogen. Instant Blue (Expedion) was used as a Coomassie staining agent. All other chemicals and enzymes were obtained from Sigma, except phosphoglycerate kinase (PGK) which was expressed and purified (see below). For homogenization of plant material, a Kenwood blender was used. Centrifugation was carried out in a Biofuge Primo R (Thermo Scientific). Purification was performed using an AKTA™ design FPLC system. SDS-PAGE gels and western blots were documented using a Fujifilm LAS-3000 Imaging System. The purity of the gels were determined using Image J software. Rubisco activity assays were performed in a Thermo Scientific Varioskan Flash Multimode reader.

Extraction of Rubisco from leaves
Unless stated otherwise, all procedures were performed rapidly at 4°C to maximize active enzyme recovery. The mid ribs of Spinach (S. oleracea) or Savoy Cabbage (B. oleracea) leaves were removed and the plant tissue was quickly frozen in liquid nitrogen and powdered in a mortar and pestle. 75 mL of extraction buffer (20 mM Hepes (pH 6.5) 5 mM MgCl$_2$, 0.33 M sorbitol, 0.2% (w/v) iso-ascorbic acid, 5 mM DDT, 0.75 mL of plant protease inhibitor cocktail (Sigma)) was added to 15 g of plant tissue, and homogenization and cell lysis carried out in a blender using 20–30 pulses, of approximately 1–2 second duration. The homogenate was filtered through two layers of Miracloth (Calbiochem) followed by centrifugation for 30 min at 3000 × g to remove intact cells and debris, and the cell extract (CE) was further centrifuged for 10 min at 8000 × g. The supernatant was subjected to (NH$_4$)$_2$SO$_4$ precipitation.

Ammonium sulphate precipitation
Two rounds of (NH$_4$)$_2$SO$_4$ precipitation were performed at 35 and 60% saturation. In each case, solid (NH$_4$)$_2$SO$_4$ was added to the desired percentage of saturation and the solution was stirred and left to equilibrate for 20 min. The precipitate was collected by centrifugation for 10 min at 10000 × g. The supernatant of the 35% saturation was subjected to the next round of precipitation. Protein pellets may be stored at 4°C for 2 days.

Hydrophobic interaction chromatography using ammonium sulphate
Purification was performed with a flow rate of 4 mL/min using a 5 mL HiTrap™ Phenyl Sepharose 6 FF (high sub) column (GE Healthcare). Buffers were filtered before use through 0.45 μm filters. (NH$_4$)$_2$SO$_4$ pellets were resuspended in HIC buffer (50 mM Tris–HCl (pH 7.6, KOH), 20 mM MgCl$_2$, 20 mM NaHCO$_3$, 0.2 mM EDTA, 2 mM DTT) containing 1 M (NH$_4$)$_2$SO$_4$, equilibrated and filtered through a 0.45 μm filter to prevent clogging of the column. After washing the HIC column with 25 mL HIC buffer, the column was equilibrated with 25 mL HIC buffer (containing 1 M (NH$_4$)$_2$SO$_4$), and then the sample was loaded. After washing the column with HIC buffer (containing 1 M (NH$_4$)$_2$SO$_4$), a series of step elutions were performed at (NH$_4$)$_2$SO$_4$ concentrations of 500, 400, 300, 200 and 0 mM (the column was washed between each elution step until the UV absorbance returned to baseline). Elution peaks were collected and desalted using 30 kDa filter units (washed 4 times with HIC buffer at 5000 × g for 10 mins), and the subsequent

samples were analysed by SDS-PAGE, Western blot, and for Rubisco activity (for protocols, see below). Protein concentrations were determined by Bradford assay [42].

Hydrophobic interaction chromatography using potassium chloride

Purification was performed with a flow rate of 4 mL/min using a 5 mL HiTrap™ Phenyl Sepharose 6 FF (high sub) column (GE Healthcare). Buffers were filtered before use through 0.45 µm filters. Crude extract samples were filtered through a 0.45 µm filter to prevent clogging of the column. After washing the HIC column with 25 mL HIC buffer, the column was equilibrated with 25 mL HIC (containing 2 M KCl), and then the sample was loaded. After washing the column with HIC buffer containing 2 M KCl, a series of step elutions were performed at KCl concentrations of 1.5 M, 1.0 M, 0.6 M, 0.3 M and 0 M (the column was washed between each elution step until the UV absorbance returned to baseline). Elution peaks were collected and desalted using 30 kDa filter units (washed 4 times with HIC buffer at 5000 × g for 10 mins), and analysed for Rubisco activity. Protein concentrations were determined by Bradford assay.

Ion exchange chromatography followed by hydrophobic interaction chromatography

Purification was performed with a flow rate of 3 mL/min using 3 × 5 mL HiTrap™ Q HP columns for increased yield (GE Healthcare). Buffers were filtered before use through 0.45 µm filters. $(NH_4)_2SO_4$ pellets were resuspended in IEC buffer (25 mM Tris–HCl (pH 7.6, KOH), 10 mM $MgCl_2$, 10 mM $NaHCO_3$, 0.1 mM EDTA, 2 mM DTT) and filtered through a 0.45 µm filter to prevent clogging of the column. After washing the IEC column with 25 mL IEC buffer containing 1 M KCl, the column was equilibrated with 25 mL of IEC buffer, and then the sample was loaded. After washing the column with IEC buffer, the sample was eluted over a linear gradient from 0 to 0.5 M KCl, over a 90 minute period. Rubisco is known to elute between ~ 0.30-0.35 M KCl.

The Rubisco peak was collected and desalted using 30 kDa filter units (washed 4 times with HIC buffer at 5000 × g for 10 mins), and analysed for Rubisco activity. The remaining sample was made up to 1 M $(NH_4)_2SO_4$ in HIC buffer, and filtered through a 0.45 µm filter. After washing the HIC column (5 mL HiTrap™ Phenyl Sepharose 6 FF (high sub)) with 25 mL HIC buffer, the column was equilibrated with 25 mL HIC (containing 1 M $(NH_4)_2SO_4$), and then the sample was loaded. After washing with HIC buffer containing 1 M $(NH_4)_2SO_4$, a series of step elutions were performed at $(NH_4)_2SO_4$ concentrations of 500, 400, 300, 200 and 0 mM (the column was washed between each elution step until

the UV absorbance returned to baseline). Elution peaks were collected and desalted using 30 kDa filter units (washed 4 times with HIC buffer at 5000 × g for 10 mins), and analysed for Rubisco activity. Protein concentrations were determined by Bradford assay.

SDS polyacrylamide gel electrophoresis (SDS-PAGE)

SDS-PAGE was performed in NuPage® Bis-Tris 4-12% gradient gels (Invitrogen) according to the manufacturer's protocol [43]. Staining was carried with InstantBlue Coomassie stain.

Western blot

After SDS-PAGE, the gel was equilibrated in 4°C Transfer buffer (Tris 25 mM, Glycine 192 mM, Methanol reagent grade 20% (v/v)) for 15 mins. Immunoblotting using a Criterion™ Blotter Cell (Bio-rad) was carried out at 70 V for 90 minutes using a Bio-rad Powerpac® 1000. After blotting, the nitrocellulose membrane was washed twice in TBST, pH 7.5 (Tris 50 mM, NaCl 150 mM , Tween 20, 0.05% (v/v)) at room temperature on an orbital shaker, and followed by washing with Blocking solution (2% (w/v) milk powder dissolved in TBST) for 2 hours at room temperature. The membrane was incubated with primary Antibody Solution (rabbit anti-RbcL (Agrisera) dissolved in 0.5% Blocking Solution at 1:5000) overnight at 4°C, followed by 5 washes with TBST, for 3 minutes each. The membrane was incubated with secondary Antibody Solution (donkey anti-rabbit IgG (H&L) HRP conjugated (Agrisera) dissolved in 0.5% Blocking Solution at 1:10000) for 2 hours, after which, the membrane was washed 3 times at 10 minute intervals with TBST. Lumi-Light substrate (Roche) was mixed with the membrane for 5 minutes and then the membrane was quickly imaged.

Rubisco activity assay

Rubisco carboxylase activity was determined using a non-radioactive microplate-based assay, which determines the product [(3-phosphoglycerate (3-PGA)] in an enzymic cycle between glycerol-3-phosphate dehydrogenase and glycerol-3-phospate oxidase, adapted from Sulpice et al. [44]. The assay was monitored through the oxidation of NADH by optical density measurements at 340 nm at 25°C. 15 µL of Rubisco was added to the initial buffer for activation (containing 100 mM Tricine (pH 8.0, KOH), 20 mM $MgCl_2$, 2 mM EDTA and 10 mM $NaHCO_3$), to give a total volume of 30 µL. The concentration of $NaHCO_3$ was confirmed to be saturating. While incubating the sample for 15 mins at room temperature, 15 uL of different concentrations of RuBP (giving final concentrations of 10 µM to 500 µM) was pipetted into a 96 well plate. The Rubisco/initial buffer mixture was then added to the wells (giving a total volume of 45 µL) and allowed to react for 60 seconds, after which, 15 μL of absolute ethanol was added to stop the reaction. To minimise the amount of assay enzymes used, 20 μL of the assay mixture was added to a 384 well plate (after 5 mins of reaction with ethanol), and 20 μL of determination buffer was added to start the reaction [final concentrations in 40 μL were: 1.875 u/mL phosphoglycerate kinase (PGK), 3 u/mL glyceraldehyde-3-phosphate dehydrogenase (GAPDH), 2.5 u/mL α-glycerol-3-phosphate dehydrogenase-/triose-P isomerase (G3PDH-TPI), 100 u/mL glycerol-3-phosphate oxidase (G3P-OX), 700 u/mL catalase, 3 mM ATP, 0.5 mM NADH, 1 mM $MgCl_2$, 60 mM Tricine (pH 8.0, KOH)]. The rates of reaction were calculated as the decrease of the absorbance in $OD.min^{-1}$ and converted to μmol by use of a 3-phosphoglyceric acid (3PGA) calibration curve.

Expression and purification of phosphoglycerate kinase

The coding region of the DNA sequence of phosphoglycerate kinase 1 (*Saccharomyces cerevisiae*) was cloned into pGEX-6P-1 expression vector (GE Healthcare). Phosphoglycerate kinase 1 expression and purification was performed as previously described [45].

Circular dichroism

CD spectra were measured using a JASCO-715 Circular Dichroism Spectropolarimeter, with the use of using JASCO PTS-604 T Temperature Controller, to regulate the temperature. Samples were allowed to incubate at 25°C and 80°C for 10 minutes before CD measurements were taken. Molecular ellipticities [θ] are reported, using an average molecular weight of 560 KDa for Rubisco, and a path length of 0.1 cm. Sample were prepared in HIC buffer at protein concentrations between 0.15-0.25 mg/mL.

Additional files

Additional file 1: Purification tables for *S. oleracea* (Table S1) and *B. oleracea* (Table S2). NB. * not calculated for these samples. ⁻ Activity was too low to measure. aCalculations based on 'cell extract'. bCalculations based on 'load'. # K_{CAT} calculated based on 8 active sites of Rubisco, with a total molecular weight of 550 K g/mol.

Additional file 2: HIC elution profile (using potassium chloride (KCl)) from cell extract of *S. oleracea*; and the SDS-PAGE gel showing protein content from the different elution steps of the HIC purification protocol. (A) HIC (using potassium chloride (KCl)) elution profile generated by loading cell extract isolated from S. oleracea;The cell extract was loaded at 2 M KCl onto a 5 mL HiTrap HIC column, washed and eluted as described in Methods**.** The gradient used for the separation (red line) and the positions of the major absorption peaks obtained by measuring OD at 280 nm (blue line) following the HIC elution profile is shown. Elutions were performed at 1.5 M, 1.0 M, 0.6 M, 0.3 M and 0 M KCl. **(B)** SDS-PAGE gel showing protein content from the different steps involved in the HIC potassium chloride purification protocol of Rubisco isolated from *S. oleracea*; Samples were separated by SDS-PAGE and subsequently stained for protein. The position of Rubisco's large (LS) and small (SS) subunits are indicated in the figure. 0.4 μg of sample was added per lane. The purity of the CE sample was calculated as 68.2 ± 3.1%,

and HIC fraction containing Rubisco had purities of 86.4 ± 1.4 (1.5 M), 87.7 ± 2.1% (1.0 M), 89.1 ± 1.7% (0.6 M), 81.3 ± 3.0% (0.3 M) and 70.1 ± 4.2% (0 M). The gels are representative of 2 sample sets.

Additional file 3: Kinetic characterisation of the IEC and HIC purified Rubisco fractions from *S. oleracea* extracts. Determination of the K_M and V_{MAX} values for the substrate RuBP for *S. oleracea* Rubisco purified using IEC and followed by further purification of the IEC fraction using HIC; Rubisco was purified using IEC, and the subsequent fraction was then loaded onto an HIC column at 1 M $(NH_4)_2SO_4$, and eluted at $(NH_4)_2SO_4$ concentrations of 500 mM, 400 mM, 300 mM, 200 mM and 0 M, as per Methods section. Average V_{MAX} and K_M values obtained for the IEC fraction was 2.8 ± 0.22 μmol.min.mg and 47 ± 13 μM. Average V_{MAX} values obtained for, 300 mM, 200 mM and 0 M was 3.0 ± 0.16, 3.1 ± 0.10 and 2.8 ± 0.13 μmol.min.mg respectively. Average K_M values obtained were 46 ± 8, 40 ± 5 and 47 ± 8 μM respectively. It is of note that although there was a peak in the UV trace of the elution profile at 400 mM $(NH_4)_2SO_4$ elution, there was not enough protein obtained to perform kinetic analysis, as the peak was very small (even when more protein was loaded). Kinetic calculations and curve-fitting was done using GraphPad Prism 6 software. Error bars shown are standard deviation with n = 6 (3 different biological replicates measured in duplicate).

Additional file 4: Circular dichroism spectra of *S. oleracea* Rubisco, purified using IEC. Rubisco purified using IEC was run at 25°C at a concentration of 0.2 mg/mL. Data are shown as an average molar ellipticity [θ] of 3 biological repeats.

Abbreviations

HIC: Hydrophobic interaction chromatography; IEC: Ion exchange chromatography; RuBP: Ribulose-1,5-bisphophate; RLPs: Rubisco like proteins; CA1P: 2 carboxy-D-arabinitol 1-phosphate; SDS-PAGE: Sodium dodecyl sulphate polyacrylamide gel electrophoresis; CD: Circular dichroism; PGK: Phosphoglycerate kinase; CE: Cell extract; 3PGA: 3-phosphoglyceric acid; GAPDH: Glyceraldehyde-3-phosphate dehydrogenase; G3PDH-TPI: mL α-glycerol-3-phosphate dehydrogenase-/triose-P isomerase; G3POX: Glycerol-3-phosphate oxidase; IPTG: Isopropyl β-D-1-thiogalactopyranoside; GST: Glutathione S-transferase.

Competing interests

The authors declare that they have no competing interests.

Authors' contributions

KOD carried out the purification study, enzymological studies, and CD analysis. PP and ZG assisted in the purification and enzymological studies. MSD and SK carried out preliminary purification studies with *B. oleracea*. LHM expressed and purified phosphoglycerate kinase. RW and LB conceived the study and participated in its design. KOD, RW and LB coordinated and drafted the manuscript. All authors read and approved the final manuscript.

Acknowledgements

This work was supported by a Royal Society University Fellowship to LB, an EPSRC Centre for Doctoral Training Studentship from the Institute of Chemical Biology (Imperial College London) awarded to KOD.

Author details

[1]Institute of Chemical Biology, Department of Chemistry, Imperial College, Flowers Building, South Kensington Campus, Exhibition Road, London SW7 2AZ, UK. [2]Department of Chemistry, Imperial College, South Kensington Campus, Exhibition Road, London SW7 2AZ, UK.

References

1. Spreitzer R, Salvucci M: **Rubisco: structure, regulatory interactions, and possibilities for a better enzyme.** *Annu Rev Plant Biol* 2002, **53**:449–475.
2. Andersson I: **Catalysis and regulation in Rubisco.** *J Exp Bot* 2008, **59**(7):1555–1568.
3. Behrenfield MJ, Randerson JT, Falkowski P: **Primary Production of the Biosphere: Integrating Terrestrial and Oceanic Components.** *Science* 1998, **281**(5374):237–240.

4.   Ellis RJ: **Biochemistry: Tackling unintelligent design.** *Nature* 2010, **463**(7278):164–165.

5.   Sage RF, Sage TL, Kocacinar F: **Photorespiration and the evolution of C4 photosynthesis.** *Annu Rev Plant Biol* 2012, **63**:19–47.

6.   Zhu XG, Portis AR, Long SP: **Would transformation of C3 crop plants with foreign Rubisco increase productivity? A computational analysis extrapolating from kinetic properties to canopy photosynthesis.** *Plant Cell Environ* 2004, **27**(2):155–165.

7.   Ellis JR: **The most abundant protein in the world.** *Trends Biochem Sci* 1979, **4**(11):241–244.

8.   Andersson I, Backlund A: **Structure and function of Rubisco.** *Plant Physiol Biochem* 2008, **46**(3):275–291.

9.   Tabita FR, Satagopan S, Hanson T, Kreel N, Scott S: **Distinct form I, II, III, and IV Rubisco proteins from the three kingdoms of life provide clues about Rubisco evolution and structure/function relationships.** *J Exp Bot* 2008, **59**(7):1515–1524.

10. Chapman MS, Cascio D, Eisenberg D, Suh SW, Smith WW: **Sliding-layer conformational change limited by the quaternary structure of plant RuBisCO.** *Nature* 1987, **329**(6137):354–356.

11. Andersson I, Knight S, Schneider G, Lindqvist Y, Lundqvist T: **Crystal-structure of the active-site of Ribulose-Bisphosphate Carboxylase.** *Nature* 1989, **337**(6204):229–234.

12. Andersson I, Taylor TC: **Structural framework for catalysis and regulation in ribulose-1,5-bisphosphate carboxylase/oxygenase.** *Arch Biochem Biophys* 2003, **414**(2):130–140.

13. Andrews TJ: **Catalysis by cyanobacterial ribulose-bisphosphate carboxylase large subunits in the complete absence of small subunits.** *J Biol Chem* 1988, **263**(25):12213–12219.

14. Lee B, Tabita FR: **Purification of recombinant ribulose-1,5-bisphosphate carboxylase/oxygenase large subunits suitable for reconstitution and assembly of active L8S8 enzyme.** *Biochemistry* 1990, **29**(40):9352–9357.

15. Whitney SM, Sharwood RE, Orr D, White SJ, Alonso H, Galmés J: **Isoleucine 309 acts as a C4 catalytic switch that increases ribulose-1,5-bisphosphate carboxylase/oxygenase (rubisco) carboxylation rate in Flaveria.** *Proc Natl Acad Sci* 2011, **108**(35):14688–14693.

16. Ishikawa C, Hatanaka T, Misoo S, Miyake C, Fukayama H: **Functional Incorporation of Sorghum Small Subunit Increases the Catalytic Turnover Rate of Rubisco in Transgenic Rice.** *Plant Physiol* 2011, **156**(3):1603–1611.

17. Spreitzer R: **Role of the small subunit in ribulose-1,5-bisphosphate carboxylase/oxygenase.** *Arch Biochem Biophys* 2003, **414**(2):141–149.

18. Uemura K, Anwaruzzaman, Miyachi S, Yokota A: **Ribulose-1,5-bisphosphate carboxylase/oxygenase from thermophilic red algae with a strong specificity for CO2 fixation.** *Biochem Biophys Res Commun* 1997, **233**(2):568–71.

19. Suarez R, Miro M, Cerda V, Perdomo JA, Galmes J: **Automated flow-based anion-exchange method for high-throughput isolation and real-time monitoring of RuBisCO in plant extracts.** *Talanta* 2011, **84**(5):1259–66.

20. Salvucci ME, Portis AR Jr, Ogren WL: **Purification of ribulose-1,5-bisphosphate carboxylase/oxygenase with high specific activity by fast protein liquid chromatography.** *Anal Biochem* 1986, **153**(1):97–101.

21. Jakob R: **2 Quick methods for isolation of Ribulose-1,5-Bisphosphate Carboxylase Oxygenase.** *Prep Biochem* 1988, **18**(3):351–360.

22. Whitney SM, Sharwood RE: **Linked Rubisco subunits can assemble into functional oligomers without impeding catalytic performance.** *J Biol Chem* 2007, **282**(6):3809–18.

23. Hemmingsen SM, Ellis RJ: **Purification and properties of ribulosebisphosphate carboxylase large subunit binding protein.** *Plant Physiol* 1986, **80**(1):269–76.

24. Carmo-Silva AE, Barta C, Salvucci ME: **Isolation of ribulose-1,5-bisphosphate carboxylase/oxygenase from leaves.** *Methods in molecular biology (Clifton, NJ)* 2011, **684**:339–47.

25. Whitney SM, Kane HJ, Houtz RL, Sharwood RE: **Rubisco oligomers composed of linked small and large subunits assemble in tobacco plastids and have higher affinities for CO2 and O2.** *Plant Physiol* 2009, **149**(4):1887–95.

26. Rosichan JL, Huffaker RC: **Source of Endoproteolytic activity associated with purified Ribulose Bisphosphate Carboxylase.** *Plant Physiol* 1984, **75**(1):74–77.

27. Kreel NE, Tabita FR: **Substitutions at Methionine 295 of Archaeoglobus fulgidus Ribulose-1,5-bisphosphate Carboxylase/Oxygenase Affect Oxygen Binding and CO2/O2 Specificity.** *J Biol Chem* 2007, **282**(2):1341–1351.

28. McCurry SD, Gee R, Tolbert NE: **Ribulose-1,5-bisphosphate carboxylase/oxygenase from spinach, tomato, or tobacco leaves.** *Methods Enzymol* 1982, **90 Pt E**(Pt E):515–21.

29. Parry MAJ, Andralojc PJ, Parmar S, Keys AJ, Habash D, Paul MJ, Alred R, Quick WP, Servaites JC: **Regulation of Rubisco by inhibitors in the light.** *Plant Cell and Environment* 1997, **20**(4):528–534.

30. Parry MAJ, Keys AJ, Madgwick PJ, Carmo-Silva AE, Andralojc PJ: **Rubisco regulation: a role for inhibitors.** *J Exp Bot* 2008, **59**(7):1569–1580.

31. Sage RF: **Variation in the k(cat) of Rubisco in C(3) and C(4) plants and some implications for photosynthetic performance at high and low temperature.** *J Exp Bot* 2002, **53**(369):609–20.

32. Uemura K, Suzuki Y, Shikanai T, Wadano A, Jensen RG, Chmara W, Yokota A: **A Rapid and Sensitive Method for Determination of Relative Specificity of RuBisCO from Various Species by Anion-Exchange Chromatography.** *Plant Cell Physiol* 1996, **37**(3):325–331.

33. Shen JB, Ogren WL: **Alteration of spinach ribulose-1,5-bisphosphate carboxylase/oxygenase activase activities by site-directed mutagenesis.** *Plant Physiol* 1992, **99**(3):1201–7.

34. Shen JB, Orozco EM Jr, Ogren WL: **Expression of the two isoforms of spinach ribulose 1,5-bisphosphate carboxylase activase and essentiality of the conserved lysine in the consensus nucleotide-binding domain.** *J Biol Chem* 1991, **266**(14):8963–8.

35. Portis AR Jr: **Rubisco activase - Rubisco's catalytic chaperone.** *Photosynth Res* 2003, **75**(1):11–27.

36. Yeoh H-H, Badger MR, Watson L: **Variations in Kinetic Properties of Ribulose-1,5-bisphosphate Carboxylases among Plants.** *Plant Physiol* 1981, **67**(6):1151–1155.

37. Kelly SM, Price NC: **The application of circular dichroism to studies of protein folding and unfolding.** *Biochim Biophys Acta* 1997, **1338**(2):161–85.

38. Kelly SM, Jess TJ, Price NC: **How to study proteins by circular dichroism.** *Biochim Biophys Acta* 2005, **1751**(2):119–39.

39. Greenfield NJ: **Applications of circular dichroism in protein and peptide analysis.** *TrAC Trends Anal Chem* 1999, **18**(4):236–244.

40. Voordouw G, van der Vies SM, Bouwmeister PP: **Dissociation of ribulose-1,5-bisphosphate carboxylase/oxygenase from spinach by urea.** *Eur J Biochem* 1984, **141**(2):313–8.

41. Li G, Mao H, Ruan X, Xu Q, Gong Y, Zang X, Zhao N: **Association of heat-induced conformational change with activity loss of Rubisco.** *Biochem Biophys Res Commun* 2002, **290**(3):1128–32.

42. Bradford MM: **A rapid and sensitive method for the quantitation of microgram quantities of protein utilizing the principle of protein-dye binding.** *Anal Biochem* 1976, **72**:248–54.

43. Laemmli UK: **Cleavage of structural proteins during the assembly of the head of bacteriophage T4.** *Nature* 1970, **227**(5259):680–5.

44. Sulpice R, Tschoep H, Von Korff M, Büssis D, Usadel B, Höhne M, Witucka-Wall H, Altmann T, Stitt M, Gibon Y: **Description and applications of a rapid and sensitive non-radioactive microplate-based assay for maximum and initial activity of D-ribulose-1,5-bisphosphate carboxylase/oxygenase.** *Plant Cell Environ* 2007, **30**(9):1163–1175.

45. Mak L, Vilar R, Woscholski R: **Characterisation of the PTEN inhibitor VO-OHpic.** *J Chem Biol* 2010, **3**(4):157–163.

5

Chlorophyll fluorescence emission can screen cold tolerance of cold acclimated *Arabidopsis thaliana* accessions

Anamika Mishra[1], Arnd G Heyer[2] and Kumud B Mishra[1*]

Abstract

Background: An easy and non-invasive method for measuring plant cold tolerance is highly valuable to instigate research targeting breeding of cold tolerant crops. Traditional methods are labor intensive, time-consuming and thereby of limited value for large scale screening. Here, we have tested the capacity of chlorophyll *a* fluorescence (ChlF) imaging based methods for the first time on intact whole plants and employed advanced statistical classifiers and feature selection rules for finding combinations of images able to discriminate cold tolerant and cold sensitive plants.

Results: ChlF emission from intact whole plant rosettes of nine *Arabidopsis thaliana* accessions was measured for (1) non-acclimated (NAC, six week old plants grown at room temperature), (2) cold acclimated (AC, NAC plants acclimated at 4°C for two weeks), and (3) sub-zero temperature (ST) treated (STT, AC plants treated at −4°C for 8 h in dark) states. Cold acclimation broadened the slow phase of ChlF transients in cold sensitive (Co, C24, Can and Cvi) *A. thaliana* accessions. Similar broadening in the slow phase of ChlF transients was observed in cold tolerant (Col, Rsch, and Te) plants following ST treatments. ChlF parameters: *maximum quantum yield of PSII photochemistry* (F_V/F_M) and *fluorescence decrease ratio* (R_{FD}) well categorized the cold sensitive and tolerant plants when measured in STT state. We trained a range of statistical classifiers with the sequence of captured ChlF images and selected a high performing *quadratic discriminant classifier* (QDC) in combination with *sequential forward floating selection* (SFFS) feature selection methods and found that linear combination of three images showed a reasonable contrast between cold sensitive and tolerant *A. thaliana* accessions for AC as well as for STT states.

Conclusions: ChlF transients measured for an intact whole plant is important for understanding the impact of cold acclimation on photosynthetic processes. Combinatorial imaging combined with statistical classifiers and feature selection methods worked well for the screening of cold tolerance without exposing plants to sub-zero temperatures. This opens up new possibilities for high-throughput monitoring of whole plants cold tolerance *via* easy and fully non-invasive means.

Keywords: High-throughput screening, Chlorophyll *a* fluorescence transients, Cold tolerance, Cold acclimation, Whole plant, *Arabidopsis thaliana*

Background

Cold tolerance is the ability of plants to withstand low temperatures and plays a crucial role in worldwide production of many important agricultural crops. In temperate climate zones, plants have developed strategies to adjust low temperature tolerance by employing a highly complex process of physiological re-arrangements, termed cold acclimation that is triggered by low but non-freezing temperatures. Numerous studies revealed that cold acclimation not only programs massive changes in transcriptome and metabolome [1-7] but also induces structural and compositional modifications of compatible solutes in various subcellular compartments [8]. Following acclimation, plants are more efficient in dealing with the impact of sustained cold or sudden temperature drops by activating a plethora of adjustments including accumulation of cryoprotective metabolites and proteins for their survival [9]. A correlation of cold acclimation capacities with habitat winter temperatures

* Correspondence: mishra.k@czechglobe.cz
[1]Global Change Research Centre, Academy of Sciences of the Czech Republic, v. v. i, Bělidla 986/4a, 603 00, Brno, Czech Republic
Full list of author information is available at the end of the article

points to high metabolic or ecological costs of these adjustments [5,10,11]. Despite intensive research the molecular mechanisms of cold tolerance are still not fully understood and remain an area for intensive research, as understanding of mechanisms responsible for cold acclimation would allow breeding of cold tolerant crops.

The primary requirement for research associated with engineering of low temperature tolerant crops is to develop efficient and cost effective methods for measuring cold tolerance. Measurement of survival or re-growth scores following triggers of cold acclimation and de-acclimation treatments are quite lengthy and may not be very accurate [12,13]. Most quantitative methods for assessing cold tolerance employ analysis of damage to plasma membranes by electrolyte leakage (EL) screens or the thylakoids *via* plastocyanin release after freeze-thaw cycles [14]. Generally, the temperature affording half of maximal damage (LT_{50}), i.e. the temperature at which 50% of electrolytes or plastocyanin are released from their respective compartments, is being evaluated and used as a proxy for cold tolerance. However, since the measurement of purified plastocyanin is comparatively difficult [15], the most popular method to study plant cold tolerance is the evaluation of EL of detached leaves by using conductivity measurements [5,16]. According to this method, detached leaves are placed in reaction tubes, which are progressively cooled down to certain temperatures. Low-temperature treated samples are then thawed at 4°C and electrical conductivity of a bathing solution is measured to calculate LT_{50} values that represent half-maximum damage to the plasma membrane [16]. EL is a widely accepted method for quantification of plant cold tolerance, although its results sometimes deviate from those obtained with alternative methods [17,18].

The response of thylakoids membranes to sustained low-temperature treatments can be measured non-invasively by the analysis of ChlF emission. ChlF is re-emitted by chlorophyll *a* molecules following light absorption, and it is modulated by photochemical and non - photochemical events in the photosynthetic pigment–protein complexes PSII and PSI of the thylakoid membranes [19,20]. The relative contribution of PSII and PSI in ChlF emission is also reported to change during cooling treatments [21]; thereby, this method can provide important insights into molecular processes of cold acclimation. The ChlF parameter F_V/F_M of detached leaves following freeze - thaw cycles has successfully been applied to quantify cold tolerance in *Arabidopsis thaliana* as well as in other plant species [reviewed in 8,22-24]. The analysis of polyphasic fluorescence rise of from initial low fluorescence (F_O) to peak F_P by JIP test [25] has also been reported to be useful for the selection of cold tolerance in wheat genotypes [26,27]. Earlier, we measured the ChlF transients of detached leaves of differentially cold tolerant *A. thaliana* accessions during progressive cooling starting

from room temperature to −15°C and found high correlation between LT_{50} measured by electrolyte leakage with ChlF parameters such as $F_O(-15°C)/F_O(4°C)$ and $[F_S/F_O]_{-15°C}$ [10]. We demonstrated that application of advanced statistics-based combinatorial imaging methods to the sequence of time resolved ChlF images could be used to categorize cold tolerance levels by training of classifiers using fluorescence emission of detached leaves that were slowly cooled at mild sub-zero-temperatures of around −4°C [10]. Vaclavik et al. [7] reported that cold acclimation induced accumulation of Gluconapin and Flavon-3-ol glycosides, respectively, in cold tolerant vs. cold sensitive *A. thaliana* accessions, and thus metabolomic patterns, could be used for the screening of cold tolerance already in the cold acclimated state i.e. without exposing plants to freezing temperatures. Although metabolomic based methods can thus have better discriminating capacity of cold tolerant and cold sensitive accessions as compared to EL methods, they are not readily applicable in high-throughput screenings. Therefore, in the search for easy methods of sensing cold tolerance, we have tested the potential of ChlF transients measured on whole plant rosettes of the model species *Arabidopsis thaliana*. We involved NAC, AC, and STT plants of nine *A. thaliana* accessions that span the north–south range of the species [5]. A short light/dark protocol was applied for recording of ChlF transients of dark-adapted plants for about 202 s using a pulse amplitude modulation (PAM) based fluorometer [28]. We found that ChlF transients averaged over whole intact plant rosettes offered valuable information for assessing impacts of cold acclimation and cold induced alterations of the photosynthetic machinery. In addition to classical analysis of ChlF transients, we employed statistical classifiers and feature selection methods on captured ChlF images to search for highly contrasting features of cold tolerant vs. cold sensitive accessions. We have trained several classifiers and chose the best performing classifier, i.e. a QDC, for further use along with a high performing SFFS feature selection methods to identify features correlated with cold tolerance. The sets of images obtained for the STT state were then tested on AC state data sets, and we found that this method, i.e. combinatorial imaging, can be applied for assessing plant cold tolerance already in the cold acclimated state, i.e. without any sub-zero temperature treatments and by fully non-invasive means.

Results

ChlF transients from whole plants are highly informative for categorization of cold tolerance in *A. thaliana* accessions

We measured the ChlF transients of whole plant rosettes in nine differentially cold tolerant *Arabidopsis thaliana* accessions in the NAC, AC, and STT state [Figure 1]. From the shape of ChlF transients it can be established that for the

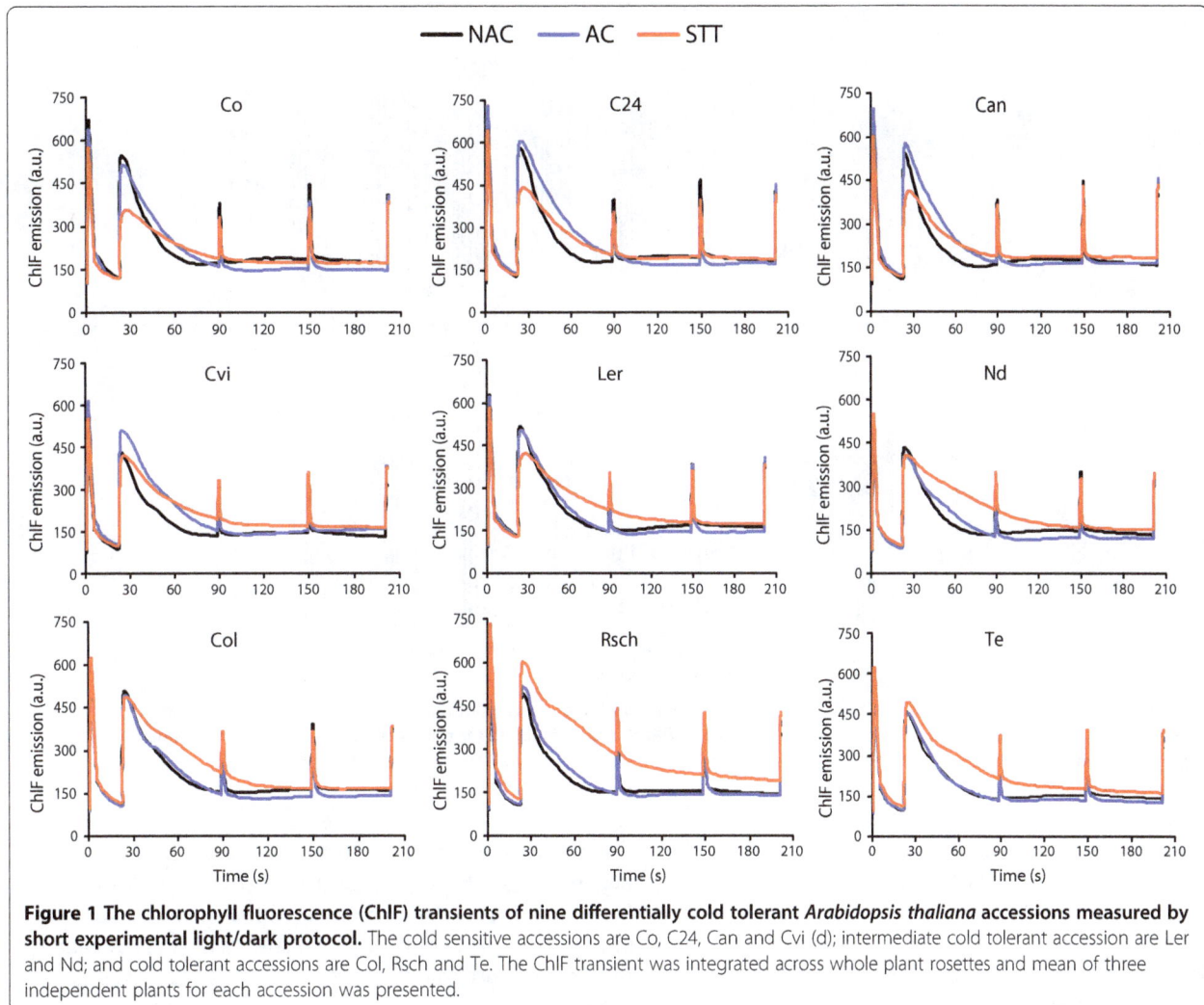

Figure 1 The chlorophyll fluorescence (ChlF) transients of nine differentially cold tolerant *Arabidopsis thaliana* accessions measured by short experimental light/dark protocol. The cold sensitive accessions are Co, C24, Can and Cvi (d); intermediate cold tolerant accession are Ler and Nd; and cold tolerant accessions are Col, Rsch and Te. The ChlF transient was integrated across whole plant rosettes and mean of three independent plants for each accession was presented.

cold sensitive accessions (Co, C24, Can, and Cvi), the fluorescence intensity following peak F_P quenched slowly in cold acclimated (AC) state, whereas the value of F_P itself decreased significantly when these plants were treated with sub-zero temperature (ST, −4°C for 8 h in dark). In cold tolerant accessions (Te, Rsch, and Col-0), however, the shape of the ChlF transients remained similar for NAC and AC states; while after STT, quenching of fluorescence intensity following peak F_P is significantly slowed down. Qualitative comparison of the ChlF transients of intermediate accessions (Ler and Nd) revealed a compound picture, and based on qualitative comparisons we tend to assign Ler as cold sensitive, while Nd can be classified cold tolerant.

ChlF parameters F_V/F_M and R_{FD} of whole plant rosettes can measure cold tolerance in subzero-temperature (ST) treated plants

Fluorescence parameters F_V/F_M and R_{FD} were evaluated from measured ChlF emission and presented as an average of three independent experiments, where values of each

parameter were estimated by integrating across whole rosette leaves. We did not find correlations between the classical ChlF parameters and cold tolerance among the investigated *A. thaliana* accessions either for NAC or for the AC state. However, the ChlF parameters F_V/F_M and R_{FD}, differed for cold sensitive, intermediate and cold tolerant accessions following STT (Figure 2). In STT state, F_V/F_M values for cold tolerant (Te, Col and Rsch), intermediate (Nd and Ler) and sensitive accessions (Co, C24, and Can) were around 0.85 ± 0.01, 0.83 ± 0.01 and 0.81 ± 0.01, respectively, while the averaged R_{FD} values were calculated as 1.98 ± 0.19, 1.48 ± 0.07 and 1.40 ± 0.19, respectively. Non-paired t-test revealed F_V/F_M to differ significantly (p <0.01) for cold tolerant and cold sensitive accessions, while there was no significant difference between cold sensitive and intermediate or intermediate and tolerant accessions. In contrast, plant vitality index R_{FD} showed significant differences between the cold sensitive and tolerant as well as between intermediate and tolerant accessions (p <0.05), but no significant difference was found between sensitive and

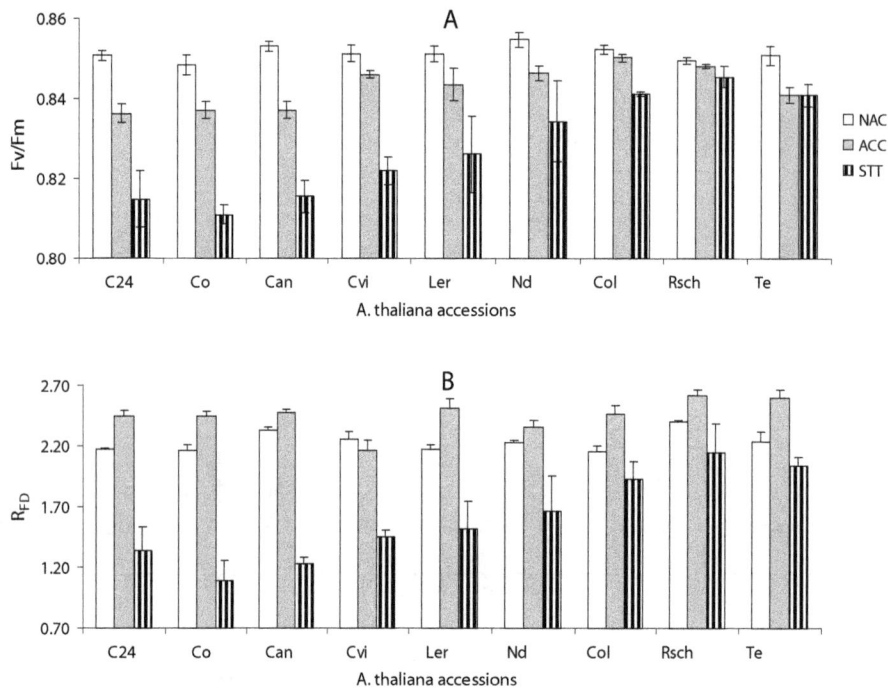

Figure 2 Chlorophyll fluorescence (ChlF) parameters: *maximum quantum yield of PSII photosystems* (F_V/F_M) [A] and *fluorescence decrease ratio* (R_{FD}) [B] of differentially cold tolerant accessions of *Arabidopsis thaliana* for non-acclimated (NAC), cold acclimated (AC), and sub-zero temperature treated (STT) states. The presented numeric values are mean of three independent plants with standard errors, and are integrated across whole plant rosettes.

intermediate tolerant accessions. For the accessions Cvi the values of ChlF parameters F_V/F_M and R_{FD} are almost similar to that of intermediate accessions Ler and Nd, thereby its tolerance level of thylakoid may be categorized intermediate, while the plasma membrane behaved cold sensitive in EL measurements.

Cold acclimation induced gain in photosynthetic performance of *A. thaliana* accessions

The ChlF parameter *effective quantum efficiency of PS II photochemistry* (Φ_{PSII}) was significantly higher (p <0.01) in AC vs. NAC plants except for Cvi (Figure 3A). Interestingly, mild sub-zero temperature treatments (–4°C) for eight hours in AC plants led to a significant decline in the value of Φ_{PSII} in all accessions (Figure 3A) without Φ_{PSII} being correlated with the cold tolerance. Two weeks of cold acclimation caused a significant (p <0.01) increase of *photochemical quenching* (qP) in all accessions, while STT led to a decline (Figure 3B). As for Φ_{PSII}, no correlation of parameter qP with cold tolerance was observed.

Combinatorial imaging

We applied statistical techniques of classifier and feature selection methods in order to identify features of cold tolerance from captured sequences of time-resolved ChlF images. This method is very powerful in identifying image

sets from large sequences of time-resolved ChlF recordings that yield the highest contrast between groups to be compared. Earlier we demonstrated the usefulness of this approach for discriminating three species of family *Lamiaceae* at very early stages of growth [29] and for investigating features of cold tolerance at non-lethal temperatures [10]. The method depends on training and performance testing of randomly selected pixels of the most contrasting ChlF image sets, i.e. recordings for the highly cold tolerant accession Te and highly sensitive Co in the STT state [for technical details, see 10,29,30]. The performance, error rate and computational time to run the algorithms of five tested classifiers, *linear discriminant classifier* (LDC), *quadratic discriminant classifier* (QDC), *k-nearest neighbors classifier* (k-NNC), *nearest neighbor classifier* (NNC) and *nearest mean classifier* (NMC), are presented in Table 1. We found that algorithms of QDC are the best performing classifier (81% correct assignment of cold tolerance group among test images) in comparatively short time (<6 s, Table 1). Therefore, QDC was chosen and applied with the SFFS feature selection method to find the most contrasting image sets for highly cold tolerant (Te) vs. highly cold sensitive (Co) accessions. Figure 4 shows the performance curve for the most efficient feature selection method SFFS in combination with simulated classifier QDC, where x-axis represents number of images. Thus, in this experiment, the SFFS algorithm reduced the full data

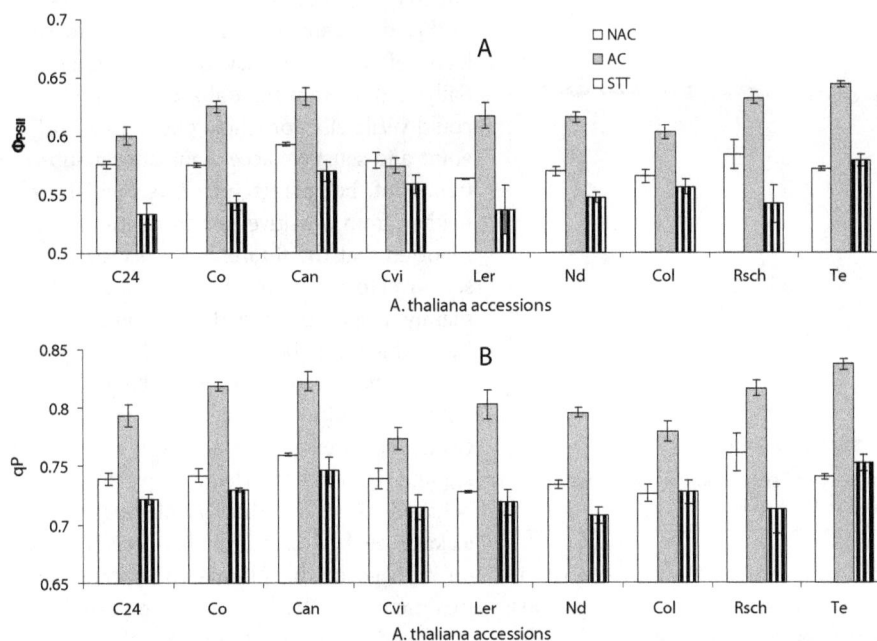

Figure 3 The *effective quantum efficiency PS II* (Φ_{PSII}) [A] and *photochemical quenching* (qP) [B] of cold sensitive (Co, C24, Cvi, Can), intermediate cold tolerant (Nd and Ler) and cold tolerant (Te, Rsch and Col) *Arabidopsis thaliana* accessions for non-acclimated (NAC), cold acclimated (ACC), and sub-zero temperature treated (STT) states. The presented numeric values are integrated across whole plant rosettes and mean of three independent plants with standard errors.

set of 218 images to three images identified as I_{21}, I_{104} and I_{107} without compromising the classification performance (~80%, Figure 4). We obtained the linear combination: $C = (0.3219)*I_{21} + (-0.5018)*I_{104} + (-0.2315)*I_{107}$, to perform best in discriminating cold tolerant and sensitive accessions. The coefficients of the linear combination were calculated according to Matouš et al. [30]. Figure 5 shows linear combinations of images for all nine accessions in the STT as well as AC state. The presented images are plotted on a false color scale where color represents the virtual fluorescence intensity of the pixel with reference to its corresponding value in the training data set. Thus, we can visualize a clear difference between cold sensitive, intermediate, and cold tolerant accessions in the STT state. When the combinatorial imaging was applied to AC plants it apparently failed to discriminate cold sensitive and intermediate accessions showing very similar patterns for leaf rosettes of the two groups of plants. However, the

tolerant group of accessions could be clearly discriminated already in the AC state.

Discussion

Towards easy methods for assessing whole plant cold tolerance in natural accessions of A. thaliana

In an attempt to establish an easy and non-invasive method for screening plant cold tolerance, we compared different concepts for measuring cold tolerance based on chlorophyll fluorescence that have been approved for detached leaves of the model species *Arabidopsis thaliana*. As far as we know, for the first time we attempted to capture the ChlF transients of intact whole plant rosettes and found interesting modulation in it for the group of cold sensitive (Co, C24, Can, Cvi), intermediate tolerant (Ler, Nd) and cold tolerant (Te, Rsch, Col-0) *A. thaliana* accessions for NAC, AC and STT state (Figure 1). The ChlF transient measured by quenching analysis of modulated fluorescence is very useful methods that can monitor slight changes in the photochemical and nonphotochemical process [31]. The slow quenching of ChlF transients following peak F_P, i.e. broadening of ChlF transients, for cold sensitive group of accessions under cold acclimated state and for tolerant group of accessions under STT state (Figure 1) is complex and can be affected by series of events such as light induced intrathylakoid H^+ concentration, non-photochemical quenching, inhibition of CO_2 assimilation processes,

Table 1 The table presents performance, error rate and computational time of the tested classifiers

Classifiers	Performance	Error rate	Computational time (s)
LDC	0.81	0.09	6.14
QDC	0.81	0.09	5.86
k-NNC (*k* = 3)	0.82	0.08	5530
NNC (*k* = 1)	0.79	0.10	5510
NMC	0.47	0.26	8.23

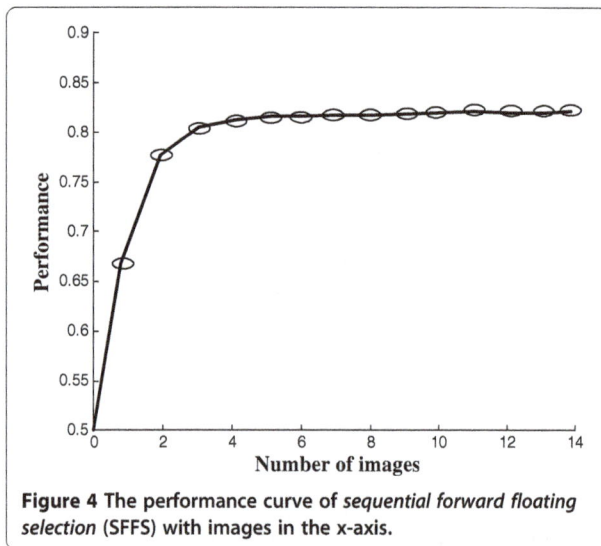

Figure 4 The performance curve of *sequential forward floating selection* (SFFS) with images in the x-axis.

ATP synthesis among others [reviewed in 32]. However, it reveals that sensitive and tolerant group of accessions possess two different strategies to utilize the absorbed irradiance following cold acclimation as well as when placed at mild subzero temperature ($-4°C$) for 8 h in darkness. Probably cold tolerant plants efficiently utilize acclimation temperature and keeping homeostasis of photosynthetic apparatus and yielded almost similar ChlF transients in NAC as well as in AC state that varied only when treated with sub-zero temperature. The reduction of F_P for the group of cold sensitive accessions in STT state indicates perturbation in PSII functional properties. Significant inhibition of F_V/F_M for sensitive vs. tolerant accessions in STT states further support slight inactivation of PSII function. Ehlert and

Hincha [23] reported that the easily recordable parameter F_V/F_M discriminates cold tolerant and sensitive detached leaves after a freeze-thaw cycle. Using a set of nine differentially cold tolerant natural accessions of *Arabidopsis* [5], we could verify also for whole plants that F_V/F_M differs for tolerant and sensitive accessions after a sub-zero temperature treatment, but plants with intermediate tolerance differed neither from sensitive nor from tolerant ones. It has been reported that the fluorescence decrease ratio, R_{FD}, is more sensitive to a variety of stress factors than F_V/F_M, which mainly responds to extreme conditions [33,34]. Indeed, R_{FD}, which can be calculated from fluorescence peak F_P and F_S measured under prevailing actinic light, was not only able to discriminate sensitive and tolerant but also tolerant and intermediate accessions. Interestingly, both F_V/F_M and R_{FD} data revealed that plastids of the accession Cvi behaved as intermediately tolerant, while the electrolyte leakage method classified this accession as sensitive based on damage to the plasma membrane during freeze-thaw treatments [5]. We have earlier reported that cold tolerance of plastids and the plasma membrane appear not strictly correlated in the accession Col-0 [8], and the differential tolerance of plastids and plasma membrane in Cvi is further evidence for independent principles underlying cold tolerance of the different types of membranes.

Considering plants not treated with sub-zero temperatures, which would be desirable for a cold tolerance screening program, neither F_V/F_M nor R_{FD} were correlated with tolerance of the nine *Arabidopsis* accessions used in this study. Since both parameters are related to intactness of PSII, it is not unexpected that they are not affected by above-zero temperatures in a cold tolerant

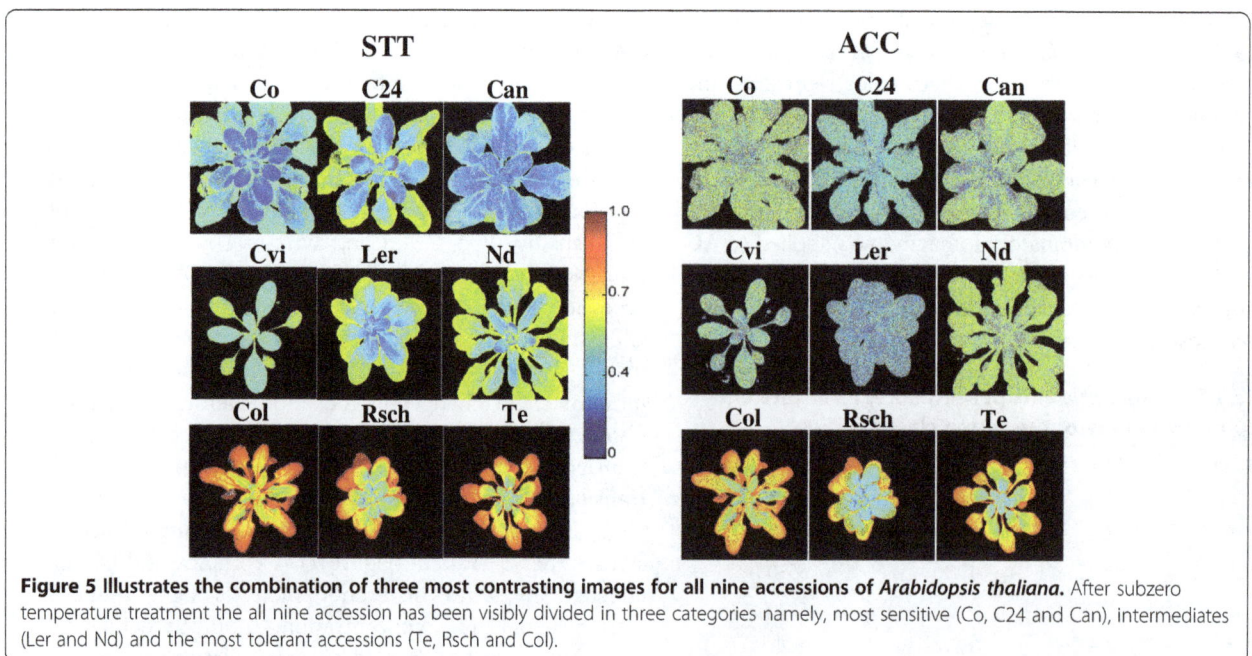

Figure 5 Illustrates the combination of three most contrasting images for all nine accessions of *Arabidopsis thaliana*. After subzero temperature treatment the all nine accession has been visibly divided in three categories namely, most sensitive (Co, C24 and Can), intermediates (Ler and Nd) and the most tolerant accessions (Te, Rsch and Col).

plant like *Arabidopsis*. However, it was noticed that parameters Φ_{PSII}, and qP that reflect effects of actinic light on chlorophyll fluorescence responded differentially to low above-zero and sub-zero temperature treatments. It has been reported that shifting plants from normal to low temperatures causes repression of genes related to photosynthesis [35], and among them are genes directly involved in light harvesting like e.g. proteins Lhca2*1 and Lhcb4*2 [36]. Down-regulation of photosynthesis genes is accompanied by a sudden suppression of photosynthesis at low temperatures [37]. However this might be a transient effect, since leaves that developed at low temperatures are reported to regain full photosynthetic activity [38]. Although a reduction in antenna size would not necessarily impact parameters Φ_{PSII} and qP, it is not self-evident why low temperatures should increase effective quantum yield. Savitch & co-workers [38] reported that low temperatures could increase photosynthetic capacity in species like e.g. wheat that maintain a high need for assimilates because of active growth at low temperatures. In the cold tolerant tomato species *Lycopersicon peruvianum*, a rise in qP during cold exposure was hypothesized to result from increased capacity of the Calvin-Benson cycle [39]. In this context it should be noted that temperature effects on light absorption and photosynthesis are often not easily separable from light effects, because in most studies light intensity is reduced at low temperature to avoid damage to the photosystems in the cold. At lower light intensity, qN is reduced and qP rises, and in turn also Φ_{PSII} is increased as long as no damage of the photosystems occurs. This effect may superimpose the temperature effect on light harvesting, and although the cold sensitive accession Cvi could be discriminated from all intermediate and tolerant accessions based on the lack of increase of Φ_{PSII} during acclimation, this parameter does not seem to be sufficient for screening cold tolerance in acclimated *Arabidopsis* plants.

Combinatorial imaging can discriminate cold acclimation induced cold tolerant plants

We have earlier reported that combinatorial imaging of ChlF transients combined with classifier and feature selection methods was able to discriminate detached leaves from cold sensitive, tolerant and intermediate accessions of Arabidopsis [10]. The main outcome of the current study is that this method can also be applied to whole plants, i.e. leaves to be compared need not be present on the same images, and thus this method is suitable for large scale screening of plant cold tolerance.

Combinatorial imaging combines sets of highly performing images from sequences of time-resolved ChlF images that provide strong discrimination capacity [10,29,30]. High performing images extract information from several thousands of pixels from hundreds of measured images (218 images per data for this experiment, and each image with

resolution 512*512 pixels). This method does not trace underlying physiological phenomena rather its algorithms select image sets having optimal contrast between sensitive and tolerant features that ultimately allow discrimination on the basis of their cold tolerance levels [30]. Therefore, parameters identified for the classifiers are very likely species and treatment specific, and training of the classifiers needs to be performed for each application of the method. While this might appear disadvantageous, it also offers great flexibility, because successful discrimination is not dependent on any specific physiological process but can exploit every feature that allows discrimination [10,29,30,40,41]. In the current study, older leaves performed better than developing ones, probably because the higher fluorescence intensity provided a better signal-to-noise ratio. However, this does not implicate that the method is not suited to the screening of plant seedlings, because it analyzes relative fluorescence intensities distributed among the image pixels and for each sample type it is different [29]. Therefore, combinatorial imaging out-performs existing methods that depend on the application of freeze-thaw treatments.

The combinatorial imaging method could discriminate sensitive and tolerant accessions in the cold acclimated state but it failed to make any difference between cold sensitive and intermediate accessions. Because training was done for STT plants, this failure could result from differential responses of the photosynthetic apparatus to low above-zero and sub-zero temperatures as outlined above, or it could relate to insufficient resolution, which might be overcome by a comparison of larger sets of images. Since this would clearly result in higher computational load, a case specific decision for higher resolution vs. shorter analysis time might be necessary.

Even if only a two-class discrimination can be achieved with the combinatorial imaging method, it would be very useful for large scale screening of cold tolerance e.g. in recombinant inbred line populations or other plant sets consisting of large numbers of individuals representing different genotypes. The method is thus well suited for quantitative trait loci (QTL) mapping or mutant screenings to investigate genetic determinants of cold tolerance that may be used for plant breeding.

Conclusions

We have demonstrated that chlorophyll fluorescence emission from whole plant rosettes of *Arabidopsis thaliana* integrates information that can be used to discriminate cold sensitive and tolerant plants in the cold acclimated state when analyzed by the advanced statistics based combinatorial imaging approach [10,29,30]. This reveals the power of combinatorial imaging for identifying features of cold tolerant and sensitive accessions at cold acclimation state (Figure 5) where well known physiological parameters of

ChlF emission (Figures 2 and 3) failed to provide any clue of discrimination. Moreover, capturing ChlF transients of whole plant rosettes following mild sub-zero temperature treatments (STT, −4°C for ~8 h in dark) is also very useful, because classical ChlF transients and the extracted parameters such as F_V/F_M and R_{FD} can categorize cold sensitive, intermediate and tolerant accessions following STT. In addition, classical ChlF analysis, in contrast to combinatorial imaging methods, can yield physiologically relevant information that could be directly exploited for breeding efforts. The screening of whole plants cold tolerance *via* fully non-invasive means following cold acclimation, i.e. without any sub-zero temperature treatments, could be highly useful in high-throughput screening of cold tolerance where it is superior to data measured from single leaf or leaf discs by alternative methods such as metabolomics [7] or electrolyte leakage [5].

Methods
Plant material and growth conditions
Six plantlets of each of the nine accessions of *Arabidopsis thaliana* [5] were grown inside a cooling chamber in pots of 0.06 m for six weeks at day/night temperatures 20°C/18°C, light 90 μmol (photons) m^{-2} s^{-1} and a relative humidity of 70%. The accessions used in this study were: Cvi (Cap Verde Islands), Can (Canary Islands), C24 (genetically related to Co, Portugal), Co (Coimbra, Portugal), Col-0 (Columbia-0, genetically related to Gü, Germany), Nd (Niederzenz, Germany), Ler (Landsberg *erecta*, Poland), Rsch (Rschew, Russia), and Te (Tenela, Finland). The experiments were executed in three independent replica with three independent sets of plants used to measure ChlF in the non acclimated (NAC, six weeks grown plants), cold acclimated (AC, NAC plants acclimated at 4°C for another 2 weeks), and sub-zero temperature treated (STT, AC plants treated with mild sub-zero temperature of −4°C for 8 hours in dark) state.

Chlorophyll fluorescence measurement
Individual plants were used for ChlF measurements at room temperature using Handy FluorCam [www.psi.cz; 23]. A short protocol of ~202 seconds modified according to Mishra et al. [10] was used for capturing time-resolved ChlF from plant rosettes. This protocol starts with the measurement of basal fluorescence (F_O) and maximum fluorescence (F_M) using measuring and saturating flashes. After a short dark period of ~20 seconds, actinic light of 40 μmol (photons) m^{-2} s^{-1} was applied to measure the fluorescence transients. Two strong flashes of saturating light were overlaid with actinic light to investigate activation of non-photochemical quenching followed by a third saturating flash 18 s after switching off the actinic light to see the relaxation of non-photochemical quenching mechanisms. Three replicate recordings were taken for NAC, AC and STT state, respectively. The images

of the measured ChlF transients were averaged across the whole rosettes for quantitative evaluation of the fluorescence parameters or plotting of ChlF transients.

Combinatorial imaging using statistical classifier and feature selection method
The method of combinatorial imaging was applied to identify the discriminant between the accessions without using the whole data set consisting of 218 images in a time series. Each time series contains some repetitive images as well as images with low contrast, which were sorted out to reduce the size of the data sets. Application of statistical techniques of classifiers was chosen to avoid biasness. We randomly classified the data for Te and Co as training and testing sets. Using the training set, features discriminating the tolerant and sensitive accession were identified, and these features were next applied using the testing set for discriminating cold tolerant and sensitive accessions. Using this method we calculated the performance of several classifiers: LDC, QDC, k-NNC, NNC and NMC, and chose the best performing classifier for further analysis. The performance of each of the investigated classifiers was quantified by a number between 0–1: value '0' means random classification (1/2 of classifications into 2 equally represented classes correct, and another 1/2 is incorrect) and value '1' meaning that the classifier was 100% successful [for details see 24]. In this experiments performance of QDC was 81% and computational time was less than 6 seconds. Therefore, QDC classifier was applied with sequential forward floating selection (SFFS) to reduce the number of images for effective classification [42]. The reduction is based on finding an image sub-set containing the most useful information for the visualization of highly contrasting features of cold tolerant vs. sensitive accessions and arranging the images in a form of descending order of their performance efficiency. The combinatorial images are developed by the combination of the three high performing images (x,y,z) multiplied with their coefficient (a,b,c) that are obtained by *linear discriminant analysis* (LDA): Combinatorial imaging (C) = (±a) $*I_x + (\pm b)*I_y + (\pm c)*I_z$.

Tool for the data analysis
The Matlab software package, version 6.5, with Pattern Reorganization toolbox (PRTools) was used for statistical analysis.

Abbreviations
ChlF: Chlorophyll *a* fluorescence; NAC: Non-acclimated; AC: Cold acclimated; ST(T): Sub-zero temperature (treated); PSII: photosystem II; PSI: Photosystem I; F_V/F_M: Maximum quantum yield of PSII photochemistry, where F_V is variable fluorescence ($F_V = F_M$-F_O), F_O is minimal fluorescence emission of a dark adapted plant with primary quinone acceptor (Q_A), oxidized and non-photochemical quenching inactive, and F_M is maximum fluorescence emission of a dark adapted plant exposed to a short pulse of a strong light

leading to a transient reduction of Q_A; $R_{FD} = (F_P\text{-}F_S)/F_S$: Fluorescence decrease ratio, where F_P is a fluorescence peak at the beginning of the transient with actinic light, and F_S is steady state fluorescence; QDC: Quadratic discriminant classifier; SFFS: Sequential forward floating selection; EL: Electrolyte leakage; LT_{50}: The half of maximal lethal temperature; EL: Electrolyte leakage; Φ_{PSII}: Effective quantum efficiency of PSII photochemistry; qP: Photochemical quenching; qN: Non-photochemical quenching; k-NNC: k- nearest neighbor classifier; NNC: Nearest neighbor classifier; NMC: Nearest mean classifier; LDA: Linear discriminant analysis.

Competing interests
The authors declare that they have no competing interests.

Authors' contributions
KBM, AGH and AM had planed and designed the experiments. KBM and AM performed the experiments. AM analyzed the data and drafted the manuscript. KBM and AGH contributed in the final version of the manuscript. All authors read and approved the final manuscript.

Acknowledgements
Participation of AM and KBM was supported by the EfCOP - IPo project ENVIMET (CZ.1.07/2.3.00/20.0246) and by the EU (FP7) EPPN project (284443). We thank Ladislav Nedbal for his inspiration.

Author details
[1]Global Change Research Centre, Academy of Sciences of the Czech Republic, v. v. i, Bělidla 986/4a, 603 00, Brno, Czech Republic. [2]Institute of Biomaterials and Biomolecular Systems, Department of Plant Biotechnology, University of Stuttgart, Stuttgart, Germany.

References
1. Jaglo-Otteson KR, Gilmour SJ, Zarka DG, Schabenberger O, Thomashow MF: Arabidopsis CBF1 overexpression induces COR genes and enhances freezing tolerance. *Science* 1998, **280**:104–106.
2. Cook D, Fowler S, Fiehn O, Thomashow MF: A prominent role for the CBF cold response pathway in configuring the low-temperature metabolomes of Arabidopsis. *Proc Natl Acad Sci U S A* 2004, **101**(45):15243–15248.
3. Kaplan F, Kopka J, Haskell DW, Zhao W, Schiller KC, Gatzke N, Sung DY, Guy CL: Exploring the temperature-stress metabolome of Arabidopsis. *Plant Physiol* 2004, **136**:4159–4168.
4. Guy C, Kaplan F, Kopka J, Selbig J, Hincha DK: Metabolomics of temperature stress. *Physiol Plant* 2008, **132**:220–235.
5. Hannah MA, Wiese D, Freund S, Fiehn O, Heyer AG, Hincha DK: Natural genetic variation of freezing tolerance in Arabidopsis. *Plant Physiol* 2006, **142**:98–112.
6. Zuther E, Schulz E, Childs LH, Hinch DK: Clinal variation in the non-acclimated and cold-acclimated freezing tolerance of *Arabidopsis thaliana* accessions. *Plant Cell Environ* 2012, **35**:1860–1878.
7. Vaclavik L, Mishra A, Mishra KB, Hajslova J: Mass spectrometry-based metabolomic fingerprinting for screening cold tolerance in *Arabidopsis thaliana* accessions. *Anal Bioanal Chem* 2013, **405**(8):2671–2683.
8. Knaupp M, Mishra KB, Nedbal L, Heyer AG: Evidence for a role of raffinose in stabilizing photosystem II during freeze–thaw cycles. *Planta* 2011, **234**(3):477–486.
9. Moellering ER, Bagyalakshmi M, Benning C: Freezing tolerance in plants requires lipid remodeling at the outer chloroplast membrane. *Science* 2010, **330**:226–228.
10. Mishra A, Mishra KB, Höermiller II, Heyer AG, Nedbal L: Chlorophyll fluorescence emission as a reporter on cold tolerance in *Arabidopsis thaliana* accessions. *Plant Signal Behav* 2011, **6**(2):301–310.
11. Degenkolbe T, Giavalisco P, Zuther E, Seiwert B, Hinchey DK, Willmitzer L: Differential remodeling of the lipidome during cold acclimation in natural accessions of *Arabidopsis thaliana*. *Plant J* 2012, **72**:972–982.
12. Rapacz M, Gasior D, Zwierzykowski Z, Lesniewska-Bocianowska A, Humphreys MW, Gay AP: Changes in cold tolerance and the mechanisms of acclimation of photosystem II to cold hardening generated by anther culture of Festuca pratensis x Lolium multiflorum cultivars. *New Phytol* 2004, **162**(1):105–114.
13. Novillo F, Alonso JM, Ecker JR, Salinas J: CBF2/ DREB1C is a negative regulator of CBF1/DREB1B and BF3/DREB1A expression and plays a central role in stress tolerance in Arabidopsis. *Proc Natl Acad Sci U S A* 2004, **101**:3985–3990.
14. Hincha DK: Cryoprotectin: a plant lipid-transfer protein homologue that stabilizes membranes during freezing. *Philos T R Soc B* 2002, **357**:909–915.
15. Hincha DK, Hofner R, Schwab KB, Heber U, Schmi JM: Membrane rupture is the common cause of damage to chloroplast membranes in leaves injured by freezing or excessive wilting. *Plant Physiol* 1987, **83**:251–253.
16. Steponkus PL, Lynch DV, Uemura M: The influence of cold-acclimation on the lipid-composition and cryobehavior of the plasma-membrane of isolated rye protoplasts. *Philos T R Soc B* 1990, **326**:571–583.
17. Azzarello E, Mugnai S, Pandolfi C, Masi E, Marone E, Mancuso S: Comparing image (fractal analysis) and electrochemical (impedance spectroscopy and electrolyte leakage) techniques for the assessment of the freezing tolerance in olive. *Trees* 2009, **23**:159–167.
18. Yamori W, Noguchi K, Terashima I: Temperature acclimation of photosynthesis in spinach leaves: analyses of photosynthetic components and temperature dependencies of photosynthetic partial reactions. *Plant Cell Environ* 2005, **28**:536–547.
19. Govindjee: Sixty-three years since Kautsky: chlorophyll a fluorescence. *Aust J Plant Physiol* 1995, **22**:131–160.
20. Baker N: Chlorophyll fluorescence: a probe of photosynthesis in vivo. *Annu Rev Plant Biol* 2008, **59**:89–113.
21. Agati G, Cerovic ZG, Moya I: The effect of decreasing temperature up to chilling values on the in vivo F685/F735 chlorophyll fluorescence ratio in *Phaseolus vulgaris* and *Pisum sativum*: The role of the photosystem I contribution to the 735 nm fluorescence band. *Photochem Photobiol* 2000, **72**:75–84.
22. Baker NR, Rosenqvist E: Applications of chlorophyll fluorescence can improve crop production strategies: an examination of future possibilities. *J Exp Bot* 2004, **55**(403):1607–1621.
23. Ehlert B, Hincha DK: Chlorophyll fluorescence imaging accurately quantifies freezing damage and cold acclimation responses in Arabidopsis leaves. *Plant Methods* 2008, **4**:12.
24. Thalhammer A, Hincha DK, Zuther E: Measuring freezing tolerance: electrolyte leakage and chlorophyll fluorescence assays. *Plant Cold Acclimation: Methods Mol Biol* 2014, **1166**:15–24.
25. Strasser RJ, Srivastava A, Govindjee: Polyphasic chlorophyll-Alpha fluorescence transcient in plants and cyanobacteria. *Photochem Photobiol* 1995, **61**(1):32–42.
26. Rapacz M, Gasior D, Koscielniak J, Kosmala A, Zwierzykowski Z, Humphreys MW: The role of the photosynthetic apparatus in cold acclimation of Lolium multiflorum. Characteristics of novel genotypes low-sensitive to PSII over-reduction. *Acta Physiol Plant* 2007, **29**(4):309–316.
27. Rapacz M, Wozniczka A: A selection tool for freezing tolerance in common wheat using the fast chlorophyll a fluorescence transient. *Plant Breed* 2009, **128**(3):227–234.
28. Nedbal L, Soukupová J, Kaftan D, Whitmarsh J, Trtílek M: Kinetic imaging of chlorophyll fluorescence using modulated light. *Photosynth Res* 2000, **66**:3.
29. Mishra A, Matouš K, Mishra KB, Nedbal L: Towards discrimination of plant species by machine vision: advanced statistical analysis of chlorophyll fluorescence transcients. *J Fluoresc* 2009, **19**:905–913.
30. Matouš K, Benediktyova Z, Berger S, Roitsch T, Nedbal L: Case study of combinatorial imaging: what protocol and what chlorophyll fluorescence image to use when visualizing infection of Arabidopsis thaliana by pseudomonas syringae? *Photosynth Res* 2006, **90**:243–253.
31. Maxwell K, Johnson GN: Chlorophyll fluorescence: a practical guide. *J Exp Bot* 2000, **51**:659–668.
32. Papageorgiou GC, Govindjee: Photosystem II fluorescence: slow changes – scaling from the past. *J Photoch Photobio B* 2011, **104**(1–2):258–270.
33. Lichtenthaler HK, Langsdorf G, Lenk S, Buschmann C: Chlorophyll fluorescence imaging of photosynthetic activity with flash-lamp fluorescence imaging system. *Photosynthetica* 2005, **43**:355–369.
34. Horgan DB, Zabkiewicz JA: Fluorescence decline ratio: comparison with quantum yield ration for plant physiological status and herbicide treatment responses. *New Zealand Plant Protection* 2008, **61**:169–173.
35. Hannah MA, Heyer AG, Hincha DK: A global survey of gene regulation during cold acclimation in *Arabidopsis thaliana*. *PLoS Genet* 2005, **1**(2):e26.
36. Fowler F, Thomashow MF: Arabidopsis transcriptome profiling indicates that multiple regulatory pathways are activated during cold acclimation in addition to the CBF cold response pathway. *Plant Cell* 2002, **14**(8):1.

37. Strand A, Hurry V, Gustafsson P, Gardeström P: **Development of *Arabidopsis thaliana* leaves at low temperature releases the suppression of photosynthesis and photosynthetic gene expression despite the accumulation of soluble carbohydrates.** *Plant J* 1997, **12**:605–614.

38. Savitch LV, Leonardos ED, Krol M, Jansson S, Grodzinski B, Huner NPA, Oquist G: **Two different strategies for light utilization in photosynthesis in relation to growth and cold acclimation.** *Plant Cell Environ* 2002, **25**:761–771.

39. Linger P, Brüggemann W: **Correlations between chlorophyll fluorescence quenching parameters and photosynthesis in a segregating Lycopersicon esculentum × L. peruvianum population as measured under constant conditions.** *Photosynth Res* 1999, **61**:145–156.

40. Codrea MC, Hakala-Yatkin M, Kårlund-Marttila A, Nedbal L, Aittokallio T, Nevalainen OS, Tyystjärvi E: **Mahalanobis distance screening of Arabidopsis mutants with chlorophyll fluorescence.** *Photosynth Res* 2010, **105**(3):273–83.

41. Mattila H, Valli P, Pahikkala T, Teuhola J, Nevalainen OS, Tyystjärvi E: **Comparison of chlorophyll fluorescence curves and texture analysis for automatic plant identification.** *Precis Agric* 2013, **14**:621–636.

42. Jain AK, Duin RPW, Mao JC: **Statistical pattern recognition: a review.** *IEEE Trans Pattern Anal Mach Intell* 2000, **22**(1):4–37.

6

Bacterially produced *Pt*-GFP as ratiometric dual-excitation sensor for *in planta* mapping of leaf apoplastic pH in intact *Avena sativa* and *Vicia faba*

Christoph-Martin Geilfus[1*], Karl H Mühling[1], Hartmut Kaiser[2] and Christoph Plieth[3]

Abstract

Background: Ratiometric analysis with H^+-sensitive fluorescent sensors is a suitable approach for monitoring apoplastic pH dynamics. For the acidic range, the acidotropic dual-excitation dye Oregon Green 488 is an excellent pH sensor. Long lasting (hours) recordings of apoplastic pH in the near neutral range, however, are more problematic because suitable pH indicators that combine a good pH responsiveness at a near neutral pH with a high photostability are lacking. The fluorescent pH reporter protein from *Ptilosarcus gurneyi* (*Pt*-GFP) comprises both properties. But, as a genetically encoded indicator and expressed by the plant itself, it can be used almost exclusively in readily transformed plants. In this study we present a novel approach and use purified recombinant indicators for measuring ion concentrations in the apoplast of crop plants such as *Vicia faba* L. and *Avena sativa* L.

Results: *Pt*-GFP was purified using a bacterial expression system and subsequently loaded through stomata into the leaf apoplast of intact plants. Imaging verified the apoplastic localization of *Pt*-GFP and excluded its presence in the symplast. The pH-dependent emission signal stood out clearly from the background. *Pt*GFP is highly photostable, allowing ratiometric measurements over hours. By using this approach, a chloride-induced alkalinizations of the apoplast was demonstrated for the first in oat.

Conclusions: *Pt*-GFP appears to be an excellent sensor for the quantification of leaf apoplastic pH in the neutral range. The presented approach encourages to also use other genetically encoded biosensors for spatiotemporal mapping of apoplastic ion dynamics.

Keywords: GFP, Genetically encoded biosensor, Plant bioimaging, Apoplast, pH, Salinity, Nitrogen forms, Stress, Signaling, *Ptilosarcus gurneyi*, Three-channel ratio imaging

Introduction

The pH in the aqueous phase of the leaf apoplast controls multiple metabolic processes and is related to signaling cascades [1,2]. Changing environmental conditions can alter the leaf apoplastic pH, consequently affecting processes that depend upon the apoplastic H^+ concentration. Among these are the proton motive force driven transport of metabolites and mineral nutrients across the membrane. Equally important is the effect of a changing pH on the protonation state of peptides or proteins [3-7]. The protonation state of amino acid residues can alter the protein's structure, leading to pH related conformational changes (misfolding) that impair the affinity to binding sites or its function [8]. For hormones that behave according to the anion trap mechanisms for weak acids (e.g. abscisic acid), the state of protonation is of particular importance for the compartmental distribution [9-11] and their affinity to receptors [12].

Biotic and abiotic environmental factors that influence the leaf apoplastic pH include, among others, the nitrogen nutrition [13,14], the onset of drought, hydric stress, salinity or anoxia [15-21] and the colonisations and

* Correspondence: cmgeilfus@plantnutrition.uni-kiel.de
[1]Institute of Plant Nutrition and Soil Science, Christian-Albrechts-Universität zu Kiel, Hermann-Rodewald-Str. 2, 24118 Kiel, Germany
Full list of author information is available at the end of the article

associations by e.g. fungal pathogens or mutualistic mycorrhiza [18,22,23]. Developmental and physiological processes like acid-growth, light-sensing, gravitropism or the nitrate assimilation in combination with a production of OH$^-$ after cytosolic nitrate reduction are related to apoplastic pH changes [24-30]. In this light, the need for methods that enable an *in planta* quantification of leaf apoplastic pH dynamics with a high spatiotemporal resolution becomes evident. Quantitative ratiometric analyses that combine H$^+$-sensitive fluorescent dyes with microscopy based imaging techniques represent a suitable approach for spatiotemporal monitoring of pH dynamics [31-34]. However, since pH-fluorophores are not sensitive over the whole physiological pH range that can exist in the leaf apoplast, the technique of ratio imaging has some limitations. Detection of the leaf apoplastic pH value in its full span that ranges from relative neutral (6.5 to 7.0) to more acidic (below 4.0 to 5.0) [1,27,35-37] is not possible, because all available ratiometric pH indicators only cover a limited range of approx. 2–2.5 pH units over which pH sensitivity is most dynamic.

For the acidic pH range, the pH-sensitive dextranated fluorescein derivative Oregon Green 488 is well suited because (i) it has a pKa of 4.7 at which its pH sensitivity is most dynamic [12] and (ii) it has a tremendous photostability in combination with a good fluorescent brightness [34]. Apoplastic pH measurements in the more neutral pH range that last over hours, however, are more problematic. Besides the requirements for a pH sensitivity in the near-neutral pH range, the dye must be photostable over hours and large enough to avoid migration from the apoplastic space across the plasmalemma membrane into the cytosol. Fluorescein isothiocyanate-based dyes that are chosen for measurements in this rage (pK$_a$ of 5.92; [38]) have a low photostability and are prone to photobleaching [39], excluding them from long term measurements. 2′,7′-Bis-(2-carboxyethyl)-5-(6)-carboxy fluorescein (BCECF), another fluorescein-based dye also appears promising as it has a pKa of 7.0 [40], but has the disadvantage that it photobleaches relatively quickly [40]. Schulte *et al.* [1] presented a fluorescent pH reporter protein from the orange seapen *Ptilosarcus gurneyi* (*Pt*-GFP) that, when expressed in *Arabidopsis thaliana*, is used as a genetically encoded pH sensor in the relatively neutral cytosol [41]. Due to its good pH responsiveness at neutral pH (pKa of 7.3), *Pt*-GFP is ideal for pH recordings in the near neutral range that prevails in the leaf apoplast of some plant species.

Unfortunately, these self-expressed biosensors can almost exclusively be inserted into plants that are readily transformable. This impairs the usage of these powerful genetically encoded ion sensors in crop research since almost all agricultural relevant plants are not straightforward to transform.

In this study an approach was elaborated that allowed to use *Pt*-GFP for ratiometric analysis of the pH in the leaf apoplast of crops such as field bean (*Vicia faba* L.) and oat (*Avena sativa* L.) that, otherwise, would need to be transformed very laborious.

It was our strategy to purify *Pt*-GFPs from a bacterial expression system and to test whether this ratiometric dual-excitation pH indicator can (i) be non-invasively loaded directly into the apoplast of intact plants through the stomata and whether (ii) *Pt*-GFPs are suitable for detecting stress-related apoplastic pH changes in the near-neutral pH range.

Material and methods
Plant cultivation
Vicia faba L., minor cv Fuego (Saaten-Union GmbH, Isernhagen, Germany) was grown under hydroponic culture conditions in a climate chamber (14/10 h day/night; 20/15°C; 60/50% humidity; Vötsch VB 514 MICON, Vötsch Industrietechnik GmbH, Balingen-Frommern, Germany) as described in detail by Geilfus and Mühling [37]. The nutrient solution had the following composition: 0.1 mM KH_2PO_4, 1.0 mM K_2SO_4, 0.2 mM KCl, 2.0 mM $Ca(NO_3)_2$ or as given in the figure legends, 0.5 mM $MgSO_4$, 60 μM Fe-EDTA, 10 μM H_3BO_4, 2.0 μM $MnSO_4$, 0.5 μM $ZnSO_4$, 0.2 μM $CuSO_4$, 0.05 μM $(NH_4)_6Mo_7O_{24}$. Hydroponic cultivation of *Avena sativa* L. was conducted in an structurally identical climate chamber with the settings and growth conditions given elsewhere [42,43]. After 10–20 d of plant cultivation, *in vivo* pH recording was performed as described below.

Bacterial expression of GFPs
Pt-GFP (Acc.No. AY015995) was expressed as described in Schulte *et al.* [1] using the bacterial expression vector pRSETb (Invitrogen GmbH; Karlsruhe, Germany). The 6xHis-tagged fluorescent protein was purified and concentrated through a Ni^{2+}/NTA-agarose column (Qiagen, Hilden, Germany) followed by gel filtration through a NAP-25 column (Pharmacia Biotech, Freiburg, Germany). The protein was stored frozen in PBS. Before apoplast loading, the *Pt*-GFP proteins were dialysed over night against Mes-buffer (5 mM MES Roth # 4256; 5 mM K_2SO_4; Merck # 5153) with a MWCO of 10 kDa.

Loading of pH indicators into the intact leaf apoplast
For means of *in planta* recording of leaf apoplastic pH values, 7.5 μg/ml of the fluorescent pH indicator *Pt*-GFP or 25 μM of the pH-sensitive dye Oregon Green 488-dextran (Invitrogen GmbH, Darmstadt, Germany) were loaded into the leaf apoplast of intact plants following the step-by-step instructions that were given elsewhere [34]. Measurements were started 2 hours after loading.

Confocal laser scanning microscopy

To visualize the *Pt*-GFP distribution within the leaf apoplast, CLSM imaging via a Leica TCS SP5 confocal laser scanning system (Leica Microsystems, Wetzlar, Germany) was carried out. For *Pt*-GFP excitation, the 488 nm beam line of an argon laser was chosen. Emission bandwith was 498–540 nm. Chloroplast autofluorescence was excited at 633 nm by a helium-neon laser (emission bandwith was 650 nm–704 nm). A planapochromatic objective (HC PLAN APO 20.0 × 0.70; Leica Microsystems) was used for image collection.

Image acquisition for *in vivo* pH-recording

Fluorescence images were collected as a time series with a Leica inverted microscope (DMI6000B; Leica Microsystems, Wetzlar, Germany) connected to a DFC camera (DFC 360FX; Leica Microsystems) via 5-fold magnification (0.15 numerical aperture, dry objective; voxel size = 0.002 mm; HCX PL FLUOTAR L, Leica Microsystems). An HXP lamp (HXP Short Arc Lamp; Osram, Munich, Germany) was used for illumination at excitation wavelengths 387/11, 440/20 and 490/10 nm. The exposure time was 25 ms for all channels. Emission was collected at 510/84 for both *Pt*-GFP channels and 535/25 for both OG channels using band-pass filter in combination with a dichromatic mirror (LP518; dichroit T518DCXR BS; Leica Microsystems). Plants were supplied with aerated nutrient solution.

Ratiometric analysis

The fluorescence ratios F_{490}/F_{387} (*Pt*-GFP) and F_{490}/F_{440} (Oregon Green 488) were obtained as a measurement of pH on a pixel-by-pixel basis. Image analysis was carried out using LAS AF software (version 2.3.5; Leica Microsystems). In order to take into account a potential variability in the leaf apoplastic pH that might exist across the imaged leaf detail, ratio image was divided in 6 ROIs per ratio image and time point. Background values were subtracted at each channel. For conversion of the fluorescence ratio data gained with the Oregon Green dye into apoplastic pH values, an *in vivo* calibration was conducted. In brief, Oregon Green dye solutions were pH buffered and loaded into the leaf apoplast. The Boltzmann fit was chosen to fit sigmoidal curves to the calibration. Fitting yielded an area of best responsiveness in the range pH 3.9–6.3 for the Oregon Green dye [34]. When the leaves were loaded with pH buffer, all regions of the apoplast showed the same ratio signal at the same buffered pH. Despite this uniformity, the absolute pH values quoted should be viewed as approximations of the apoplastic pH [44], because we cannot exclude the possibility that the buffer reaches equilibrium with the steady-state pH environment within the leaf. Nevertheless, this does not preclude a biological interpretation of leaf apoplastic pH responses to experimental treatments, because it was demonstrated that manipulation of the PM proton pump ATPase (PM-H$^+$-ATPase) activity with fusicoccin or vanadate lead to the expected effects on the apoplastic pH as measured by a ratiometric dye [37]. For pseudo-color display, the ratio was color-coded ranging from purple (no signal) over blue (lowest detectable pH signal) to pink (highest detectable pH signal). The *Pt*-GFP ratio signal was calibrated following the same procedure using citric acid/sodium citrate (3.5 ≤ pH ≤ 5.5; 10 mM), MES (5.5 ≤ pH ≤ 6.5;

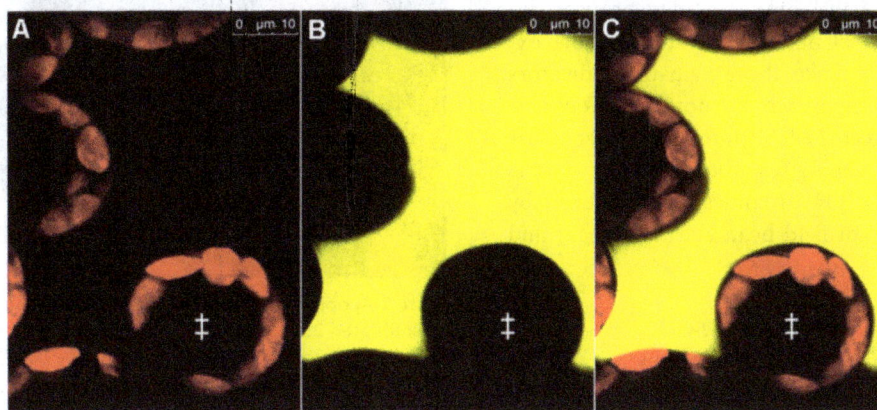

Figure 1 Apoplastic distribution of the *Pt*-GFP in a *Vicia faba* leaf infiltrated 2–4 minutes prior image acquisition. *Pt* GFP is exclusively located in the apoplast. Confocal image in **(A)** shows adaxial view on palisade cell chloroplasts (exited at 633 nm; pseudo-red). Image in **(B)** shows same detail with *Pt*-GFP (excited at 488 nm; pseudo-yellow). **(C)** Overlay of **(A)** and **(B)** demonstrates that the *Pt* GFP is only located in the apoplast. No *Pt* GFP signal is emitted from between the chloroplasts, indicating that the *Pt*-GFP did not enter the cytosol. Moreover, the inside of the palisade cells remained black, proving that *Pt*-GFP did not enter the vacuole or other symplastic organelles, as otherwise signals would be detectable from the cells. Symplastic *Pt* GFP location was negated in several leaves derived from different plants. ‡, palisade cells.

50 mM), PIPES ($6.0 \leq pH \leq 7.5$; 50 mM), HEPES ($7.0 \leq pH \leq 8.5$; 50 mM) and TRIS-base/MES ($8.5 \leq pH \leq 10.5$; 50 mM).

Results and discussion

Plants respond to stress through complex signaling networks that regulate and coordinate transcriptional and physiological processes initiating adaptations that help the plant to endure under unfavorable conditions [45] and references therein; [46] and references therein. Choi et al. [46] report in accordance to other authors [2,36,47,48] that apoplastic pH is one of the key factors in transmitting information regarding stress to distant unaffected plant organs [12,18]. After all, alterations in pH have an impact on protein folding, hormone distribution, channel and transporter activity or on membrane integrity and traffic [12,49].

There is increasing evidence that pH dynamics in the apoplast are involved in stress perception and systemic communication [2,20,45,50]. To better understand the role of transient pH dynamics, the need for indicators that allow the detection of apoplastic pH in its full span ranging from relative neutral to more acidic becomes evident. While the more acid range can easily be covered by the dextranated dye Oregon Green 488, there is lack in photostable dyes that allow ratiometric dual-excitation measurements in the relatively neutral range.

The fluorescent pH reporter protein from the orange seapen Ptilosarcus gurneyi (Pt-GFP) may provide a solution since it has a very broad pH-responsiveness that also covers relatively neutral pH values and an excellent dynamic ratio range [1]. However, it has the drawback that this genetically encoded pH sensors can only be used in plants that can be genetically transformed. For this reason, the abundance of self expressed biosensors which is currently available cannot readily be used for some agricultural crop plants without considerable expenditure. In order to make these valuable indicators available, we tested an approach for non-invasively loading bacterially produced Pt-GFP into the leaf apoplast of intact plants. In order to test the suitability of this strategy, leaf apoplastic pH dynamics were induced by salt stress at the roots of field bean (Vicia faba L.) and oat (Avena sativa L.).

Localisation of leaf apoplastic loaded Pt-GFP

Studies on leaf apoplastic ion concentrations require an ion indicator that is reliably localized in the apoplast and not unintentionally in cellular compartments, the cytosol or the vacuole. Otherwise, signals from the e.g. neutral cytosol or the very acidic vacuole would affect the apoplastic pH estimations. Additionally, it must be ensured that the bacterially produced Pt-GFP proteins can be inserted into the leaf apoplast by liquid mass flow

through stomatal pores, following the protocol to which was referred in the material and methods-section. To visualize the indicator's localization after apoplast loading, CLSM imaging was carried out. Confocal imaging (Figure 1) revealed that the Pt-GFP could easily be loaded into the apoplast following the protocol described by Geilfus and Mühling [34]. Moreover, images proved that Pt-GFP had not unintentionally entered the symplast at 2–4 minutes after loading, as otherwise signals would be detectable from the cells. It is very likely that the size of the Pt-GFP that is approximately 105 kDa in its native form [1] ensured that it did not access the symplast by crossing the plasma lemma from the apoplast. Images in Figure 2 demonstrated that in longer periods of time in which experiments are conducted, e.g. 2.5 hours after loading, the Pt-GFP still maintained to be exclusively apoplastically located. Moreover, it can be

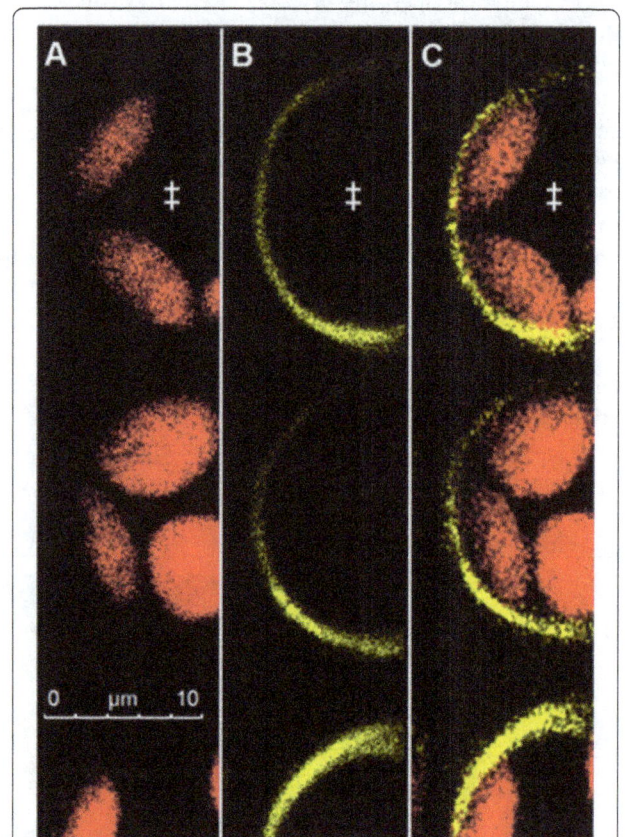

Figure 2 Apoplastic distribution of the Pt-GFP in a Vicia faba leaf infiltrated 2.5 h prior image acquisition. Pt GFP is exclusively located in the apoplast. Confocal image in **(A)** shows adaxial view on palisade cell chloroplasts (exited at 633 nm; pseudo-red). Image in **(B)** shows same detail with Pt-GFP (excited at 488 nm; pseudo-yellow). **(C)** Overlay of **(A)** and **(B)** verifies the apoplastic distribution of the Pt-GFP that is attached outside of the palisade cells and, by this means, outlines the cell boundaries at 2.5 hours after loading. No Pt-GFP signal is emitted from between the chloroplasts, proofing that no Pt-GFP had entered the cytosol. Symplastic Pt GFP location was negated in several leaves derived from different plants. ‡, palisade cells.

seen that the apoplast is not flooded (and thus not anoxic) during measurements and that *Pt*-GFP behaves like Oregon green because it is attached outside of the palisade cells as it was previously observed for the fluorescent dye ([37], Figure number eight therein).

Background and photostability

In planta measurements of apoplastic ion dynamics using microscopy-based ratio analysis require a signal-to-background ratio that is large enough to coherently reflect changes in the analyte concentration in the natural environment of the specimen. Background is all the light in the optical system that is not specifically emitted from the pH sensors and, if not considered, might introduce

errors in quantitation. Background signals sum up from autofluorescence coming from the measuring devices (i.e., lens elements), the specimen (i.e., chloroplasts or cell wall compounds such as oxidized phenols), the shot background associated with sampling of the signal [32,51], and the avoidable background arising from residual light in the laboratory (i.e., computer LEDs, monitor screens). In order to evaluate whether the signal-to-background ratio of the *Pt*-GFP is large enough for ratio analysis, the specimen without the dye was illuminated (background signal intensity) and was compared with the specimen plus dye (signal intensity). In result, only negligible background was detectable (Figure 3; less than 1‰ of the weakest fluorescence signals). The emission signal stood out

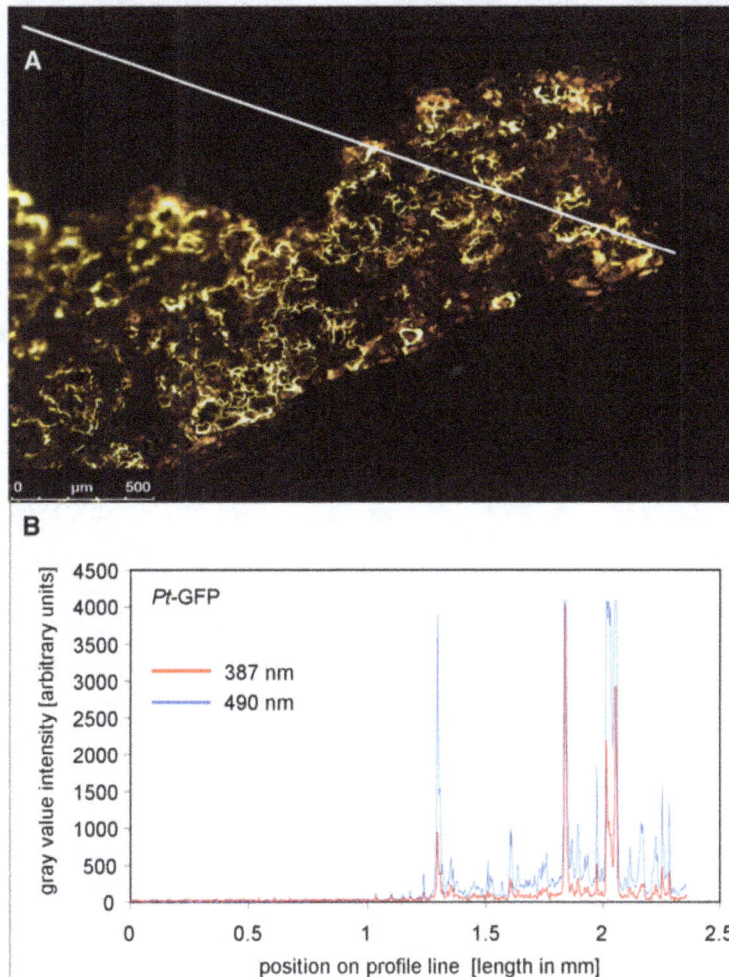

Figure 3 *Pt*-GFP emission signals are markedly higher than unspecific background signals. (A) Overlay of ex 490 nm (pseudo-green) and ex 387 nm (pseudo-blue) fluorescence images shows adaxial leaf apoplast that was partially loaded with the *Pt*-GFP. Under illumination, dye-loaded areas appear yellowish because pseudo-green and pseudo-blue are mixed within the overlay. Areas that were not loaded with the pH reporter protein appear black when illuminated and serve to compare the amount of unspecific signals (background) to signals that are specifically emitted from the proteins. For this comparison, the intensity of grey values was chosen as a measure for signal intensity. A profile of the intensity values was taken from the white line in **(A)** and is presented in **(B)**. Profiles are displayed for both fluorescence excitation channels (F_{490} and F_{387}). Only negligible signals were emitted from the area without dye and signals were much higher in the dye-loaded areas. Twelve separate images captured from different plants proved the low background intensity.

clearly from the background. This test revealed that the emitted analyte signal was strong enough to allow ratiometric analysis.

Besides a large signal-to-background ratio, the *Pt*-GFP needs to have a high photostability allowing to record pH dynamics over a period of several hours without

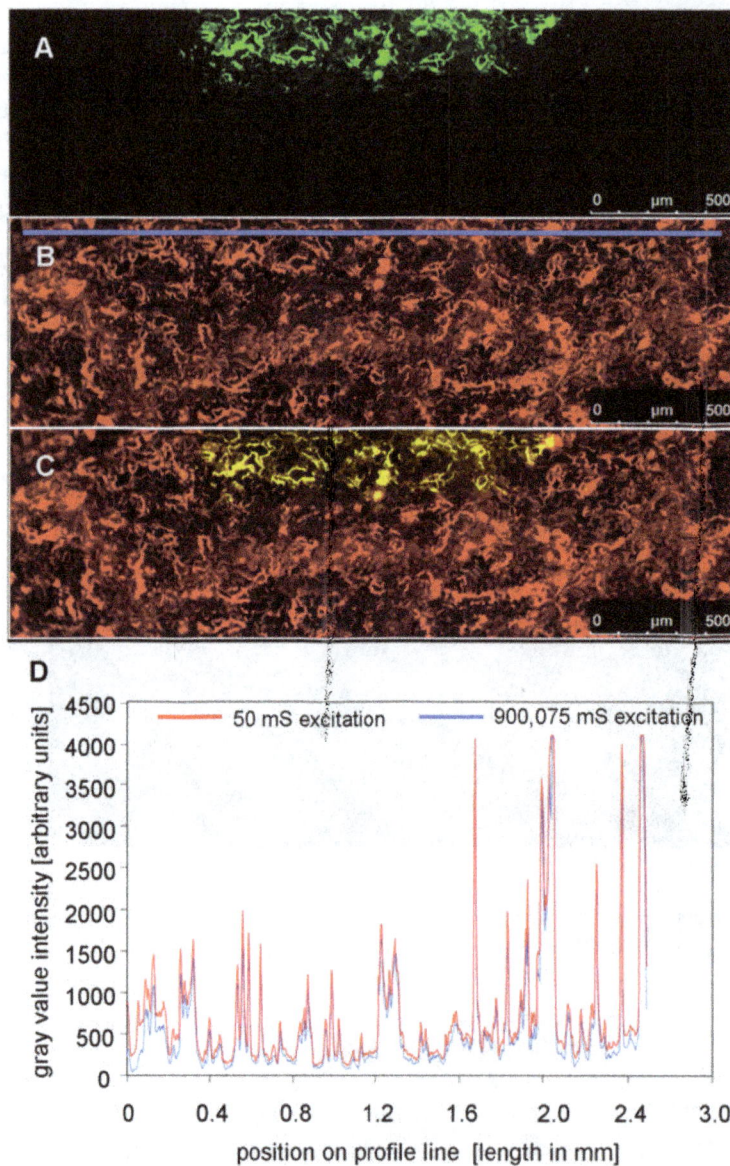

Figure 4 *Pt*-**GFP is photostable. (A)** The leaf apoplast of *Vicia faba* was loaded with *Pt*-GFP. To test whether *Pt*-GFP is prone to bleaching, a selected area (pseudo-green) was designated to be continuously excited by 490 nm illumination over a period of 15 min (=900,000 ms). The outer edges of the specimen were protected against continuous illumination by foreclosing the field diaphragm (non-bleached are appears black). Prior bleaching was started, initial signal intensity of the specimen was documented (image not shown). **(B)** After 900,000 ms continuous excitation, the field diaphragm was opened for collecting an image at ex 490 nm (exposure time was 25 mS). The image is presented in pseudo-red and contains the part of the specimen that was continuously illuminated (in total 3*25 ms illumination from three image acquisitions plus 900,000 ms from bleaching treatment) plus the area of the specimen that was not bleached (exposed in total to 2*25 ms illumination from two acquisitions). **(C)** Merged overlay of **(A)** and **(B)**. The yellow area (mixing pseudo-red and pseudo-green yields orange) represents the part that was continuously exposed to light treatment (in total 900,075 ms) and, thus, contains the possibly bleached proteins. Pseudo-red area represents the non-bleached part of the leaf with only 50 ms illumination in total (due to image acquisition cycles). Image **(B)** was used to create a profile of the emission intensity values from the area tagged by the blue line as a measure for the photostability. This line covers the bleached and non-bleached areas. The intensity values are presented in **(D)**. A comparison of the intensity values derived from the bleached and non-bleached areas revealed that no significant bleaching occurred after 15 min of continuous illumination. Eight separate bleaching experiments proved photostability of *Pt*-GFP.

(photo-) bleaching. In order to test whether *Pt*-GFP was prone to bleaching, the leaf apoplast was loaded with *Pt*-GFP proteins and continuously excited by 490 nm illumination over a period of 15 min. Subsequently, signal emission intensity was compared between bleached and non-bleached areas. This comparison revealed that no dye-bleaching had occurred to a relevant extent after 15 min of continuous illumination (Figure 4), which is satisfactory because in the present work leaves were illuminated about approximately 4000 ms in maximum. *Pt*-GFP was extremely photostable under continuous F_{490}-light exposure (same was true for the F_{387}-light channels; data not shown). This means that *Pt*-GFP is suitable for hours of pH recording in the apoplast of intact plants.

Pt-GFP as leaf apoplastic pH indicator

The suitability and responsiveness of apoplastically loaded *Pt*-GFP to pH changes was evaluated by a comparison to the established fluorescent pH indicator Oregon Green 488-dextran. In order to enable a proper comparability, measurements were conducted simultaneously side by side within the apoplast of the same field bean leaf. For this purpose, the *Pt*-GFP was loaded adjacent to a region loaded with Oregon Green (Figure 5A-C). The leaf veins clearly separated the GFP proteins from the dye and thus prevented the mixture of both H^+-indicators (Figure 6). Thereby, leaf regions loaded with different indicators could be monitored within a single image frame and ratios were calculated according to the optimal wavelength of the respective indicator (Figure 5D-E). Following this loading strategy, the dynamics of two different ions/analytes could be visualized simultaneously in the identical leaf and *in planta* by "three-channel ratio imaging". For the sake of comparison, pH changes in the leaf apoplast were specifically induced in a controlled manner by adding chloride into the nutrient solution harbouring the roots. This was done because chloride is carried from the roots to the shoot where it probably primes a systemic transient alkalinization in the leaf apoplast [21,34]. As visualized by the Oregon Green dye, the addition of 50 mM Cl^- via L-cysteinium chloride into the nutrient solution resulted in the expected transient leaf apoplastic alkalinization (Figure 7, black kinetic) that might reflect the action of a Cl^-/nH^+ symporter. A chloride symport across the PM [52-54] possibly results in a decrease of the leaf apoplastic $[H^+]$ caused by the co-transfer of protons together with chloride anions from the apoplast into the cytosol. However, the *Pt*-GFP ratios did not reflect this alkalinization from pH 4.3 to 5.0 that was indicated by the

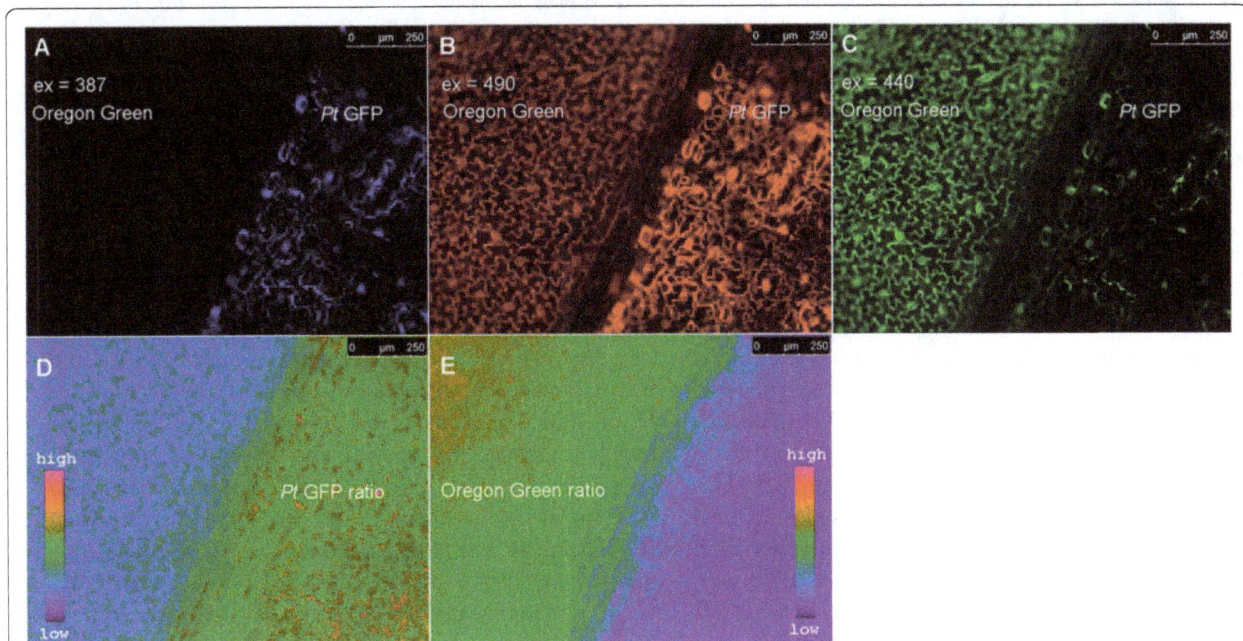

Figure 5 Principle of ratiometric analysis using two pH indicators within a single image frame. Fluorescence images shows adaxial view of *Vicia faba* leaf as excited at **(A)** F_{387}, **(B)** F_{490} and **(C)** F_{440}. Leaf apoplast as loaded with the pH-indicator protein *Pt*-GFP (right of leaf vein) and the pH-indicator dye Oregon Green-dextran 488 (left of leaf vein). Images were captured approximately 3 hours after loading. The fluorescence ratios **(D)** F_{490}/F_{387} (*Pt*-GFP) and **(E)** F_{490}/F_{440} (Oregon Green 488-dextran) were obtained as a measurement of pH. Emission was collected at 510/84 for both Pt-GFP channels and 535/25 for both OG channels. For this reason, the F_{490} channel was captured two times, once with emission 510/84, then with emission 535/25 (only the 535/25 emission image is shown in this figure). The ratios were coded by hue on a spectral colour scale ranging from purple (no signal) to blue (lowest signal) to pink (highest signal). Following this new loading strategy, leaf regions loaded with different indicators could be monitored and ratios were calculated according to the optimal wavelength of the respective indicator.

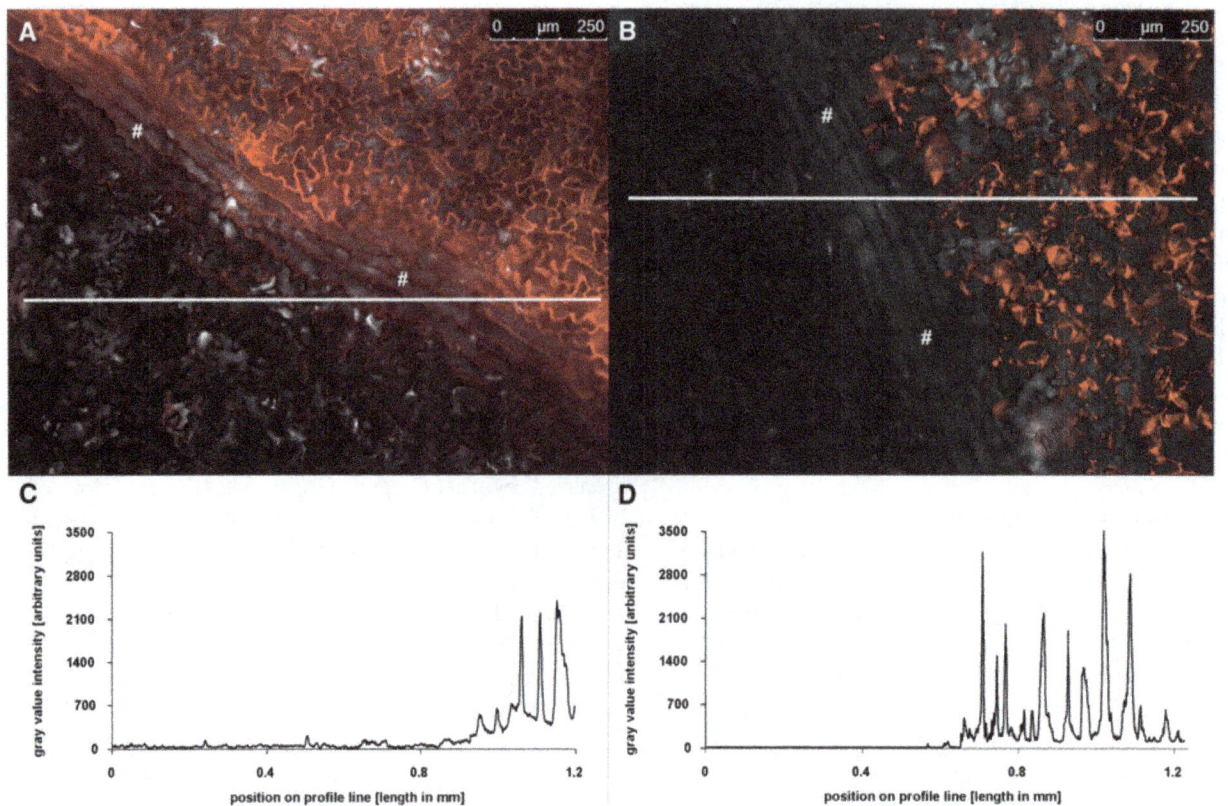

Figure 6 Leaf vein as a structural barrier that separates apoplastically located dyes. Adaxial leaf apoplast partially loaded with **(A)** Oregon Green 488-dextran or with **(B)** *Pt*-GFP. Overlay of pseudo-red fluorescence image at F_{490} and corresponding bright field image captured approximately 3 hours after loading. Dye-loaded areas in **(A)** and **(B)** appear red. Areas that were not loaded with the fluorescent pH reporter appear grey when illuminated and serve as a suitable area to compare the amount of unspecific signals (background) to specific signals being emitted from the pH reporters. For this, the intensity of grey values was chosen as a measure for emitted signal intensity. A profile of the intensity values was taken from the white line in **(A)** and is presented in **(C)**. The same was done for the *Pt*-GFP: The intensity values were taken from the white line in **(B)** and are presented in **(D)**. Only negligible signals were emitted from the areas without pH indicator that were separated by the leaf vein from the loaded apoplast. Signals were markedly higher in the dye-loaded areas. #, leaf vein. Results were confirmed by 10 replicates captured from different plants.

peak in the Oregon Green ratios (Figure 7, grey kinetic). It seems that the apoplastic pH in field bean leaves was below the range of best responsiveness for the engineered *Pt*-GFP which ranges from approx. 4.0 to 8.0 as measured by Schulte *et al.* [1] *in vitro* with a fluorescence spectrometer and organic buffers adjusted to the desired pH. This raises the question as to whether *Pt*-GFP is still functional and sensitive to high proton concentrations when used *in vivo*. In order to conduct an *in planta* calibration with the aim to test whether the *Pt*-GFP reacts *in vivo* on pH increments from pH 4.5 to 5.0, we buffered the *V. faba* apoplast to pH values ranging from 4.5 to 10.5 in increments of 0.5 pH units. It turned out that a pH below 5 can not be measured *in vivo* with the *Pt*-GFP (Figure 8), finally explaining the discrepancies in the comparative measurement presented in Figure 7. Based on the *in vivo* calibration, only pH changes ranging from values > 5 to 8 can be monitored.

Nitrate nutrition alkalizes the leaf apoplast of *Vicia faba* L.
In a next experiment, the leaf apoplastic pH was alkalized by increasing the nitrate concentration in the nutrient solution from 4 mM up to 15 mM nitrogen (Figure 9). This nitrogen form-related nutrition increased the permanent apoplastic pH from approx. 4.5 up to approx. 5.0 (compare initial pH in Figures 7 and 9). In this way, the leaf apoplastic pH was lifted to the range of best responsiveness for *Pt*-GFP. The subsequent addition of 20 mM Cl⁻ via L-cysteinium chloride into the nutrient solution resulted in the expected transient apoplastic alkalinization as reflected by the Oregon Green fluorescent dye (Figure 9, black kinetik). This transient alkalinization was also indicated by the *Pt*-GFP (Figure 9, grey kinetik). However, the absolute pH values measured with Oregon Green and *Pt*-GFP differed at the maximum peak height. It is possible that e.g. the cell wall did somehow modify the responsiveness of one of the dye system to pH. Another explanation

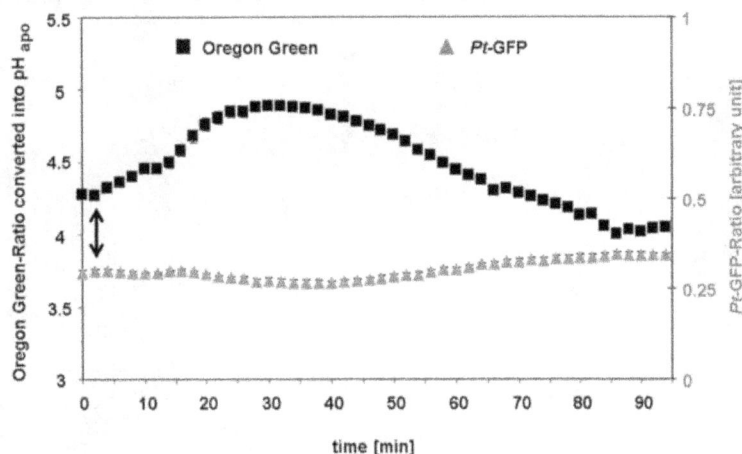

Figure 7 Unsuitability of *Pt*-GFP in the acid leaf apoplast of *Vicia faba*. Comparison between the responsiveness of the pH indicator protein *Pt*-GFP (grey) and the pH indicator dye Oregon Green (black) to apoplastic pH changes as induced by the addition of 50 mM Cl⁻ via L-cysteinium chloride to the roots of *Vicia faba*. Time point of chloride addition is indicated by the arrow. pH, as quantified at the adaxial face of *Vicia faba* leaves is plotted over time. Fluorescence ratio data obtained by *Pt*-GFP were below the linear range of the *in vivo* pH calibration and, therefore, could not be converted into pH data. Leaf apoplastic pH quantification was averaged (n = 6 ROIs per ratio image and time point; mean ± SE of ROIs). Representative kinetics of eight equivalent recordings of plants gained from 8 independent experiments.

could be that the indicators are localized in different regions of the apoplast were slightly different pH values prevail [20,21]. Nevertheless, this does not detract from the interpretation of the effects of Cl⁻-treatment on leaf apoplastic pH because both indicators uniformly recorded the chloride-induced transient leaf apoplastic alkalinization.

The experiments presented in Figures 7 and 9 demonstrated that under conditions of 4 mM nitrate fertilization the leaf apoplastic pH was too acidic, so that the *Pt*-GFP did not act in a range of good responsiveness, possibly due to a fluorescence quench at all wavelengths that caused irreversible conformational changes due to too low pH [1]. Increasing nitrate concentration in the nutrient solution of the beans and the associated alkalinization of the extra cellular space that is partially known to be caused by a nitrate cotransport with H⁺ across the PM [13,14] increased the leaf apoplastic pH to a range that can be monitored with the *Pt*-GFP.

Pt-GFP as apoplastic pH indicator in *Avena sativa* L.

Once oat (*Avena sativa* L.) with its less acidic apoplast [35,55] was chosen for analyzing the formation of the NaCl-induced leaf apoplastic pH peaks, the acidotropic [56] Oregon Green 488 dye seemed to be the wrong choice: Regardless of the fact that the leaf apoplastic pH response was challenged by the addition of 25 mM Cl⁻ given via L-cysteinium chloride into the nutrient solution, the expected transient alkalinization was not

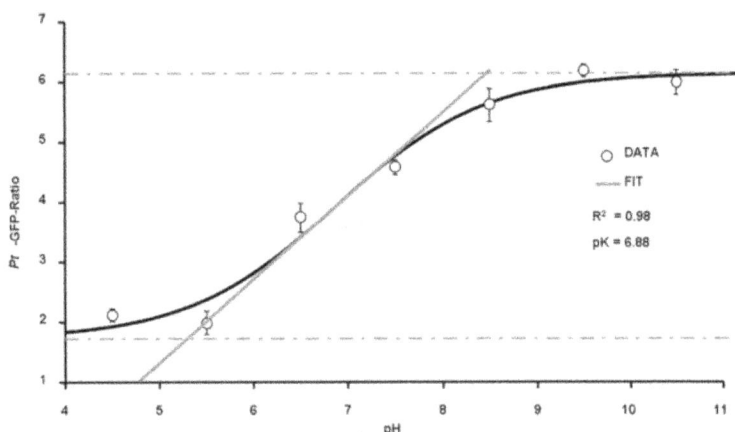

Figure 8 *In vivo* calibration of *Pt*-GFP fluorescence ratio (F_{490}/F_{387}). The Boltzmann fit was chosen for fitting sigmoidal curves to calibration ratio data. Fitting resulted in an optimal dynamic range for pH measurements between 5.3 and 8.4. *In vivo* calibration was conducted on six different plants, each biological replicate was technically replicated. Data are mean of n = 6 ± SE.

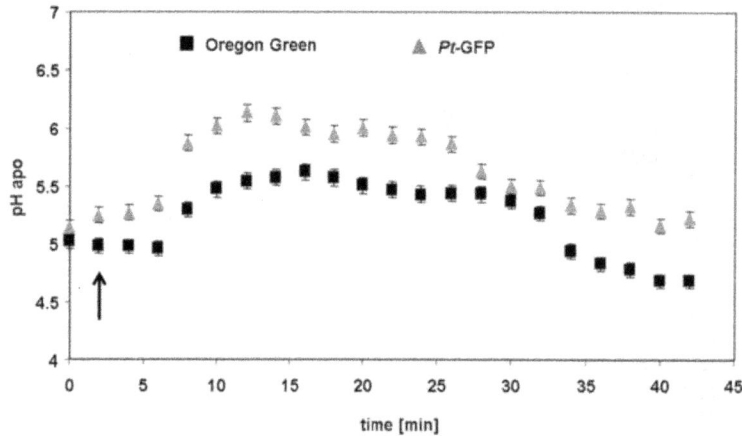

Figure 9 Nitrate nutrition alkalized the leaf apoplast into *Pt*-GFP's range of responsiveness. Comparison between the responsiveness of *Pt*-GFP (grey) and Oregon Green (black) to apoplastic pH changes as induced by the addition of 25 mM Cl^- via L-cysteinium chloride to the roots of *Vicia faba*. Time point of chloride addition is indicated by the arrow. Plants were cultivated with 15 mM nitrate in the nutrient solution given as $Ca(NO_3)_2$. pH as quantified at the adaxial face of *Vicia faba* leaves is plotted over time. Leaf apoplastic pH was averaged (n = 6 ROIs per ratio image and time point; mean ± SE of ROIs). Representative kinetics of six equivalent recordings of plants gained from independent experiments (n = 6 biological replicates).

reflected by the Oregon Green ratio (Figure 10). It appears that the Oregon Green ratios has reached a 'plateau phase' that cannot be exceeded. In contrast, the *Pt*-GFP ratio clearly reflected the transient alkalinization by peaking from an initial pH of approx. 6.25 up to a neutral pH of almost 7.25 (Figure 10). This clearly showed that knowledge about the prevailing pH is mandatorily necessary for choosing the suitable pH indicator. If, for example, Oregon Green would be selected as pH indicator under near neutral conditions, an alkalizing response of the biological system to the stimulus could not be detected and would be misinterpreted as being absent. The experiment has shown that the chloride induced transient alkalinizations which are thought to play

a role in the root-to-shoot communication of salt stress [19] are also present in the leaf apoplast of oat. This new finding delivers further indication that the salt-stress induced pH response may occur universally in different plant species.

Bypassing the cellular secretion pathway

The expression of genetically encoded pH sensors in plants that are targeted to the apoplast is accompanied by the problem that the proteins have to pass the cellular secretion pathway. From here they may also emit fluorescence. These signals from the cytosol, the ER or from Golgi vesicles average with signals from the acid apoplast and may thus lead to incorrect apoplastic pH

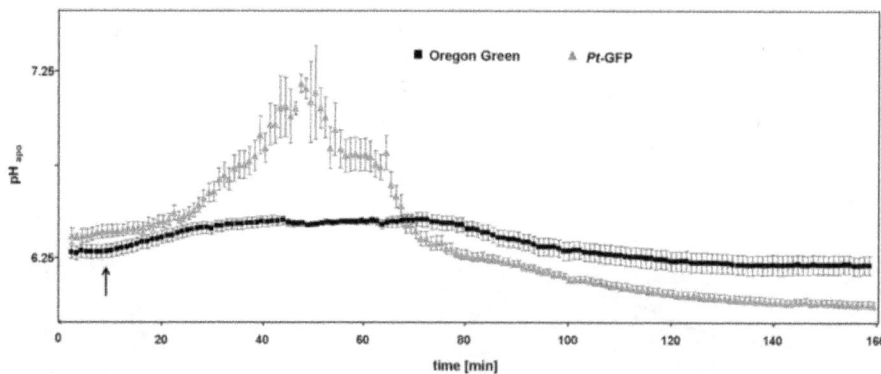

Figure 10 Near neutral leaf apoplast in *Avena sativa* requires *Pt*-GFP for detecting alkalizing effects. Leaf apoplastic pH response of *Avena sativa* as provoked by the addition of 25 mM Cl^- via L-cysteinium chloride into the nutrient solution. Time point of chloride addition as indicated by the arrow. pH, as quantified at the adaxial leaf face is plotted over time. *Pt*-GFP, grey curve; Oregon Green, black curve. Plants were cultivated with 15 mM nitrate in the nutrient solution given as $Ca(NO_3)_2$. Leaf apoplastic pH was averaged (n = 6 ROIs per ratio image and time point; mean ± SE of ROIs). Representative kinetics of eight equivalent recordings of plants gained from independent experiments (n = biological replicates).

values [1]. Since there is a pronounced proton gradient between cytosol and apoplast, this problem cannot be neglected. The presented approach of infiltrating bacterially produced and purified pH-sensitive *Pt*-GFP proteins through the stomatal pores into the leaf apoplast bypasses this problem. Moreover, the new approach describes how to use fluorescent reporter proteins for the measurement of leaf apoplastic ion relations in plants that are not readily transformable such as some of the agricultural relevant crop plants. In the light of the increasing availability of self-expressed biosensors [12,36,46,57-64], the presented approach should be seen as example and could also be applied to other genetically encoded sensor proteins or synthetic dyes for spatiotemporal mapping of ion relationships in the intact apoplast.

Concluding remarks

In summary, confocal laser scanning imaging showed that the tested loading technique is suitable for inserting bacterially produced *Pt*-GFPs into the apoplast of intact plants and negates the unintended presence of the *Pt*-GFP proteins in the symplast. The emission signal stood out clearly from the background and the *Pt*-GFP was characterized by a high photostability allowing ratiometric dual-excitation measurements over hours. *Pt*-GFP appeared to be a very good choice for the *in planta* quantification of leaf apoplastic pH-dynamics in plants that exhibit a relative neutral apoplastic pH. By using this approach, it was found that chloride-induced alkalinizations are not only formed in field beans [20] but also in oat. A strategy is presented that explains how to use bacterially expressed biosensors for the ratiometric *in planta* quantification of apoplastic ion kinetics in real time.

Abbreviations

CLSM: Confocal laser scanning microscopy; GFP: Green fluorescent protein; Pt: *Ptilosarcus gurneyi*; PM: Plasma membrane; ROI: Region of interest; OG: Oregon Green.

Competing interests

The authors declare that they have no competing interests.

Author details

[1]Institute of Plant Nutrition and Soil Science, Christian-Albrechts-Universität zu Kiel, Hermann-Rodewald-Str. 2, 24118 Kiel, Germany. [2]Botanisches Institut, Christian-Albrechts-Universität zu Kiel, Am Botanischen Garten 3-9, 24118 Kiel, Germany. [3]Zentrum für Biochemie und Molekularbiologie, Christian-Albrechts-Universität zu Kiel, Am Botanischen Garten 3-9, 24118 Kiel, Germany.

References

1. Schulte A, Lorenzen I, Bötcher M, Plieth C: **A novel fluorescent pH probe for expression in plants.** *Plant Methods* 2006, **2**:7.
2. Monshausen GB, Miller ND, Murphy AS, Gilroy S: **Dynamics of auxin-dependent Ca2+ and pH signalling in root growth revealed by integrating high-resolution imaging with automated computer vision-based analysis.** *Plant J* 2011, **65**:309-318.
3. Sanders D: **Generalized kinetic analysis of ion driven cotransport systems: II. random ligand binding as a simple explanation for non-Michaelis kinetics.** *J Membr Biol* 1986, **90**:67-87.
4. Lemoine R, Delrot S: **Proton-motive-force-driven sucrose uptake in sugar beet plasma membrane vesicles.** *FEBS Lett* 1989, **249**:129.
5. Bush DR: **Proton-coupled sugar and amino acid transporters in plants.** *Annu Rev Plant Physiol Plant Mol Biol* 1993, **44**:513-542.
6. Marschner H: *Mineral nutrition of higher plants.* 2nd edition. London: Academic; 1995.
7. Maathuis FJ, Sanders D: **Plasma membrane transport in context – making sense out of complexity.** *Cur Opin Plant Biol* 1999, **2**:236-243.
8. Ben-Shimon A, Shalev DE, Niv MY: **Protonation States in molecular dynamics simulations of peptide folding and binding.** *Curr Pharm Des* 2013, **19**(23):4173-4181.
9. Hartung W: **Die intrazelluläre Verteilung von Phytohormonen in Pflanzenzellen.** *Hahenheimer Arbeiten* 1983, **129**:64-80.
10. Hartung W, Solvik S: **Physicochemical properties of plant growth regulators and plant tissues determine their distribution and redistribution: stomatal regulation by abscisic acid in leaves.** *New Phytol* 1991, **119**:361-382.
11. Daeter W, Slovik S, Hartung W: **The pH-gradients in the root system and the abscisic acid concentration in xylem and apoplastic saps.** *Philos T Roy Soc B* 1993, **341**:49-56.
12. Swanson SJ, Choi WG, Chanoca A, Gilroy S: **In vivo imaging of Ca2+, pH, and reactive oxygen species using fluorescent probes in plants.** *Annu Rev Plant Biol* 2011, **62**:273-297.
13. Gerendás J, Ratcliffe RG: **Intracellular pH regulation in maize root tips exposed to ammonium at high external pH.** *J Exp Bot* 2000, **51**(343):207-219.
14. Mühling KH, Läuchli A: **Influence of chemical form and concentration of nitrogen on apoplastic pH of leaves.** *J Plant Nutr* 2001, **24**:399-411.
15. Van Volkenburgh E, Boyer JS: **Inhibitory effects of water deficit on maize leaf elongation.** *Plant Physiol* 1985, **77**:190-194.
16. Hartung W, Radin JW, Hendrix DL: **Abscisic acid movement into the apoplastic solution of water-stressed cotton leaves: role of apoplastic pH.** *Plant Physiol* 1988, **86**:908-913.
17. Wilkinson S, Davies WJ: **Xylem sap pH increase: a drought signal received at the apoplastic face of the guard cell that involves the suppression of saturable abscisic acid uptake by the epidermal symplast.** *Plant Physiol* 1997, **113**:559-573.
18. Felle HH, Hermann A, Hückelhoven R, Kogel KH: **Root-to-shoot signalling: apoplastic alkalinization, a general stress response and defence factor in barley (Hordeum vulgare).** *Protoplasma* 2005, **227**:17-24.
19. Felle HH: **pH regulation in anoxic plants.** *Ann Bot* 2005, **96**:519-532.
20. Geilfus CM, Mühling KH: **Transient alkalinization in the leaf apoplast of Viciafaba L. depends on NaCl stress intensity: an in situ ratio imaging study.** *Plant Cell Environ* 2012, **35**(3):578-587.
21. Geilfus CM, Mühling KH: **Microscopic and macroscopic monitoring of adaxial– abaxial pH gradients in the leaf apoplast of Vicia faba L. as primed by NaCl stress at the roots.** *Plant Sci* 2014, **223**:109-115.
22. Felle HH, Herrmann A, Hanstein S, Hückelhoven R, Kogel KH: **Apoplastic pH signalling in barley leaves attacked by the powdery mildew fungus Blumeria graminis f.sp. hordei.** *Mol Plant Microbe In* 2004, **17**:118-123.
23. Schäfer P, Pfiffi S, Voll LM, Zajic D, Chandler PM, Waller F, Scholz U, Pons- Kühnemann J, Sonnewald S, Sonnewald U, Kogel KH: **Manipulation of plant innate immunity and gibberellin as factor of compatibility in the mutualistic association of barley roots with Piriformospora indica.** *Plant J* 2009, **59**:461-474.
24. Hager A, Menzel H, Krauss A: **Versuche und Hypothese zur Primärwirkung des Auxins beim Streckenwachstum.** *Planta* 1971, **100**:47-75.
25. Raven JA, Farquhar GD: **Leaf apoplast pH estimation in Phaseolus vulgaris.** In *Plant membrane transport: the current position.* Edited by Dainty J, De Michelis MI, Marre E, Rasi-Caldogno F. Amsterdam: Elsevier; 1989:607-610.
26. Ullrich WR: **Transport of nitrate and ammonium through plant membranes.** In *Nitrogen Metabolism in Plants.* Edited by Mengel K, Pilbeam DJ. Oxford, UK: Oxford University Press; 1992:121-137.
27. Mühling KH, Plieth C, Hansen UP, Sattelmacher B: **Apoplastic pH of intact leaves of Vicia faba as influenced by light.** *J Exp Bot* 1995, **16**:377-382.
28. Mühling KH, Läuchli A: **Light-induced pH and K+ changes in the apoplast of intact leaves.** *Planta* 2000, **212**:9-15.

29. Monshausen GB, Bibikova TN, Messerli MA, Shi C, Gilroy S: Oscillations in extracellular pH and reactive oxygen species modulate tip growth of Arabidopsis root hairs. *Proc Natl Acad Sci U S A* 2007, **104**(52):20996–21001.

30. Toyota M, Gilroy S: Gravitropism and mechanical signaling in plants. *Am J Bot* 2012, **100**(1):111–125.

31. Fricker MD, Plieth C, Knight H, Blancaflor E, Knight MR, White NS, Gilroy S: Fluorescent and luminescent techniques to probe ion activities in living plant cells. In *Fluorescent and luminescent probes*. Edited by Mason WT. London: Academic; 1999:569–596.

32. Fricker MD, Parsons A, Tlalka M, Blancaflor E, Gilroy S, Meyer A, Plieth C: Fluorescent probes for living plant cells. In *Plant Cell Biology: A Practical Approach*. 2nd edition. Edited by Hawes C, Satiat- Jeunemaitre B. Oxford: University Press; 2001:35–84.

33. Gilroy S: Fluorescence microscopy of living plant cells. *Annu Rev Plant Physiol Plant Mol Biol* 1997, **48**:165–190.

34. Geilfus CM, Mühling KH: Real-time imaging of leaf apoplastic pH dynamics in response to NaCl stress. *Front Plant Sci* 2011, **2**:13.

35. Grignon C, Sentenac H: pH and ionic conditions in the apoplast. *Annu Rev Plant Physiol Plant Mol Biol* 1991, **42**:103–128.

36. Gao D, Trewavas AJT, Knight MR, Sattelmacher B, Plieth C: Selfreporting Arabidopsis thaliana expressing pH- and [Ca 2+]-indicators unveil ion dynamics in the cytoplasm and in the apoplast under abiotic stress. *Plant Physiol* 2004, **134**:898–908.

37. Geilfus CM, Mühling KH: Ratiometric monitoring of transient apoplastic alkalinizations in the leaf apoplast of living Vicia faba plants: chloride primes and PM-H^+-ATPase shapes NaCl-induced systemic alkalinizations. *New Phytol* 2013, **197**(4):1117–1129. doi:10.1111/nph.12046.

38. Hoffmann B, Kosegarten H: FITC-dextran for measuring apoplast pH and apoplastic pH gradients between various cell types in sunflower leaves. *Physiol Plant* 1995, **95**:327–335.

39. Mahmoudi AR, Shaban E, Ghods R, Jeddi-Tehrani M, Emami S, Rabbani H, Zarnani AH, Mahmoudian J: Comparison of photostability and photobleaching properties of FITC- and dylight 488 conjugated Herceptin. *Int J Green Nanotech* 2011, **3**(3):264–270.

40. Han JY, Burgess K: Fluorescent Indicators for Intracellular pH. *Chem Rev* 2010, **110**:2709–2728.

41. Shen J, Zeng Y, Zhuang X, Sun L, Yao X, Pimpl P, Jiang L: Organelle pH in the Arabidopsis endomembrane system. *Mol Plant* 2013. doi:10.1093/mp/sst079.

42. Geilfus CM, Zörb C, Neuhaus C, Hansen T, Lüthen H, Mühling KH: Differential transcript expression of wall-loosening candidates in leaves of maize cultivars differing in salt resistance. *J Plant Growth Regul* 2011. doi:10.1007/s00344-011-9201-4.

43. Zörb C, Geilfus CM, Mühling KH, Ludwig-Müller J: The influence of salt stress on ABA and auxin concentrations in two maize cultivars differing in salt resistance. *J Plant Physiol* 2013, **170**(2):220–224.

44. Bibikova TN, Jacob T, Dahse I, Gilroy S: Localized changes in apoplastic and cytoplasmic pH are associated with root hair development in Arabidopsis thaliana. *Development* 1998, **125**:2925–2934.

45. Felle HH: pH: signal and messenger in plant cells. *Plant Biol* 2001, **3**:577–591.

46. Choi WG, Swanson SJ, Gilroy S: High-resolution imaging of Ca2+, redox status, ROS and pH using GFP biosensors. *Plant J* 2012, **70**:118–128.

47. Bacon MA, Wilkinson S, Davies WJ: pH-regulated leaf cell expansion in droughted plants is abscisic acid dependent. *Plant Physiol* 1998, **118**:1507–1511.

48. Wilkinson S: pH as a stress signal. *Plant Growth Regul* 1999, **29**:89–99.

49. Casey JR, Grinstein S, Orlowski J: Sensors and regulators of intracellular pH. *Nat Rev Mol Cell Biol* 2010, **11**:50–61.

50. Monshausen GB: Visualizing Ca^{2+} signatures in plants. *Curr Opin Plant Biol* 2012, **15**(6):677–682.

51. Fricker MD, Errington RJ, Wood JL, Tlalka M, May M, White NS: Quantitative confocal fluorescence measurements in living tissues. In *Signal Transduction - Single Cell Research*. Edited by Van Duijn B, Wiltink A. Heidelberg: Springer- Verlag; 1997:569–596.

52. Sanders D, Hansen UP: Mechanism of Cl – transport at the plasma membrane of Chara corallina. part II. transinhibition and the determination of H+/Cl – binding order from a reaction kinetic model. *J Membrane Biol* 1981, **58**:139–153.

53. Felle HH: The H^+/Cl^- -symporter in root-hair cells of Sinapis alba. an electrophysiological study using ion-selective microelectrodes. *Plant Physiol* 1994, **106**:1131–1136.

54. Lorenzen I, Aberle T, Plieth C: Salt stress-induced chloride flux: a study using transgenic Arabidopsis expressing a fluorescent anion probe. *Plant J* 2004, **38**:539–544.

55. Rayle DL: Auxin-induced hydrogen-ion secretion in Avena coleoptiles and its implications. *Planta* 1973, **114**(1):63–73.

56. DiCiccio JE, Steinberg BE: Lysosomal pH and analysis of the counter ion pathways that support acidification. *J Gen Physiol* 2011, **137**:385–390. doi:10.1085/jgp.201110596.

57. Miesenböck G, de Angelis DA, Rothman JE: Visualizing secretion and synaptic transmission with pH-sensitive green fluorescent protein. *Nature* 1998, **394**:192–195.

58. Hanson GT, McAnaney TB, Park ES, Rendell MEP, Yarbrough DK, Chu SY, Xi LX, Boxer SG, Montrose MH, Remington SJ: Green fluorescent protein variants as ratiometric dual emission pH sensors. 1. structural characterization and preliminary application. *Biochemistry* 2002, **41**:15477–15488.

59. Bizzarri R, Arcangeli C, Arosio D, Ricci F, Faraci P, Cardarelli F, Beltram F: Development of a novel GFP-based ratiometric excitation and emission pH indicator for intracellular studies. *Biophys J* 2006, **90**:3300–3314. doi:10.1529/biophysj.105.074708.

60. Meyer AJ, Brach T, Marty L, Kreye S, Rouhier N, Jacquot JP, Hell R: Redox-sensitive GFP in Arabidopsis thaliana is a quantitative biosensor for the redox potential of the cellular glutathione redox buffer. *Plant J* 2007, **52**:973–986.

61. Young B, Wightman R, Blanvillain R, Purcel SB, Gallois P: pH-sensitivity of YFP provides an intracellular indicator of programmed cell death. *Plant Methods* 2010, **6**:27.

62. Gjetting KS, Ytting K, Karkov C, Schulz A, Fuglsang AT: Live imaging of intra- and extracellular pH in plants using pHusion, a novel genetically encoded biosensor. *J Exp Bot* 2012, **63**(8):3207–3218.

63. Gjetting KS, Schulz A, Fuglsang AT: Perspectives for using genetically encoded fluorescent biosensors in plants. *Front Plant Sci* 2013, **4**:234.

64. Martinière A, Desbrosses G, Sentenac H, Paris N: Development and properties of genetically encoded pH sensors in plants. *Front Plant Sci* 2013, **4**:523. doi:10.3389/fpls.2013.00523.

7

An easy-to-use primer design tool to address paralogous loci and T-DNA insertion sites in the genome of *Arabidopsis thaliana*

Gunnar Huep[†], Nils Kleinboelting[†] and Bernd Weisshaar[*]

Abstract

Background: More than 90% of the *Arabidopsis thaliana* genes are members of multigene families. DNA sequence similarities present in such related genes can cause trouble, e.g. when molecularly analysing mutant alleles of these genes. Also, flanking-sequence-tag (FST) based predictions of T-DNA insertion positions are often located within paralogous regions of the genome. In such cases, the prediction of the correct insertion site must include careful sequence analyses on the one hand and a paralog specific primer design for experimental confirmation of the prediction on the other hand.

Results: GABI-Kat is a large *A. thaliana* insertion line resource, which uses in-house confirmation to provide highly reliable access to T-DNA insertion alleles. To offer trustworthy mutant alleles of paralogous loci, we considered multiple insertion site predictions for single FSTs and implemented this 1-to-N relation in our database. The resulting paralogous predictions were addressed experimentally and the correct insertion locus was identified in most cases, including cases in which there were multiple predictions with identical prediction scores. A newly developed primer design tool that takes paralogous regions into account was developed to streamline the confirmation process for paralogs. The tool is suitable for all parts of the genome and is freely available at the GABI-Kat website. Although the tool was initially designed for the analysis of T-DNA insertion mutants, it can be used for any experiment that requires locus-specific primers for the *A. thaliana* genome. It is easy to use and also able to design amplimers with two genome-specific primers as required for genotyping segregating families of insertion mutants when looking for homozygous offspring.

Conclusions: The paralog-aware confirmation process significantly improved the reliability of the insertion site assignment when paralogous regions of the genome were affected. An automatic online primer design tool that incorporates experience from the in-house confirmation of T-DNA insertion lines has been made available. It provides easy access to primers for the analysis of T-DNA insertion alleles, but it is also beneficial for other applications as well.

Keywords: *Arabidopsis thaliana*, T-DNA, Insertion mutants, Paralog, Primer design, GABI-Kat

Background

Arabidopsis thaliana is widely and very successfully used as a model organism in basic plant research. After the completion of its genome sequence in the year 2000 [1], several large mutant collections have been established. In most cases, T-DNA insertional mutagenesis mediated by *Agrobacterium tumefaciens* has been used for the generation of knock out alleles for reverse genetic approaches [2,3]. The insertion of T-DNA in the plant genome occurs almost randomly [4-6], and different methods for the identification of insertion sites in specific lines have been established. The most frequently used method is based upon flanking sequence tags (FSTs). FSTs are short DNA sequences, which flank the T-DNA insertion site and contain genome sequence information adjacent to the insertion site. They are generated with PCR-based methods after digestion of genomic

* Correspondence: bernd.weisshaar@uni-bielefeld.de
[†]Equal contributors
Center for Biotechnology & Department of Biology, Bielefeld University, Universitaetsstrasse 25, D-33615 Bielefeld, Germany

DNA and adapter ligation [7]. Their sequences can be compared to the *A. thaliana* genome sequence using BLAST [8] to predict the insertion position(s) of the T-DNA in a given line. The GABI-Kat collection is the world's second largest FST-based T-DNA insertion line collection for *A. thaliana* [9]. The FST data along with insertion site predictions and information about confirmed T-DNA insertion alleles is accessible at Simple-Search, which is the user interface to the database at the website of the project [10]. GABI-Kat lines can be accessed via SimpleSearch, and links to the stock centres are provided if the line that contains the relevant allele has been donated to the American and/or European stock centres for *A. thaliana* seeds [11]. In case of direct orders at GABI-Kat, predicted insertions are requested rather than simply seed of GABI-Kat lines. Upon a user request, T2 plants of the respective line are grown and the insertion site prediction is confirmed at GABI-Kat by PCR with an insertion site-specific primer and a T-DNA border primer, followed by sequencing of the amplicon. The experimental data for confirmed insertion alleles is presented on the SimpleSearch website [12]. This includes the amplicon sequences as well as the sequences of the primers used in the confirmation PCR. A more detailed overview on the features of the SimpleSearch site is summarised in [9].

A major problem during the FST-based insertion site prediction in T-DNA insertion lines occurs when the FST sequence cannot be assigned unambiguously to a *single* specific locus in the *A. thaliana* genome. Such events are inevitable, because even in a small genome like the one from *A. thaliana* only about 10% of all genes encode unique proteins. All other genes have at least one additional homologue [13,14]. One reason for the occurrence of homologues within the genomes of eukaryotes lies in genome duplication events leading to paralogous genes. This has already been studied in detail in the original *A. thaliana* genome sequence analysis [1], and searchable databases are available which support analyses of duplication events and paralogous gene families [15]. Beside genes (and without considering transposable elements), also non-genic sequences of the *A. thaliana* genome occur in higher copy numbers. Several mechanisms have been discussed for the duplication events (reviewed for example in [16]). Regardless of the exact mechanism, the ultimate result of duplication events is that even after evolutionary diversification of the duplicated sequences, larger stretches of similar sequences occur at different positions in the *A. thaliana* genome. In this article we will refer to regions with more than one copy of similar DNA sequences in the genome as "paralogous regions", regardless if genic or non-genic regions are concerned. In this sense, we will also use the term "paralog" for individual regions in paralogous regions, even if those regions are non-genic and if the

genetic origin (i.e. duplication event) of the respective region is not clear.

In all large FST-based T-DNA insertion line collections, the FSTs have so far been used to predict a single locus in the genome as the corresponding insertion site. If the locus is located within a paralogous region, the prediction and the decision for one of the paralogs is error-prone. However, given that the sequences of the paralogous loci are known, the confirmation process at GABI-Kat, which considers the DNA sequence of the confirmation amplicon, is able to resolve ambiguities concerning the correct insertion position in most cases. Only if the paralogous regions contain (almost) identical sequences a definite assignment to a single locus is not possible.

We present data from example cases in which the correct insertion locus was identified only after PCR-based confirmation using optimised primers, even though FST-based insertion site prediction was unable to assign a unique best-fitting locus. Insertion site predictions were redone using the TAIRv10 genome sequence and BLAST, and multiple predictions derived from single FSTs were combined into "paralog groups". When attempting to confirm a prediction from such a group, specific primers (as far as possible) unique for relevant insertion sites were designed. We developed a primer design method that identifies possible primers using a multiple alignment, which enables the discrimination between the different paralogous regions. An optimised, easy to use version of the tool is available on the website of GABI-Kat and allows users to design primers at their own locus or genome position of interest.

Results and discussion
FSTs and T-DNA insertion site predictions
The GABI-Kat database contains insertion site predictions from about 135,000 FSTs, which were generated for the 93,504 lines in the T-DNA insertion line collection. During the generation of GABI-Kat FSTs, genomic DNA of individual T1 plants was digested with *Bfa*I, adaptors were added, and fragments containing T-DNA borders as well as sequences of plant origin next to the T-DNA were amplified with a T-DNA- and an adaptor-specific primer [7]. The length of the resulting amplicons is dependent on the position of the *Bfa*I recognition sites in the genome relative to the insertion site. In case of more than one T-DNA insertion in a given line, more than one amplicon might be generated in a single reaction. Due to different sizes of these amplicons, an insertion corresponding to a longer amplicon is measured at the tail of the FST sequence, and an insertion corresponding to a shorter amplicon at the head of the FST sequence. Usually the shorter amplicon causes a stronger signal because of higher fragment abundance after PCR. We refer to these

FSTs, which allow to correctly predict several insertion sites in one line from different regions of one FST, as "composite FSTs". Such FSTs have also been described for the SALK collection [17].

We often observed that GABI-Kat FSTs contain sequence parts from borders of two distinct T-DNA insertions present in a single line. When addressing insertion site predictions in paralogous regions of the genome, predictions from "composite FSTs" had to be considered as well because they share the feature "additional BLAST hit from one FST". The optimised analysis pipeline (see below) that has been established at GABI-Kat detects, in addition to paralogous hits, also hits from "composite FSTs". The additional predictions derived from GABI-Kat FSTs by using this optimised pipeline have been made available with the GABI-Kat database release No 27 [12].

Initially, the insertion predictions were deduced from the FSTs in a 1-to-1 relation. Only the best BLAST hit from a given single FST was evaluated for the prediction of a single insertion site [9,18]. In order to address paralogous regions of the genome, we recalculated the insertion site predictions for all FSTs. For this new assessment, a 1-to-N relation of FST to insertion site predictions was implemented in the internal GABI-Kat database. To be able to filter for the most relevant predictions, three categories (designated 0, 1 and 2) were defined and assigned to the different types of insertion predictions deduced from a single FST based on the BLAST e-value. Category 0 was assigned to the prediction deduced from the best BLAST hit. This was the one that had been selected as the only prediction (1-to-1) before the extended analysis (1-to-N) was performed. Additional predictions from the evaluated FST were assigned to category 1 if the BLAST e-values were lower than 1e-3, and to category 2 if the e-values were 1e-3 or higher (for details see Methods). The additional predictions were taken into account during the confirmation process at GABI-Kat if necessary. This was especially important if the e-values of the BLAST hits for a given FST region were highly similar or even identical because paralogous genomic loci were affected. Only one of the insertion site predictions derived from a single region of an FST corresponds to the correct insertion locus. Details about the results from the "1-to-N" type FST evaluation and insertion site prediction are listed in Table S1, which is included in the document "Additional file 1".

We observed that the prediction of category 0 could be wrong due to small errors in the FST sequence, even if the BLAST analysis results in a unique best hit. Consequently, analysis of only this locus would have made confirmation impossible. The access to several BLAST hits from one FST region allowed creating groups of paralogous insertion site predictions of categories 0 and 1, which were derived from subsets of the FSTs of the

respective line. In total, about 11,000 paralog groups were detected in the GABI-Kat FST dataset. If a paralogous prediction was addressed for confirmation experimentally, several predictions in the respective group were analysed during the confirmation process, if necessary. Until now, more than 1,200 groups with paralogous predictions in the GABI-Kat collection have been solved experimentally. If a prediction other than the best prediction has been confirmed, this prediction was made available in SimpleSearch in addition to the "category 0 prediction" which was included anyway.

Primers in paralogous regions of the genome

Even when most parts of paralogous sequences are highly similar or even identical, the individual sequences often differ at certain positions. Based upon sequence alignments of the genomic DNA sequences of the individual paralogs in groups of paralogous predictions, we developed a primer design algorithm that allows designing specific primers (see Methods). Uniqueness for the individual paralog was preferably constructed into their 3′-ends. Such primers can enable the determination of the correct insertion site prediction by PCR and sequencing, even if only one base pair differs in the paralogous regions surrounding the set of paralogous insertion site predictions. A direct comparison of PCR results with the different paralog-specific primers allows the discrimination between the paralogs, sometimes taking into account that mispriming usually leads to weaker PCR products.

In addition to paralogous regions in the genome, random mispriming sites in the genome might occur for primers. In our experience and with our PCR conditions, even short sequence stretches at the 3′-end of primers can lead to unspecific PCR products (see [19] and references therein). We have regularly observed examples of primers, which were able to amplify unspecific PCR products when only 12 bases of their 3′-end had a perfect match in the genome. For example, in GABI-Kat line 011B05 we tried to confirm the predicted insertion at position 51,137 on chromosome 3. The primer that was used for this purpose (5′-CTCAATTTATGTGT GACTGCAAGC-3′) had the unique, perfect annealing site from position 50,794 to 50,817 on chromosome 3. Unexpectedly, we observed an amplicon of roughly 1.3 kb. BLAST analysis of the sequence of this amplicon resulted in a hit in the gene At4g33170 with a BLAST e-value of 0.0 and a derived T-DNA insertion site at position 15,997,766 on chromosome 4. Analysis of the primer sequence showed a perfect match of the last 12 bp at the positions 15,996,481 to 15,996,492 on chromosome 4. The insertion in the line 011B05 was subsequently confirmed with an At4g33170-specific primer, essentially by using the "wrong" confirmation sequence as an FST for insertion site prediction. More extreme examples of mis-priming occur in rare cases. This is taken into account in our primer design by minimizing the number of possible 12 bp-matches within the genome (see Methods).

Application of the paralog primer method

An example that illustrates the advantages of the paralog primer method is the confirmation of the insertion in *At5g41740* in the GABI-Kat line 683F05 (see Figure 1). The best insertion site prediction and the only one that had been available in the "1-to-1 prediction dataset" in this line was *At5g41750* with a BLAST e-value of 5e-27. This gene is annotated to encode a "disease resistance protein (TIR-NBS-LRR class) family" in TAIRv10. After newly calculating the insertion site predictions and setting up the "paralog groups", the second best prediction for the same FST was *At5g41740* with a BLAST e-value of 1e-24 and the same annotation. The experimental analysis with paralog specific primers designed using the tools described above resulted in an amplicon with the *At5g41740*-specific primer and no product with the At5g41750-specific primer. BLAST analysis of the sequence of the confirmation amplicon resulted in fully aligning sequences for both genes. However, when comparing the score values, *At5g41740* reached 775 while *At5g41750* ended up with 672. Manual inspection of the alignments confirmed a few SNP positions that distinguish the two loci with *At5g41740* representing the correct locus. The confirmed insertion in *At5g41740* is available in SimpleSearch with a search for the respective AGI gene/locus code, and the data from the experimental analysis as well as a link to the stock centre NASC are displayed.

Another example is the confirmation of the insertion in *At1g07930* in the line 902F05. In this line the best predictions were *At1g07920* and *At1g07940* with a BLAST e-value of 1e-95 in both cases. A third and slightly worse prediction for the same FST was *At1g07930* with a BLAST e-value of 7e-91. Only for *At1g07930* a PCR product could be obtained with paralog-specific primers. BLAST analysis of the sequence of the amplicon confirmed *At1g07930* as the correct paralog via the score values similar to the example above (data available at the Simple-Search website for line 902F05).

Besides examples of GABI-Kat lines with second- or third-best insertion site predictions being confirmed, there are several cases of lines which have two or more predictions with identical reliability according to the

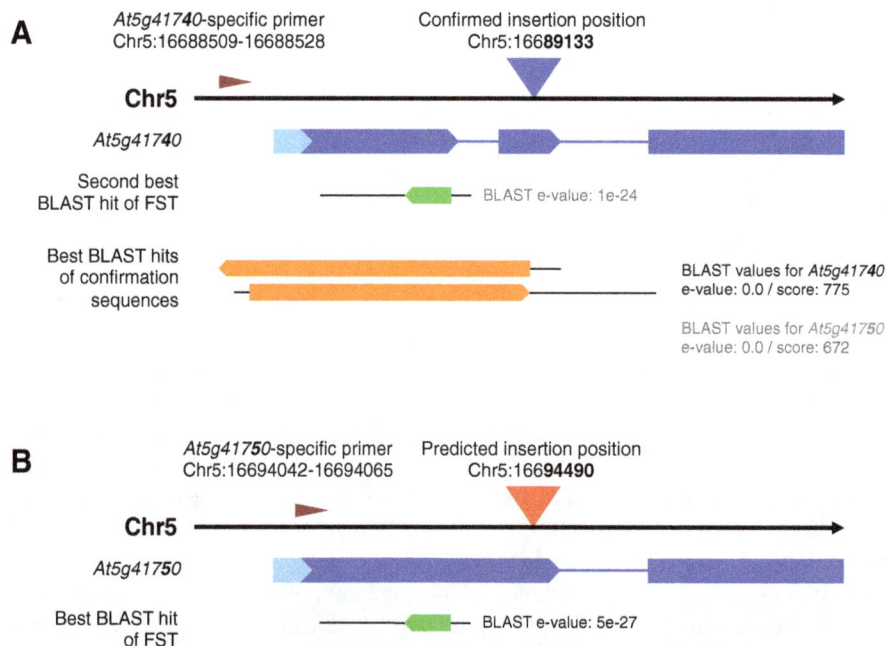

Figure 1 Confirmation of the insertion allele of *At5g41740* in the GABI-Kat line 683F05. In line 683F05 two paralogous loci were predicted as possible insertion sites. Initiated by a user request for *At5g41750*, both loci were examined with primers specific for the respective paralogous locus. As a result, the worse prediction could be confirmed successfully **(A)** and the PCR for the better prediction failed **(B)**. FST-based insertion site predictions are shown with blue (confirmed) and red (failed) triangles. Primer positions are given using dark red arrowheads. The affected genes are symbolised by blue arrows. Light blue arrows are 5′-UTRs, CDSs are shown in dark blue. Introns are indicated by thin blue lines. BLAST hits of the FSTs are shown as green arrows, BLAST hits of the confirmation sequences are shown as orange arrows. Sequence parts of FSTs and confirmation sequences that do not fit to the indicated positions are indicated as thin black lines. The genomic positions in the figure are given according to the TAIRv10 genome sequence dataset of *A. thaliana*.

BLAST e-values of the FSTs. Primers designed using the tools described above allowed the determination of the correct insertion locus in these lines. Examples are the GABI-Kat lines 583H04 and 742E06. In line 583H04, *At1g29350* was identified as a wrong prediction and the insertion in *At1g29370* was confirmed. In line 742E06, *At2g38210* was the wrong prediction and *At2g38230* was confirmed.

Access to the easy-to-use primer design tool

In order to offer public access to the primer design algorithms developed at GABI-Kat, we implemented an easy-to-use tool into SimpleSearch that includes the paralog-specific design if necessary. The design method is chosen as described in Methods and an overview on the primer design process is shown in Figure 2. Details on the selection of suitable primers using the paralog-specific design are summarised in Figure 3. The publicly available primer tool was implemented within the visualisation part and displays the location of the designed

primers. It can be accessed directly with the URL [20] or via the menu on the GABI-Kat website [10]. Insertions found in SimpleSearch also provide a link to the primer design for their respective position and a button in the visualisation allows quick access to the primer design for the currently selected position in the genome (Figure 4). It differs from previously available tools for the analysis of paralogous regions in a number of important aspects. Other tools, for example the very useful tool Primer-BLAST [19], checked the redundancy of the combination of both primer annealing sites for the amplimers defined by a primer pair. It also uses MegaBLAST of the complete target zone to the genome sequence of the addressed organism to avoid primer design in redundantly matching parts of the target zone. In contrast to this, our tool checks the 3′-ends of every single primer for redundancy, which is essential for the analysis of T-DNA insertions because in this experimental setup only one genome-specific primer is used. Primer-BLAST uses a BLAST of the complete target zone sequence and

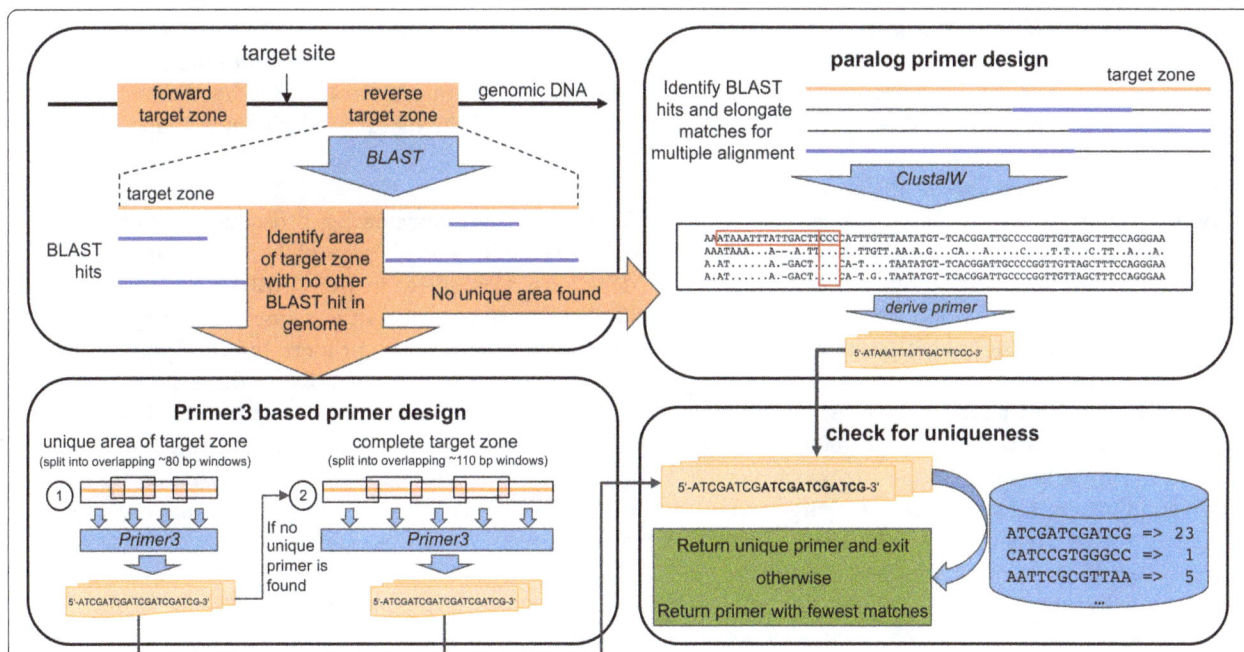

Figure 2 Overview on the primer design tool. Initially, each of the two target zones with a distance to the chosen target site (distance and target site defined by the user) is examined by a BLAST vs. the *A. thaliana* genome sequence. If there is an area that has no other BLAST hit somewhere in the genome, the Primer3-based approach is used (bottom left box), otherwise the paralog primer design is used (upper right box). A number of candidate primers is designed in both approaches which are then checked for uniqueness (bottom right box). If a unique primer with no additional 12 bp-hit at the 3′-end in the genome is found, the primer design is stopped and the primer is returned as a result. Otherwise the primer with the fewest matches is returned. When designing additional primers, the next best primers are returned. The Primer3-based primer design (bottom left box) uses multiple runs of Primer3 in overlapping windows and altering temperatures to generate a large set of candidate primers. First, only the unique area of the target zone with no additional BLAST hit in the genome is considered. If no unique primer is found, the process is performed again with the complete target zone. The paralog primer design first creates a multiple alignment with all sequences showing a BLAST hit to the target zone using ClustalW. The algorithm searches for mismatches in the multiple alignment (see Figure 3 for details). To reduce the runtime of ClustalW for sequences with many hits, the target zone is also split into overlapping windows and alignments are computed separately (not shown in Figure). The primer sequences shown in the figure are to be regarded as example sequences that cannot fit to *all* features of the scheduled workflow.

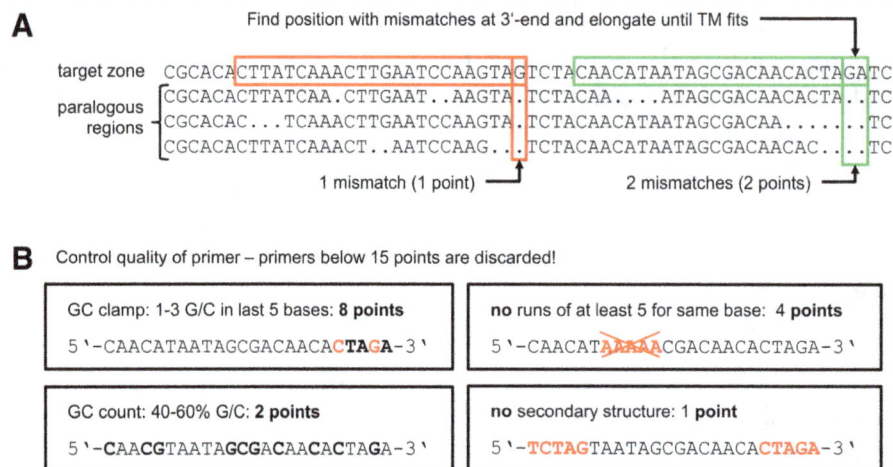

Figure 3 Paralog primer design. **(A)** Possible primers are identified by finding mismatches within a multiple alignment. Each mismatch is worth one point, which is later used to determine the quality of the primer. **(B)** Further criteria are: (i) GC-clamp: there should be 1 to 3 G or C bases in the last 5 bases of the primer (8 points); (ii) primers should not have runs of more than 5 identical bases (4 points); (iii) the GC content should be between 40% and 60% (2 points); (iv) the last 5 bases of the primer should not form a secondary structure with the first 5 bases or rather not be the reverse complement (1 points); all primers need to have at least 15 points in order to be considered as candidate.

optimisation of the BLAST alignments by the Needleman-Wunsch global alignment algorithm to perform a specificity check that deselects primer pairs with amplimers on other targets than the submitted template [19]. Our approach to rank primer candidates uses a precomputed index of all occurrences of 12 bp sequences in the *A. thaliana* genome and considers the last 12 bp of a primer candidate. This 12 bp strech was in our experience the crucial part of the primer. For using the tool, a genomic nucleotide position central to the locus to be addressed (designated "target position") must be selected. Upon starting the tool, primers are automatically designed around this target position with a default minimal and maximal distance of 300 to 800 bp to the target position on each side. We refer to this sequence range surrounding the target position as "target zone". The distance to the target position can be changed to values between 100 to 1500 bp with a minimum range on each side of the target position of 100 bp. The larger the target zone, the better are the chances to obtain a unique primer. For the primers, the default annealing temperature has been set to 60.5°C, but it can also be set by the user to a value between 50 and 72°C. The primers can either be used in combination with T-DNA border primers in order to confirm a T-DNA insertion of interest, or simply to create amplicons from the targeted genomic locus. If the automatically designed primers are not acceptable to the user for some reason, the design tool can be executed repeatedly to acquire further primer combinations.

As an additional usability feature, the tool determines and reports the size of the amplimer with respect to the pseudochromosome sequence, and presents a summary of information related to genotyping insertion alleles if a (predicted) insertion site is spanned by the amplimer. The PHP code of the tool is available upon request.

Conclusions

We describe the primer design procedure that has been used successfully and in large scale for confirmation of T-DNA insertion alleles in the GABI-Kat project. Since 2007 [21] users can access the sequences of experimentally proven confirmation primers for confirmed insertion alleles via SimpleSearch. Now, the primer design procedure established at GABI-Kat has been integrated into the publicly available SimpleSearch interface. The tool can be useful for confirming T-DNA insertion alleles, including those from SALK or other insertion mutant collections. At GABI-Kat, usually only one insertion is confirmed per line. After this first confirmation, the lines are donated to NASC and can further on only be ordered from there. Access to the GABI-Kat primer design tool might therefore help in the analysis of additional insertions, which are listed as predictions in SimpleSearch. Moreover, the tool allows easy design of amplimers for the genotyping of insertion alleles because the amplimers spanning the insertion site differentiate between the wt allele (amplicon produced) and the insertion allele (no amplicon; see [14]). The tool presented in this work differs from the already available tools in several aspects, as discussed above. Also, we simplified the primer design process by providing an easy-to-use user interface, which only requires several mouse clicks and no copy-paste of target sequences. Furthermore, existing

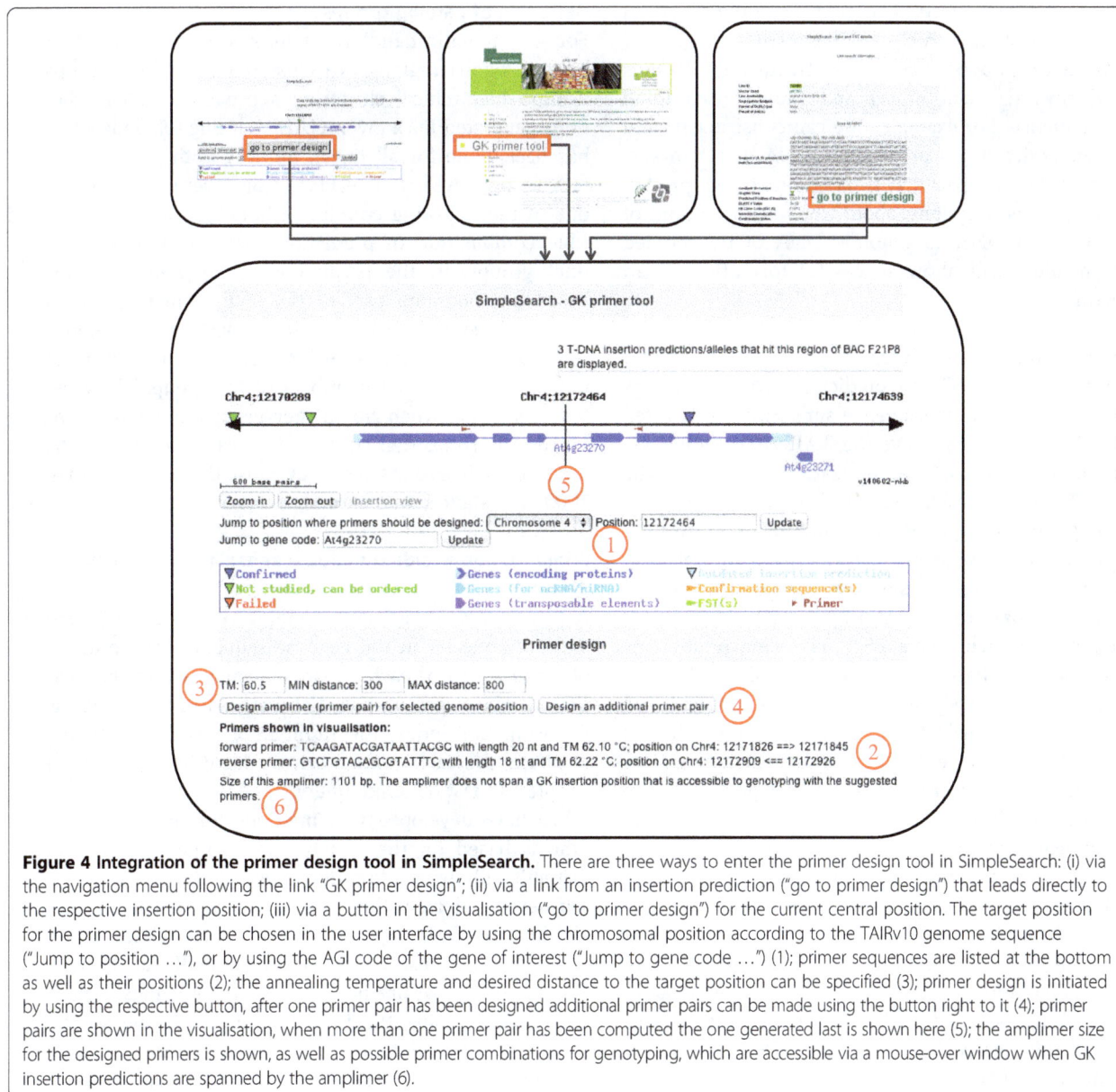

Figure 4 Integration of the primer design tool in SimpleSearch. There are three ways to enter the primer design tool in SimpleSearch: (i) via the navigation menu following the link "GK primer design"; (ii) via a link from an insertion prediction ("go to primer design") that leads directly to the respective insertion position; (iii) via a button in the visualisation ("go to primer design") for the current central position. The target position for the primer design can be chosen in the user interface by using the chromosomal position according to the TAIRv10 genome sequence ("Jump to position …"), or by using the AGI code of the gene of interest ("Jump to gene code …") (1); primer sequences are listed at the bottom as well as their positions (2); the annealing temperature and desired distance to the target position can be specified (3); primer design is initiated by using the respective button, after one primer pair has been designed additional primer pairs can be made using the button right to it (4); primer pairs are shown in the visualisation, when more than one primer pair has been computed the one generated last is shown here (5); the amplimer size for the designed primers is shown, as well as possible primer combinations for genotyping, which are accessible via a mouse-over window when GK insertion predictions are spanned by the amplimer (6).

tools usually require the definition of several parameters by their users for the design of primers, which is often laborious and confusing for the users, especially when the underlying algorithms are unknown to them. Our tool ensures most convenient primer design, because it only requires the absolute minimum of parameter definition. This is mainly the genome position to be addressed which is easily accessible through GenBank or SimpleSearch or even already known. In addition, the distance to the target position to be considered for primer design and a value for the desired melting temperature of the primers is required. A main advantage is that problems with difficult genomic positions are taken care of automatically. With the new tool we hope to contribute to the simplification of the analysis of T-DNA insertions as well as of other PCR-based applications in *A. thaliana*.

Methods

Plant growth conditions as well as molecular biological methods used for generation of the data used in this study have been described in [3]. General database aspects have been described in [9,21]. All FSTs generated in GABI-Kat are publicly available through ENA/GenBank and SimpleSearch.

Terminology

We use the term "amplicon" to refer to the DNA fragment that has been physically formed after PCR and which can be sequenced. The term "amplimer" refers to the theoretical construct, which consists of a primer pair (located on opposite strands and with their 3′-ends directed towards each other) and the corresponding source sequence. For example, one can construct several amplimers addressing a predicted T-DNA insertion site, but only the primer pair that is based on correct predictions and/or assumptions about the configuration of the fusion of T-DNA to genomic DNA of the studied insertion allele will allow successful formation of an amplicon.

Categorisation of insertion sites

The evaluation of FSTs to predict insertion sites was based on hits (weighted sequence similarities) generated by BLAST [8] and by using the TAIRv10 genome sequence and annotation dataset [22]. Initially, only the best BLAST hit for each FST had been stored [18]. To address paralogous predictions, and also "composite FSTs" that contain sequence parts derived from at least two independent insertion sites, we changed the internal GABI-Kat database to allow storing several insertion site predictions for each single FST. The new predictions were categorised based on the e-value of the BLAST hit. Category 0 is assigned to the prediction deduced from the best BLAST-hit as long as its e-value is below (better than) 1e-3. "Best" hits with larger (less significant) e-values were assigned to category 2. There can only be one best hit for each FST, and this hit is selected by using the top-of-the-list hit in the BLAST output, which results in the prediction of category 0. Note that the best hit might end up at the top position by chance if the respective FST region hits several parts of the genome with identical e-values and scores. If there are additional hits from the same FST with e-values below 1e-3, the deduced predictions are classified as category 1. If different regions of the same FST have hits in different parts of the genome, these regions are handled individually to cover the cases of several insertions being deduced from one composite FST. From each of these regions (and in addition to the single category 0 prediction for the complete FST) a maximum of 3 hits are used to produce predictions of category 1; further BLAST hits are ignored. This restriction is necessary to reduce the amount of lab work caused by FST regions that are not only paralogous but repetitive. For further filtering, the deduced insertion sites (i) need to have a distance of 1000 bp to each other to be considered as a new insertion prediction, if they are closer to each other they are assigned to the same prediction, and (ii) are discarded if the e-value difference to the best hit for one FST region is larger than a factor of 1e10. In general, our subsequent analyses considered only the predictions of categories 0 and 1.

Definition of paralog groups

For a systematic handling of insertion predictions that hit paralogous regions, we clustered them into groups using a hierarchical clustering approach for all predictions generated for a given GABI-Kat line (obviously this has been done for all lines). Starting with groups containing one prediction each, groups were combined if one of the following conditions hold true for each possible combination of predictions between the two distinct groups: (i) the prediction for different loci was based on the same part of the FST sequence (with a minimal overlap of 30 bp); (ii) the 400 bp of sequence next to the predicted insertion site have an identity of more than 79% to all members of the group. The clustering stopped when no further groups could be combined. All groups that contained at least two predictions for distinct insertions (i.e. they must display more than 1000 bp distance) were stored.

Primer design avoiding multiple annealing sites for the 3′-end

A primer was regarded as "unique" if its 12 bp-3′-end had only one hit in the genome sequence. The number of occurrences of each 12 bp sequence within the genome has been precomputed and stored in our database. By using this index, the number of possible matching positions for each primer can be identified easily by a simple and fast database query.

We have developed two methods for primer design. One is based on the widely used primer design tool Primer3 [23] with additional filtering, the other uses a self-developed algorithm that searches for mismatches within a multiple alignment of sequence-related target zones. For in-house confirmation at GABI-Kat, primers for insertions that are not located in paralogous regions are designed using the first method, while the second method is applied to insertions in paralogous regions. The public primer design tool works for all positions within the A. thaliana genome, is not dependent on (predicted) GABI-Kat insertion sites and automatically chooses the best method for the genomic locus addressed. In order to decide which method is suited best for the target zone in question, a BLAST of the sequences of this zone (that is, the sequences surrounding the target position limited by the value set for the distance to the target position) is performed against the A. thaliana genome sequence with an e-value cutoff of 1e-5, which is high enough to detect hits of down to 24 bp. If there is a sequence part within the target zone that has no other hit in the A. thaliana genome and is at least 100 bp long, problems with paralogous regions should not occur and the Primer3-based method is used. If such a unique sequence part cannot be determined in the target zone sequence, the paralog primer design method is chosen.

Primer3-based method

In order to find primer candidates within a target zone, the part of the target zone without additional paralogous regions in the genome (identified during the decision which primer design method should be used) is used first. This part of the target zone sequence is divided into overlapping windows of at least 80 bp and for each of these windows a primer is designed by Primer3. Windows overlap with 30 bp to ensure that possible primers at the ends of the windows are considered as well. To further increase the number of possible primers the selected melting temperature is altered in steps of 0.4°C to maximally 1.2°C below or above the defined melting temperature. All primers are checked for uniqueness of their 3′-end as described above. As soon as a primer is detected that has only one 12 bp-hit within the genome, the primer design finishes successfully. If the initial search within the unique part of the target zone did not yield a result, the procedure is repeated within the whole target zone sequence with a window size of at least 110 bp and alternating melting temperatures in the same way as described above. If no unique primer could be found, the one with the fewest 12 bp-hits among all primers designed during the whole process is regarded as the best possible primer for this target zone.

Paralog primer method

In target zones that do have paralogous regions throughout their sequence somewhere in the genome, the Primer3-based approach often does not lead to satisfying results. Our algorithm first identifies all potentially paralogous regions within the genome by an initial BLAST with an e-value cutoff of 1e-5 and a minimum required length of 50 bp. All hits are elongated to fit the length of the target zone, and a multiple alignment is computed using ClustalW [24]. In order to reduce the runtime of ClustalW, a sliding window approach with overlapping windows of sizes around 220 bp (and an overlap of 30 bp) is used to compute the multiple alignments. In these multiple alignments, the algorithm searches for positions with a maximum number of mismatches to the sequence of the target zone. This position is defined as the 3′-end of a possible primer and is elongated to match the desired melting temperature. After that, the primer candidate is checked for GC-clamp (1–3 G/C in the last 5 nucleotides), base repeats (less than 5 identical nucleotides in a row), GC-content (between 40 and 60%) and secondary structures (last 5 nucleotides should not appear as reverse complement in the primer). The more of these conditions hold, the better the primer – primers not fulfilling some of these criteria are discarded (see Figure 3). The annealing temperature of the primer is computed using the same formula used in Primer3

(according to [25]) to achieve results comparable to those from the Primer3-based method:

$$T_m[°C] = 81.5 - 11.6 + 0.41(\%GC) - \frac{600}{length}$$

A large number of candidates are generated and further checked for uniqueness as described above. If a unique primer is found it is returned as result. If this is not possible, the primer with the fewest matches among all examined primers is returned.

Additional file

Additional file 1: Table S1. The file contains a table (Table S1), which shows statistics about insertion site predictions in the GABI-Kat collection before and after the 1-to-N analysis of the FSTs. An explanation for the data presented in Table S1 is included in the file as well.

Competing interests
The authors declare that they have no competing interests.

Authors' contributions
GH, NK and BW conceived and designed research, analysed and interpreted the data and wrote the manuscript. GH conducted wet-lab experiments. NK did database programming and bioinformatics. All authors read and approved the manuscript.

Authors' information
Gunnar Huep and Nils Kleinboelting are joint first authors.

Acknowledgements
The authors thank Andreas Kloetgen and Tina Zekic for their contributions to the implementation of the primer design, Yong Li, Mario Rosso, Prisca Viehoever, the MPI for Plant Breeding Research and all former co-workers for their contribution to GABI-Kat, and Ute Buerstenbinder, Eliane Quittschau, Helene Schellenberg, Nina Schmidt, Andrea Voigt for technical assistance. The work described in this article is funded by the German Federal Ministry of Education and Research (BMBF) in the context of the German plant genomics program GABI (Förderkennzeichen 0313855). We acknowledge support of the publication fee by Deutsche Forschungsgemeinschaft and the Open Access Publication Funds of Bielefeld University.

References
1. Initiative TAG: **Analysis of the genome sequence of the flowering plant** *Arabidopsis thaliana*. *Nature* 2000, **408**(6814):796–815.
2. Alonso JM, Ecker JR: **Moving forward in reverse: genetic technologies to enable genome-wide phenomic screens in Arabidopsis.** *Nat Rev Genet* 2006, **7**(7):524–536.
3. Rosso MG, Li Y, Strizhov N, Reiss B, Dekker K, Weisshaar B: **An** *Arabidopsis thaliana* **T-DNA mutagenized population (GABI-Kat) for flanking sequence tag based reverse genetics.** *Plant Mol Biol* 2003, **53**(1):247–259.
4. Szabados L, Kovacs I, Oberschall A, Abraham E, Kerekes I, Zsigmond L, Nagy R, Alvarado M, Krasovskaja I, Gal M, Berente A, Redei GP, Haim AB, Koncz C: **Distribution of 1000 sequenced T-DNA tags in the Arabidopsis genome.** *Plant J* 2002, **32**:233–242.
5. Li Y, Rosso MG, Ulker B, Weisshaar B: **Analysis of T-DNA insertion site distribution patterns in Arabidopsis thaliana reveals special features of genes without insertions.** *Genomics* 2006, **87**(5):645–652.
6. Kim S, Veena, Gelvin S: **Genome-wide analysis of Agrobacterium T-DNA integration sites in the Arabidopsis genome generated under non-selective conditions.** *Plant J* 2007, **51**(5):779–791.

7.   Strizhov N, Li Y, Rosso MG, Viehoever P, Dekker KA, Weisshaar B: **High-throughput generation of sequence indexes from T-DNA mutagenized Arabidopsis thaliana lines.** *BioTechniques* 2003, **35**(6):1164–1168.

8.   Altschul SF, Gish W, Miller W, Myers EW, Lipman DJ: **Basic local alignment search tool.** *J Mol Biol* 1990, **215**:403–410.

9.   Kleinboelting N, Huep G, Kloetgen A, Viehoever P, Weisshaar B: **GABI-Kat SimpleSearch: new features of the Arabidopsis thaliana T-DNA mutant database.** *Nucleic Acids Res* 2012, **40**:D1211–D1215.

10. **GABI-Kat Website.** [http://www.gabi-kat.de].

11. Scholl RL, May ST, Ware DH: **Seed and molecular resources for Arabidopsis.** *Plant Physiol* 2000, **124**(4):1477–1480.

12. **SimpleSearch.** [http://www.gabi-kat.de/simplesearch.html].

13. Armisén D, Lecharny A, Aubourg S: **Unique genes in plants: specificities and conserved features throughout evolution.** *BMC Evol Biol* 2008, **8**:280.

14. Bolle C, Huep G, Kleinbolting N, Haberer G, Mayer K, Leister D, Weisshaar B: **GABI-DUPLO: a collection of double mutants to overcome genetic redundancy in Arabidopsis thaliana.** *Plant J* 2013, **75**(1):157–171.

15. Ding G, Sun Y, Li H, Wang Z, Fan H, Wang C, Yang D, Li Y: **EPGD: a comprehensive web resource for integrating and displaying eukaryotic paralog/paralogon information.** *Nucleic Acids Res* 2008, **36**:255–262.

16. Rutter MT, Cross KV, Van Woert PA: **Birth, death and subfunctionalization in the Arabidopsis genome.** *Trends Plant Sci* 2012, **17**(4):204–212.

17. O'Malley RC, Ecker JR: **Linking genotype to phenotype using the Arabidopsis unimutant collection.** *Plant J* 2010, **61**(6):928–940.

18. Li Y, Rosso MG, Strizhov N, Viehoever P, Weisshaar B: **GABI-Kat SimpleSearch: a flanking sequence tag (FST) database for the identification of T-DNA insertion mutants in Arabidopsis thaliana.** *Bioinformatics* 2003, **19**(11):1441–1442.

19. Ye J, Coulouris G, Zaretskaya I, Cutcutache I, Rozen S, Madden TL: **Primer-BLAST: a tool to design target-specific primers for polymerase chain reaction.** *BMC Bioinformatics* 2012, **13**(134):.

20. **GABI-Kat primer design.** [http://www.gabi-kat.de/db/primerdesign.php].

21. Li Y, Rosso MG, Viehoever P, Weisshaar B: **GABI-Kat SimpleSearch: an Arabidopsis thaliana T-DNA mutant database with detailed information for confirmed insertions.** *Nucleic Acids Res* 2007, **35**:D874–D878.

22. Huala E, Dickerman AW, Garcia-Hernandez M, Weems D, Reiser L, LaFond F, Hanley D, Kiphart D, Zhuang M, Huang W, Mueller LA, Bhattacharyya D, Bhaya D, Sobral BW, Beavis W, Meinke DW, Town CD, Somerville C, Rhee SY: **The Arabidopsis Information Resource (TAIR): a comprehensive database and web-based information retrieval, analysis, and visualization system for a model plant.** *Nucleic Acids Res* 2001, **29**(1):102–105.

23. Untergasser A, Cutcutache I, Koressaar T, Ye J, Faircloth BC, Remm M, Rozen SG: **Primer3–new capabilities and interfaces.** *Nucleic Acids Res* 2012, **40**(15):e115.

24. Larkin MA, Blackshields G, Brown NP, Chenna R, McGettigan PA, McWilliam H, Valentin F, Wallace IM, Wilm A, Lopez R, Thompson JD, Gibson TJ, Higgins DG: **Clustal W and Clustal X version 2.0.** *Bioinformatics* 2007, **23**(21):2947–2948.

25. Sambrook J, Fritsch EF, Maniatis T: *Molecular Cloning: A Laboratory Manual.* 2nd edition. New York, NY: Cold Spring Harbor Laboratory Press; 1989.

8

Metabolite profiling of wheat (*Triticum aestivum* L.) phloem exudate

Lachlan James Palmer[1*], Daniel Anthony Dias[2], Berin Boughton[2], Ute Roessner[2], Robin David Graham[1] and James Constantine Roy Stangoulis[1]

Abstract

Background: Biofortification of staple crops with essential micronutrients relies on the efficient, long distance transport of nutrients to the developing seed. The main route of this transport in common wheat (*Triticum aestivum*) is via the phloem, but due to the reactive nature of some essential micronutrients (specifically Fe and Zn), they need to form ligands with metabolites for transport within the phloem. Current methods available in collecting phloem exudate allows for small volumes (µL or nL) to be collected which limits the breadth of metabolite analysis. We present a technical advance in the measurement of 79 metabolites in as little as 19.5 nL of phloem exudate. This was achieved by using mass spectrometry based, metabolomic techniques.

Results: Using gas chromatography–mass spectrometry (GC-MS), 79 metabolites were detected in wheat phloem. Of these, 53 were identified with respect to their chemistry and 26 were classified as unknowns. Using the ratio of ion area for each metabolite to the total ion area for all metabolites, 39 showed significant changes in metabolite profile with a change in wheat reproductive maturity, from 8–12 to 17–21 days after anthesis. Of these, 21 were shown to increase and 18 decreased as the plant matured. An amine group derivitisation method coupled with liquid chromatography MS (LC-MS) based metabolomics was able to quantify 26 metabolites and semi-quantitative data was available for a further 3 metabolites.

Conclusions: This study demonstrates that it is possible to determine metabolite profiles from extremely small volumes of phloem exudate and that this method can be used to determine variability within the metabolite profile of phloem that has occurred with changes in maturity. This is also believed to be the first report of the presence of the important metal complexing metabolite, nicotianamine in the phloem of wheat.

Keywords: Aphid stylectomy, Exudate, Grain loading, GC-MS, LC-MS, Metabolomics, Method development, Phloem, Wheat

Background

Deficiencies of Fe and Zn in humans have been identified as a serious issue of concern for developing countries. In a 2002 World Health Organisation report it was estimated that in 2000, 1.6 million people died as a direct result of Fe and Zn deficiency and a further 60 million healthy life years were lost [1]. Approximately 60% of the health life years lost occurred in developing countries within Africa and South-East Asia [1]. Biofortification of staple crops has been identified as a possible way of combating the issue of micronutrient deficiency [2] and attempts

to increase the levels of mineral and vitamin micronutrients in the harvested and edible plant parts using genetic or agronomic techniques is currently underway [3]. An important part of the mineral biofortification process is the transport of these elements from the source to the sink (i.e. from soil, through to the roots, stems and leaves, and then to the seed). Within a plant, the long distance transport pathways of the xylem and phloem are the major routes for nutrient movement to developing seeds [4]. In the case of wheat, the phloem is very important as there is a xylem discontinuity at the base of the grain [5] which results in all macro and micro nutrients first transferring to the phloem before unloading into the grain. During the transport of Fe and Zn in the phloem these minerals must be complexed due to their reactive nature

* Correspondence: Lachlan.palmer@flinders.edu.au
[1]School of Biological Science, Flinders University, Bedford Park, South Australia 5042, Australia
Full list of author information is available at the end of the article

[6]. A variety of metabolites have been theorised to complex Fe, Zn and other essential minerals within the phloem [7]. Of these, nictoianamine and cystine are proposed to play a major role in the modelled transport of Fe and Zn [7], and in rice, nicotianamine has been found to complex Zn in the phloem [8].

Phloem is a complex matrix which consists of water, sugars, amino acids, organic acids, secondary metabolites, peptides and hormones along with ions and a number of macromolecules, including proteins, small RNAs and mRNAs [9,10]. Recent reviews have highlighted the importance of phloem composition in long distance transport and signalling throughout the plant [9,10] and these reviews have also examined the difficulty and issues related to collection of phloem for analysis. There are three main techniques in which phloem can be collected for direct analysis: 1) cutting the stem and collecting the liquid that exudes; 2) making use of an Ethylenediamine-tetraacetic acid (EDTA) solution to allow a freshly cut plant part to continue to exude; 3) using insect stylectomy to collect phloem exudate (see [11] for further details). The first two methods have limitations when applied to cereal crops. Cutting the stem for collecting phloem is limited to a small selection of plant species such as castor bean [12] and cucurbits [13] and is not possible for cereals. In wheat, phloem will not exude from cuts made to the stem or leaves under field or glasshouse conditions, however phloem will exude from the grain pedicel after the removal of the seed [14]. This limits the accessibility to wheat phloem and also involves interference with the developing ear. EDTA facilitated exudation also has its limitations, owing to the difficulty in quantifying phloem volume for accurate concentration measurements and also because EDTA facilitated exudation may be contaminated by components from damaged cells other than the phloem and the apoplastic space [11]. Insect stylectomy using aphids and planthoppers has been used to access the phloem of cereal crops for the analysis of some metabolites within the phloem [8,15]. The main limitation of stylectomy based collection is the small volumes involved. With exudation rates ranging from 4.2 to 354 nl h^{-1} [16] volumes collected are in the low μl to nl range [16,17]. Due to these small volumes, accurate measurement of phloem collections has been difficult which has limited the scope of metabolomic profiling of the phloem. In most reports of metabolites in phloem collected by stylectomy, collections were made over several hours to enable sufficient volumes to be collected for analysis, as measured using 0.5 μl micro capillaries [8,18]. In more recent work, an alternative technique for measuring phloem volume has been used to measure diurnal variability in amino acid concentrations in volumes as little as 2.1 nl [15]. In this current research, we demonstrate the use of accurate volume measurements for the quantitative analysis of amine-containing metabolites detected by LC-MS and semi-quantitative analysis of the metabolite profile of wheat phloem using GC-MS. We also present the results of semi-quantitative analysis of changes in the metabolite profile during the grain loading period.

Results
GC-MS metabolite profiling

GC-MS metabolomic profiling was tested as this has been used previously to profile the metabolites in plant tissues. For example, tissue level changes as a tolerance response to Fe deficiency in peas [19]. Additional file 1: Table S1 details all 79 metabolites identified by GC-MS and for some metabolites, multiple derivatives are created and these are shown in Additional file 1: Table S2 as they were included in calculations of the ion ratio. Of the 79 metabolites identified it was found that 40 had non-normal distributions and attempts were made to transform the data prior to statistical analysis. Of the 40 metabolites, 2 were not able to be transformed to produce a normal distribution (Additional file 1: Table S3) and so were not included in statistical testing.

Of the 79 metabolites detected, there were 26 unknown compounds found and these were not identified using either in-house or commercial libraries nor the GOLM Metabolome Database [20], so they are listed with the following notation. UN1_10.61_158 = Unknown 1 with a retention time of 10.61 minutes with a unique ion at 158 m/z. The area of a particular fragment ion was selected and was subsequently adjusted for each sample by dividing it by the volume of phloem collected and then the ratio of this area to the total area for all identified ions in the sample was calculated and used for statistical comparison between different stages during grain loading.

The results from independent student t-tests on metabolites showing significant changes between peak grain loading (9–11 DAA) and the end of grain loading (18–20 DAA) are shown in Tables 1 and 2. The results listed in Table 1 show the 18 metabolites had a statistically significant decrease in the phloem as grain loading progressed, from 9–11 DAA to 18–20 DAA. Ornithine had the greatest reduction showing a 4.6 fold decrease. 3-amino-piperidin-2-one, UN08 and Glutamine also declined by 3.5-, 3.4- and 3.4-fold respectively (Table 1). There were another 7 metabolites that had more than a two-fold decrease as grain loading progressed (Table 1).

There were 21 metabolites that had a significant increase in the phloem as grain loading progressed from 8–12 DAA to 17–21 DAA (Table 2). Of these, shikimic acid had the greatest increase (2.9 fold), while quinic acid, succinate and glycine also had more than a 2 fold increase (2.5, 2.3 and 2.2 respectively, Table 2).

Table 1 The mean and standard error of the difference and Fold change for metabolites, profiled using GC-MS, that significantly decreased (p <0.05) in the phloem from 9–11 DAA (n = 15) to 18–20 DAA (n = 16)

Metabolite	Transformation	Levene's test for equality of variances sig.	t-test for equality of means sig. (2-tailed)	Mean difference	Std. error difference	Fold change
3-amino-piperidin-2-one 2TMS	SQRT	0.227	0.000	−0.34838	0.07402	−3.4
Alanine 2TMS	SQRT	0.000[un]	0.009	−0.23127	0.08092	−1.9
Arginine 3TMS	Ln	0.286	0.000	−0.84617	0.20727	−2.3
Glutamate 3TMS	None	0.003[un]	0.005	−1.11392	0.35445	−2.3
Glutamine 3TMS	CBRT	0.296	0.001	−0.34581	0.09220	−2.6
Histidine 3TMS	Ln	0.994	0.032	−0.62065	0.27622	−1.9
Homoserine 3TMS	None	0.613	0.000	−0.08141	0.01851	−1.9
Lysine 4TMS	None	0.180	0.006	−0.57860	0.19365	−1.4
Ornithine 3TMS	None	0.001[un]	0.007	−0.17965	0.05689	−4.6
Pyroglutamate 2TMS	None	0.444	0.018	−3.21496	1.28551	−1.3
Serine 3TMS	None	0.073	0.000	−4.85198	1.16680	−1.6
Trehalose 8TMS	Ln	0.952	0.000	−1.25509	0.19795	−3.5
UN01_10.61_158	CBRT	0.360	0.000	−0.07041	0.01492	−2.3
UN07_17.62_275	Ln	0.177	0.002	−0.68564	0.20462	−2.0
UN08_17.96_360	CBRT	0.673	0.001	−0.19065	0.05177	−3.4
UN09_18.15_275	CBRT	0.363	0.003	−0.08408	0.02541	−2.0
UN11_19.48_299	CBRT	0.003[un]	0.003	−0.08195	0.02451	−2.0
UN26_14.48_229	Ln	0.034[un]	0.001	−0.61709	0.15206	−1.9

Also shown are the p values for Levene's test of equality of variance, where values are less than 0.05 (equal variances not assumed = [un]), equal sample variances were not assumed when calculating t-test. (xTMS = Trimethylsilyl derivative where x = the number of TMS groups; yMX = methoxyamine derivatised product where y = 1 or 2).

LC-MS

Quantification of amine group containing metabolites in the phloem exudates was conducted according to Boughten et al. [21]. The concentrations of the metabolites identified are detailed in Table 3. Two metabolites were not able to be quantified due to issues with standard stability affecting the calibration curve and the response relative to the ISTD is presented instead. To the authors knowledge this is the first time that nicotianamine (NA) has been directly quantified. For metabolites where an authentic standard was not available or could not be generated, the ratio of ion area to internal standard was used to generate a relative response. For the LC-MS analysis, phloem collection volumes used were between 43.4 nl and 180.34 nl, with an mean of 95.7 nl ± 61.43. Due to the exploratory nature of this analysis, only four samples were analysed, two from each maturity, and no significant changes were observed (data not shown). Therefore only the average for all samples is presented in Table 3.

Discussion

The analysis of the metabolite profile of phloem has been restricted in the past due to limitations in the amount collected and the availability and sensitivity of analytical techniques. With exudate flow rates in wheat ranging from 0.07 to 5.9 nl min^{-1} [16], volumes that are collected per sample are normally less than a µl. For previous work examining metabolite profiles in the phloem, volume measurement has been done in a variety of ways [22-24]. Phloem volumes or sample amounts are estimated using the length of the liquid within the microcap [24], the weight of sample collected [23] or by collecting to the volume of the microcap [22]. For a series of papers examining metabolites in phloem [25-27], phloem volumes of between 10 and 60 nl were collected using 0.5 µl microcaps but to our knowledge there is no mention of how volume was estimated in the microcap. Even in the original method, these papers all refer to where volumes of 5 nl were collected using a 0.5 µl microcap [28] and there are no details of how the sample volume is derived. The measurement of sample volume is a key part of improving the accuracy of analysis and enables better detection of changes in metabolite profile.

One of the main issues for the accurate measurement of sub-µl volumes is evaporation and work carried out previously has demonstrated an increase in the osmotic potential when phloem samples are collected in air [29,30]. To counter the effects of evaporation, collection of exudate under oil is the accepted method [11] but in more recent work by our laboratory, we have demonstrated that accurate measurement under oil has technical difficulties due to the potential for measurement errors arising from

Table 2 The mean and standard error of the difference and Fold change for metabolites, profiled using GC-MS, that significantly increased (p <0.05) in the phloem from 9–11 DAA (n = 15) to 18–20 DAA (n = 16)

Metabolite	Transformation	Levene's test for equality of variances sig.	t-test for equality of means sig. (2-tailed)	Mean difference	Std. error difference	Fold change
Citric acid 4TMS	None	0.512	0.000	0.23250	0.05283	1.8
Fructose_MX1	None	0.002[un]	0.000	0.27792	0.05953	1.8
Fumarate 2TMS	None	0.000[un]	0.004	0.00942	0.00289	1.6
Gluconic acid-1,5-lactone 4TMS	None	0.066	0.002	0.34643	0.09990	1.7
Glucose MX1	None	0.008[un]	0.002	0.63124	0.17488	1.7
Glycine 3TMS	None	0.026[un]	0.000	0.17284	0.03384	2.2
Hexadecanoate 1TMS	SQRT	0.004[un]	0.017	0.27188	0.10404	1.9
Methionine 1TMS	None	0.337	0.024	0.00554	0.00233	1.3
Octadecanoate 1TMS	SQRT	0.028[un]	0.021	0.21070	0.08527	1.8
Phenylalanine 2TMS	None	0.654	0.018	1.07470	0.42754	1.5
Putrescine 4TMS	None	0.697	0.000	0.82294	0.13397	1.9
Quinic acid 5TMS	SQRT	0.010[un]	0.007	0.32413	0.10951	2.5
Shikimic acid 4TMS	SQRT	0.060	0.000	0.23331	0.05785	2.9
Succinate 2TMS	None	0.000[un]	0.010	0.05240	0.01802	2.3
Tyrosine 3TMS	None	0.180	0.000	1.09938	0.24564	1.8
UN04_15.56_185	None	0.000[un]	0.019	0.10668	0.04183	1.8
UN10_19.08_217	None	0.106	0.007	0.44254	0.15290	1.4
UN16_25.71_339	None	0.016[un]	0.015	0.00806	0.00302	1.7
UN17_27.24_375	None	0.003[un]	0.000	0.01570	0.00379	1.8
UN18_28.91_437	None	0.011[un]	0.001	0.01317	0.00358	1.6
UN20_32.34_503	InvCBRT	0.020[un]	0.049	−0.62079	0.29828	1.8

Also shown are the p values for Levene's test of equality of variance, where values are less than 0.05 (equal variances not assumed = [un]), equal sample variances were not assumed when calculating t-test. (xTMS = Trimethylsilyl derivative where x = the number of TMS groups; yMX = methoxyamine derivatised product where y = 1 or 2).

the optical nature of the measurement and the surface shape of the oil used in the collection step [16]. Recent work has made advances in volume measurement: work examining the diurnal effect on the concentration of amino acids in the phloem, made use of a correction factor for air based droplet volume measurements [15]. This technique used measurements made under oil, which reduces the effect of evaporation, as a comparison for collections measured in air and enable a correction factor to be derived. We have further refined this technique of volume measurement using digital photography and software to further reduce the effect of evaporation and to quantify the accuracy of the volume measurement [16]. This method has been used successfully to quantify inorganic components in the phloem of wheat and to detect significant changes with maturity [31]. Oil has also been identified as a potential contaminant for certain metabolite analyses [15], and we have also found that for GC-MS profiling, paraffin oil is a significant contaminant suppressing metabolite signal from the phloem causing an increase in the baseline due to the elution of multiple hydrocarbons present in paraffin oil (refer to Additional

file 1 for trace images). This observation was consistent with the GC-MS analysis of petroleum based oil as a contaminant [32].

When using phloem samples that were measured in air we were able to identify, using GC-MS, 79 different metabolites (Additional file 1: Table S1) and of these, 38 showed significant variability in phloem composition with a change in maturity (Tables 1 and 2). Of these metabolites with significant changes, 21 increased and 18 decreased in the phloem as the plant aged. For samples analysed using an LC-MS method specific for amine containing compounds, we were able to identify 30 metabolites and of these quantify the concentration of 27 metabolites within the phloem (Table 3). For the GC-MS metabolite profiling analysis, full quantification of all metabolites is not feasible. This is due to the large number of metabolites detected by GC-MS, making it unfeasible to establish calibration curves for all metabolites detected and so peak area is used for semi-quantitative analysis. Due to the use of peak area, metabolites cannot be compared to one another as the peak area is dependent on the derivitisation process, and all metabolites have a different

Table 3 Amine group containing metabolites identified in the phloem of wheat as measured by LC-MS collected at two maturities (n =4), unless otherwise stated mean and standard error of all quantitated metabolites are reported as mMol L^{-1}

Metabolite	Mean	SE
Ammonia (ISTD RR)	33.8	1.275
Glutathione (ISTD RR)	0.58	0.0905
Glutamine	170.8	50.895
Valine	75.5	17.52
Histidine	64.7	25.24
Serine	63.1	25.455
Glutamate	36.3	15.32
Arginine	39.8	3.74
Alanine	42.0	6.045
Proline	21.2	8.125
Phenylalanine	58.1	18.365
Threonine	49.7	15.515
Isoleucine	44.3	11.275
Leucine	43.7	16.34
Tryptophan	44.1	8.405
Lysine	48.0	10.315
Glycine	39.9	18.275
Tyrosine	27.7	9.495
Aspartic acid	18.7	7.205
Methionine	18.8	8.31
Asparagine	7.4	2.77
β-Alanine	1.7	0.595
GABA	1.1	0.45
Ornithine	1.2	0.58
Citrulline (μmol L^{-1})	232.3	39.14
Nicotianamine (μmol L^{-1})*	255.4	96.71
Cysteine (μmol L^{-1})	70.9	13.91
4-Hydroxy-proline (μmol L^{-1})	38.5	9.675

Metabolites that were measured but did not have an external standard are reported as a ratio of response in relation to the internal standard (ISTD RR). *n = 3, outlier was removed.

response factor. For specific metabolites, it may be possible to set up calibration curves for quantification of phloem concentrations by GC-MS and so allow further exploration of results of interest identified from metabolite profiling.

One result of interest is the significant changes in sugars other than sucrose within the phloem. For most work on the phloem, only sucrose is reported [22,33,34]. It has been established that sucrose is the dominant sugar involved in sugar transport [35] and it is assumed that hexoses (glucose and fructose) are not normally present, and when they are, they are mostly seen when using EDTA facilitated exudation [35]. Glucose and fructose have been shown to exist (5.5% and 1.5%) in the phloem of perennial ryegrass (Lolium perenne L.) collected from aphid stylectomy [36]. A change in sugar composition was found when plants were defoliated with a decrease of more than 80% for sucrose concentration and decreases of 42% and 47% in glucose and fructose concentrations, respectively [36]. This may indicate a source other than leaf for hexoses within the phloem. The results from the GC-MS analysis highlighted a significant increase in glucose and fructose levels within the phloem during the grain loading period (Table 2).

Of particular interest are the results presented from the LC-MS analysis quantifying NA and demonstrating the presence of glutathione. These two metabolites have been found to play important roles in complexing essential micronutrients during long distance transport in the phloem [7,8,37,38]. In a theoretical model of metabolite and micronutrient speciation in the phloem it was found that 54.4% of Zn was likely to be bound to NA and the remaining Zn complexed with amino acid complexes of cysteine and cysteine with histidine (41.2% and 2.8% respectively) [7]. In the case of Fe, 99% of the ferrous ions would be complexed by NA and only 19.3% of ferric ions would be complexed by NA while the remaining ferric ions would be complexed with glutamate and citrate (70% and 9.2% respectively) [7]. A potential fault with this model is that it does not incorporate 2'-deoxymugineic acid (DMA) as a potential candidate for complex formation. In work performed on the phloem of rice, the concentration of DMA was found to be between 152 μMol L^{-1} [39] and 150 μMol L^{-1} [8]. This was much higher than the NA values reported of 66 μMol L^{-1} [39] and 76 μMol L^{-1} [8]. In rice it has been found that the main Zn complex ligand was NA whilst for Fe, DMA was the main metabolite responsible for complexing this metal [8]. Glutathione has been found to play a role in Cd transportation as part of the detoxification process [40]. Glutathione was tested as a chelating agent in the speciation model but was found to complex less than 2% of Zn though it was mentioned that if Cd was included this may affect the model dynamics [7]. In this work we have reported for the first time the concentration of NA in the phloem of wheat with an average concentration of 255.4 μMol L^{-1} ± 96.71 which is within the range reported in castor bean [37,38,41] but is much higher than what has been reported in rice [8,39]. Previous work in our lab has shown a significant increase in Zn and Mg in the phloem during grain loading [31] and it's also likely that there is an associated increase in metal complexing metabolites. This relationship has been demonstrated in castor bean where the NA phloem concentrations reported 4 and 8 days after imbibition were 206 μMol L^{-1}, [38] which is close to what was observed in this work

(255.4 μMol L^{-1} ± 96.71). Further exploration of NA flux at different maturities may give further insight into the role of NA in essential micronutrient transport.

It is anticipated that further method development would enable the analysis and quantification of DMA and glutathione within the phloem which would assist in the development of a wheat specific speciation model for the long distance transport of essential micronutrients. The amine binding LC-MS method used here gave in-conclusive results for DMA (data not shown) and a different method is required.

The ability to obtain a profile of a broad range of metabolites is a powerful tool not only for understanding the transport of essential micronutrients to the grain and other vegetative tissues, but also for examining responses to toxic or deficient conditions. An example of this is where it was possible to identify a metabolite complex responsible for boron mobilisation in the phloem [42] that could lead to efficiency in boron utilization [43]. Metabolite profiling has also been used widely on plant tissues to examine tissue level dynamics such as changes involved in tolerance to nutrient deficient conditions such as Fe deficiency in pea [19], and also metabolite variation during the ripening process in capsicum [44]. The methods outlined in this study add further capability to researchers interested in metabolite changes both on a whole plant level during maturation and also when plants are under abiotic or nutrient based stresses.

Conclusion

This study demonstrates the production of a complex metabolite profile from extremely small volumes of phloem exudate using GC-MS. We also demonstrate that this method of metabolite profiling can be used to determine significant maturity based variability within the metabolite profile of the phloem. To our knowledge this is the first report of the presence of and quantification of NA in the phloem of wheat.

Materials and methods

Plant material

Wheat (*Triticum aestivum* L. genotype 'Samnyt 16') seedlings were grown in 70x100 mm pots in Debco™ Green Wizard potting mix within a growth room. Growth room conditions were 13/11 h light/dark at 20°C/10°C with a minimum of 400 μmol m^{-2} s^{-1} light at the leaf surface. Plants were transferred to a greenhouse where aphids were applied and kept there for a maximum of 48 h.

Aphid stylectomy

Aphid stylectomy procedures were adapted from the method established by Downing and Unwin [45]. A short video of the method is presented in Additional file 2 and a summary of the method is as follows. Aphids were taken from an anholocyclic *Sitobion miscanthi* (Indian grain aphid) culture maintained at Flinders University on wheat plants kept under greenhouse conditions. Only apterous aphids were used in the experiments.

Aphids were secured to wheat plants (immediately below the head on the peduncle), a minimum of 12 h prior to stylectomy, using specially prepared cages (refer to Additional file 3 for construction specifications). Plants were watered to saturation at time of aphid caging. Stylectomy was performed using high-frequency micro-cauterisation under a Leica microscope (M165 C or MZ16) using an electrolytically-sharpened tungsten needle in combination with a micromanipulator. Exudate samples were collected using glass micro-capillaries (30–0017, Harvard Apparatus) pulled using a capillary puller (Narishige). The relative humidity during collections ranged from 41% to 50%.

Microscope measurement of nanolitre phloem exudate volumes

Exudate volumes were measured using the method published in [16]. In brief, exudate flow rates were estimated from photo sequences taken, in air, using a Leica microscope (M165C or MZ16) with an attached camera (DFC295 or DFC280) and the multi-time module from the Leica Application Suite software (v3.6.0). Photo sequences were taken immediately after obtaining an exuding stylet, approximately every 15 minutes during collection and immediately prior to the end of the collection. Photo sequences consisted of five photos with a one second interval between photos. The droplet radius for each photo in a sequence was measured using the interactive measurement module within the Leica Application Suite and an estimate of droplet volume calculated. Using the time interval between each two photos in a sequence the exudation flow rate was estimated from the change in volume between photos. The average of the estimated flow rate from all sequences in each collection was multiplied by the respective collection length to give an estimate of the collection volume. A correction factor, as determined previously, was applied to correct for evaporation [16]. Samples were deposited into 200 μl glass vial inserts (Agilent) containing 5 μl of Millipore Milli-Q™ water (>18.2 MΩ cm^{-1}), centrifuged for 20 seconds at 14,000 rpm in a 1.5 ml microcentrifuge tube and insert was transferred to 2 ml auto sampler vials (Agilent) and stored at −80°C. After samples had been collected samples were lyophilized in a freeze dryer prior to shipment for metabolomics profiling and quantitation at Metabolomics Australia, School of Botany, The University of Melbourne. Table 4 shows the average and standard error for the time that sample collection was started, average volume and average maturity (DAA) for the samples allocated to the maturity groupings analysed using LC-MS and GC-MS.

Table 4 N, Mean and SE, for both maturity groups (DAA group), for sample collection start times, phloem exudate volumes and number of days after anthesis for samples analysed by GC-MS and LC-MS

Analysis	Variable	DAA Group	N	Mean	SE
GC-MS	collection start	8-12 DAA	16	16:02:33.75	0:06:50.17
		17-21 DAA	20	15:35:45.00	0:11:25.50
	Corrected vol	8-12 DAA	16	90.6	13.26
		17-21 DAA	20	62.3	6.91
	DAA	8-12 DAA	16	9.9	0.22
		17-21 DAA	20	18.5	0.17
LC-MS	collection start	8-12 DAA	2	15:04:30	0:08:30
		17-21 DAA	2	14:58:00	1:24:00
	Corrected vol	8-12 DAA	2	140.6	39.79
		17-21 DAA	2	50.9	7.48
	DAA	8-12 DAA	2	10.0	0.00
		17-21 DAA	2	18.5	0.50

GC-MS analysis of phloem exudate

GC-MS analysis of phloem exudate samples was carried out using a method modified from [46]. The lyophilized phloem samples were re-dissolved in 5 μL of 30 mg mL^{-1} methoxyamine hydrochloride in pyridine and derivatised at 37°C for 120 min with mixing at 500 rpm. The samples were then treated for 30 min with 10 μL N,O-bis-(trimethylsilyl)trifluoroacetamide (BSTFA) and 2.0 μL retention time standard mixture [0.029% (v/v) n dodecane, n-pentadecane, n-nonadecane, n-docosane, n-octacosane, n-dotriacontane, n-hexatriacontane dissolved in pyridine] with mixing at 500 rpm. Each derivatised sample was allowed to rest for 60 min prior to injection.

Samples (1 μL) were injected into a GC-MS system comprised of a Gerstel 2.5.2 autosampler, a 7890A Agilent gas chromatograph and a 5975C Agilent quadrupole MS (Agilent, Santa Clara, USA). The MS was adjusted 171 according to the manufacturer's recommendations using tris-(perfluorobutyl)-amine (CF43). The GC was performed on a 30 m VF-5MS column with 0.2 μm film thickness and a 10 m Integra guard column (Agilent J&W GC Column). The injection temperature was set at 250°C, the MS transfer line at 280°C, the ion source adjusted to 250°C and the quadrupole at 150°C. Helium was used as the carrier gas at a flow rate of 1.0 mL min^{-1}. For the polar metabolite analysis, the following temperature program was used; start at injection 70°C, a hold for 1 min, followed by a 7°C min^{-1} oven temperature, ramp to 325°C and a final 6 min heating at 325°C. Both chromatograms and mass spectra were evaluated using the Chemstation Data Analysis program (Agilent, Santa Clara, USA). Mass spectra of eluting compounds were identified and validated using the public domain mass spectra library of Max-Planck-Institute for Plant Physiology, Golm,

Germany (http://csbdb.mpimp-golm.mpg.de/csbdb/dbma/msri.html) and the in-house Metabolomics Australia mass spectral library. All matching mass spectra were additionally verified by determination of the retention time by analysis of authentic standard substances. Resulting relative response ratios, that is, selected ion area of each metabolite was normalized to phloem volume for each identified metabolite. For metabolites which had multiple TMS derivatives, normalized ion areas were presented.

LC-MS analysis of phloem exudate

Quantification of amine containing metabolites in nl phloem exudate was modified from the method reported in Boughton et al. [21]. To the lyophilized phloem sample 4 μl of Borate buffer (200 mM at pH 8.8 with 1 mM ascorbic acid, 10 mM tris(2-carboxyethyl)phosphine (TCEP) and 25 μM 2-aminobutyric acid (added as an ISTD) was added, then 1 μl of 10 mM 6-Aminoquinolyl-N-hydroxysuccinimidylcarbamate (AQC, dissolved in 100% acetonitrile). The glass inserts were sealed with parafilm then left to rest at room temperature for 30 minutes. Analysis and quantification by LC-MS was done as previously reported [21].

Statistical analysis

Statistics including Student's t-test were calculated using IBM SPSS statistics software (version 22). A normal distribution was determined to achieved when the z-scores for the skew and kurtosis were less than 1.96 which the cut off indicating a significant skew or kurtosis at the 95% level [47]. Of the metabolites identified, 40 were found to have significant skew or kurtosis to their distribution. These metabolites were transformed using the following transformations in increasing power; 16 using square root (SQRT), 8 using cube root (CBRT), 11 using the natural logarithm (Ln) and 3 using the inverse cube root (InvCBRT). A further 2 metabolites were unable to be transformed successfully. For the Student's t-test as calculated using SPSS, results are produced with and without the assumption of equal variance between treatments and the Levene's statistic gives an indication if this assumption is met or rejected.

Additional files

Additional file 1: Further information on GC-MS analysis, including example of separation traces with and without oil contamination, full list of metabolites detected, replicated derivatives and un-normalisable metabolites.

Additional file 2: Short video outlining the method of aphid stylectomy.

Additional file 3: Specifications and method for constructing aphid cages.

Abbreviations
GC-MS: Gas chromatography - mass spectrometry; LC-MS: Liquid chromatography - mass spectrometry; EDTA: Ethylenediaminetetraacetic acid; DAA: days after anthesis; ISTD: Internal standard.

Competing interests
The authors declare that they have no competing interests.

Authors' contributions
LJP helped in conceiving study, collected phloem samples, did ICP-MS analysis, statistical analysis and drafted the manuscript. DAD assisted with GC-MS method development, acquisition and data analysis of phloem samples. BB assisted with LC-MS method development, running of sample and data analysis. UR helped in the design of the study, and GC-MS and LC-MS method development. RG helped in the design of the study and drafting of manuscript. JS participated in conception and design of the study and assisted with drafting of manuscript. All authors read and approved the final manuscript.

Acknowledgements
The authors wish to acknowledge HarvestPlus for helping to fund this work and Metabolomics Australia (School of Botany), funded through Bioplatforms Australia Pty Ltd., a National Collaborative Research Infrastructure Strategy (NCRIS) with co-investment from the Victorian State Government and The University of Melbourne.

Author details
[1]School of Biological Science, Flinders University, Bedford Park, South Australia 5042, Australia. [2]Metabolomics Australia, School of Botany, The University of Melbourne, Parkville, Melbourne 3010, Victoria, Australia.

References
1. World Health Organization: *The World Health Report: 2002: Reducing the Risks, Promoting Healthy Life.* Geneva, Switzerland: World Health Organization; 2002.
2. Mayer JE, Pfeiffer WH, Beyer P: **Biofortified crops to alleviate micronutrient malnutrition.** *Curr Opin Plant Biol* 2008, **11**:1–5.
3. Murgia I, Arosio P, Tarantino D, Soave C: **Biofortification for combating 'hidden hunger' for iron.** *Trends Plant Sci* 2012, **17**:47–55.
4. Atwell BBJ, Kriedmann PE, Turnbull CGN: *Plants in Action: Adaptation in Nature, Performance in Cultivation.* South Yarra, Australia: MacMillan Education Australia; 1999.
5. Zee SY, O'Brien TP: **A special type of tracheary element associated with "xylem discontinuity" in the floral axis of wheat.** *Aust J Biol Sci* 1970, **23**:783–791.
6. Blindauer CA, Schmid R: **Cytosolic metal handling in plants: determinants for zinc specificity in metal transporters and metallothioneins.** *Metallomics* 2010, **2**:510–529.
7. Harris WR, Sammons RD, Grabiak RC: **A speciation model of essential trace metal ions in phloem.** *J Inorg Biochem* 2012, **116**:140–150.
8. Nishiyama R, Kato M, Nagata S, Yanagisawa S, Yoneyama T: **Identification of Zn–nicotianamine and Fe-2"-deoxymugineic acid in the phloem sap from rice plants (*Oryza sativa* L.).** *Plant Cell Physiol* 2012, **53**:381–390.
9. Turgeon R, Wolf S: **Phloem transport: cellular pathways and molecular trafficking.** *Annu Rev Plant Biol* 2009, **60**:207–221.
10. Dinant S, Bonnemain J-L, Girousse C, Kehr J: **Phloem sap intricacy and interplay with aphid feeding.** *C R Biol* 2010, **333**:504–515.
11. Dinant S, Kehr J: **Sampling and analysis of phloem sap.** In *Plant Mineral Nutrients. Volume 953.* Edited by Maathuis FJM. New York: Humana Press; 2013:185–194. Methods in Molecular Biology.
12. Hall SM, Baker DA: **The chemical composition of *Ricinus* phloem exudate.** *Planta* 1972, **106**:131–140.
13. Richardson PT, Baker DA, Ho LC: **The chemical composition of cucurbit vascular exudates.** *J Exp Bot* 1982, **33**:1239–1247.
14. Fisher DB, Gifford RM: **Accumulation and conversion of sugars by developing wheat grains : VI. gradients along the transport pathway from the peduncle to the endosperm cavity during grain filling.** *Plant Physiol* 1986, **82**:1024–1030.
15. Gattolin S, Newbury HJ, Bale JS, Tseng H-M, Barrett DA, Pritchard J: **A diurnal component to the variation in sieve tube amino acid content in wheat.** *Plant Physiol* 2008, **147**:912–921.
16. Palmer LJ, Palmer LT, Pritchard J, Graham RD, Stangoulis JCR: **Improved techniques for measurement of nanolitre volumes of phloem exudate from aphid stylectomy.** *Plant Methods* 2013, **9**:18.
17. Mittler TE: **Studies on the feeding and nutrition of Tuberolachnus salignus (Gmelin) (Homoptera, Aphididae): II. The nitrogen and sugar composition of ingested phloem sap and excreted honeydew.** *J Exp Biol* 1958, **35**:74–84.
18. Winter H, Lohaus G, Heldt HW: **Phloem transport of amino acids in relation to their cytosolic levels in barley leaves.** *Plant Physiol* 1992, **99**:996–1004.
19. Kabir AH, Paltridge NG, Roessner U, Stangoulis JCR: **Mechanisms associated with Fe-deficiency tolerance and signaling in shoots of Pisum sativum.** *Physiol Plant* 2013, **147**:381–395.
20. Kopka J, Schauer N, Krueger S, Birkemeyer C, Usadel B, Bergmüller E, Dörmann P, Weckwerth W, Gibon Y, Stitt M, Willmitzer L, Fernie AR, Steinhauser D: **GMD@CSB.DB: the Golm metabolome database.** *Bioinformatics* 2005, **21**:1635–1638.
21. Boughton BA, Callahan DL, Silva C, Bowne J, Nahid A, Rupasinghe T, Tull DL, McConville MJ, Bacic A, Roessner U: **Comprehensive profiling and quantitation of amine group containing metabolites.** *Anal Chem* 2011, **83**:7523–7530.
22. Fukumorita T, Chino M: **Sugar, amino acid and inorganic contents in rice phloem sap.** *Plant Cell Physiol* 1982, **23**:273–283.
23. Girousse C, Bonnemain J-L, Delrot S, Bournoville R: **Sugar and amino acid composition of phloem sap of Medicago sativa: a comparative study of two collecting methods.** *Plant Physiol Biochem* 1991, **29**:41–48.
24. Sandström J, Telang A, Moran NA: **Nutritional enhancement of host plants by aphids — a comparison of three aphid species on grasses.** *J Insect Physiol* 2000, **46**:33–40.
25. Lohaus G, Moellers C: **Phloem transport of amino acids in two Brassica napus L. genotypes and one B. carinata genotype in relation to their seed protein content.** *Planta* 2000, **211**:833–840.
26. Winzer T, Lohaus G, Heldt H-W: **Influence of phloem transport, N-fertilization and ion accumulation on sucrose storage in the taproots of fodder beet and sugar beet.** *J Exp Bot* 1996, **47**:863–870.
27. Lohaus G, Burba M, Heldt HW: **Comparison of the contents of sucrose and amino acids in the leaves, phloem sap and taproots of high and low sugar-producing hybrids of sugar beet (Beta vulgaris L.).** *J Exp Bot* 1994, **45**:1097–1101.
28. Riens B, Lohaus G, Heineke D, Heldt HW: **Amino acid and sucrose content determined in the cytosolic, chloroplastic, and vacuolar compartments and in the phloem sap of spinach leaves.** *Plant Physiol* 1991, **97**:227–233.
29. Downing N: **Short communications: measurements of the osmotic concentrations of stylet sap, haemolymph and honeydew from an aphid under osmotic stress.** *J Exp Biol* 1978, **77**:247–250.
30. Pritchard J: **Aphid stylectomy reveals an osmotic step between sieve tube and cortical cells in barley roots.** *J Exp Bot* 1996, **47**:1519–1524.
31. Palmer LJ, Palmer LT, Rutzke MA, Graham RD, Stangoulis JCR: **Nutrient variability in phloem: examining changes in K, Mg, Zn and Fe concentration during grain loading in common wheat (Triticum aestivum L.).** *Physiol Plant* 2014. doi:10.1111/ppl.12211.
32. Serra Bonvehi J, Orantes Bermejo FJ: **Detection of adulterated commercial Spanish beeswax.** *Food Chem* 2012, **132**:642–648.
33. Lohaus G, Hussmann M, Pennewiss K, Schneider H, Zhu JJ, Sattelmacher B: **Solute balance of a maize (Zea mays L.) source leaf as affected by salt treatment with special emphasis on phloem retranslocation and ion leaching.** *J Exp Bot* 2000, **51**:1721–1732.
34. Fisher D: **Changes in the concentration and composition of peduncle sieve tube sap during grain filling in normal and phosphate-deficient wheat plants.** *Funct Plant Biol* 1987, **14**:147–156.
35. Liu DD, Chao WM, Turgeon R: **Transport of sucrose, not hexose, in the phloem.** *J Exp Bot* 2012, **63**:4315–4320.
36. Amiard V, Morvan-Bertrand A, Cliquet J-B, Jean-Pierre B, Huault C, Sandström JP, Prud'homme M-P: **Carbohydrate and amino acid composition in phloem sap of Lolium perenne L. before and after defoliation.** *Can J Bot* 2004, **82**:1594–1601.
37. Stephan UW, Scholz G: **Nicotianamine: mediator of transport of iron and heavy metals in the phloem?** *Physiol Plant* 1993, **88**:522–529.

38. Schmidke I, Stephan UW: **Transport of metal micronutrients in the phloem of castor bean (Ricinus communis) seedlings.** *Physiol Plant* 1995, **95**:147–153.

39. Kato M, Ishikawa S, Inagaki K, Chiba K, Hayashi H, Yanagisawa S, Yoneyama T: **Possible chemical forms of cadmium and varietal differences in cadmium concentrations in the phloem sap of rice plants (Oryza sativa L.).** *Soil Sci Plant Nutr* 2010, **56**:839–847.

40. Mendoza-Cózatl DG, Butko E, Springer F, Torpey JW, Komives EA, Kehr J, Schroeder JI: **Identification of high levels of phytochelatins, glutathione and cadmium in the phloem sap of Brassica napus. A role for thiol-peptides in the long-distance transport of cadmium and the effect of cadmium on iron translocation.** *Plant J* 2008, **54**:249–259.

41. Stephan U, Schmidke I, Pich A: **Phloem translocation of Fe, Cu, Mn, and Zn in Ricinus seedlings in relation to the concentrations of nicotianamine, an endogenous chelator of divalent metal ions, in different seedling parts.** *Plant Soil* 1994, **165**:181–188.

42. Stangoulis J, Tate M, Graham R, Bucknall M, Palmer L, Boughton B, Reid R: **The mechanism of boron mobility in wheat and canola phloem.** *Plant Physiol* 2010, **153**:876–881.

43. Stangoulis JCR, Brown PH, Bellaloui N, Reid RJ, Graham RD: **The efficiency of boron utilisation in canola.** *Funct Plant Biol* 2001, **28**:1109–1114.

44. Aizat WM, Dias DA, Stangoulis JCR, Able JA, Roessner U, Able AJ: **Metabolomics of capsicum ripening reveals modification of the ethylene related-pathway and carbon metabolism.** *Postharvest Biol Technol* 2014, **89**:19–31.

45. Downing N, Unwin DM: **A new method for cutting the mouth-parts of feeding aphids, and for collecting plant sap.** *Physiol Entomol* 1977, **2**:275–277.

46. Temmerman L, Livera AMD, Bowne JB, Sheedy JR, Callahan DL, Nahid A, Souza DPD, Schoofs L, Tull DL, McConville MJ, Roessner U, Wentworth JM: **Cross-platform urine metabolomics of experimental hyperglycemia in type 2 diabetes.** *J Diabetes Metab* 2012, **S6**:002.

47. Field A: *Discovering Statistics Using SPSS.* London, UK: SAGE Publications; 2007.

9

Improved methodology for assaying brassinosteroids in plant tissues using magnetic hydrophilic material for both extraction and derivatization

Jun Ding[1,2†], Jian-Hong Wu[3†], Jiu-Feng Liu[1], Bi-Feng Yuan[1] and Yu-Qi Feng[1*]

Abstract

Background: Brassinosteriods (BRs) are a group of important phytohormones that have major effects on plant growth and development. To fully elucidate the function of BRs, a sensitive BR assay is required. However, most of the previously reported methods are tedious and time-consuming due to multiple pretreatment steps. Therefore, it is of great significance to develop a method to increase the throughput and detection sensitivity of BR analysis.

Results: We established a novel analytical method of BRs based on magnetic solid phase extraction (MSPE) combined with in situ derivatization (ISD). TiO_2-coated magnetic hollow mesoporous silica spere(TiO_2/MHMSS) was served as a double identity- a microextraction sorbent and "microreactor" for the capture and derivatization of BRs in sequence. BRs were first extracted onto TiO_2/MHMSS through hydrophilic interaction. The BR-adsorbed TiO_2/MHMSS was then employed as a "microreactor" for the derivatization of BRs with 4-(N,N-dimethyamino)phenylboronic acid (DMAPBA). The MSPE-ISD method was simple and fast, which could be accomplished within 10 min. Furthermore, the derivatives of BRs showed better MS response because they were incorporated with tertiary amino groups. Uniquely, endogenous BRs were detected in only 100 mg fresh weight plant tissue.

Conclusion: Our proposed MSPE-ISD method for the determination of endogenous BRs is rapid and sensitive. It can be applied to the analysis of endogenous BRs in 100 mg fresh plant tissue (*Brassica napus* L. (*B. napus* L)). The proposed strategy for plant sample preparation may be extended to develop analytical methods for determination of a wide range of analytes with poor MS response in other complex sample matrices.

Keyword: Brassinosteroid, Magnetic solid phase extraction, In situ derivatization, Hydrophilic interaction, Liquid chromatography-mass spectrometry

Background

Brassinosteroids (BRs), a class of polyhydroxy steroid phytohormones, play critical roles in the growth and development of plants, including the germination of seeds, rhizogenesis, flowering, senescence, photomorphogenesis etc. [1,2]. Extensive studies also suggest that BRs can synergize with other phytohormones to function in the processes of reproduction, embryogenesis, hypocotal elongation and so on [3-5]. The investigations of BR functions rely heavily on monitoring of the temporal and spatial variation of the BR concentrations. Therefore, an effective BR analytical method is necessary.

In recent years, the technological breakthroughs in instrumentation have improved the selectivity and sensitivity of analytical methods with the advent of high-performance liquid chromatography-tandem mass spectrometry (HPLC-MS/MS) [6]. However, for the analysis of plant samples, the compromised sensitivity is frequently caused by the signal suppression from complex sample matrix during mass spectrometry (MS) analysis. Moreover, the trace amounts of BRs in complex plant matrixes and their

* Correspondence: yqfeng@whu.edu.cn

†Equal contributors

[1]Key Laboratory of Analytical Chemistry for Biology and Medicine (Ministry of Education), Department of Chemistry, Wuhan University, Wuhan 430072, China

Full list of author information is available at the end of the article

inherently low MS response makes reliable qualitative and quantitative analysis of BRs challenging. The current pretreatment methods of BRs to remove the sample matrix required the combination of two or more sample preparation processes, including SPE [7,8], LLE [9], MSPE [10] etc. Besides, BRs lack ionization groups, thus the MS responses of BRs are far from satisfaction. To improve MS responses of BRs, a pre-column derivatization process was employed to incorporate ionized moieties into BRs before LC-MS analysis [7,11]. Obviously, the multiple sample preparation processes with the following derivatization procedure made BR analysis labor-consuming and time-consuming. Therefore, it is essential to develop a fast and sensitive BR assay.

In situ derivatization (ISD) is a relatively new technique, which can couple with multiple sample preparation methods to simplify the connection of the extraction and derivatization [12,13]. So far, single-drop microextraction (SDME) [14,15], solid-phase extraction (SPE) [16,17], hollow fiber liquid–liquid–liquid extraction (HF-LLLME) [18,19], polymer monolith microextraction (PMME) [20], solid phase microextraction (SPME) [21,22] and stir bar sorptive extraction (SBSE) [23], have effectively combined with ISD for the analysis of a variety of compounds. Herein, the extraction media served as a double identity—an extractant and microreactor. After analytes were loaded onto the extraction media, the chemical derivatization reaction can occur directly on the surface of the sorbents by adding derivatization reagent. In the process, a redundant desorption/re-dissolution step was prevented and the errors associated with the multistep sample preparation process were reduced. Most importantly, the enrichment of target analytes in the extractant would benefit the fast derivatization reaction due to the local relatively high concentration. Despite of the advantages of ISD, considerable pretreatment time was still required to achieve satisfactory extraction efficiency due to the inherent limitation of the current extraction methods themselves.

Magnetic solid phase extraction (MSPE), a new mode of extraction technique based on magnetic or magnetizable nanoparticles, has been widely used in sample preparation in recent years [24-27]. The sorbents can be dispersed uniformly in sample solution by vortex, instead of being packed into the SPE cartridge. Moreover, magnetic sorbents can be readily agglomerated and re-dispersed in a sample solution by the application and removal of an external magnetic field, which makes the phase separation very convenient. From the view of mass transfer, the dispersive extraction mode also provides a large contact area between the extractant phase and sample solution, which is favorable for the mass transfer of analytes and therefore results in shorter extraction time [28]. In virtue of these properties, MSPE coupled with

ISD is a promising technique for the fast and sensitive pretreatment of BRs.

BRs contain multiple polyhydroxy groups and thus exhibit hydrophilic property. In light of this property, hydrophilic magnetic materials were chosen as sorbents, and a fast and convenient MSPE-ISD method based on hydrophilic interaction was developed for the determination of endogenous BRs in plant tissues. By employing hydrophilic magnetic material as both a microextraction sorbent and "microreactor", the MSPE-ISD method integrates extraction and derivatization together, which largely simplifies the analytical process. First, BRs were extracted onto the surface of a magnetic sorbent through hydrophilic interaction in the acetonitrile extract of the plant sample; in the meantime, hydrophobic interferents from the extract were removed. Subsequently, magnetic sorbents served as a "microreactor", where the captured BRs were rapidly and efficiently derivatized with 4-dimethylphenyl boronic acid (DMPBA). The BR derivatives could be desorbed from the sorbents with water as the desorption solvent for further UPLC-ESI-MS/MS analysis. The proposed MSPE-ISD procedure could be accomplished within 10 min, and endogenous BRs could be detected in 100 mg fresh weight plant tissues.

Results and discussion
Optimization of MSPE-ISD

The proposed MSPE-ISD method for the analysis of BRs utilized hydrophilic interaction to fulfill both the extraction and ISD process. In hydrophilic interaction chromatography (HILIC), the high content of acetonitrile is normally used as the sampling solution. It was already reported that the extraction efficiencies of BRs in acetonitrile were satisfactory [29], which provides an opportunity to separate them from the hydrophobic interferents based on hydrophilic interaction. Moreover, the cis-diol groups in the BR structure can react efficiently with boronate derivatization reagent [10,30]. Based on these backgrounds, a series of magnetic hydrophilic materials were chosen as sorbents, and DMAPBA was selected as the derivatization reagent. Several parameters affecting the extraction and derivatization efficiencies were investigated.

We first examined the performance of different types of sorbents on the extraction of BRs. Plant extract with acetonitrile contains large amounts of interfering matrix, such as pigments and hydrophobic compounds, which may jeopardize the following in situ derivatization and UPLC-ESI-MS/MS analysis. As BRs exhibit relative hydrophilic properties due to their multiple hydroxyl groups, we selected magnetic hydrophilic sorbents (TiO_2/MHMSS, Fe_3O_4, Fe_3O_4@mSiO_2, Fe_3O_4/SiO_2, Fe_3O_4/TiO_2) for the extraction of BRs and the removal of hydrophobic matrix and pigments in the hydrophilic solid-phase extraction mode. Comparison of the performance of these

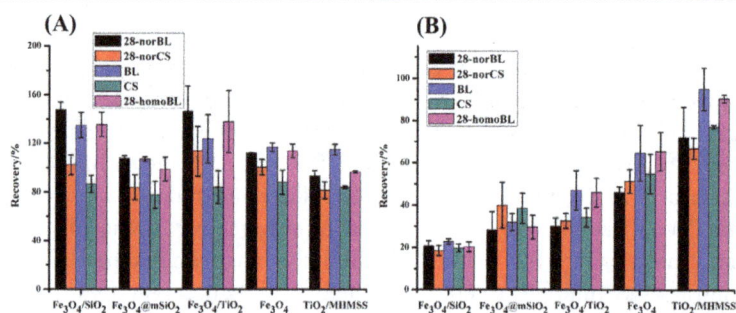

Figure 1 Effect of different magnetic sorbents on the BR recoveries in acetonitrile solution sample (n = 2) (A) and plant extract (n = 3) (B). BRs were spiked in the sampling solution at 1 ng/mL each.

hydrophilic sorbents on the extraction of BRs was conducted by examining the recoveries of BRs in acetonitrile or in plant extract. As shown in Figure 1, the five types of hydrophilic sorbents exhibited no significant difference in the recoveries of BRs spiked in acetonitrile (Figure 1A), whereas remarkable differences in the recoveries of BRs spiked in the plant extract were observed (Figure 1B). The extraction efficiencies for BRs are in the order of TiO_2/MHMSS > Fe_3O_4 > Fe_3O_4/TiO_2 > Fe_3O_4@mSiO_2 > Fe_3O_4/SiO_2, which may be ascribed to their differences in hydrophilic properties and the number of adsorption sites for BRs and polar compounds from the plant matrix. For BRs spiked in acetonitrile, all the five sorbent exhibited great extraction efficiencies towards BRs due to no matrix effect. However, for the plant extract, the massive matrix interferents competed with BRs for the adsorption sites, leading to low extraction efficiencies of sorbents to different extent. To assure sufficient recoveries of BRs, TiO_2/MHMSS was chosen as the hydrophilic sorbent for the following experiments.

We further optimized the sampling solution in MSPE-ISD. Volume percentages of acetonitrile in the range of 0 to 100% were investigated. As shown in Figure 2A, the highest recoveries were obtained with 100% acetonitrile; once, the proportion of water was greater than 5% (v/v), the recoveries dropped dramatically, suggesting that the

extraction efficiencies of BRs by hydrophilic sorbent of TiO_2/MHMSS are strongly dependent on the acetonitrile content (more than 95%) in the sampling solution.

To obtain high extraction efficiencies of BRs by TiO_2/MHMSS, the water in the extraction solution should be removed as much as possible. However, in the process of the grinding of plant tissues in liquid nitrogen and acetonitrile extraction, water was inevitably brought into the plant extract. Here, we evaluated the dehydration strategies by either direct evaporation of plant extract and reconstitution with acetonitrile or the addition of NaCl and anhydrous $MgSO_4$ into the plant extract. As shown in Figure 2B, a better dehydration effect was obtained by direct evaporation of plant extract and reconstitution with acetonitrile. Hence, the plant extract was evaporated under a mild nitrogen atmosphere and then re-dissolved in acetonitrile for the following experiments.

Because the plant extract was very complex, DMAPBA would also react with cis-diol-containing interferents in the plant extract. Therefore, to ensure high derivatization efficiency, the DMAPBA amount was investigated in both acetonitrile solution spiked with BR standards and plant extract spiked with BR standards (1 ng for each) (Figure 3). In acetonitrile, the peak areas of the BR derivatives dropped as the molar ratios increased (Figure 3A), whereas in the plant extract, the maximal peak areas

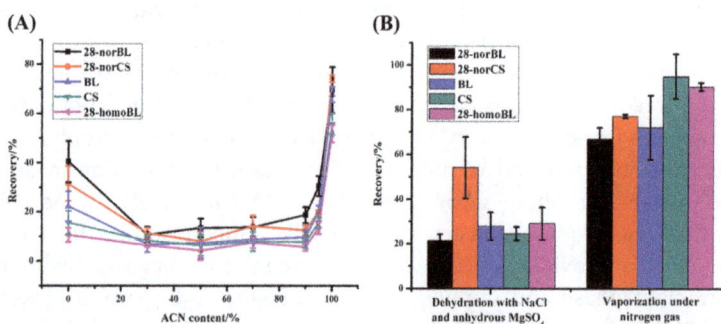

Figure 2 Effect of the water content in the sampling solution on the BR recoveries (n = 3) (A) and comparison of the effects of different dehydration strategies on plant extract (n = 3) (B). BRs were spiked in the sampling solution at 1 ng/mL each.

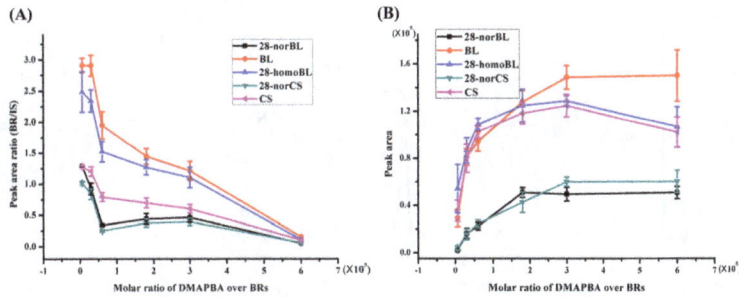

Figure 3 Effect of the DMAPBA amount on the derivatization efficiencies of BRs in acetonitrile (n = 3) (A) and in plant extract (n = 3) (B). BRs were spiked in the sampling solution at 1 ng/mL each.

appeared at molar ratios (DMAPBA molar quantity by five BR molar quantity) above 300,000 (Figure 3B). We reason that in acetonitrile solution, as the molar ratio increased, more DMAPBA would enter the LC-MS/MS system during sample injection, which may suppress the ionization efficiencies of the BR derivatives. In plant extract, the existing BR analogues and other cis-diol-containing compounds might consume large amounts of DMAPBA; therefore, a greater amount of DMAPBA was required to guarantee high derivatization efficiencies of BRs. In this regard, 500 μg DMAPBA (molar ratio 500,000/1) was selected.

TiO$_2$/MHMSS amounts were examined in the range of 10 to 100 mg. As shown in Figure 4A, the signal of the BR derivatives significantly increased as the TiO$_2$/

MHMSS amounts increased from 10 mg to 50 mg, and most of the signal of the BR derivatives remained nearly constant with greater amounts of TiO$_2$/MHMSS. Therefore, 50 mg TiO$_2$/MHMSS sorbent was used in the following experiments.

To obtain fast mass transfer between the sorbent and plant extract, the sampling, derivatization and desorption process were all performed under vortexing. Sampling time ranging from 30 seconds to 10 minutes was investigated. As shown in Figure 4B, the sampling time had no obvious effect on the extraction efficiencies; therefore, 30 seconds sampling was chosen. Similarly, the derivatization time and desorption time were also optimized (Figure 4C and D). The results showed that 30 seconds was enough for both procedures.

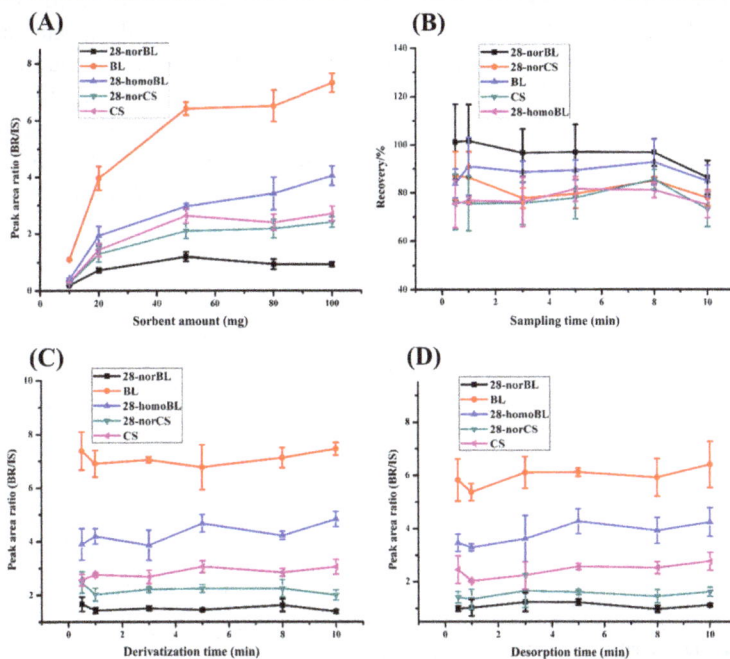

Figure 4 Effect of the sorbent amount on the extraction and derivatization efficiencies (n = 3) (A), effects of extraction time (n = 3) (B), derivatization time (n = 3) (C) and desorption time (n = 3) (D) on the extraction and derivatization efficiencies. BRs were spiked in the sampling solution at 1 ng/mL each, and DMAPBA (500 μg/mL) was added for derivatization.

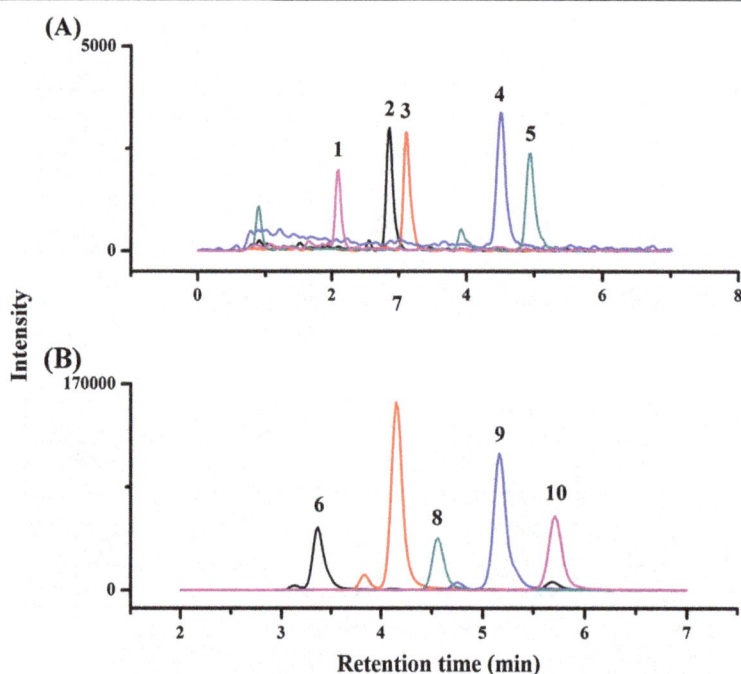

Figure 5 MRM chromatograms of the five BRs obtained without (A) or with MSPE-ISD (B). Peaks: 1. 28-norBL; 2. 28-norCS; 3. BL; 4. CS; 5. 28-homoBL; 6. 28-norBL derivative; 7. BL derivative; 8. 28-norCS derivative; 9. 28-homoBL derivative; 10. CS derivative.

On the basis of the above-described discussion, the optimal extraction conditions were as follows: 50 mg TiO_2/MHMSS as the sorbents, BRs in acetonitrile (1 mL) as the sampling solution, 500 µg/mL DMAPBA in acetonitrile (1 mL) as the derivatization solution, H_2O (0.5 mL) as the desorption solution, 30 s for the extraction, derivatization and desorption time. In the optimal conditions, the MSPE-ISD process could be accomplished within 10 minutes.

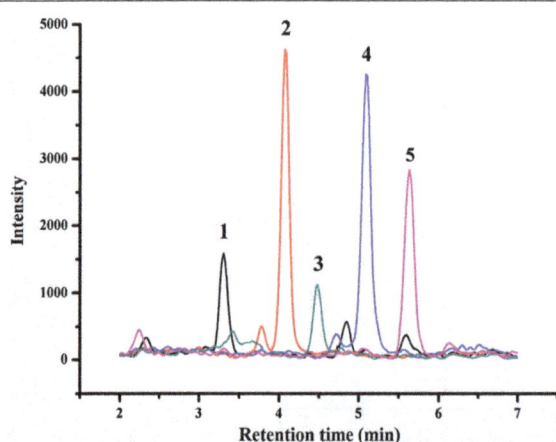

Figure 6 MRM chromatograms of the BRs in plant tissue (100 mg FW) treated with the MSPE-ISD method. The five BRs were all spiked at 0.1 ng/mL. Peaks: 1. 28-norBL; 2. BL; 3. 28-norCS; 4. 28-homoBL; 5. CS.

Sensitivity evaluation

To evaluate the performance of MSPE-ISD, we compared the detection sensitivity of BRs with or without MSPE-ISD. After treatment, equal amount of BRs or BR-derivatives were injected into LC-MS/MS system. As shown in Figure 5, significant enhancement of the peak areas of five BRs could be achieved by MSPE-ISD. Specifically, after labeled with DMAPBA by ISD method, the peak areas of the BR derivatives increased by 18-48-fold compared to that of BRs, which demonstrated the MS responses of BRs greatly increased.

Method validation

Because only two IS standards are commercially available and this is not sufficient to normalize the extraction and derivatization process, matrix-matched calibration curves were chosen as reference curves in the current

Table 1 Linearities, LODs and LOQs of the BR derivatives

Analyte	Linear range	Regression data			LODs	LOQs
	(ng/mL)	Slope	Intercept	R value	(ng/L)	(ng/L)
28-norBL	0.01-5	6.9291	0.0448	0.9867	4.86	16.20
BL	0.01-5	9.6684	0.0560	0.9981	1.94	6.48
28-homoBL	0.01-5	7.4722	0.0669	0.9923	4.49	14.97
28-norCS	0.01-5	6.0189	0.0262	0.9872	5.12	17.07
CS	0.01-5	7.5004	−0.0103	0.9992	4.23	14.08

Table 2 Accuracy and precision (intra- and inter-day) for the determination of BRs in *O. sativa* L seedlings (100 mg FW)

Analyte	Intra-day precision (RSD, %, n = 3)			Inter-day precision (RSD, %, n = 3)			Recovery (%, n = 4)		
	Low	Medium	High	Low	Medium	High	Low	Medium	High
	(0.5 ng/g)	(1 ng/g)	(10 ng/g)	(0.5 ng/g)	(1 ng/g)	(10 ng/g)	(0.5 ng/g)	(1 ng/g)	(10 ng/g)
28-norBL	7.6	12.4	0.4	12.6	14.7	5.7	115.4	109.7	111.3
BL	5.7	7.4	8.0	1.3	10.5	4.1	102.9	111.0	113.7
28-homoBL	14.6	8.3	0.0	16.0	14.0	16.3	113.0	120.9	119.7
28-norCS	10.9	7.7	3.9	8.6	10.0	14.5	110.6	94.2	96.4
CS	12.5	7.5	5.8	2.1	11.7	8.0	109.1	108.0	112.9

study. The calibration curves were constructed by plotting the analyte/IS peak area ratio versus the concentrations with triplicate measurements from 100 mg rice shoots. MRM chromatograms of the BRs in plant tissue spiked at 1 ng/g are shown in Figure 6. As shown in Table 1, satisfactory correlation coefficients were obtained with R values ranging from 0.9867 to 0.9992. Moreover, the sensitivity of the method was evaluated by examining the limit of detection (LOD) and the limit of quantification (LOQ). The LOD was defined as the lowest detectable concentration with a signal-to-noise ratio of at least 3, and the LOQ was defined as the lowest quantifiable concentration with a signal-to-noise ratio of at least 10. The LODs and LOQs were in the range of 1.94 to 5.12 ng/L and 6.48 to 17.07 ng/L, respectively.

The reproducibility and accuracy of the proposed method were evaluated by intra- and inter-day precisions and recoveries. *O. sativa* L shoot extracts were spiked with BR standards (BL, CS, 28-norBL, 28-norCS, and 28-homoBL) at three concentration levels (0.5 ng/g, 1 ng/g, and 10 ng/g). Three parallel extractions of a sample solution over 1 day gave the intra-day RSDs, and the inter-day RSDs were determined by extracting sample solutions that had been independently prepared for 3 continuous days. As shown in Table 2, acceptable precision was obtained, with RSD values below 16.3%, indicating good reproducibility of the proposed method.

The recoveries were also obtained using *O. sativa* L extracts. The endogenous concentrations of BRs in *O. sativa* L extract were calculated based on the calibration curves. The spiked BR amounts were calculated by subtracting the endogenous concentration of each BR in the extract from the total concentration of BRs. Therefore, the recoveries were obtained by comparing the concentration of measured spiked BRs with the corresponding spiked values. As shown in Table 2, the relative recoveries were in the range of 94.2% to 119.7%, demonstrating that the accuracy of the proposed method was satisfactory.

Effect of plant tissue amount on BR detection

With increased amounts of plant tissue, the endogenous BR contents also increased, which would facilitate BR detection. However, increased amounts of plant tissue may introduce more matrix interferents and therefore cause a negative impact on both extraction and detection. In this vein, an appropriate sample amount should be selected. Different amounts of plant tissue (50-500 mg) were treated by the MSPE-ISD method, and IS derivatives were

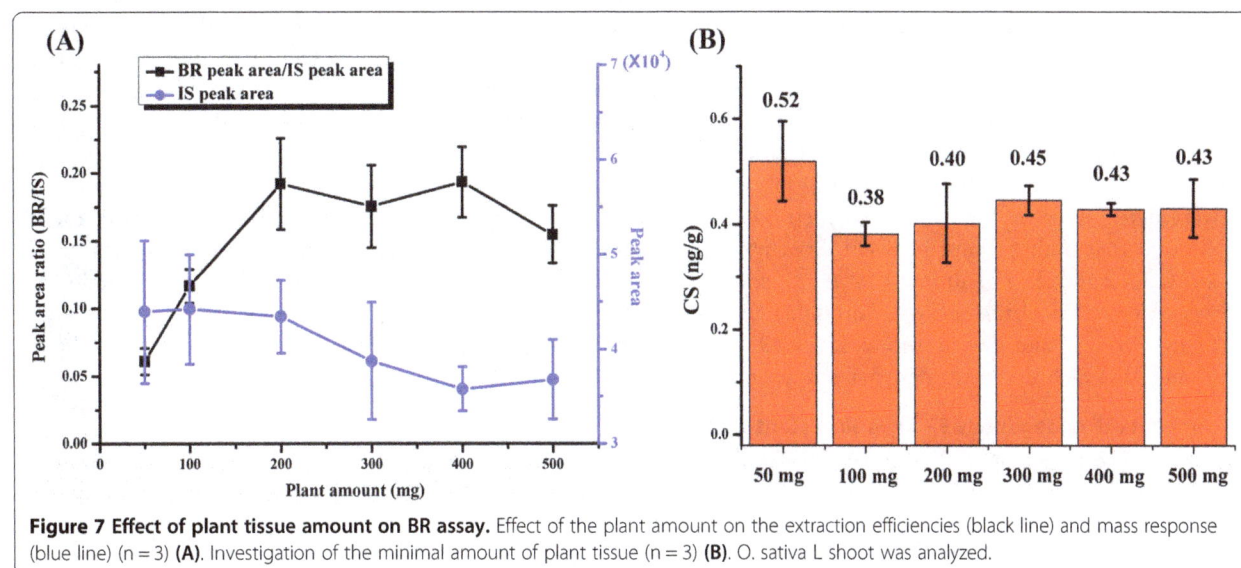

Figure 7 Effect of plant tissue amount on BR assay. Effect of the plant amount on the extraction efficiencies (black line) and mass response (blue line) (n = 3) **(A)**. Investigation of the minimal amount of plant tissue (n = 3) **(B)**. O. sativa L shoot was analyzed.

Table 3 Matrix effect of plant tissue analyzed by MSPE-ISD

Analytes	Matrix effect/%
28-norBL	85.9
BL	67.4
28-homoBL	76.0
28-norCS	93.1
CS	77.8

Matrix effect = BR/IS peak area ratio of the real sample over the BR/IS peak area ratio of a standard sample.
Plant extract (100 mg FW) and acetonitrile were both spiked with five BRs at 0.1 ng/mL.

added prior to the UPLC-ESI-MS/MS analysis (Figure 7A). When matrix effects are negligible, the peak area of the IS derivatives should keep constant with the increase of plant amount, and the ratio of BR peak area to IS derivative peak area should increase linearly with the increase of plant amount. However, the matrix effects on the extraction and detection were obviously observed when using plant samples greater than 100 mg. The matrix effect of 100 mg of plant tissue was 67.4 to 93.1%, indicating that most of the hydrophobic matrix that might have a negative effect on ESI-MS ionization of BR derivatives had been removed using 100 mg plant tissue (Table 3).

In some cases, a limited amount of plant tissue can be obtained for phytohormone analysis. To investigate the minimal amount of plant tissue required for endogenous BR detection, different amounts (from 50 to 500 mg) of *O. sativa* L shoots were used for the analysis of endogenous BRs by the MSPE-ISD method. As shown in Figure 7B, the results showed that the quantification of endogenous BRs was not affected by different amounts of *O. sativa* L shoot, but the signal-to-noise ratio of CS was near the LOQ when the amount was less than 50 mg. Therefore, 100 mg was used for the real sample analysis.

Analysis of BRs in plant tissues

The BR contents in five plant samples (the control and drought *O. sativa* L shoot, *O. sativa* L. cv. 9311-A shoot, *O. sativa* L. cv. 9311-B shoot and *Brassica napus* L. shoot) were determined by the MSPE-ISD method. The results showed that both CS and BL were detectable in *Oryza sativa* L. (control) and *Brassica napus* L. shoots, and CS was detected in *Oryza sativa* L. (drought), *Oryza sativa* L. cv. 9311-A and *Oryza sativa* L. cv. 9311-B shoots (Table 4), which demonstrates that our proposed

method is suitable for the sensitive analysis of low contents of BRs in plant tissue.

Furthermore, we designed a biological experiment to test the proposed method. BRs were reported to take part in plant photomorphogenesis [31]. In light, the related genes in BR biosynthesis pathway were inhibited, while in dark these genes got activated. To investigate the effect of light periods on the BR levels, we grew *O. sativa* L under three different light periods (all dark, 8 h light/16 h dark, 16 h light/8 h dark) and observed different growth patterns of these *O. sativa* L shoots. In dark, the seedlings showed an etiolation pattern that did not produce chlorophyll but instead elongated upwards. In light, the seedlings were all green and relatively short. The endogenous BR contents of the three samples were analyzed by our proposed method. As shown in Table 5, the BR contents showed no difference between the seedlings of the 8 h light/16 h dark and 16 h light/8 h dark conditions. Remarkably, the CS content was reduced sharply in the all dark condition, whereas 0.04 ng/g BL was observed. BL was the final product of the BR biosynthesis pathway and was reported to be the most active among all of the BRs. The quantitative results of BRs revealed that BR synthesis gene got activated, and CS was converted into BL in the absence of light, which was coincided with the reported physiological function of BRs, demonstrating the feasibility and accuracy of the proposed BR assay.

Method comparison

We summarized the representative articles published in the last four years for BR analysis using different methods in Table 6 [7,9,30,32], and the analytical time, LODs and the amount of samples were compared. The proposed MSPE-ISD-UPLC-MS/MS assay could be finished within an hour, and only 100 mg fresh weight of plant tissues were required for the quantification of endogenous BRs. Compared with the published methods, the proposed method showed significant advantages in both the sensitivity and the analysis speed.

Conclusion

In this study, we developed an MSPE-ISD method for the determination of endogenous phytohormones in plant tissues. Using TiO_2/MHMSS as both an extraction sorbent and microreactor, the extraction and derivatization processes and magnetic separation were successfully combined. The method largely simplified the sample

Table 4 Amounts of endogenous BR in various plant tissues

Analyte	*O. sativa* L. (control)	*O. sativa* L. (drought)	*B. napus* L. shoot	*O. sativa* YTA shoot	*O. sativa* YTB shoot
CS	0.09 ± 0.01	0.11 ± 0.04	0.17 ± 0.02	0.19 ± 0.02	0.26 ± 0.02
BL	0.04 ± 0.01	n.d.	0.13 ± 0.01	n.d.	n.d.

Unit: ng/g; n.d., not detectable.

Table 5 Amounts of endogenous BR in *O. sativa* L shoots under three different light conditions

Analyte	*O. sativa* L shoot with 16 h light/8 h dark	*O. sativa* L shoot with 8 h light/16 h dark	*O. sativa* L shoot with all dark
CS	0.10 ± 0.02	0.12 ± 0.01	0.04 ± 0.00
BL	n.d.	n.d.	0.04 ± 0.01

Unit: ng/g; n.d., not detectable.

preparation procedure and the BR assay can be accomplished within 1 hour. In the meantime, the MS response of BRs was significantly improved due to derivatization with 4-DMAPBA, which can benefit the quantification of BRs with a small amount of plant tissue (100 mg fresh weight in the current study). We then successfully determined the concentration of endogenous BRs in various plant tissues. The developed MSPE-ISD technique may also have potential for the determination of a wide range of analytes in other complex biological and environmental sample matrices.

Methods

Chemicals and reagents

Standard BRs and stable isotope-labeled standards (IS), including 28-norbrassinolide (28-norBL, purity > 98%), 28-norcastasterone (28-norCS, purity >98%), 28-homobrassinolde (28-homoBL, purity >95%), brassinolide (BL, purity >95%), castasterone (CS, purity >98%), $[^2H_3]BL$ and $[^2H_3]CS$, were purchased from Olchemim Ltd. (Olomouc, Czech Republic). All of the BRs standards and stable isotope-labeled standards were dissolved in acetonitrile to obtain stock solutions at the concentration of 200 ng/mL for each. Working solutions were obtained by appropriate dilution of the stock solutions.

Chromatographic grade acetonitrile was obtained from Tedia Co. (Fairfield, OH, USA). Ultrapure water was purified by a Milli-Q water purification system (Millipore, Milford, MA, USA). 4-(N,N-dimethyamino) phenylboronic acid (DMAPBA) was purchased from J&K Scientific Ltd (Beijing, China). Cetyltrimethylammonium bromide (CTAB), sodium silicate nonahydrate ($Na_2SiO_3 \cdot 9H_2O$), iron nitrate nonahydrate ($Fe(NO_3)_3 \cdot 9H_2O$), ethylene glycol

(EG), ammonium hexfluorotitanate (($NH_4)_2TiF_6$), boric acid (H_3BO_3) and ethyl acetate were all of analytical grade and supplied by Sinopharm Chemical Reagent Co., Ltd (Shanghai, China). Titania spheres (Titansphere, 5 μm) were purchased from GL Sciences Inc. (Tokyo, Japan). Silica spheres (SiO_2, 200-300 mesh) were obtained from Qingdao Haiyang Chemical Co., Ltd (Qingdao, China).

Plant materials

Nine types of plant leaves, including rice (*Oryza sativa* L. (*O. sativa* L)) and rape (*B. napus* L), were analyzed in this study. Three-month-old wild-type B. napus L leaves were harvested from the ground. Two rice mutant shoots (*Oryza sativa* ssp. *Indica* cv. YueTai A (YTA) (Sterile Lines) (*O. sativa* YTA) and *Oryza sativa* ssp. *indica* cv. YueTai B (maintainer line) (*O. sativa* YTB)) were grown in the field for 3 months and harvested. Wild-type *O. sativa* L shoots, under three different light periods (all dark, 8 h light/16 h dark, 16 h light/8 h dark), were grown in a cultivation room at 25°C (night) and 30°C (day) for 2 weeks. The drought and control groups of *O. sativa* L were both germinated and grown in the cultivation room at 25°C (night) and 30°C (day) for 2 weeks. The seedlings grown without water were called the drought group, and the seedlings which were watered on time were called the control group. All plant materials were immediately frozen in liquid nitrogen after harvest and were then stored at -80°C.

Preparation of hydrophilic magnetic sorbents

TiO_2-coated magnetic hollow mesoporous silica spheres (MHMSS) were prepared according to a previously reported method with minor modification [33]. Briefly, CTAB (19.6 g) and $Na_2SiO_3 \cdot 9H_2O$ (23.2 g) were dissolved in water (337 mL) to form a clear solution at 30°C. Then, ethyl acetate (35 mL) was quickly added, followed by vigorous stirring for 30 seconds. After standing at 30°C for 5 hours, the mixture was refluxed at 90°C for 48 hours. Finally, the mixture was filtered and washed several times with ethanol. The filtered HMSS was dried in a vacuum oven and then calcined at 550°C for 5 hours. Magnetic nanoparticles were introduced

Table 6 Comparison of different BR analytical methods

Pretreatment method	Separation/detection	Analyte	LOD	Amount of plant tissues	Analysis time
LLE-MSPE-derivatization [9]	LC-FLD	24-epiBL	0.12 ng	50 g	More than 3 hours
SPE-ultrafiltration-SPE-derivatization [7]	Online trapping-UPLC-MS/MS	28-epihomoBL	0.2 pg	400 mg	7 hours
MCX SPE-MAX SPE-derivatization [32]	UPLC-MS/MS	BL, CS, teasterone (TE), typhasterol (TY)	1.5-3.9 pg	1 g	1 day
On-line two-dimensional microscale SPE-on column derivatization-HPLC-MS/MS [30]	On-line-HPLC-MS/MS	24-epiBL, 24-epiCS, 6-deoxo-24-epiCS,TE, TY	1.4-6.6 pg	225 mg	40 minutes
MSPE coupled with ISD (this work)	UPLC-MS/MS	28-norBL, 28-norCS, 28-homoBL, BL, CS	0.1-0.3 pg	100 mg	1 hour

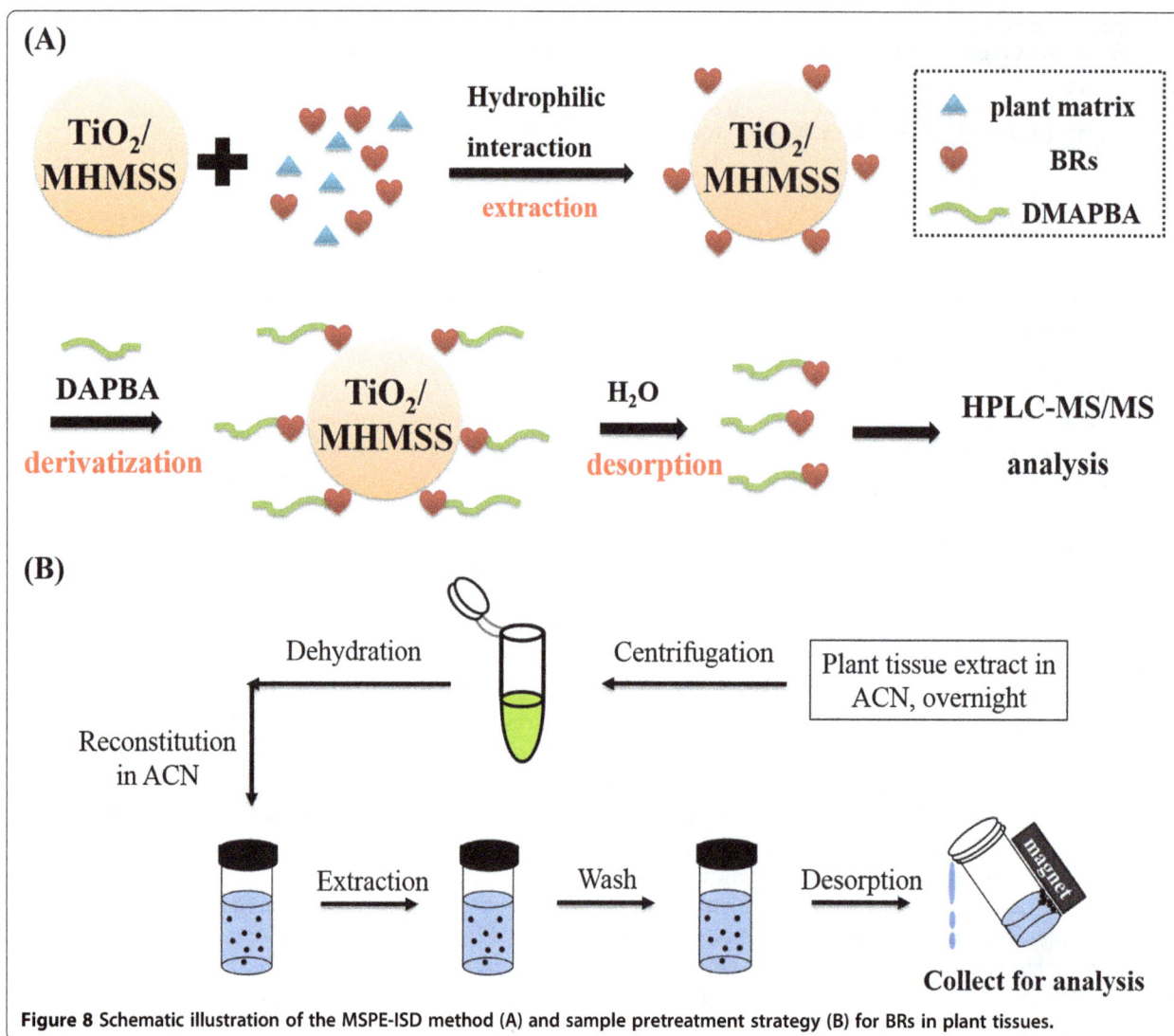

Figure 8 Schematic illustration of the MSPE-ISD method (A) and sample pretreatment strategy (B) for BRs in plant tissues.

into the hollow core of HMSS through a vacuum impregnation of $Fe(NO_3)_3$. HMSS (2.4 g) was soaked in $Fe(NO_3)_3 \cdot 9\,H_2O$ aqueous solution (24 g/L, 200 mL). The suspension was heated in a microwave oven until boiling and then cooled in an ice water mixture, allowing the Fe^{3+} to enter the hollow core of the HMSS. The process was repeated several times until the water completely dried. Subsequently, the product was washed with 10 mL ethanol twice and dried again. The product was impregnated with 1 mL ethylene glycol up to incipient wetness. The impregnated sample was then subjected to heat treatment under nitrogen atmosphere at 450°C for

Table 7 Optimized MRM parameters of seven BR derivatives by UPLC-ESI-MS/MS

Analyte	Quantification					Confirmation				
	Q1 (m/z)	Q3 (m/z)	Q1 pre bias/V	CE	Q3 pre bias/V	Q1(m/z)	Q3(m/z)	Q1 pre bias/V	CE	Q3 pre bias/V
28-norBL	596.4	190.1	−30	−44	−20	596.4	246.1	−30	−34	−17
BL	610.4	190.2	−32	−41	−13	610.4	122.1	−32	−40	−24
28-homoBL	624.4	190.2	−32	−41	−20	624.4	418.0	−32	−39	−23
28-norCS	580.4	190.1	−30	−49	−21	580.4	562.4	−30	−30	−28
CS	594.4	190.1	−30	−44	−20	594.4	576.4	−30	−32	−22
^2H_3BL	613.4	190.4	−34	−51	−13	613.4	345.4	−34	−38	−16
^2H_3CS	597.4	194.4	−32	−46	−13	597.4	579.6	−32	−32	−22

2 hours. Finally, TiO_2 was loaded onto the obtained MHMSS through the liquid phase deposition method. MHMSS (2.0 g) was added into a solution (200 mL) containing 0.1 M $(NH_4)_2TiF_6$ and 0.3 M H_3BO_3 in a PTFE container. After keeping under vacuum conditions for 1 h, the mixture was heated at 35°C for 12 h under continuous shaking. The resulting composite was washed with water thoroughly and dried at 60°C in a vacuum oven for 6 h. The resultant TiO_2/MHMSS was obtained by heat treatment under nitrogen up to 300°C at the rate of 1 K/min and was then kept at 300°C for 2 h.

Nano-scale Fe_3O_4 was prepared through the solvothermal method according to a previously reported method [34]. $FeCl_3 \cdot 6H_2O$ (5.0 g) was dissolved in EG (100 mL) to form a clear solution. Then, NaAc (15.0 g) and ED (50 mL) were added to the solution. After vigorously stirring for 30 min, the homogeneous mixture was sealed in a Teflon-lined stainless-steel autoclave and was heated to 200°C for 8 hours. The product was magnetically collected and washed with water/ethanol several times and vacuum-dried at 60°C for 6 h.

MSPE-ISD procedure for the determination of BRs in plant tissue

The schematic illustrations of MSPE-ISD (A) and the sample pretreatment strategy (B) are depicted in Figure 8. Plant tissue (100 mg fresh weight (FW)) was smashed with a mortar and pestle in liquid nitrogen. The powdered sample was extracted at −20°C overnight with acetonitrile (1 mL) containing $[^2H_3]$BL and $[^2H_3]$CS (0.4 ng each) as IS for quantification. After centrifugation at 3,500 g for 10 min, the supernatant was collected in a 1.5-mL vial followed by evaporation to dryness under a mild nitrogen gas stream. The residue was re-dissolved with acetonitrile (1 mL) for the following MSPE-ISD process.

TiO_2/MHMSS (50 mg) was added to a 15-mL glass vial and activated with acetonitrile before use. Subsequently, the aforementioned plant extract (1 mL) was added into the vial and vortexed vigorously for 30 seconds to form a homogenous dispersive solution. The supernatant was separated and discarded by applying a magnet. Acetonitrile (1 mL) was added to wash the residual matrix interferences on the surface with 30 seconds of vortexing and then disposed of. The washing process was repeated twice. Subsequently, DMAPBA-acetonitrile solution (500 μg/mL, 1 mL) was added to the vial for ISD by vortexing for 30 seconds. Finally, water (0.5 mL) was added to the mixture solution to elute BR derivatives from the sorbents by 30 seconds of vortexing. The desorption solution was magnetically separated and evaporated to dryness under a mild nitrogen gas flow at 35°C. The residue was dissolved in acetonitrile/H_2O (50 μL, 1/1 v/v), and then 20 μL was used for the analysis by UPLC-ESI-MS/MS.

UPLC-ESI-MS/MS analysis

The mass spectrometry analysis was performed on a UPLC-ESI (+)-MS/MS system consisting of a Shimadzu LC-30AD HPLC system (Tokyo, Japan) with two 30AD pumps, an SIL-30AC auto sampler, a CTO-30A thermostat column compartment, a DGU-20A$_{5R}$ degasser, and a Shimadzu MS-8040 mass spectrometer (Tokyo, Japan) with an electrospray ionization source (Turbo Ionspray). The separation of BRs was achieved on a Shim-pack ODS column (75 × 2.0 mm id, 1.6 μm, Shimadzu, Tokyo, Japan). The column oven temperature was set at 40°C. Mobile phases A and B were 0.1% formic acid in water and acetonitrile, respectively. An isocratic elution of 85% B at 0.2 mL/min for 7 minutes was employed. The injection volume was 20 μL.

All BRs were quantified by multiple reaction monitoring (MRM) in the positive mode. The optimal ESI source conditions were as follows: DL temperature 250°C, heat block temperature 400°C, nebulizing gas 3 L/min and drying gas 15 L/min. The MRM mass spectrometric parameters are summarized in Table 7. Data were acquired by Labsolutions software (version 5.53 sp2, Shimadzu, Tokyo, Japan).

Competing interests
The authors declare that they have no competing interests.

Authors' contributions
JD, JHW and YQF conceived and designed the method. JD, JHW and JFL carried out the experiments. JD, JHW and YQF wrote the manuscript. BFY and YQF revised the manuscript. All authors read and approved the final manuscript.

Acknowledgments
The authors are thankful for the financial support from the National Natural Science Foundation of China (91217309, 91017013), and the Fundamental Research Funds for the Central Universities.

Author details
[1]Key Laboratory of Analytical Chemistry for Biology and Medicine (Ministry of Education), Department of Chemistry, Wuhan University, Wuhan 430072, China. [2]Chinese Acad Sci, Key Lab Plant Germplasm Enhancement & Specialty A, Wuhan Bot Garden, Wuhan 430074, China. [3]College of Chemical Engineering, Wuhan Textile University, Wuhan 430200, China.

References
1. Bajguz A, Hayat S: Effects of brassinosteroids on the plant responses to environmental stresses. *Plant Physiol Biochem* 2009, **47**(1):1–8.
2. Krishna P: Brassinosteroid-mediated stress responses. *J Plant Growth Regul* 2003, **22**(4):289–297.
3. Stephen D, Christain SH: Hormone signalling crosstalk in plant growth regulation. *Curr Biol* 2011, **21**(9):R365–R373.
4. Bishop GJ, Yokota T: Plants steroid hormones, brassinosteroids: current highlights of molecular aspects on their synthesis/metabolism, transport, perception and response. *Plant Cell Physiol* 2001, **42**(2):114–120.
5. Robert-Seilaniantz A, Grant M, Jones JDG: Hormone Crosstalk in Plant Disease and Defense: More Than Just JASMONATE-SALICYLATE Antagonism. In *Annual Review of Phytopathology, Vol 49*. Edited by VanAlfen NK, Bruening G, Leach JE. 317–343. vol. 49.
6. Nováková L, Vlčková H: A review of current trends and advances in modern bio-analytical methods: chromatoraphy and sample preparation. *Anal Chim Acta* 2009, **656**(1–2):8–35.

7. Huo F, Wang X, Han Y, Bai Y, Zhang W, Yuan H, Liu H: **A new derivatization approach for the rapid and sensitive analysis of brassinosteroids by using ultra high performance liquid chromatography-electrospray ionization triple quadrupole mass spectrometry.** *Talanta* 2012, **99**:420–425.

8. Swaczynova J, Novak O, Hauserova E, Fuksova K, Sisa M, Kohout L, Strnad M: **New techniques for the estimation of naturally occurring brassinosteroids.** *J Plant Growth Regul* 2007, **26**(1):1–14.

9. Zhang Z, Zhang Y, Tan W, Li G, Hu Y: **Preparation of styrene-co-4-vinylpyridine magnetic polymer beads by microwave irradiation for analysis of trace 24-epibrassinolide in plant samples using high performance liquid chromatography.** *J Chromatogr A* 2010, **1217**(42):6455–6461.

10. Xin PY, Yan JJ, Fan JS, Chu JF, Yan CY: **A dual role of boronate affinity in high-sensitivity detection of vicinal diol brassinosteroids from sub-gram plant tissues via UPLC-MS/MS.** *Analyst* 2013, **138**(5):1342–1345.

11. Wang X, Ma Q, Li M, Chang CL, Bai Y, Feng YQ, Liu HW: **Automated and sensitive analysis of 28-epihomobrassinolide in Arabidopsis thaliana by on-line polymer monolith microextraction coupled to liquid chromatography-mass spectrometry.** *J Chromatogr A* 2013, **1317**:121–128.

12. Atapattu SN, Rosenfeld JM: **Solid phase analytical derivatization as a sample preparation method.** *J Chromatogr A* 2013, **1296**:204–213.

13. Rosenfeld JM: **Solid-phase analytical derivatization: enhancement of sensitivity and selectivity of analysis.** *J Chromatogr A* 1999, **843**(1–2):19–27.

14. Park Y-K, Choi K, Ahmed AYBH, Alothman ZA, Chung DS: **Selective preconcentration of amino acids and peptides using single drop microextraction in-line coupled with capillary electrophoresis.** *J Chromatogr A* 2010, **1217**(20):3357–3361.

15. Sharma N, Jain A, Singh VK, Verma KK: **Solid-phase extraction combined with headspace single-drop microextraction of chlorophenols as their methyl ethers and analysis by high-performance liquid chromatography-diode array detection.** *Talanta* 2011, **83**(3):994–999.

16. Poole CF: **New trends in solid-phase extraction.** *Trac-Trends Anal Chem* 2003, **22**(6):362–373.

17. Salvador A, Moretton C, Piram A, Faure R: **On-line solid-phase extraction with on-support derivatization for high-sensitivity liquid chromatography tandem mass spectrometry of estrogens in influent/effluent of wastewater treatment plants.** *J Chromatogr A* 2007, **1145**(1–2):102–109.

18. Ito R, Kawaguchi M, Honda H, Koganei Y, Okanouchi N, Sakui N, Saito K, Nakazawa H: **Hollow-fiber-supported liquid phase microextraction with in situ derivatization and gas chromatography-mass spectrometry for determination of chlorophenols in human urine samples.** *J Chromatogr B Analyt Technol Biomed Life Sci* 2008, **872**(1–2):63–67.

19. Meng L, Liu X, Wang B, Shen G, Wang Z, Guo M: **Simultaneous derivatization and extraction of free cyanide in biological samples with home-made hollow fiber-protected headspace liquid-phase microextraction followed by capillary electrophoresis with UV detection.** *J Chromatogr B Analyt Technol Biomed Life Sci* 2009, **877**(29):3645–3651.

20. Zhang H-J, Huang J-F, Lin B, Feng Y-Q: **Polymer monolith microextraction with in situ derivatization and its application to high-performance liquid chromatography determination of hexanal and heptanal in plasma.** *J Chromatogr A* 2007, **1160**(1–2):114–119.

21. Awan MA, Fleet I, Thomas CLP: **Determination of biogenic diamines with a vaporisation derivatisation approach using solid-phase microextraction gas chromatography-mass spectrometry.** *Food Chem* 2008, **111**(2):462–468.

22. Campins-Falco P, Herraez-Hernandez R, Verdu-Andres J, Chafer-Pericas C: **On-line determination of aliphatic amines in water using in-tube solid-phase microextraction-assisted derivatisation in in-valve mode for processing large sample volumes in LC.** *Anal Bioanal Chem* 2009, **394**(2):557–565.

23. Magi E, Di Carro M, Liscio C: **Passive sampling and stir bar sorptive extraction for the determination of endocrine-disrupting compounds in water by GC-MS.** *Anal Bioanal Chem* 2010, **397**(3):1335–1345.

24. Ding J, Gao Q, Li XS, Huang W, Shi ZG, Feng YQ: **Magnetic solid-phase extraction based on magnetic carbon nanotube for the determination of estrogens in milk.** *J Sep Sci* 2011, **34**(18):2498–2504.

25. Li XS, Wu JH, Xu LD, Zhao Q, Luo YB, Yuan BF, Feng YQ: **A magnetite/oxidized carbon nanotube composite used as an adsorbent and a matrix of MALDI-TOF-MS for the determination of benzo[a]pyrene.** *Chem Commun* 2011, **47**(35):9816–9818.

26. Ding J, Gao QA, Luo D, Shi ZG, Feng YQ: **n-Octadecylphosphonic acid grafted mesoporous magnetic nanoparticle: Preparation, characterization, and application in magnetic solid-phase extraction.** *J Chromatogr A* 2010, **1217**(47):7351–7358.

27. Gao Q, Luo D, Ding J, Feng YQ: **Rapid magnetic solid-phase extraction based on magnetite/silica/poly(methacrylic acid-co-ethylene glycol dimethacrylate) composite microspheres for the determination of sulfonamide in milk samples.** *J Chromatogr A* 2010, **1217**(35):5602–5609.

28. Zhao Q, Wei F, Xiao N, Yu Q-W, Yuan B-F, Feng Y-Q: **Dispersive microextraction based on water-coated Fe3O4 followed by gas chromatography mass spectrometry for determination of 3-monochloropropane-1,2-diol in edible oils.** *J Chromatogr A* 2012, **1240**:45–51.

29. Ding J, Mao LJ, Yuan BF, Feng YQ: **A selective pretreatment method for determination of endogenous active brassinosteroids in plant tissues: double layered solid phase extraction combined with boronate affinity polymer monolith microextraction.** *Plant Methods* 2013, **9**:13.

30. Wu Q, Wu D, Shen Z, Duan C, Guan Y: **Quantification of endogenous brassinosteroids in plant by on-line two-dimensional microscale solid phase extraction-on column derivatization coupled with high performance liquid chromatography–tandem mass spectrometry.** *J Chromatogr A* 2013, **1297**:56–63.

31. Ashraf M, Akram NA, Arteca RN, Foolad MR: **The physiological, biochemical and molecular roles of Brassinosteroids and salicylic acid in plant processes and salt tolerance.** *Crit Rev Plant Sci* 2010, **29**(3):162–190.

32. Xin PY, Yan JJ, Fan JS, Chu JF, Yan CY: **An improved simplified high-sensitivity quantification method for determining brassinosteroids in different tissues of rice and Arabidopsis.** *Plant Physiol* 2013, **162**(4):2056–2066.

33. Wu JH, Li XS, Zhao Y, Gao QA, Guo L, Feng YQ: **Titania coated magnetic mesoporous hollow silica microspheres: fabrication and application to selective enrichment of phosphopeptides.** *Chem Commun* 2010, **46**(47):9031–9033.

34. Guo S, Li D, Zhang L, Li J, Wang E: **Monodisperse mesoporous superparamagnetic single-crystal magnetite nanoparticles for drug delivery.** *Biomaterials* 2009, **30**(10):1881–1889.

10

A novel method to follow meiotic progression in *Arabidopsis* using confocal microscopy and 5-ethynyl-2′-deoxyuridine labeling

Patti E Stronghill[1*], Wajma Azimi[2] and Clare A Hasenkampf[1]

Abstract

Background: Meiosis progression in the more recent past has been investigated using 5-bromo-2′-deoxyuridine (BrdU) uptake by S-phase meiocytes undergoing DNA replication. BrdU uptake is detected by reaction with BrdU antibody followed by epifluorescent microscopy examination of chromosome spreads and/or squashes. We here report using confocal microscopic examination of intact meiocytes in conjunction with the new thymidine analog 5-ethynyl-2′-deoxyuridine (EdU). The simplicity of the EdU detection coupled with confocal examination of anthers provides a more exact temporal description of meiotic prophase I progression in *Arabidopsis* and opens up the possibility of examining the coordination of microsporocyte development with the other tissues of the anther.

Results: Using our time course protocol, we have determined the duration of wild type *Arabidopsis* leptotene to be 5 h, zygotene -6 h, pachytene -10 h and a diplotene duration of approximately 1 h. We estimate G2 duration to be approximately 7 h based on the timing of the initial appearance of EdU signal in early leptotene meiocytes. In addition we have found that DNA replication in meiocytes is not done synchronously with the associated tapetal layer of cells. The EdU labeling suggests that S-phase replication of meiocyte DNA precedes the duplication of tapetal cell DNA.

Conclusions: The increased number of meiotic staging criteria that can be assessed in our confocal analysis, as compared to chromosome spreading or squashing, makes the identification of even the early and late portions of the prophase I substages attainable. This enhanced staging coupled with the ability to easily generate large data sets at hourly time points makes it possible to more exactly determine substage duration and to detect modest temporal abnormalities involving meiocyte entrance into and/or exit from leptotene, zygotene and pachytene. Confocal analysis also makes it possible to study the relationships between different cell types within the flower bud as meiosis proceeds.

Keywords: *Arabidopsis*, Meiosis, S-phase, Prophase I, Confocal microscopy, EdU labeling, Meiocyte filament, Tapetal cells, Time course, Multi-criteria meiotic staging, DNA replication

Background

S-phase is the portion of both the mitotic and meiotic cell cycle when DNA replicates. Van't Hof [1], used tritiated thymidine labeling of meristematic root tips to investigate the duration of the mitotic cycle. He reported that mitotic cycle duration was varied amongst plants and that plants with relatively long mitotic cycles also possessed longer S-phases. Some years later tritiated thymidine labeling of S-phase meiocytes during DNA replication was used to investigate substage duration in plant pollen mother cells from wheat (*Triticum aestivum*), and rye (*Secale cereale*) [2]. Systematic studies have revealed that the S-phase that precedes meiosis is longer than that preceding mitosis for a given species [3]. Since then tritiated thymidine labeling has been replaced by the introduction of thymidine-based analogs into the replicating DNA of S-phase meiocytes. The first nucleoside analog to be used was 5-bromo-2′-deoxyuridine; this labeling was followed by the immuno-detection of BrdU and epifluorescent imaging. The timing of meiosis progression in *Arabidopsis* was first examined using chromosome spreading and BrdU

---

* Correspondence: stronghill@utsc.utoronto.ca
[1]Department of Biology, University of Toronto, 1265 Military Trail, Scarborough, Canada
Full list of author information is available at the end of the article

uptake into S-phase male meiocytes [4-6]. The duration of prophase I substages were determined as follows; leptotene 6.0 h, zygotene/pachytene 15.3 h, diplotene to tetrad 2.7 h [5]. Subsequently, using the percentage of total cells found in the two different stages, the individual durations of zygotene and pachytene were estimated to be 4.8 h and 10 h respectively [7].

Recently a new thymidine analog 5-ethynyl-2′-deoxyuridine (EdU), has been developed by Invitrogen (California, USA). EdU labeling of DNA is detected with a fluorophore tagged azide that forms a covalent bond with the terminal acetylene group on the ethynyl component of EdU. This new technology has been extensively used in animals to study cell proliferation [8-10]. EdU incorporation has been used in conjunction with dispersed *Arabidopsis* chromosomes to determine the relative time course of the appearance of several key meiotic proteins [11]. We have developed a protocol whereby EdU label was taken up into *Arabidopsis* inflorescences by the vasculature, followed by confocal analysis of intact filaments of meiotic cells surrounded by a layer of tapetal cells. Because we used these intact filaments we were able to use features of both types of cells to define each meiotic substage and determine the duration of each substage with high precision.

Results and discussion
Precise estimates of all meiotic substage durations
Confocal microscopy permits multi-criteria staging
We stained our preparations with propidium iodide; it stains both DNA and RNA. Thus we can visualize chromosomes and the RNA-rich nucleoli. Chromosome spreads (using acid or detergent disruption of cell wall/membrane) and chromosome squashes (using physical pressure disruption of cell wall/membrane) both rely mainly on chromosome 'thickness' for meiotic stage identification [12]. This 'thickness' relates to whether homologous chromosomes are paired or not and also relates to the increasing degree of chromosome condensation that occurs as meiosis progresses [6]. Typically, spreading and to a lesser degree squashing lead to the undesirable liberation of the nucleolus [11,13]. With our style of preparation, chromosome thickness remains an important staging tool but we also have nucleoli size, shape, and position within the meiocyte nuclei. the tapetal cell nuclei characteristics and callose deposition between meiocytes as additional diagnostic wfeatures. Our confocal meiocyte preparations either completely or partially extrude meiotic filaments from anthers thus leaving meiocytes intact and nucleoli undisturbed (Figure 1). Both the nucleoli's shape and position within the nucleus change as meiosis proceeds; this provides valuable staging information. Partial and non-extruded filaments maintain their association with the tapetal cell layer surrounding them. Tapetal cell nuclei are mainly

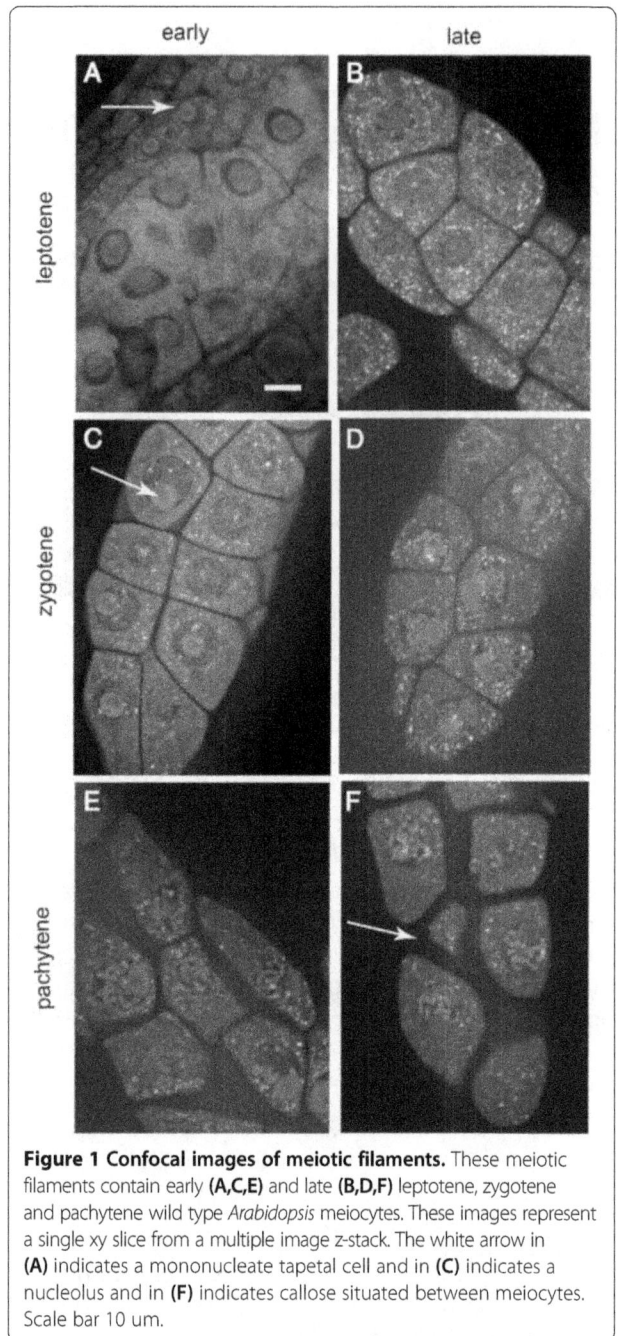

Figure 1 Confocal images of meiotic filaments. These meiotic filaments contain early **(A,C,E)** and late **(B,D,F)** leptotene, zygotene and pachytene wild type *Arabidopsis* meiocytes. These images represent a single xy slice from a multiple image z-stack. The white arrow in **(A)** indicates a mononucleate tapetal cell and in **(C)** indicates a nucleolus and in **(F)** indicates callose situated between meiocytes. Scale bar 10 um.

mononucleate in leptotene, a mixture in zygotene and binucleate in pachytene. Callose deposition between meiocytes (Figure 1) becomes thicker as prophase I progresses. Collectively these criteria make very precise meiotic staging possible. Figure 1 provides examples of meiotic filaments at both the early and late stages of leptotene, zygotene and pachytene. The criteria used to stage leptotene, zygotene, pachytene, diplotene and diakinesis meiocytes are summarized in Table 1. Using multiple staging criteria that do not rely on 'pairing status' is particularly important in meiotic mutant with

Table 1 Meiotic staging criteria for wild type *Arabidopsis thaliana* prophase I substages

Meiotic staging criteria	Sub-stages of prophase I				
	Leptotene	Zygotene	Pachytene	Diplotene	Diakinesis
Tapetal cell nuclei	Mononuclear	Mononuclear (early) Few binuclear (mid) Many binuclear (late)	Binuclear	Binuclear	Binuclear
Callose wall thickness	Very thin	Thin	Thick	Thicker	Thickest
Nucleolus position	Central (early) Peripheral (late)	Peripheral	Peripheral (early) Peri-central (late)	Central	Central
Nucleolus shape	Round	Round (early) flattened (late)	Flattened (early) round (late)	Round	Round
Chromosome distribution in the nucleus	Circular and takes up entire nuclear space	Crescent shape in one hemisphere of nuclear space	Roughly circular but does not take up entire nuclear space	Takes up more of nuclear space than in pachytene	Five distinct bivalents distributed in nuclear space
Chromosome thickness	Thin	Mixture of thin and thick	Thick	Mixture of thin and thick (early) Thick (late)	Very thick

pairing disruptions. By using multiple criteria, it should be possible to determine if the increased duration of a particular meiotic substage in a mutant meiocyte is due to premature entry, slow progression or both.

Hourly time point sampling, large sample size and precise staging increases accuracy in the determination of meiotic stage duration

In our time course analysis inflorescences incorporate EdU, and then are collected and fixed hourly. Subsequent preparation of meiocytes for detection of EdU labeling, using confocal microscopy, is relatively simple. This preparation produces a large number of analyzable nuclei as buds from several inflorescences, from the same time point, can be labeled together with EdU and subsequently prepared and viewed on the same slide. Examples of EdU labeling of wild type prophase I meiocytes (counterstained with propidium iodide) are shown in Figure 2. EdU, a thymidine analog stains only DNA whereas propidium iodide (PI) stains DNA, RNA and the cell wall. PI stains

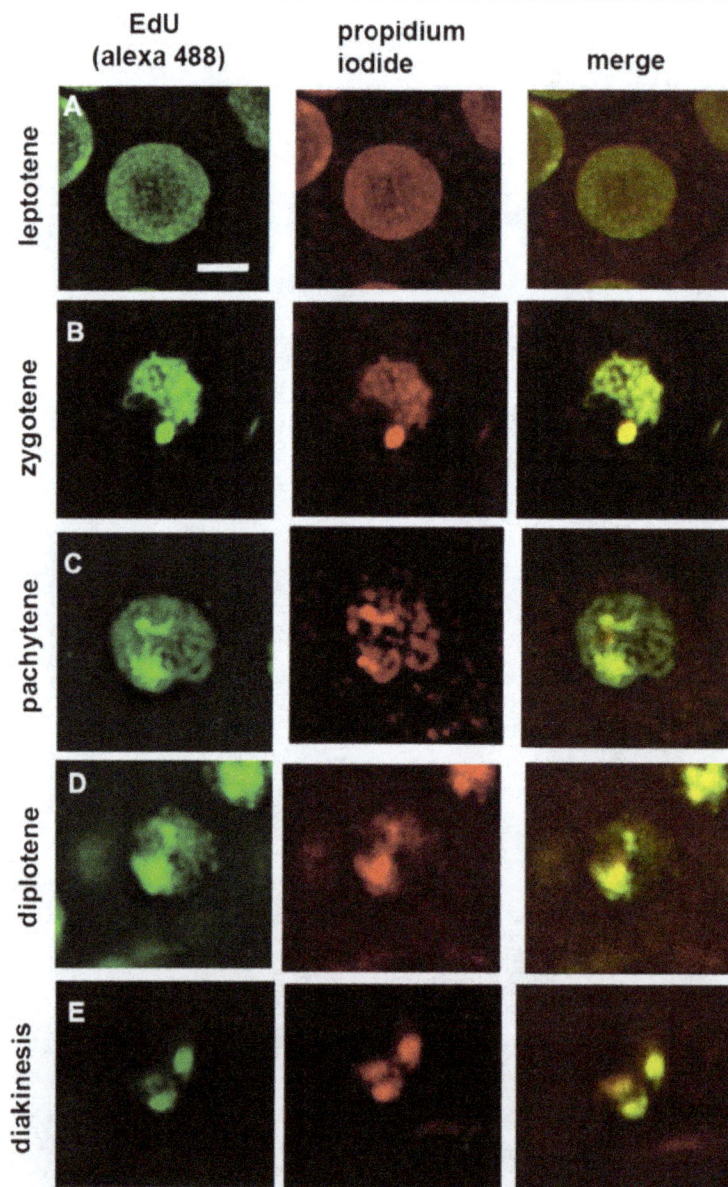

Figure 2 Confocal images of EdU labeled meiocytes. Detection of 5- ethynyl-2'-deoxyuridine(EdU) labeled. *Arabidopsis* pollen mother cells with Alexa Fluor 488 azide (green). The following prophase I substages are included: **A)** leptotene, **B)** zygotene, **C)** pachytene, **D)** diplotene, **E)** diakinesis. DNA is counterstained with propidium iodide (PI) (red). Each image is a single xy slice from a confocal z-stack. The PI stained nucleolus is not seen in these single slices but is always evident and used for staging when the entire z stack is analyzed . It should also be noted that only 3 of 5 bivalents can be seen in the diakinesis optical slice shown. Scale bar 5 um.

the nucleolus because of its RNA content. Because we are able to distinguish early from late leptotene, zygotene, and pachytene and are able to take samples for cytology at hourly time points after the addition of EdU, we can obtain precise estimations for the duration of each of the prophase I substages in wild type and an rough estimation of G2 duration (Figure 3).

EdU incorporation occurs during S-phase. The first time point (after EdU is added to anthers) in which EdU shows up in meiotic cells of a given stage is critical in determining the duration of each stage. These 'first appearance' time points are provided in Table 2. The first time point that EdU signal was observed in early leptotene meiocytes was 8 hours post EdU pulse initiation. This finding implies that premeiotic G2 duration is approximately 7 hours, since meiocytes that incorporated EdU into their replicating DNA, at the end of S-phase, required approximately 7 hours to progress through G2 and enter into early leptotene. The minimum time required for EdU signal to appear in early zygotene meiocytes was 13 hours after pulse initiation. The difference in timing between first appearance of EdU signal in early leptotene and early zygotene meiocytes defines leptotene duration to be 5 hours in *Arabidopsis* meiocytes (L*er* ecotype). The minimum time required for first appearance of EdU labeled early pachytene meiocytes was 19 hours post EdU pulse initiation; the difference in timing between first appearance of EdU signal in early zygotene and early pachytene meiocytes defines zygotene duration to be 6 hours. The minimum time required for first appearance of EdU labeled early diplotene meiocytes was 29 hours post EdU pulse initiation; the difference in timing between first appearance of EdU signal in early pachytene and early diplotene meiocytes defines pachytene duration to be 10 hours. Only one hour separated the first appearance of EdU signal in early diplotene and early diakinesis meiocytes. Therefore diplotene duration appears to be approximately one hour. Hourly time point-data collection is an absolute requirement for measuring the durations of diplotene, diakinesis and the remaining stages of meiosis, which are all relatively short. In previous work we determined leptotene, zygotene and pachytene durations to be 6.0, 4.8 and 10 hours respectively by an indirect method. This indirect method was based on examination of a very large number of meiotic nuclei to ascertain the percentages of meiocytes observed in each substage. These percentages were applied to the total duration of meiosis as determined by BrdU analysis to yield approximate substage durations [7]. The substage durations of leptotene (5 h), zygotene (6 h) and pachytene (10 h) presented in this paper are close to our previous values but are more precise due to the direct method of measurement. All of the data on the percentage of EdU labeled meiocyte observed, for each prophase I substage, at each time point, is given in Additional file 1: Table S1.

From this work, we have found that EdU labeling is well suited for confocal microscopy analysis of whole meiotic filaments as the fluorophore tagged azide penetrates the tissue, all the way into meiocytes more easily than the antibody required for BrdU detection. The EdU signal obtained was easily observed in *Arabidopsis* meiocytes (Figure 2). The preparation of meiocytes for confocal examination is simpler than for chromosome spread preparations, making the analysis of hourly sampled EdU labeled meiocytes feasible. In addition the number of analyzable meiocytes for each time point is much greater. Consequently the data set from our confocal microscopy- based, time course analysis is significantly larger than for other methods and allows us to more precisely study meiotic progression.

Relationship between tapetal and meiocyte DNA replication

The primary purpose of our study was to develop a precise, simplified method for labeling DNA during S-phase for the purpose of determining the duration of subsequent meiotic stages in normal and aberrant situations. However since confocal microscopy can be used to create optical sectioning through multiple cell layers, we decided to explore the use of this feature with EdU labeling to address questions relating to coordination of events between cell layers of the anthers. We therefore examined EdU incorporation in the tapetum, the cell layer that surrounds meiotic cells of the anther. The tapetal cells are known to undergo a relatively synchronous mitosis at the transition between zygotene and pachytene; we did a preliminary look at when tapetal cells replicate their DNA.

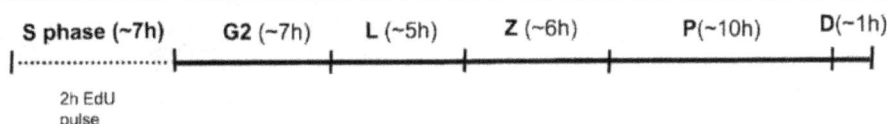

S phase (~7h)   G2 (~7h)   L (~5h)   Z (~6h)   P (~10h)   D(~1h)

2h EdU
pulse

Figure 3 A time course schematic of *Arabidopsis* L*er* Substage Prophase I Durations. A meiosis prophase I time course schematic that extends from EdU pulse initiation in S-phase to diakinesis. Note: 1) that the uptake of EdU can occur at the start, middle or end of S-phase 2) S-phase duration [5] 3) rough estimate of G2 duration Leptotene (L), Zygotene (Z), Pachytene, (P), Diplotene (D).

Table 2 A summary of *Arabidopsis* L*er* prophase I substage durations

Meiotic stage	Time from EdU pulse initiation to first labeling (h)	Stage duration (h)
Leptotene	8	5
Zygotene	13	6
Pachytene	19	10
Diplotene	29	1
Diakinesis	30	

To study tapetal cells we examined those of our cytological preparations that had only partially extruded the filaments of meiotic cells from the anthers and were associated with the surrounding tapetal layer. EdU labeling pattern was observed in 50 partially and non-extruded meiotic filaments from three different time course experiments. Our observations are summarized in Table 3. Four patterns are theoretically possible based on whether or not the tapetal and/or meiotic cells were labeled. Pattern 1 - Meiotic and tapetal cells are not labeled. This pattern would be generated in meiotic filaments that did not incorporate EdU in the time frame of our experiment, producing unlabeled meiotic and tapetal cells. This pattern was observed. Pattern 2 - Meiotic and tapetal cells are both labeled. This pattern could be generated in filaments that contained meiotic cells undergoing S-phase when the EdU was present as well as tapetal cells undergoing S-phase at the same time or after the meiotic cells. This pattern was observed. Pattern 3- Meiotic cells are labeled, but tapetal cells are not labeled. This pattern could be generated in filaments if meiotic cells underwent S-phase when the EdU was present, but tapetal cell DNA replication only occurred before EdU addition. This pattern was not observed. Pattern 4- Meiotic cells are not labeled but tapetal cells are labeled. This pattern could be generated if the meiotic cells had already completed S-phase before EdU was present but tapetal cell DNA replication had not yet occurred. We did see this pattern and in fact when we saw this pattern, all of the cells in the tapetal layer were labeled with EdU (Figure 4). The existence of this last pattern demonstrates that tapetal cells undergo DNA replication after the meiotic cells have completed their last DNA replication.

Furthermore we noticed that for the meiotic cells of one anther chamber, either none of the cells were labeled, or all were labeled. This is not surprising since the meiotic cells are synchronized from pre-meiotic S-phase onward [7]. The tapetal cell layer also had this same 'all or none' labeling pattern. Therefore either the tapetal cells all undergo synchronous DNA replication after the meiotic cells have completed S-phase, or they are cycling fast enough such that all tapetal cells have undergone an additional unsynchronized replication in the time period of EdU exposure. Future experiments, designed specifically to address this question will be done.

Conclusions

Confocal examination of EdU labeled meiocytes allows for a more precise analysis of meiotic substage duration in wild type and presents the opportunity to study possible temporal effects on meiotic progression in meiotic mutants. This claim is based on the increased precision of meiotic staging, hourly sampling and the much larger sampling size that our confocal method allows. Our confocal time course method promises to be a powerful tool that can be used, in future, to identify and pinpoint even subtle abnormalities in stage duration and meiotic progression. Furthermore coordination of DNA replication in the various anther cell types can be evaluated. The method also could be easily adapted to the study of meiosis in megasporocytes within plant gynoecium. Entire intact EdU labeled anthers can be analyzed with confocal microscopy to allow for studies of coordination of meiosis within the development of the other tissues of anthers.

Table 3 Combinations of EdU labeling patterns observed in partially extruded anthers

EdU labeling patterns observed in meiotic filaments (n = 50)			
Pattern #	Meiocytes	Tapetal cells	# of filaments examined
1	Not labeled	Not labeled	16
2	Labeled	Labeled	14
3	Labeled	Lot labeled	0
4	Not labeled	Labeled	20

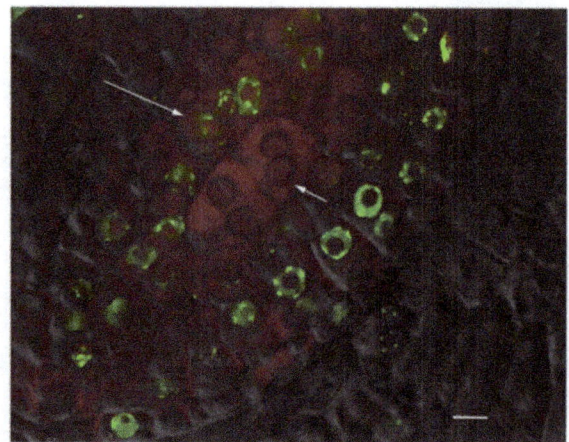

Figure 4 Edu labeling of a meiotic filament. A representative image of a meiotic filament comprised of non-EdU labeled, leptotene meiocytes (short arrow) surrounded by a layer of EdU labeled tapetal cell nuclei (long arrow). EdU labeling (green) and PI counterstain (red) This image is a single xy slice from a confocal z-stack. Scale bar 10 um.

Method

Plant material

Seeds of *A. thaliana* cv. Landsberg *erecta* (L*er*) were obtained from the *Arabidopsis* Biological Resource Center (http://www.arabidopsis.org/abrc). Plants were grown under long day conditions (16 h light/8 h dark); the day/night temperatures were 22°C/19°C. Light intensity was ~170 umoles/m^2/s. Plants were ready to be used in the time course experiment when the stems of primary inflorescences were approximately 10 cm long.

EdU S-phase labeling

Stems of young inflorescences (1–3 opened flowers) were quickly cut under tap water to a length of approximately 9 cm. The cut ends were immediately submerged in a small quantity of 10 mM EdU labeling solution from a Click-IT assay kit (Invitrogen, California, USA). These inflorescences were placed under grow lights at 21°C for 2 h (light intensity ~170 um/m^2/s) to allow the EdU to reach the buds via their vasculature and subsequently to be incorporated into the DNA of S-phase cells. After the 2 h pulse the EdU was exchanged with tap water several times to stop addition of EdU to the nucleotide pool. After these rinses inflorescences in tap water were returned to the grow lights and at hourly intervals two inflorescences were sampled. Negative controls consisted of inflorescences prepared exactly as experimental ones except stems were placed in water only (no EdU).

Inflorescence fixation

For each time-point the inflorescences were processed as follows. Larger buds (beyond meiosis) were removed from the inflorescence and a couple of small holes were made in the remaining attached buds (to facilitate formaldehyde uptake), taking care not to damage the anthers within. The inflorescences were then placed in 3.7% formaldehyde in meiocyte buffer A [15 mM pipes-NaOH (pH 6.8), 80 mM KCl, 20 mM NaCl, 0.5 mM EGTA, 2 mM EDTA, 0.15 mM spermine tetra-HCl, 0.05 mM spermidine, 1 mM dithiothreitol, 0.32 M sorbitol] (pH =8.2). Vacuum infiltration was done (70 psi) at room temperature for 30 min to assist fixative penetration. After fixation the inflorescences were rinsed 3 × 10 min in meiocyte buffer A and then stored in meiocyte buffer A at 4°C until used, typically within 24 hours.

EdU detection

Inflorescences from the first time point were brought to room temperature and washed twice with 3% Bovine Serum Albumin (BSA) in Phosphate Buffered Saline (PBS) (pH =7.4) and then transferred to a depression slide containing the same solution. The sepals and petals were removed from 0.3-0.5 mm buds leaving stamens and pistils intact. These buds with the two outer whorls removed, were then transferred to an epitube containing 3% BSA in PBS. This step was repeated for the inflorescences from all the remaining time points. After brief centrifugation the 3% BSA in PBS was exchanged twice with 0.5% triton X in PBS and allowed to incubate at room temperature for 60 min. Next the intact pistil-stamens were rinsed twice in 3% BSA in PBS, then placed in EdU Click-It colour reaction cocktail as per manufacturer's instructions (Invitrogen) for one hour at room temperature in the dark. The intact pistil-stamens were then rinsed twice in 1 ml BSA in PBS and kept in the dark until used for cytology.

Cytological preparation

The intact pistil-stamens for the first time point, were removed from 3% BSA in PBS. Using a dissecting microscope, anthers were removed from the stamens, scored with a scalpel at their midpoint and transferred to approximately 15 ul of propidium iodide in Vectashield (Vector Laboratories, Burlington, Canada) that had been placed in the center of a clean slide. A 22 × 22 mm No. 1 coverslip was placed over the sample area and a slight pressure applied and the edges of the coverslip were sealed with nail polish. This procedure was repeated for the pistil-stamens for all the remaining time points.

Microscopic analysis

Slides were observed using a Quorum Wave-FX spinning disk confocal microscope and a Leica 63X oil immersion Plan-apo objective (NA 1.4). The solid state lasers used were 491 nm and 561 nm. The emission filters used were bandpass 525/50 nm and bandpass 620/60 nm respectively. The CCD camera used to capture images was a Hamamatsu Orca R2. The software used for image capture and analysis were Metamorph version 7.7.9.0 and Velocity version 6.1.1. Photoshop CS2 was used to label images and create figure montages.

Meiotic staging

Our confocal microscopy sample preparation method extruded meiotic filaments (a structure comprised of ~30 meiocytes) from anther locules. The filaments were either completely or partially extruded. Partially extruded filaments had the advantage of preserving some tapetal–meiocyte associations and in these instances tapetal cells were also used in the staging process. We have taken advantage of the attributes of both the meiocytes and the tapetal cells in the staging of meiotic cells [14,15]. The amount of callose deposition between meiocytes and the nuclear morphology of tapetal cells also served as criteria in meiotic staging. As well intact meiocytes retained the nucleolus (propidium iodide labels the nucleolus), whose shape and position within the nucleus changes as meiosis proceeds, served as an additional valuable staging tool. Collectively this information made it possible to more precisely identify meiotic stages. Furthermore we used our precise multi-criteria staging to identify the early and late portions of leptotene, zygotene and pachytene substages. The criteria we used for precise meiotic staging are outlined in Table 1.

Additional file

> **Additional file 1: Table S1.** Time course data for all meiotic filaments examined.

Abbreviations

BrdU: 5-bromo-2′-deoxyuridine; EdU: 5-ethynyl-2′-deoxyuridine; PMC: Pollen mother cell; CCD: Charge coupled device.

Competing interests

The authors declare that they have no competing interests.

Authors' contributions

The methods were conceived by WA, CH and PS. The experiments were performed by PS. The data was analyzed by PS and CH. The paper was written by PS and CH. All authors read and approved the final manuscript.

Acknowledgements

This project was funded partially by a NSERC grant to CAH.

Author details

[1]Department of Biology, University of Toronto, 1265 Military Trail, Scarborough, Canada. [2]Kingston General Hospital, Kingston, Canada.

References

1. Van't Hof J: **Relationships between mitotic cycle duration, S period duration and the average rate of DNA synthesis in the root meristem cells of several plants.** *Exp Cell Res* 1965, **39**:48–58.
2. Bennet MD, Chapman V, Riley R: **The duration of meiosis in pollen mother cells of wheat, rye and** *Triticale*. *Proc Roy Soc Lond B* 1971, **178**:250–275.
3. Strich R: **Meiotic DNA Replication.** *Curr Topics in Dev Biol* 2004, **61**:29–60.
4. Armstrong S, Franklin FCH, Jones GH: **Nucleolus-associated telomere clustering and pairing precede meiotic chromosome synapsis in** *Arabidopsis thaliana*. *J Cell Sci* 2001, **114**:4207–4217.
5. Armstrong SJ, Franklin FCH, Jones GH: **A meiotic time course for** *Arabidopsis thaliana*. *Sex Plant Repro* 2003, **16**:141–149.
6. Armstrong SJ, Jones GH: **Meiotic cytology and chromosome behaviour in wild type** *Arabidopsis thaliana*. *J Exp Biol* 2003, **380**:1–10.
7. Stronghill P, Hasenkampf CA: **Analysis of Substage Associations in Prophase I of Meiosis in Floral Buds of Wild Type** *Arabidopsis thaliana* **(Brassicaceae).** *Amer J Bot* 2007, **94**:2063–2067.
8. Warren M, Puskarcyzk K, Chapman SC: **Chick embryo proliferation studies using EdU labeling.** *Dev Dynamics* 2009, **238**:944–949.
9. Yu Y, Arora A, Roifman CM, Grunebaum E: **EdU incorporation is an alternative non-radioactive assay to [(3)H]thymidine uptake for in vitro measurement of mice T-cell proliferations.** *J Immunol Methods* 2009, **350**:29–35.
10. Mead TJ, Lefebvre V: **Proliferation Assays (BrdU and EdU) on Skeletal Tissue Sections.** *Focus* 2014, **1130**:233–243.
11. Armstrong SJ: **A time course for the analysis of meiotic progression in** *Arabidopsis thaliana*. *Methods in Mol Biol* 2013, **990**:119–123.
12. Ross KJ, Fransz P, Jones GH: **A light microscopic atlas of meiosis in** *Arabidopsis thaliana*. *Chromosome Res* 1996, **7**:507–516.
13. Sanchez- Moran E, Santos J-L, Jones GH, Franklin FCH: **ASY1 mediates AtDMC1-dependent interhomolog recombination during meiosis in** *Arabidopsis*. *Genes & Dev* 2007, **21**:2220–2233.
14. Owen HA, Makaroff CA: **Ultrastructure of microsporogenesis and microgametogenesis in** *Arabidopsis thaliana*. *Protoplasma* 1995, **185**:7–21.
15. Pathan N, Stronghill P, Hasenkampf C: **Transmission electron microscopy and serial reconstructions reveal novel meiotic phenotypes for the** *ahp2* **Mutant of** *Arabidopsis thaliana*. *Genome* 2013, **3**:139–145.

11

GrainScan: a low cost, fast method for grain size and colour measurements

Alex P Whan[1*], Alison B Smith[2], Colin R Cavanagh[1], Jean-Philippe F Ral[1], Lindsay M Shaw[1], Crispin A Howitt[1] and Leanne Bischof[3]

Abstract

Background: Measuring grain characteristics is an integral component of cereal breeding and research into genetic control of seed development. Measures such as thousand grain weight are fast, but do not give an indication of variation within a sample. Other methods exist for detailed analysis of grain size, but are generally costly and very low throughput. Grain colour analysis is generally difficult to perform with accuracy, and existing methods are expensive and involved.

Results: We have developed a software method to measure grain size and colour from images captured with consumer level flatbed scanners, in a robust, standardised way. The accuracy and precision of the method have been demonstrated through screening wheat and *Brachypodium distachyon* populations for variation in size and colour.

Conclusion: By using GrainScan, cheap and fast measurement of grain colour and size will enable plant research programs to gain deeper understanding of material, where limited or no information is currently available.

Keywords: Wheat, Brachypodium distachyon, Seed size, Seed colour, Image analysis

Introduction

Measurement of seed characteristics is a vital aspect of cereal research. Grain size represents one of the major components of yield, it contributes to seedling vigour [1,2], and larger grains may lead to an increase in milling yield [3-5]. Seed colour is also important for breeding of cereal varieties because it affects the quality and appeal of processed grain, and is also associated with dormancy in multiple species [6,7].

Grain size

Grain (or seed) size is an important component of both basic plant research, since seed formation and development is a fundamental aspect of plant reproduction, and cereal breeding, as a component of yield and vigour. Existing methods of determining seed size tend to either favor speed of measurement while sacrificing resolution, or are so involved that high throughput measurement is challenging. In the context of cereal breeding, seed weight is an important trait related to seed size, and therefore measuring the weight of a standard number or volume of seeds is practical and informative. Measures such as thousand-grain weight or hectolitre weight are commonly used since they are fast, and not prone to error. However, they give no measure of variation within a sample. Detailed measurement of seed shape characteristics such as length and width traditionally depends on laborious techniques such as manual measurement of individual seeds [8]. The single kernel characterization system (SKCS, [9]) is a relatively low throughput, destructive technique that measures hardness as well as seed size. Systems such as SeedCount (Next Instruments, NSW, Australia) utilize image analysis to give measures of size for individual seeds within a sample, allowing for a detailed understanding of variation, as well as an accurate estimation of the sample mean. However the time required for sample preparation especially for large numbers of samples (SeedCount samples need to be placed in wells in a sample tray), along with the initial cost of such systems can be prohibitive (~ $AUD15000).

Grain colour

The association between red seed colour and increased dormancy has been recognized in wheat for over a

* Correspondence: alex.whan@csiro.au
[1]CSIRO Plant Industry, GPO Box 1600, Canberra ACT 2601, Australia
Full list of author information is available at the end of the article

century. Nilsson-Ehle [10], cited in [11] suggested that three genes were controlling red pigmentation in wheat, and subsequently three homoeologous loci have been mapped to the long arm of chromosome group 3 [12] encoding a Myb-type transcription factor having pleiotropic effects on both dormancy and expression of genes in the flavonoid biosynthesis pathway [13]. With increased copy number of red genes (3A, 3B, 3D) there is an additive effect on increasing dormancy in wheat, however other genetic loci such as those on 4AL and 3AS have been found to explain a greater percentage of the genetic variation [14]. White wheat may be more desirable because of increased milling efficiency and consumer preferences for some end products, such as Udon noodles [15].

No simple methods for measuring seed colour (other than human estimation) are available. Colour estimation is generally performed on a modal scale by eye, resulting in loss of colour gradation information (inability to classify gene number). Unless the colour difference is stark, there is a high likelihood of inconsistent estimation [16]. For classification of wheat as genetically either red or white, seeds can be soaked in NaOH to increase the contrast between the two [17], however this is relatively low throughput, and does not take into account further colour variation due to environmental or other genetic factors.

Accurate, widely interpretable measurement of colour is technically challenging, and a field unfamiliar to many biologists. Because perception of colour is affected by the environment in which it is observed, standardised measurement is critical. Such a requirement generally involves somewhat laborious sample preparation and high cost analytical equipment. Chroma meters are standard tools for accurate colour determination in many industries, and can be applied to cereal products along the processing chain, including grain, flour, dough and the final processed product. For standardised, comparable colour measurements, chroma meters measure in the CIELAB colour space, a device independent colour space which includes all perceivable colours. CIELAB is made up of three channels: L^*, which ranges from 0 to 100 and represents the lightness of the colour; a^*, negative or positive values of which represent green or magenta, respectively; and b^*, representing blue (negative) or yellow (positive). These channels can then be used individually to quantify specific colour attributes, which may be linked to biological factors [18]. While the measurements given by chroma meters are highly controlled and standardised, when applied to grain, there are several drawbacks. Because of the small area that is measured, only a limited number of grains are visible by the observer, and a single average value is reported. This, therefore, provides no information regarding variation within a sample of grain. An alternative method is the SeedCount system, which also provides colour information based on the CIELAB colour space, as well as other grain characteristics such as size and disease state.

There is increasing use of image analysis in plant science and agriculture, especially in the field of phenomics [19,20]. While demonstrating great potential in accelerating detailed plant measurements, many of the available methods depend on very costly infrastructure, limiting widespread adoption. Developments in the availability of image analysis for plant measurement applications have made low cost alternatives available, including: RootScan, which analyses root cross sections [21]; Tomato Analyzer, which measures a range of features including shape and disease state in tomatoes and other fruits [22]; and the web application PhenoPhyte, which allows users to quantify leaf area and herbivory from above ground plant images [23]. ImageJ is general purpose image analysis software that is freely available [24], and has been used to analyse seed shape and size parameters in a range of plant species including wheat, rice and *Arabidopsis* [25-28]. SmartGrain [29] is another image analysis system that is free to use, and is also based on images captured by consumer level flatbed scanners to extract seed characteristics. SmartGrain builds ellipses on identified grains to establish seed area, perimeter, width and length, but does not measure colour information. Seed shape can also be analysed with the software SHAPE [30], which produces elliptic Fourier descriptors of 2- and 3-dimensional characteristics from photographs of vertically and horizontally oriented seed, which has the advantage of potentially identifying different loci affecting seed shape, but due to the nature of the image capture, requires manual handling and preparation of individual seeds [31].

Here, we present GrainScan [32], a low cost, high-throughput method of robust image capture and analysis for measurement of cereal grain size and colour. GrainScan utilizes reflected light to accurately capture colour information described in a device independent colour space (CIELAB), allowing comparison of colour data between scanning devices.

Results and discussion

To test the accuracy of GrainScan, wheat seeds from a diverse mapping population were measured with Grain-Scan, SmartGrain and Seedcount. These comparisons were used because SmartGrain and SeedCount are specifically designed for grain analysis, and each includes components that provide similar functionality to elements of GrainScan.

Size traits

The distribution of size traits measured by GrainScan for individual images could be reasonably approximated by a Guassian distribution (Figure 1). Because of the number

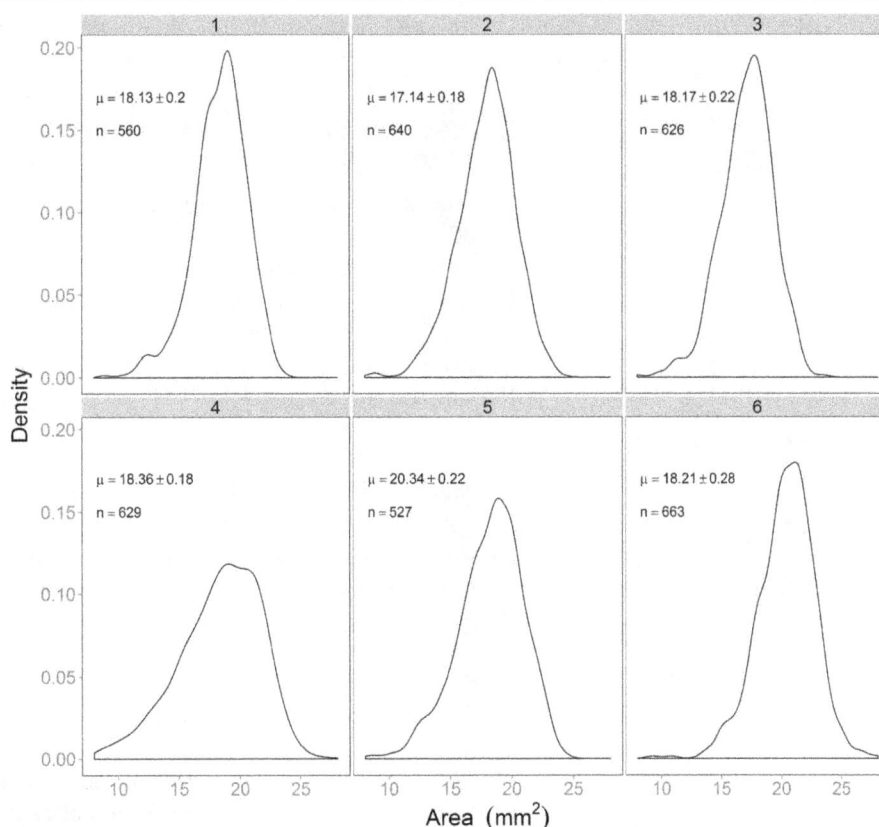

Figure 1 Density distributions of grain area for six randomly chosen samples of wheat grain. The mean and confidence interval, along with the number of seeds included in each scan is noted on each panel.

of seeds measured in each scan, there was a high level of confidence in the mean trait value for each image.

Comparison of screening methods

Summary data for each size trait as measured by Grain-Scan, SmartGrain and SeedCount is shown in Table 1. Mean values and ranges for size traits across the population were similar between methods. The REML estimates

Table 1 Summary statistics (minimum, mean and maximum) of raw packet means for each trait and method

	GrainScan	SmartGrain	SeedCount
Area-min	11.68	10.22	10.00
Area-mean	17.99	15.96	16.07
Area-max	24.52	21.34	22.05
Length-min	5.40	5.25	5.36
Length-mean	6.71	6.51	6.71
Length-max	7.99	7.70	7.94
Width-min	2.65	2.47	2.58
Width-mean	3.41	3.24	3.39
Width-max	3.91	3.74	3.88

Seed area is measured in mm², length and width are in mm.

of the correlations between the packet effects for different methods are shown in Figure 2. Each correlation gives a measure of the agreement in the ranking of effects between methods. In the context of a breeding program this measure would relate to the similarity between methods in terms of genotype rankings and thence selection. A correlation near +1 suggests identical rankings for the two methods; a correlation near -1 suggests a complete reversal of rankings and a correlation near 0 suggests very little relationship between the rankings. Figure 2 shows that GrainScan correlates highly with both methods for all size traits, but most strongly with SeedCount. The strength of the correlations is also reflected in the pairwise plots of the packet effect BLUPs in Figure 2.

The average accuracy (correlation between true and predicted packet effects, Table 2) for GrainScan was very high (0.981 – 0.996) and similar to SeedCount (0.991 – 0.994) for both replicated and unreplicated packets, while the average accuracy for trait measurements from SmartGrain was lower (0.871 – 0.947).

Measurements took approximately twice as long using SeedCount compared to scanning for analysis by GrainScan or SmartGrain (210 seconds and 101 seconds, respectively). This time only considered the image capture, which for

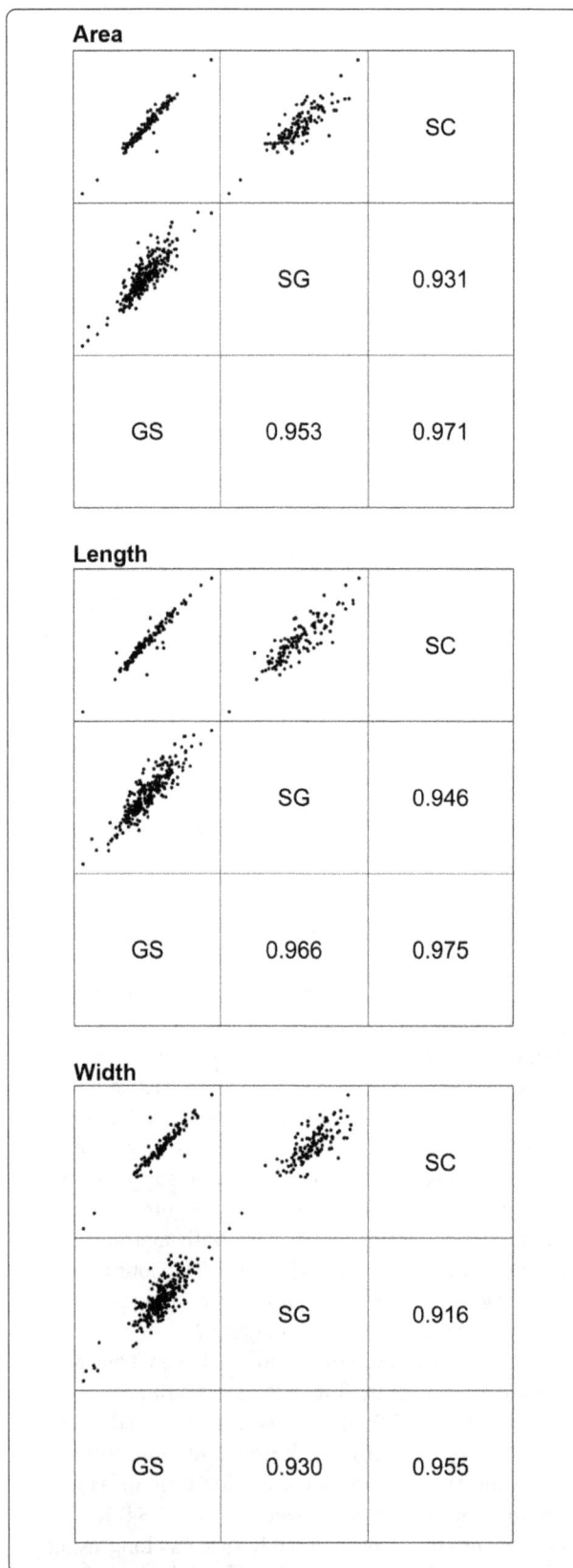

Figure 2 Correleation of BLUPs for size traits. Pairwise plot of BLUPs of packet effects (above the diagonal) and REML estimates of correlations between packet effects (below the diagonal) for size traits from GrainScan, SmartGrain and SeedCount. Method labels are on the diagonal: SC (SeedCount), SG (SmartGrain) and GS (GrainScan).

SeedCount included image processing time, while for the other methods, image processing was done as a batch after all images were captured. However, the difference in time was mainly due to the time taken to lay out seeds as required in the sample tray for SeedCount, as opposed to scattering in the glass tray for the flatbed scanning. Because wheat grains are rounded, when they are scattered on the glass, they can roll into different orientations. GrainScan provides a facility to detect grain creases (described below), which can be used to filter out data from grains that are not oriented crease down. In our comparison of methods we have used measurements from all visible seeds, since it represents the complete GrainScan output.

Colour traits
GrainScan colour determination
GrainScan can output colour channel intensity in the standardised CIELAB colourspace. To test whether the crease region on a seed image distorted colour measurements in GrainScan measurements, three ways of calculating colour were tested with GrainScan. Each method measured colour on different parts of the detected seed – the entire seed area (abbreviated GS), the entire seed area of seeds where no crease was detected (abbreviated GSncd) or only the non-crease area of seeds where a crease was detected (abbreviated GSwc). Mean values and ranges (Table 3) agreed very closely between each method, and REML estimates of the correlations between packet effects were all greater than 0.99 (Figure 3). Therefore, for the grain images included in this analysis, the crease area does not effect colour determination, however the option to detect grain crease and differentiate colour measurements based

Table 2 Average accuracies for each size trait for each method

	Unreplicated packets	Replicated packets	Trait
GrainScan	0.993	0.996	
SmartGrain	0.900	0.945	
SeedCount	0.992	0.994	Area
GrainScan	0.981	0.990	
SmartGrain	0.903	0.947	
SeedCount	0.994	0.995	Length
GrainScan	0.990	0.994	
SmartGrain	0.871	0.928	
SeedCount	0.991	0.994	Width

Averages are computed separately for unreplicated and replicated packets.

Table 3 Summary statistics of raw packet means for colour traits for each method

	GS	GSCD	GSNC	Minolta	SC
L-min	48.82	49.72	47.36	47.11	43.50
L-mean	57.44	57.67	56.29	51.86	49.78
L-max	66.09	66.27	64.34	58.20	54.80
a*-min	6.25	6.07	6.92	5.50	3.30
a*-mean	9.08	9.00	9.50	6.81	4.74
a*-max	11.46	11.13	12.03	7.94	6.50
b*-min	21.46	21.55	21.95	13.73	15.90
b*-mean	27.69	27.79	27.86	16.89	18.66
b*-max	31.72	31.89	32.18	20.76	21.60

on crease presence is included in the GrainScan interface, a facility that is not available in the other methods considered. While crease detection has only been considered for wheat seeds in this comparison, we anticipate successful detection for any species with a defined crease.

Comparison of screening methods

Mean values for colour measurement varied between GrainScan, Minolta and SeedCount (Table 3). REML estimates of correlations between packet effects for colour traits between methods are shown in Figure 3. All methods correlated highly (>0.96) for L* (lightness). GrainScan and SeedCount were strongly correlated for a* (0.96), but less so with Minolta (0.78 and 0.75, respectively). For b*, GrainScan and Minolta were strongly correlated (0.97), compared to SeedCount (0.90 and 0.87 respectively).

Average accuracies (Table 4) were higher for Seed-Count (0.988 – 0.995) than GrainScan for all channels (0.874 – 0.988) for both replicated and unreplicated packets. This improved accuracy for colour determination may be due to improved control and uniformity of lighting conditions inside the SeedCount equipment.

Based on these comparisons, GrainScan is an excellent alternative to costly, low throughput methods for standardised colour measurement. GrainScan could be used to determine the presence of genetic variation for colour traits within a population, and where large enough, be sufficiently accurate to conduct complete analysis. Because of its low investment requirement, both in labour and equipment, GrainScan could also be used as an initial investigative tool to determine the value of further investigation with higher cost tools.

Brachypodium distachyon

Traits measured for *B.distachyon* seeds were area, perimeter, width and length. Despite the marked difference in shape between seeds from wheat and *B. distachyon*, GrainScan successfully identified seeds, and allowed

estimation of mean size as well as variation within a sample (Figure 4, Table 5). The distributions of grain size suggested the possibility of bimodality in these samples, although the sample sizes were much lower than those for wheat. Because of the reduced number of seeds per image, standard errors were higher than those for wheat, highlighting the benefit of scanning larger number of seeds. Since GrainScan can accurately measure seed size across two species with largely differing seed shapes, it is therefore likely that GrainScan can be successfully implemented for many different plant species that also have regular, approximately elliptical morphology.

Conclusion

GrainScan enables robust, standardized and detailed study of grain size, shape and colour at very low cost and relatively high throughput. We have demonstrated that size measurements from GrainScan are reproducible between scans, agree well with accepted image analysis techniques, and result in similar rankings of sample material. Because of the dramatically lower cost, and higher throughput of GrainScan compared to other standardized colour measurement methods, GrainScan facilitates detailed study of grain colour in large populations.

GrainScan is freely available as an executable application (http://dx.doi.org/10.4225/08/536302C43FC28).

Method

Image capture

Wheat images were scanned using an Epson Perfection V330 (Seiko Epson Corporation, Suwa, Japan) and *B. distachyon* images with a Canon CanoScan LiDE 700 F (Canon Inc, Tokyo, Japan), which are both consumer grade flatbed scanners (<$250 AUD). To standardise image capture, scanning was managed through VueScan (Hamrick Software, http://www.hamrick.com), which allows for a wide range of flatbed scanner manufacturers. All images were scanned at 300 dpi with no colour adjustment or cropping applied. For wheat scanning, grains were spread onto a glass bottomed tray for ease of collection, while for *B. distachyon*, seeds were spread on an overhead transparency film both to avoid scratching the scanner glass and to allow the seeds to be easily collected. Since the wheat seed was bulked from field trial material, a non-uniform subsample of seed was scattered from a seed packet. The operator assessed the appropriate amount of seed to avoid excessive touching of grains. The number of seeds per image ranged from 382 to 985 with a mean value of 654. For *B.distachyon*, seeds were assessed from single spikes from individual plants and all seeds from a spike were measured. The average number of seeds per scan was 18. To maximise contrast at the border of each seed, either a piece of black cardboard, or a matte black box was upturned over the scanning surface, minimizing reflection

L*

				SC	
				Min	0.963
			GSncd	0.990	0.967
		GSwc	0.997	0.993	0.970
	GS	>0.999	0.997	0.992	0.970

a*

				SC	
				Min	0.748
			GSncd	0.778	0.957
		GSwc	0.997	0.780	0.960
	GS	>0.999	0.997	0.780	0.960

b*

				SC	
				Min	0.869
			GSncd	0.965	0.892
		GSwc	0.995	0.969	0.896
	GS	>0.999	0.995	0.970	0.896

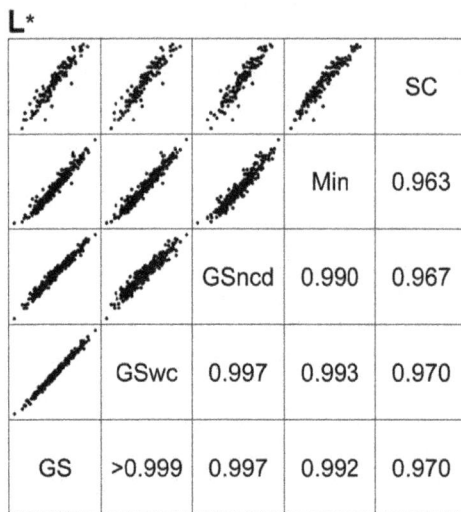

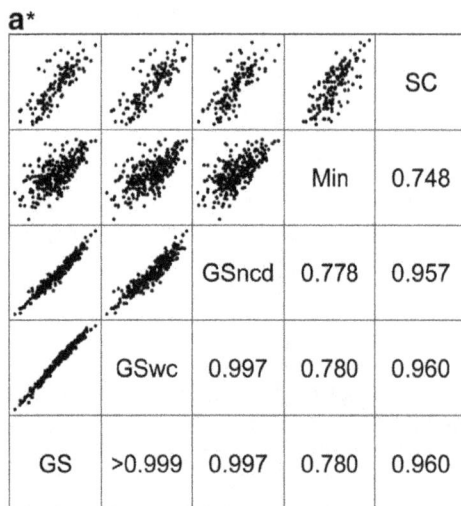

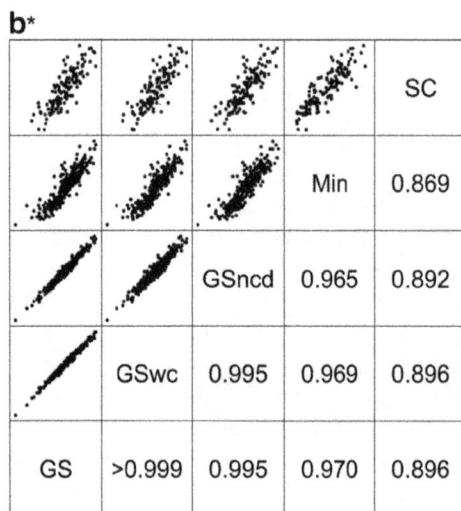

Figure 3 Correlation of BLUPs for colour traits. Pairwise plot of BLUPs of packet effects (above the diagonal) and REML estimates of correlations between packet effects (below the diagonal) for colour traits from GrainScan, SmartGrain and SeedCount. Panels represent each colour trait (**L*, a* and b***) as labelled. Labels for each method are on the diagonal of each panel: SC (SeedCount), Min (Minolta Colorimeter), GSncd (GrainScan - only those grains where no crease was detected), GSwc (GrainScan – only the non-crease areas of seeds where a crease was detected) and GS (total grain area of all seeds detected by GrainScan).

and shadow. All wheat images used to compare methods are available online [33].

To allow standardisation of colour measurements to the CIELAB colourspace, a Munsell ColorChecker Mini card (X-Rite Corp., MI, USA) was scanned under the same settings as the seed, and used within GrainScan to generate conversion parameters for the colour information measured by the flatbed scanner.

Image analysis

The image analysis workflow in GrainScan is as follows. A grayscale image is derived from the scanned colour image by averaging the Red and Green channels, since these provide the greatest contrast for seeds considered. Preprocessing is applied to simplify the image prior to segmentation. The functions used in this simplification are mostly connected component (or attribute) morphological operators [34]. These operators are used in preference to older structuring element based morphological functions because they are contour-preserving and there is more selectivity in the way the image is modified. The preprocessing steps include Gaussian smoothing to reduce noise, an attribute closing based on width (0.3 × *Min grain width, a variable accessible to the user*) to fill in the grain crease, a morphological thinning based on elongation to

Table 4 Average accuracies for each colour trait for each method

	Unreplicated packets	Replicated packets	Trait
GrainScan	0.978	0.988	L*
gsCreaseDown	0.979	0.989	
gsNoCrease	0.974	0.986	
SeedCount	0.994	0.995	
GrainScan	0.874	0.930	a*
gsCreaseDown	0.871	0.928	
gsNoCrease	0.867	0.926	
SeedCount	0.992	0.994	
GrainScan	0.926	0.960	b*
gsCreaseDown	0.925	0.960	
gsNoCrease	0.925	0.959	
SeedCount	0.988	0.992	

Averages were computed separately for unreplicated and replicated packets.

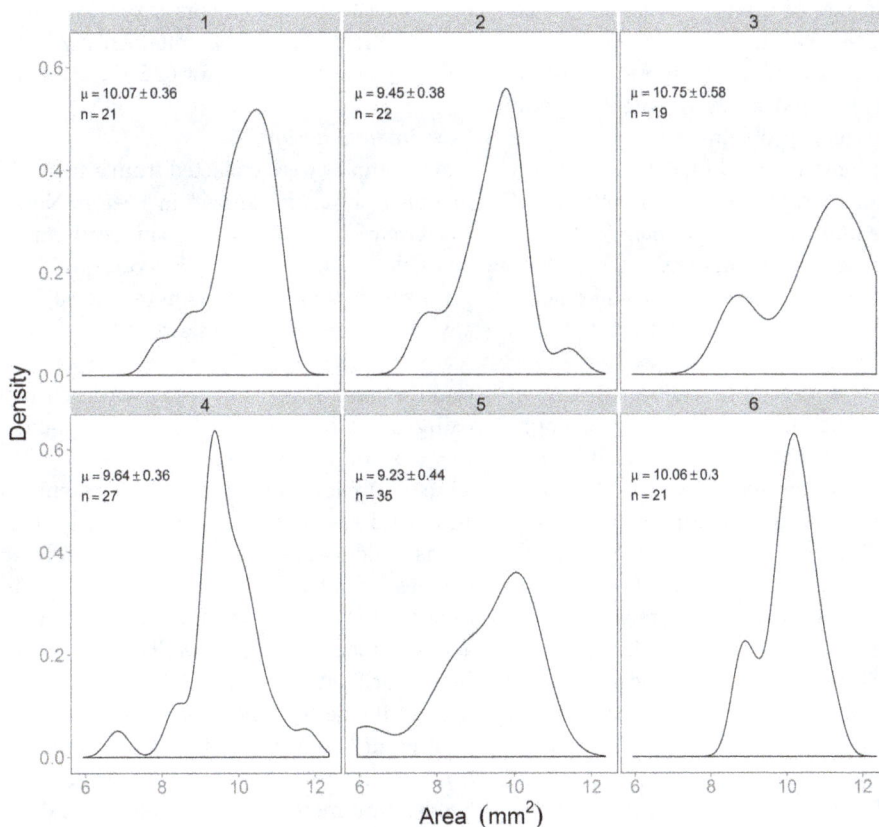

Figure 4 Density distributions of grain area for six randomly chosen samples of *Brachypodium*. The mean and confidence interval, along with the number of seeds included in each scan is noted on each panel.

remove any scratches in the background, an attribute opening based on width ($0.7 \times Min\ grain\ width$) to remove thin debris and an attribute opening based on length ($0.7 \times Min\ grain\ length$) to remove thick debris.

Because flatbed scanners have uniform lighting and the scanner background provides good contrast with the grain colour, there is no need for sophisticated segmentation techniques. The grains can be separated from the background through simple global thresholding. This threshold is determined using an automated thresholding method, based on a bivariate histogram of input grey level versus gradient, as it is more reliable than methods based on the simple image histogram and is used in image normalisation [35]. Touching grains are separated using a common binary object splitting technique based on finding the troughs between regional maxima in the smoothed distance transform. To remove any small regions created by the grain splitting step, a filtering based on the connected component area ($0.5 \times Min\ grain\ width \times Min\ grain\ length$) is then performed.

Individual grains are labelled and measurements made of their size and colour. The dimension measurements are area, perimeter, and surrogates for length and width

Table 5 Summary statistics for *B.distachyon* size traits

Trait	Min	Mean	Max
Area	7.80	10.00	11.17
Perimeter	20.32	22.94	25.13
Length	7.70	8.71	9.55
Width	1.22	1.47	1.64

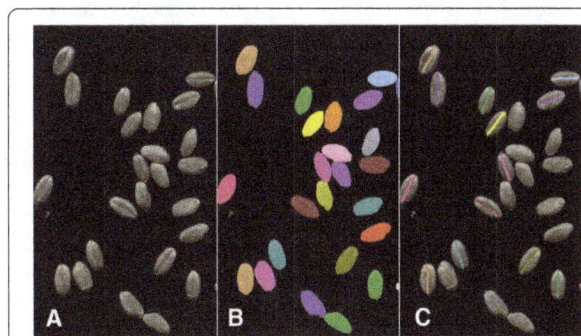

Figure 5 Examples of GrainScan input and output. Panel **A**: Scanned wheat grain for GrainScan input. Panel **B**: GrainScan output highlighting segmented grains as determined by the software. Different colours indicate different grains. Panel **C**: Optional crease detection output highlighting regions identified as grain crease.

– the major and minor axes of the best fit ellipse (called majellipse and minellipse respectively). These surrogates are quick to compute and tend to be more robust to noise (small bumps and dents) in the segmented grain boundary which can cause problems with algorithms that measure the exact length and width. The dimension units are converted from pixels to millimetres (mm) based on the input *Scanner resolution* in dots per inch (dpi).

The software has two independent options in the analysis of colour. One option is to make the colour measurements for each grain in CIELAB values rather than the raw RGB values measured by the scanner. To use the colour calibration option, the image of a calibrated colour checker card must first be analysed using the Colour-Calibration software. This software locates the card, segments each of the colour swatches, extracts the mean RGB values for each swatch, and determines the transformation matrix, RGB2Lab, by linear regression between the measured RGB values and the supplied CIELAB values for each swatch. For convenience, the transformation matrix is saved as two images, one containing the 3×3 matrix and one the 3x1 offset (with filename suffixes of *RGB2Labmat.tif and *RGB2Laboff.tif respectively). By inputting this transformation matrix into the GrainScan software, colour measurements made within each labelled grain can be converted from raw RGB values to calibrated L*, a*, and b* values.

The second colour analysis option is to detect the grani crease and to make additional colour measurements in the non-crease region and if present, the crease region. The crease detection is performed on each grain by finding the shortest path along the long axis of the grain after mean filtering preferentially along this axis to suppress intensity variability unrelated to the crease.

The resulting dimension and colour measurements are saved to a Results sub-directory in Comma Separated Variable (CSV) format. To permit visual inspection of the segmentation results, the labelled grain image and optionally the labelled crease image are saved (with filename suffixes of *.grainLbl.tif and *.creaseLbl.tif respectively). Overlay images with each labelled grain, or crease, overlaid in a different colour on the input image are also saved (with filename suffixes of *.grainOvr.jpg and *.creaseOvr.jpg respectively, Figure 5).

Comparison to other methods

To compare the image analysis algorithm for size parameters, scanned images were processed with both GrainScan and SmartGrain [29]. Output from these systems was compared to results from a SeedCount system, which was used as a standard for size parameters. SeedCount measurements were taken according to manufacturer's instructions. To compare between colour measurements determined by GrainScan and SeedCount, output was

compared to measurements taken by a Minolta CR-400 chroma meter (Konica Minolta Sensing, Osaka, Japan), an industry standard device for CIE L^*, a^* and b^* values.

Experimental design

Grain samples were collected from a field trial of a diverse mapping population grown in Leeton, New South Wales. For GrainScan and SmartGrain, seed was scanned from 300 field plots, each of which corresponded to a different genotype. It is important to note that no field replicates of any of the genotypes were available in this study. Prior to scanning, seed was cleaned by a vacuum separator to remove chaff. Packets of seed from each plot were tested using an experimental design in which a proportion ($p = 0.4$) of the packets was tested with replication. Thus 120 packets were tested twice and the remaining 180 were tested once. This equated to a total of 420 scans which were conducted by a single operator in 14 batches. Each batch comprised 30 scans done sequentially. Replication was achieved for a packet by tipping out seeds and scanning to obtain the first image, then tipping the seeds back into the packet for a subsequent scan. The second image for any packet was always obtained from a different batch to the first image. Thus the design was a p – replicate design [36] with batches as blocks. The SeedCount method was tested on 150 packets, 45 of which were tested with replication, making a total of 195 images. The experimental design was similar to GrainScan and SmartGrain in the sense of involving batches (13 batches with 15 images per batch). Colorimeter (Minolta) measurements were not taken according to a p-replicate design with a blocking structure, but were in duplicate for the 300 packets that were included for GrainScan and SmartGrain.

Data analysis

Analyses were conducted using the ASReml-R package [37] in the R statistical computing environment [38]. For the size data, the analysis commenced with the fitting of a separate mixed model for each trait and method. Since the SeedCount and the SmartGrain methods produce a single value per packet, mean values of the GrainScan data were used to allow comparisons between methods. Each model included random effects for packets and batches. The separate analyses for each method were used to obtain a measure of accuracy for each, defined in terms of the correlation between the predicted packet effects and the true (unknown) packet effects. The data for the different methods were then combined in a multi-variate analysis. The mixed model included a separate mean for each method, random packet effects for each method, random batch effects for each method and a residual for each method. The variance model used for the random packet effects was a factor analytic model [39] which allows for a separate variance for each method and separate correlations between pairs of methods. The other variance models were commensurate with the structure of the experiment. In particular we note that correlations between the GrainScan and SmartGrain methods were included for the batch and residual effects, since these methods were used on the same experimental units (images). The multi-variate analysis provides residual maximum likelihood (REML) estimates of the correlations between the true (unknown) packet effects for different methods. It also provides best linear unbiased predictions (BLUPs) of the packet effects for each method.

For colour measurements, comparisons were made between the complete GrainScan output, GrainScan output for seeds where no crease was detected (abbreviated GSncd), GrainScan output for the non-crease portion of seeds where a crease was detected (abbreviated GSwc), SeedCount and Minolta colorimeter. Since SeedCount and the Minolta methods produce a single value per packet, mean values of the GrainScan data were used to make comparisons between methods.

Initially a separate mixed model analysis was conducted for the data for each trait for each method apart from Minolta. Measurements using the latter were not derived using a design or replication structure as per the other methods and so could not be assessed in the same way. Each model included random effects for packets and batches. The data for the different methods (including Minolta) were then combined in a multivariate analysis. The mixed model was analogous to that used for the seed size analyses.

Brachypodium size analysis was only performed with GrainScan, so no comparisons with other methods were performed.

Abbreviations
GSncd: GrainScan no crease detected; GSwc: GrainScan with a detected crease; REML: Residual maximum likelihood; BLUP: Best linear unbiased predictor.

Competing interests
The authors declare that they have no competing interests.

Authors' contributions
AW assisted in developing the method, experimental design, conducting the analysis and drafting the manuscript. AS analysed the data and assisted in drafting the manuscript. CC assisted in experimental design, analysis and drafting the manuscript. JR assisted in coordinating the experiment and drafting the manuscript. LS assisted in developing the method, coordinating the experiment and drafting the manuscript. CH assisted in coordinating the experiment and drafting the manuscript. LB developed the image analysis method and assisted in drafting the manuscript. All authors read and approved the final manuscript.

Acknowledgements
The authors acknowledge Geoff Ellacott and Freddie Loyman for their assistance in scanning grain images. AS gratefully acknowledges the financial support of the Grains Research and Development Corporation of Australia.

Author details
[1]CSIRO Plant Industry, GPO Box 1600, Canberra ACT 2601, Australia. [2]National Institute for Applied Statistics and Research Australia, Univeristy of Wollongong, Wollongong NSW 2522, Australia. [3]CSIRO Computational Informatics, North Ryde NSW 2113, Australia.

References
1. Lafond GP, Baker RJ: Effects of genotype and seed size on speed of emergence and seedling vigor in nine spring wheat cultivars1. *Crop Sci* 1986, **26**:341.
2. Demirlicakmak A, Kaufmann ML, Johnson LPV: The influence of seed size and seeding rate on yield and yield components of barley. *Can J Plant Sci* 1963, **43**:330–337.
3. Berman M, Bason ML, Ellison F, Peden G, Wrigley CW: Image analysis of whole grains to screen for flour-milling yield in wheat breeding. *Cereal Chem* 1996, **73**:323–327.
4. Novaro P, Colucci F, Venora G, D'Egidio MG: Image analysis of whole grains: a noninvasive method to predict semolina yield in durum wheat. *Cereal Chem* 2001, **78**:217–221.
5. Marshall DR, Mares DJ, Moss HJ, Ellison FW: Effects of grain shape and size on milling yields in wheat .2. experimental studies. *Aust J Agric Res* 1986, **37**:331–342.
6. Anderson JA, Sorrells ME, Tanksley SD: RFLP analysis of genomic regions associated with resistance to preharvest sprouting in wheat. *Crop Sci* 1993, **33**:453–459.
7. Gu X-Y, Kianian SF, Foley ME: Multiple loci and epistases control genetic variation for seed dormancy in weedy rice (Oryza sativa). *Genetics* 2004, **166**:1503–1516.
8. Ramya P, Chaubal A, Kulkarni K, Gupta L, Kadoo N, Dhaliwal HS, Chhuneja P, Lagu M, Gupt V: QTL mapping of 1000-kernel weight, kernel length, and kernel width in bread wheat (Triticum aestivum L.). *J Appl Genet* 2010, **51**:421–429.
9. Martin CR, Rousser R, Brabec DL: Development of a single-kernel wheat characterization system. *Trans ASAE* 1993, **36**(5):1399–1404.
10. Nilsson-Ehle H: Zur Kenntnis der mit der keimungsphysiologie des weizens in zusammenhang stehenden inneren faktoren. *Z Für Planzenzüctung* 1914, **2**:153–187.
11. Groos C, Gay G, Perretant M-R, Gervais L, Bernard M, Dedryver F, Charmet G: Study of the relationship between pre-harvest sprouting and grain color by quantitative trait loci analysis in a white × red grain bread-wheat cross. *Theor Appl Genet* 2002, **104**:39–47.
12. Nelson JC, Deynze AEV, Sorrells ME, Autrique E, Lu YH, Negre S, Bernard M, Leroy P: Molecular mapping of wheat: homoeologous group 3. *Genome* 1995, **38**:525–533.
13. Himi E, Noda K: Red grain colour gene (R) of wheat is a Myb-type transcription factor. *Euphytica* 2005, **143**:239–242.
14. Mares D, Mrva K, Cheong J, Williams K, Watson B, Storlie E, Sutherland M, Zou Y: A QTL located on chromosome 4A associated with dormancy in white- and red-grained wheats of diverse origin. *Theor Appl Genet* 2005, **111**:1357–1364.
15. Liu S, Bai G, Cai S, Chen C: Dissection of genetic components of preharvest sprouting resistance in white wheat. *Mol Breed* 2011, **27**:511–523.
16. Peterson CJ, Shelton DR, Martin TJ, Sears RG, Williams E, Graybosch RA: Grain color stability and classification of hard white wheat in the US. *Euphytica* 2001, **119**:101–106.
17. Lamkin WM, Miller BS: Note on the use of sodium-hydroxide to distinguish red wheats from white common, club, and durum cultivars. *Cereal Chem* 1980, **57**:293–294.
18. Humphries JM, Graham RD, Mares DJ: Application of reflectance colour measurement to the estimation of carotene and lutein content in wheat and triticale. *J Cereal Sci* 2004, **40**:151–159.
19. Furbank RT, Tester M: Phenomics - technologies to relieve the phenotyping bottleneck. *Trends Plant Sci* 2011, **16**:635–644.
20. White JW, Andrade-Sanchez P, Gore MA, Bronson KF, Coffelt TA, Conley MM, Feldmann KA, French AN, Heun JT, Hunsaker DJ, Jenks MA, Kimball BA, Roth RL, Strand RJ, Thorp KR, Wall GW, Wang GY: Field-based phenomics for plant genetics research. *Field Crops Res* 2012, **133**:101–112.

21. Burton AL, Williams M, Lynch JP, Brown KM: **RootScan: Software for high-throughput analysis of root anatomical traits.** *Plant Soil* 2012, **357**:189–203.

22. Gonzalo MJ, Brewer MT, Anderson C, Sullivan D, Gray S, van der Knaap E: **Tomato fruit shape analysis using morphometric and morphology attributes implemented in Tomato Analyzer software program.** *J Am Soc Hortic Sci* 2009, **134**:77–87.

23. Green JM, Appel H, Rehrig EM, Harnsomburana J, Chang J-F, Balint-Kurti P, Shyu C-R: **PhenoPhyte: a flexible affordable method to quantify 2D phenotypes from imagery.** *Plant Methods* 2012, **8**:45.

24. Abramoff MD, Magalhães PJ, Ram SJ: **Image processing with ImageJ.** *Biophotonics Int* 2004, **11**:36–42.

25. Li D, Wang L, Wang M, Xu Y-Y, Luo W, Liu Y-J, Xu Z-H, Li J, Chong K: **Engineering OsBAK1 gene as a molecular tool to improve rice architecture for high yield.** *Plant Biotechnol J* 2009, **7**:791–806.

26. Igathinathane C, Pordesimo LO, Batchelor WD: **Major orthogonal dimensions measurement of food grains by machine vision using ImageJ.** *Food Res Int* 2009, **42**:76–84.

27. Breseghello F, Sorrells ME: **QTL analysis of kernel size and shape in two hexaploid wheat mapping populations.** *Field Crops Res* 2007, **101**:172–179.

28. Herridge RP, Day RC, Baldwin S, Macknight RC: **Rapid analysis of seed size in Arabidopsis for mutant and QTL discovery.** *Plant Methods* 2011, **7**:11.

29. Tanabata T, Shibaya T, Hori K, Ebana K, Yano M: **SmartGrain: high-throughput phenotyping software for measuring seed shape through image analysis.** *Plant Physiol* 2012, **160**:1871–1880.

30. Iwata H, Ukai Y: **SHAPE: a computer program package for quantitative evaluation of biological shapes based on elliptic fourier descriptors.** *J Hered* 2002, **93**:384–385.

31. Williams K, Munkvold J, Sorrells M: **Comparison of digital image analysis using elliptic Fourier descriptors and major dimensions to phenotype seed shape in hexaploid wheat (Triticum aestivum L.).** *Euphytica* 2013, **190**:99–116.

32. Whan A, Bolger M, Bischof L: **GrainScan - Software for analysis of grain images.** 10.4225/08/536302C43FC28.

33. Whan A, Cavanagh C: **Scanned wheat grain images.** 10.4225/08/52F9AE7262532.

34. Salembier Clairon PJ, Wilkinson M: **Connected operators: A review of region-based morphological image processing techniques.** *IEEE Signal Process. Mag* 2009, **26**:136–157.

35. Sintorn I-M, Bischof L, Jackway P, Haggarty S, Buckley M: **Gradient based intensity normalization.** *J Microsc* 2010, **240**:249–258.

36. Cullis BR, Smith AB, Coombes NE: **On the design of early generation variety trials with correlated data.** *J Agric Biol Environ Stat* 2006, **11**:381–393.

37. Butler D, Cullis B, Gilmour A, Gogel BJ: *ASReml-R Reference Manual.* Brisbane: Queensland Department of Primary Industries and Fisheries; 2007.

38. R Core Team: *R: A Language and Environment for Statistical Computing.* Vienna, Austria: R Foundation for Statistical Computing; 2013.

39. Smith A, Cullis B, Thompson R: **Analyzing variety by environment data using multiplicative mixed models and adjustments for spatial field trend.** *Biometrics* 2001, **57**:1138–1147.

12

Multiscale imaging of plants: current approaches and challenges

David Rousseau[1*], Yann Chéné[2], Etienne Belin[2], Georges Semaan[2], Ghassen Trigui[3], Karima Boudehri[3], Florence Franconi[4] and François Chapeau-Blondeau[2]

Abstract

We review a set of recent multiscale imaging techniques, producing high-resolution images of interest for plant sciences. These techniques are promising because they match the multiscale structure of plants. However, the use of such high-resolution images is challenging in the perspective of their application to high-throughput phenotyping on large populations of plants, because of the memory cost for their data storage and the computational cost for their processing to extract information. We discuss how this renews the interest for multiscale image processing tools such as wavelets, fractals and recent variants to analyse such high-resolution images.

Keywords: Mutiscale imaging, Multiscale filtering, Wavelets, Fractal, ImageJ plugins

Introduction

Finding the good practices to perform high-throughput phenotyping of large populations of plants is a current challenge to meet the high-throughput capacity of genotyping and push forward the knowledge on the development of plants in different environments. Because they allow contactless and noninvasive measurements, imaging techniques are regarded as tools of highest interest in this context, to provide anatomical or physiological objective traits and outperform the limit of human vision either in terms of sensitivity, accuracy or throughput. Conversely, plant sciences constitute a new field of application for computer vision which traditionally, when applied in life sciences, used to focus more on biomedical imaging. Among the specificities of computer vision for plant sciences that are not found in biomedical imaging, is the possibility to monitor, continuously over the whole life cycle, the process of growth on structures possessing complex 3D multiscale organisation with a part visible in the air (shoot) and a part hidden in the soil (root).

There has been a significant increase in interest in plant imaging and image analysis methods in recent years, but most of the techniques proposed focus on measurements at a single scale - cell, organ, whole plant, etc. This is in contrast to modelling efforts which have stressed multiscale approaches. Such numerical models have been proposed at the scale of the entire structure of plants from iterated replication processes using L-systems (see [1,2] for reviews). Such replication processes have been shown able to reproduce the fractal organization of plant structures as measured on entire real plants. These can also serve to model the root systems [3] and have recently been used to validate image processing algorithms for root segmentation [4]. Multiple plant modeling coupled to agronomical models have also been developed [5] and allow the numerical validation of image processing algorithms at the scale of canopy. Replication processes have also been modeled at the cellular scale with possibilities of explanatory physical mechanisms for the shape of the plant at higher scales [6]. As another instance, the so-called dead leaves model takes inspiration from the foliage of plants, with leaves of different sizes and illumination which are reproduced at various scales with occlusions [7-9]. Such models have been shown to produce fractal patterns with controllable properties, and in return they offer models for the multiscale constitution of plants.

Due to the increase in size and resolution of the imaging sensors and to the development of efficient registration methods, the number of scales accessible in imaging is now ready to meet the multiscale structure of plants.

*Correspondence: david.rousseau@univ-lyon1.fr
[1] Université de Lyon, Laboratoire CREATIS, CNRS, UMR5220, INSERM, U1044, Université Lyon 1, INSA-Lyon, Villeurbanne, France
Full list of author information is available at the end of the article

In this review article, we present a set of recent high-resolution imaging techniques which cover the plant scales from molecules in the cell up to the field, and we detail how this renews the interest of scale-analysis tools for image processing.

Multiscale high-resolution imaging in plant sciences

We give in Table 1 a list of imaging techniques which have been shown in the recent literature to cover multiple scales of interest for plant sciences. At the smallest scales, single molecules up to the the entire cell are now also accessible for plants [10] with super-resolution imaging techniques [11] outperforming the classical diffraction limits such as photoactivated localization microscopy (PALM), stochastic optical resolution microscopy (STORM) [12], stimulated-emission depletion microscopy (STED) [13], three-dimensional structured illumation microscopy (3D-SIM) [14] and total internal reflection fluorescence microscopy (TIRF) [15]. At a higher range of scales, some recent microscopic imaging techniques now allow to discriminate cells of an entire organ. This is illustrated in Figure 1 with an example of optical coherence tomography (OCT) of a seedling of *Arabidopsis thaliana* during elongation with a resolution enabling to discriminate the cells of the seedling and the entire seedling. Other microscopic imaging techniques also have this capability and have been applied to plants like X-ray phase contrast imaging (X-ray PCT) [16] for microstructure analysis of the voids in an entire seed, light sheet fluorescence microscopy (LSFM) [17], multiangles confocal microscopy [18] to observe the entire seedling growth cell by cell, or optical projection tomography (OPT) [19] to image an entire leaf with possibility of cell resolution. At larger scales, inside the soil, imaging techniques give access to nodules on the root system up to the entire root system. This has been recently demonstrated in 3D in soil with absorption-based micro X-ray computed tomography [20-22], and with high-resolution imaging in 2D with rhizotron using reflectance imaging [23], or with bioluminescence imaging [24,25]. At the same metric scales but in the air, imaging techniques give access to a leaf in the shoot up to the entire shoot. This has been recently demonstrated with a variety of 3D imaging systems (see [26,27] for a recent reviews). At still larger metric scales, in field conditions, one can capture with high-resolution imaging setups embedded on an airplane or unmanned aerial vehicle (UAV) [28,29] the entire shoot from top view up to the canopy constituted by assemblies of shoots.

The list of imaging techniques given in Table 1 is not exhaustive (see [32-34] for recent reviews). This family of new imaging systems bring some challenges that would be interesting to be discussed in the field of instrumentation when applied to plants. To point only one, the new microscopies of Table 1 have been introduced for applications of broad interest in life sciences and often demonstrated on organisms which serve as models for biology, such as C-elegans, zebra fish, mice, *Drosophilae* fly or *Arabidopsis thaliana*... Consequently, the non-invasiveness property of the light used to acquire images of such a variety of organisms is mainly expressed as nonphototoxic if it does not kill the organism on a time scale linked with the time required by instrumentation for image acquisition. For specific applications on plants however, imaging the development can necessitate long time-lapsed acquisitions. For instance, imbibition and germination of a seed take hours while elongation of a seedling several days. At these stages of development illustrated in Figure 1, plants are supposed to grow in dark conditions in the soil with no light exposure, as light strongly modifies the physiology of seedlings since it activates the process of photosynthesis. It would therefore be important to revisit, as recently done for seedling in [35], the concept of phototoxicity, by adapting wavelength, energy and duration of the light used by the family of mutiscale microscopies when applied to plants.

In this review, we rather put the stress on current approaches and challenges brought by new imaging systems at the level of image processing. The point here is that techniques of Table 1 have in common, although working at very different metric and biological scales, to produce images requiring a huge capacity of data storage. This is due to the increasing size, resolution and dynamic of imaging sensors, but also to the coupling of imaging systems with motorized scanning systems. By this coupling, multiple views can be acquired and registered to produce high-resolution imaging. Multiview imaging is common practice in remote sensing. This is now extending to the scale of a single plant with rotating plates, or

Table 1 Multiple scale high-resolution imaging in plant sciences

Biological scales	Metric scales	Imaging techniques
From molecule to cell	10 nm to 10 μm	PALM-STORM [12], STED [13], 3DSIM [14]
From cell to organs	0.1 μm to dcm	OCT [30], LSFM [17], X-ray PCT [16], confocal [18], OPT [19]
From nodules to root system	μm to m	Rhizotron [24,25], X-ray μCT [20-22]
From leaf to entire shoot	mm to 10 m	depth-imaging, LIDAR [26,27]
From shoot to canopy	m to hm	remote sensing, UAV imaging [28,29]

Acronyms are explicated in Section "Multiscale high-resolution imaging in plant sciences".

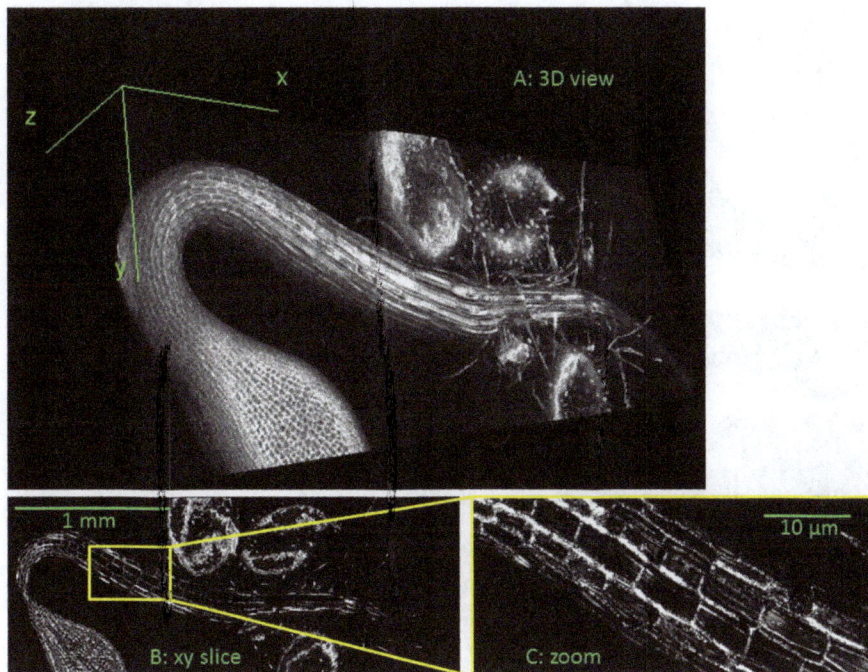

Figure 1 Images of a seedling of *Arabidopsis thaliana* acquired with optical coherence tomography (see [31] for another illustration of OCT with plants). Panel **A**: 3D view of an entire seedling. Panel **B**: 2D view in XY. Panel **C**: zoom in the solid rectangle of the 2D view of panel **B**.

at the scale of the cells with microscope scanners. For instance, the OCT system used to produce Figure 1 is associated to microstage translation systems, in such a way that the imaging technique can, after registration, capture in 3D and at the cell resolution, hundreds of such entire seedlings in a single run, resulting in some Giga bytes of data. This is 10^6 more than what has to be stored for one single plant imaged with a standard imaging resolution. Such large images can still be opened by a software like ImageJ but image processing, even some basic ones, can become very slow. This high memory cost, specially in the perspective of high-throughput phenotyping for large population of plants, calls for adapted approaches. We propose a review of the most prominent of them in the following.

Image processing tools for multiscale imaging
Combining modalities with different scales
A problematic of current interest in multiscale imaging is to combine imaging modalities providing different scales and contrasts. This association has for instance been illustrated in plant sciences with electron microscopy combined with confocal microscopy [36], or magnetic resonance imaging (MRI) combined with positon emission tomography (PET) [37] or again depth imaging combined with thermal imaging [26]. In these examples one of the modality has a relative high spatial resolution (electron microscopy, MRI, depth imaging) which provides an anatomical information while the other modality (respectively confocal, PET, thermal imaging) gives a more functional information. The functional modality can be used to locate a region a interest to be further analysed from an anatomical point of view or the other way round. This is a useful way to reduce the amount of data to be explored at high resolution. Also, the high-resolution modality can be used to analyze separately different anatomical compartments, not clearly contrasted in the functional modality. This is illustrated in Figure 2 where a 3D image of sugar beet dry seed has been acquired with a high-resolution X-ray tomograph and a MRI sequence providing gray levels proportional to the content of lipid in the seed. This gives an image of the embryo of the dry seed. As shown in Figure 2, the high-resolution modality can be used to identify the position of the cotyledon and the radicle in the embryo. If the two modalities are registered, the landmark corresponding to the beginning of the separation between cotyledon and radicle can be applied onto the MRI images and then allowing a comparison of the lipid content of these two sub-organs of the seed. Specifically here, this shows the expected higher content of lipid in the cotyledon than in the radicle. The registration step is a key image processing step in this combination of modalities. Image registration is a problematic of image processing by itself [38] with various approaches (conventionnally classified as rigid versus non rigid, automatic versus manual, ...) which have in common the calculation of a transformation

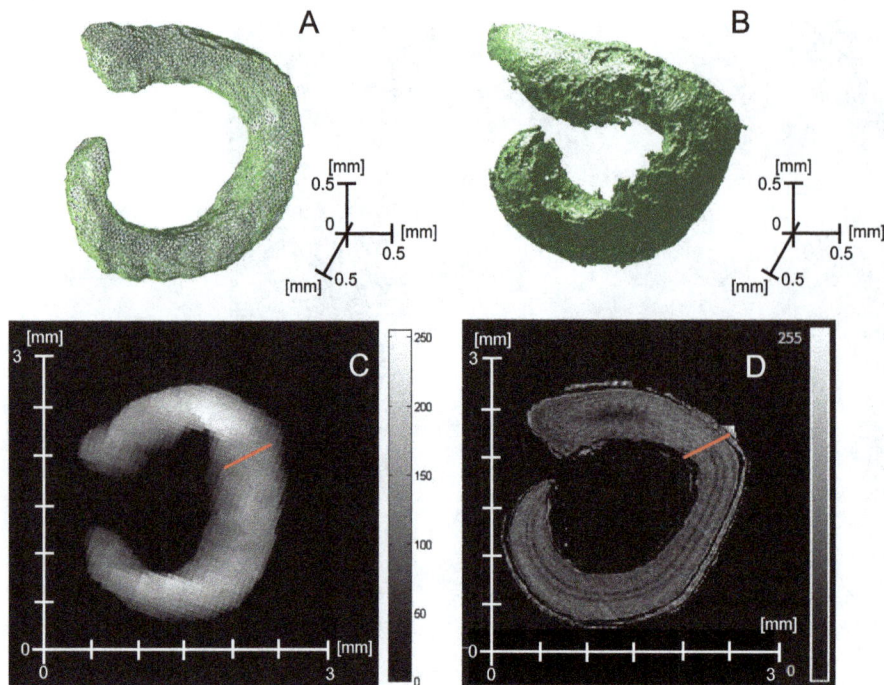

Figure 2 Bimodal imaging of the embryo of a dry seed of sugar beet with a low spatial resolution of 0,187 mm per isotropic voxel in MRI (**A** external 3D view and **C** medial 2D slice) and high spatial resolution of 7,84 μm per isotropic voxel in X-ray tomography (**B** external 3D view and **D** medial 2D slice). The MRI is a spin-echo sequence giving gray-level propotional to the lipid content of the embryo. The red line in panel **D** is positionned manually on the X-ray at the separation between cotyledon and radicle. Red line in panel **C** is positioned after registration of both imaging modalities with the ImageJ plugin TrakEM2 of Table 2.

matrix to be applied on one of the modality so as to be able to surperimpose both modalities with a locally accurate match all over the images. The development of high-resolution multiscale images has called for the design of approaches adapted to the computational cost due to the large size of the images to be registered. Instead of performing the computation of the registration on the whole image, the transformation matrix is computed on a region of interest containing landmarks and then applied on the entire image (this is available in the ImageJ Plugin TurboReg pointed in Table 2). These landmarks can be selected manually or detected automatically with scale invariant feature transforms (SIFT) [39] or variants

implemented in the ImageJ Plugin TrakEM2 pointed in Table 2. Random local deformation can occur with electron microscopy due to slicing or with MRI due to the so-called blooming effect or also with thermal imaging due to the presence of mixed pixels on edges of structures. The compensation of these local deformations randomly occuring in one of the imaging modalities remains an open challenge for image registration.

Selecting scales

The selection of structures appearing in the images at given scales can be realized with filters. The design of these filters has to incorporate some prior knowledge on

Table 2 Multiple scale image processing tools available under the free and open software ImageJ

Image processing task	ImageJ plugin weblink
Image registration	http://fiji.sc/TrakEM2
Landmark detection	http://fiji.sc/Feature_Extraction
Wavelet filtering	http://bigwww.epfl.ch/demo/fractsplines/java.html
Multiscale blob extraction	http://bigwww.epfl.ch/sage/soft/Log3D/
Multiscale vessellness extraction	http://www.longair.net/edinburgh/imagej/tubeness/
Nonlocal mean denoising	https://code.google.com/p/ij-non-local-means/
Fractal analysis	http://rsb.info.nih.gov/ij/plugins/fraclac/
Multiscale color analysis	http://www.signal-image.net/2010/04/color-inspector-3d/

the shape of the objects to be found at each scale. Among the strategies for the design of bank of filters, wavelets have shown to be a very powerful approach for application in plant sciences, see [40] for a review of the late 90's, which continue to be investigated to select patterns on leaves [41-44] or on canopies [45,46]. The wavelet approach is versatile since a large panel of wavelet functions have been designed. A wavelet is a wave-like oscillation with an amplitude that begins at zero, increases, and then decreases back to zero on a scale which can be defined by the user. Such functions are expected to constitute good filters when they share common features with the shape of the objects to be extracted. Some familly of filters have specifically been developed to extract given shapes. Let us shortly underline the vesselness filter [47] which enhances area in images where the gradient in the image is almost null in one direction and much higher in the other perpendicular directions. This situation is found with any tubular structures. Therefore, although initially developed to enhance biomedical images with vascular vessels the vesslness filter is also very much suited to enhance tubular structures met in plant sciences such as cell walls, leaf veins or branches in trees. Based on the same philosophy, enhancing areas with high gradient in all directions extract the blob-like structures [48] (cells, nodules, spherical fruits, …), or also enhancing areas with high gradient in only one direction in space extracts surface-like structures (cellular layers, plant leaves,…). These familly of filters are available under ImageJ as mentioned in Table 2.

When no prior knowledge on the shapes or scales of the objects of interest in the image is available, it is necessary to use self-adaptive methods to automatically select the appropriate scales of interest. Such methods are known as wavelet packets decomposition. However, in this case, the choice of the wavelet and of the range of scales to be analysed still have to be performed by the user. Another self-adaptive method, of more recent introduction, is the empirical mode decomposition also called Hilbert-Huang transform, where the scale analysis is purely based on the data itself. Data-dependent modes, corresponding to the local frequency data, are extracted by the analysis to decompose the signal, instead of a decomposition on preexisting elemental functions such as wavelets. Introduced for monodimensional signals [49], empirical mode decomposition has then been extended to images [50] and successfully applied to texture characterization [51]. The dominant modes of the decomposition single out the main scales in the signals or images under analysis, and keeping only the dominant modes offers natural methods for parsimonious representation and for data compression. Efficient compression schemes have been developed for landscapes captured in remote sensing for scales from canopy to field [52]. Such compression approaches by

scale selection remain open for investigation for the other scales of Table 1.

Another active field of image processing associated to the selection of scales is image denoising. Benchmark are found in the litterature [53] so as to identify the best techniques. However, such benchmarks are mostly organized on natural images not specifically suited for a given scientific field. It is very likely that the ranking of best practices may vary depending on the specific type of images. A specificity of multiscale images in plant sciences is the presence of replicated structures. This is visible in Figure 1 with cells or in Figure 3 with leaves. This replication process found in plant architecture constitutes a prior which is not found in all natural images. This observation motivates the choice of the so-called nonlocal mean [54] as interesting denoising methods. Nonlocal mean denoising is realized by averaging pixel content weighted by how similar these pixels are to the target pixel. In its principle, this non local averaging process, available under ImageJ plugin given in Table 2, will be very efficient if a lot of pixels are similar to the target pixel like in the self-similar structures found in plant sciences.

Characterizing multiscale signatures

Instead of selecting specific scales of interest, another approach is to characterize the global organization over multiple scales in the images. Nontrivial regularities developing in a self-similar way across a significant range of scales usually identify the existence of a fractal organization. Fractal concepts have been shown relevant to the description of plants, of their roots and shoots, which often show self-similar organizations across scales [55-59]. Especially, such organizations lead to high surface areas at the interfaces with the environment, ensuring for the plant efficient capture of nutrients and energy.

Self-similarity accross scales, i.e. fractal features, can thus be found in various properties of images from plants. For instance, they have been reported in the spatial organization of gray-level luminance images from outdoor scenes of woods and plants [60,61]. This is manifested by scale-free power-law evolutions present in the frequency spectrum of luminance images or also in their spatial correlation functions. Also, the colorimetric organization of natural images including landscapes with plants has been reported to carry self-similarity and fractal properties [62-64]. More recently, multiscale analysis has been undertaken for plant images obtained from another imaging technique delivering depth images of a physical scene [65]. The depth map images from outdoor scenes of woods and plants as in [60,61] were shown in [65] to also reveal self-similarity and fractal properties. Such multiscale image analyses revealing and characterizing fractal properties in plants are important to contribute to their understanding, since the fractal and multiscale

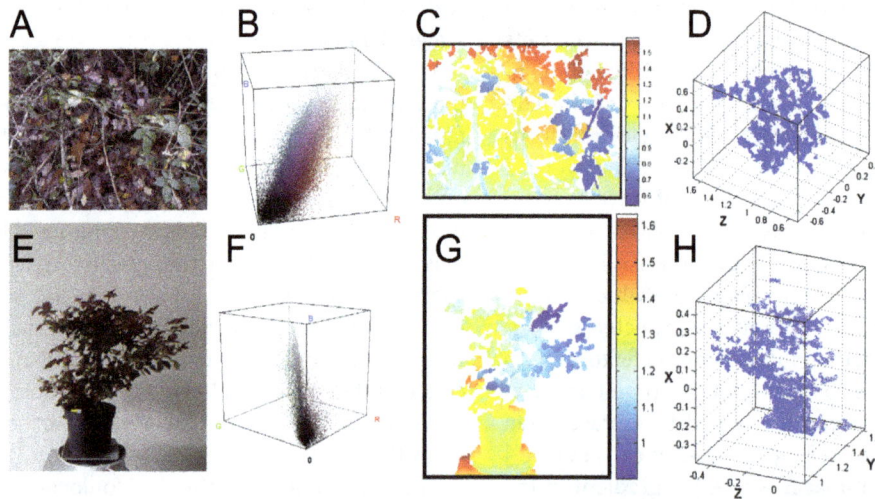

Figure 3 Bimodal RGB-depth representation of a forestery scene (first row) and a single plant (second row). Panels **A** and **E**: RGB luminance. Panels **B** and **F**: corresponding RGB histogram. Panels **C** and **G**: depth map expressed in meter. Panels **D** and **H**: corresponding point cloud of the depth map.

organization of plants has a direct impact on their functioning, for instance for efficient interactions with their environment as evoked above [58,66-68]. Also, fractal characterization of plants is useful to devise synthetic models of plants with sufficient realism [1].

For illustration, we proceed to the scale analysis of several images from a forestry scene and from a single plant, as shown in Figure 3, acquired with a bimodal RGB-depth camera [69]. Figure 3 shows four possible ways of vizualizing such data, with an RGB luminance image,

with a 3D RGB histogram, with depth map or with a 3D depth point cloud. We analyze the scale organization in each of these four representations. The spatial frequencies of the RGB luminance images in Figure 3 are analyzed with the power spectrum computed via the periodogram method, through the squared modulus of the two-dimensional Fourier transform, expressed in polar coordinates in the plane of spatial frequencies from a single plant, as shown in Figure 3, acquired with a bimodal RGB-depth camera [65]. An average is then realized over

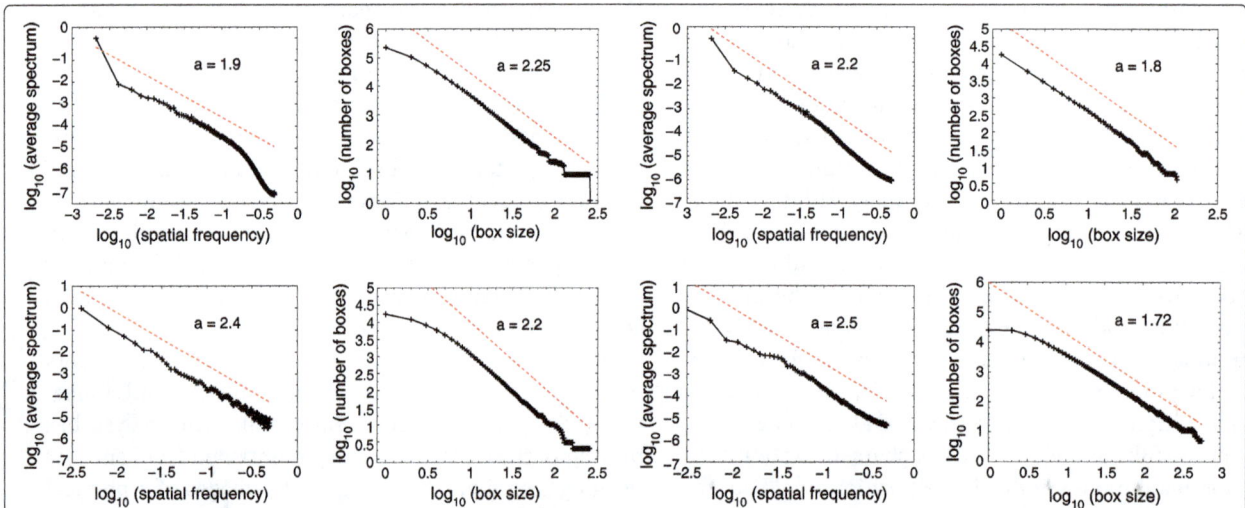

Figure 4 First and second rows: multiscale analysis of RGB-depth images of first and second rows of Figure 3. First column: average spectrum of RGB luminance image as a function of spatial frequency on a log-log plot. Second column: box counting in the RGB histogram as a function of the box size on a log-log plot. Third column: average spectrum of depth map image as a function of spatial frequency on a log-log plot. Fourth column: box counting in the point cloud of the depth map as a function of the box size on a log-log plot. In each graph, the dotted line with its slope indicated represents a model to appreciate a power-law evolution to match the data. The slopes reveal noninteger exponents for the power-law evolutions matching the data over a significant range of scales. This indicates nontrivial self-invariance of the data across scales, i.e. a fractal organization.

the angular coordinate to yield the orientationally averaged spectrum. This power spectrum is computed on a gray-level version of the RGB image of Figure 3 and on the depth image of Figure 3 as a function of the spatial frequency. The results shown in Figure 4 demonstrate for both the forestry scene and for the single plant, and with both imaging modalities, scale-invariant power-law signatures over a significant range of scales, represented by the spatial frequency. Also, in Figure 4 we implemented the box counting method [65] on the point clouds constituted by the RGB histogram of Figure 3 and by the depth image of Figure 3. The box counting values are obtained in terms of scales represented by the side length of the various boxes. For each side length, we compute the number of boxes with this side length which are needed to cover all the point cloud. Here again the results shown in Figure 4 demonstrate, for both scenes and both modalities, power-law signatures over a significant range of scales, represented by the size of the covering boxes. As shown in Figure 4, the measures computed in luminance space, in RGB space as well as in depth space, all display scale-invariant power-law signatures over significant ranges of scales. Such fractal signatures are interesting in the context of multiscale imaging since they constitute an efficient and concise way to characterize a complex organization. Fractal image processing tools have been widely applied to characterize plants (see [70] for a review) at the scales of leaf [67,71], canopy [72]. So far the fractal characterization of plants at the microscopic scale is open for investigation. Fractal analyses of root systems have been undertaken but mainly from plants taken out of the soil [73-75]. The new high-resolution X-ray CT reported in [20-22] therefore opens new perspectives for the fractal characterization of the root system directly in 3D and in the soil.

Conclusion

High-resolution multiscale imaging in plant sciences was until recently limited to the domain of remote sensing. It is now also possible to capture entire roots or shoots of plants, at various stages of development, with cellular or subcellular spatial resolution. These high-resolution imagings are producing huge amounts of data, specially when they are applied to large populations of plants in high-throughput phenotyping. In this framework, we have highlighted here some current approaches connected to the multiscale analysis of plants and pointed toward efficient computational implementation under the free and open software ImageJ in Table 2. Open problems emerge for image compression and image characterization. Multiscale approaches are specifically relevant for the new microscopies such as those presented in Table 1; these are more recent and have received so far, in a multiscale

perspective, less attention than remote-sensing imaging or than proximal detection in the field.

Competing interests
The authors declare that they have no competing interests.

Authors' contributions
DR conceived and designed this review, carried out the acquisition of OCT and RGB-depth data, conceived and interpreted the whole data, wrote and revised the manuscript. YC realized the acquisition and data analysis of RGB-depth images. EB helped in the analysis of RX-MRI data and revised the manuscript. GS helped in the acquisition and carried out the analysis of MRI data. GT and KB carried out the acquisition and analysis of RX data. FF carried out the MRI acquisition. FCB helped in the analysis of RGB-depth images and contributed in the writing and revision of the manuscript. All authors read and approved the final manuscript.

Acknowledgements
This work received supports from the French Government supervised by the Agence Nationale de la Recherche in the framework of the program Investissements d'Avenir under reference ANR-11-BTBR-0007 (AKER program).

Author details
[1] Université de Lyon, Laboratoire CREATIS, CNRS, UMR5220, INSERM, U1044, Université Lyon 1, INSA-Lyon, Villeurbanne, France. [2] Laboratoire Angevin de Recherche en Ingénierie des Systèmes (LARIS), Université d'Angers, 62 avenue Notre Dame du Lac, 49000 Angers, France. [3] GEVES, Station Nationale d'Essais de Semences (SNES), rue Georges Morel, 49071 Beaucouzé, France. [4] La Plateforme d'Ingénierie et Analyses Moléculaires (PIAM), Université d'Angers, 49000 Angers, France.

References
1. Prusinkiewicz P, Lindenmayer A. The Algorithmic Beauty of Plants. Berlin: Springer; 2004.
2. Godin C. Representing and encoding plant architecture: a review. Ann Forest Sci. 2000;57(5):413–38.
3. Leitner D, Klepsch S, Bodner G, Schnepf A. A dynamic root system growth model based on L-systems. Plant and Soil. 2010;332(1-2):177–92.
4. Benoit L, Rousseau D, Belin É, Demilly D, Chapeau-Blondeau F. Simulation of image acquisition in machine vision dedicated to seedling elongation to validate image processing root segmentation algorithms. Comput Electron Agric. 2014;104:84–92.
5. Dufour-Kowalski S, Pradal C, Donès N, Barbier De Reuille P, Boudon F, Chopard J, et al. OpenAlea: An open-software plateform for the integration of heterogenous FSPM components. In: 5th International Workshop on Functional-Structural Plant Models, FSPM07, November, 2007. Napier, New Zealand: The Horticulture and Food Research Institute of New Zealand Ltd.; 2007. p. 1–2.
6. Mirabet V, Das P, Boudaoud A, Hamant O. The role of mechanical forces in plant morphogenesis. Annu Rev Plant Biol. 2011;62:365–85.
7. Lee AB, Mumford D, Huang J. Occlusion models for natural images: A statistical study of a scale-invariant dead leaves model. International Journal of Computer Vision. 2001;41:35–59.
8. Bordenave C, Gousseau Y, Roueff F. The dead leaves model: A general tessellation modeling occlusion. Adv Appl Probability. 2006;38:31–46.
9. Gousseau Y, Roueff F. Modeling occlusion and scaling in natural images. SIAM Journal of Multiscale Modeling and Simulation. 2007;6:105–34.
10. Langhans M, Meckel T. Single-molecule detection and tracking in plants. Protoplasma. 2014;251(2):277–91.
11. Elgass K, Caesar K, Schleifenbaum F, Stierhof Y-D, Meixner AJ, Harter K. Novel application of fluorescence lifetime and fluorescence microscopy enables quantitative access to subcellular dynamics in plant cells. PLoS One. 2009;4(5):5716.
12. Gutierrez R, Grossmann G, Frommer WB, Ehrhardt DW. Opportunities to explore plant membrane organization with super-resolution microscopy. Plant Physiol. 2010;154(2):463–6.

13. Sparkes I, Graumann K, Martinière A, Schoberer J, Wang P, Osterrieder A. Bleach it, switch it, bounce it, pull it: using lasers to reveal plant cell dynamics. J Exp Bot. 2010351. doi:10.1093/jxb/erq351.

14. Fitzgibbon J, Bell K, King E, Oparka K. Super-resolution imaging of plasmodesmata using three-dimensional structured illumination microscopy. Plant Physiol. 2010;153(4):1453–63.

15. Wan Y, Ash WM, Fan L, Hao H, Kim MK, Lin J. Variable-angle total internal reflection fluorescence microscopy of intact cells of Arabidopsis thaliana. Plant Methods. 2011;7(1):27.

16. Cloetens P, Mache R, Schlenker M, Lerbs-Mache S. Quantitative phase tomography of Arabidopsis seeds reveals intercellular void network. Proc Nat Acad Sci USA. 2006;103(39):14626–30.

17. Costa A, Candeo A, Fieramonti L, Valentini G, Bassi A. Calcium dynamics in root cells of Arabidopsis thaliana visualized with selective plane illumination microscopy. PLoS ONE. 2013;8:75646.

18. Fernandez R, Das P, Mirabet V, Moscardi E, Traas J, Verdeil J, et al. Imaging plant growth in 4D: robust tissue reconstruction and lineaging at cell resolution. Nat Methods. 2010;7:547–53.

19. Leea K, Avondob J, Morrisonc H, Blotb L, Starkd M, Sharpec J, et al. Visualizing plant development and gene expression in three dimensions using optical projection tomography. Plant Cell. 2006;18:2145–56.

20. Mairhofer S, Zappala S, Tracy SR, Sturrock C, Bennett M, Mooney SJ, et al. RooTrak: automated recovery of three-dimensional plant root architecture in soil from X-ray microcomputed tomography images using visual tracking. Plant Physiol. 2012;158(2):561–9.

21. Zappala S, Mairhofer S, Tracy S, Sturrock CJ, Bennett M, Pridmore T, et al. Quantifying the effect of soil moisture content on segmenting root system architecture in X-ray computed tomography images. Plant and Soil. 2013;370(1-2):35–45.

22. Mairhofer S, Zappala S, Tracy S, Sturrock C, Bennett MJ, Mooney SJ, et al. Recovering complete plant root system architectures from soil via X-ray micro-computed tomography. Plant Methods. 2013;9(8):1–7.

23. Salon C, Jeudy C, Bernard C, Mougel C, Coffin A, Bourion V, et al. Google Patents. EP Patent App. 2014EP20,130,173,626. http://www.google.com/patents/EP2679088A1?cl=fr.

24. Dinneny JR. Luciferase Reporter System for Roots and Methods of Using the Same. Google Patents. US Patent App. 2014;13/970:960. http://www.google.com/patents/US20140051101.

25. Rellán Álvarez R. Growth and luminescence observatory for roots (glo-roots): A platform for the analysis of root structure and physiology in soil. In: Plant and Animal Genome XXII Conference. Plant and Animal Genome; 2014.

26. Chéné Y, Belin E, Chapeau-Blondeau F, Boureau T, Caffier V, Rousseau D. Anatomo-functional bimodality imaging for plant phenotyping: An insight through depth imaging coupled to thermal imaging. In: Plant Image Analysis: Fundamentals and Applications. Boca Raton: CRC Press; 2014. Chap. 9.

27. Paulus S, Behmann J, Mahlein A-K, Plümer L, Kuhlmann H. Low-cost 3d systems: Suitable tools for plant phenotyping. Sensors. 2014;14(2):3001–18.

28. Jones HG, Vaughan RA. Remote Sensing of Vegetation: Principles, Techniques, and Applications. Oxford: Oxford University Press; 2010.

29. Zarco-Tejada PJ, González-Dugo V, Berni JA. Fluorescence, temperature and narrow-band indices acquired from a UAV platform for water stress detection using a micro-hyperspectral imager and a thermal camera. Remote Sensing Environ. 2012;117:322–37.

30. Hettinger JW, de la Pena Mattozzi M, Myers WR, Williams ME, Reeves A, Parsons RL, et al. Optical coherence microscopy. a technology for rapid, in vivo, non-destructive visualization of plants and plant cells. Plant Physiol. 2000;123(1):3–16.

31. Meglinski I, Buranachai C, Terry L. Plant photonics: application of optical coherence tomography to monitor defects and rots in onion. Laser Phys Lett. 2010;7(4):307.

32. Dhondt S, Wuyts N, Inzé D. Cell to whole-plant phenotyping: the best is yet to come. Trends Plant Sci. 2013;18(8):428–39.

33. Ehrhardt DW, Frommer WB. New technologies for 21st century plant science. The Plant Cell Online. 2012;24(2):374–94.

34. Sozzani R, Busch W, Spalding EP, Benfey PN. Advanced imaging techniques for the study of plant growth and development. Trends in Plant Sci. 2014;19(5):304–10.

35. Benoit L, Belin É, Durr C, Chapeau-Blondeau F, Demilly D, Ducournau S, et al. Computer vision under inactinic light for hypocotyl-radicle separation with a generic gravitropism-based criterion. Comput Electron Agric. 2015;111:12–7.

36. Bell K, Mitchell S, Paultre D, Posch M, Oparka K. Correlative imaging of fluorescent proteins in resin-embedded plant material1. Plant Physiol. 2013;161(4):1595–603.

37. Jahnke S, Menzel MI, Van Dusschoten D, Roeb GW, Bühler J, Minwuyelet S, et al. Combined mri–pet dissects dynamic changes in plant structures and functions. The Plant J. 2009;59(4):634–44.

38. Zitova B, Flusser J. Image registration methods: a survey. Image Vision Comput. 2003;21(11):977–1000.

39. Lowe DG. Distinctive image features from scale-invariant keypoints. Int J Comput Vision. 2004;60(2):91–110.

40. Dale MRT, Mah M. The use of wavelets for spatial pattern analysis in ecology. Journal of Vegetation Science. 1998;9(6):805–14.

41. Gu X, Du J-X, Wang X-F. Leaf recognition based on the combination of wavelet transform and Gaussian interpolation. In: Advances In Intelligent Computing. Berlin: Springer; 2005. p. 253–62.

42. Prasad S, Kumar P, Tripathi R. Plant leaf species identification using curvelet transform. In: 2nd International Conference on Computer and Communication Technology (ICCCT). IEEE; 2011. p. 646–52.

43. Casanova D, de Mesquita Sa Junior JJ, Martinez Bruno O. Plant leaf identification using Gabor wavelets. Int J Imaging Sys Technol. 2009;19(3): 236–43.

44. Bours R, Muthuraman M, Bouwmeester H, van der Krol A. Oscillator: A system for analysis of diurnal leaf growth using infrared photography combined with wavelet transformation. Plant Methods. 2012;8(1):29.

45. Epinat V, Stein A, de Jong SM, Bouma J. A wavelet characterization of high-resolution NDVI patterns for precision agriculture. Int J Appl Earth Observation Geoinformation. 2001;3(2):121–32.

46. Strand EK, Smith AM, Bunting SC, Vierling LA, Hann DB, Gessler PE. Wavelet estimation of plant spatial patterns in multitemporal aerial photography. Int J Remote Sensing. 2006;27(10):2049–54.

47. Frangi AF, Niessen WJ, Vincken KL, Viergever MA. Multiscale vessel enhancement filtering. In: Medical Image Computing and Computer-Assisted Intervention-MICCAI'98. Berlin: Springer; 1998. p. 130–7.

48. Sage D, Neumann FR, Hediger F, Gasser SM, Unser M. Automatic tracking of individual fluorescence particles: Application to the study of chromosome dynamics. IEEE Trans Image Process. 2005;14:1372–83.

49. Huang NE, Shen Z, Long SR, Wu MC, Shih HH, Zheng Q, et al. The empirical mode decomposition and the Hilbert spectrum for nonlinear and non-stationary time series analysis. Proc R Soc London Ser A: Math Phys Eng Sci. 1998;454(1971):903–95.

50. Nunes JC, Bouaoune Y, Delechelle E, Niang O, Bunel P. Image analysis by bidimensional empirical mode decomposition. Image Vision Comput. 2003;21(12):1019–26.

51. Nunes JC, Guyot S, Deléchelle E. Texture analysis based on local analysis of the bidimensional empirical mode decomposition. Machine Vision Appl. 2005;16(3):177–88.

52. Myers WL, Patil GP. Pattern-Based Compression of Multi-Band Image Data for Landscape Analysis. Berlin: Springer; 2006.

53. Buades A, Coll B, Morel J-M. A review of image denoising algorithms, with a new one. Multiscale Model Simul. 2005;4(2):490–530.

54. Buades A, Coll B, Morel J-M. A non-local algorithm for image denoising. In: Computer Vision and Pattern Recognition, 2005. CVPR 2005. IEEE Computer Society Conference On. vol. 2. IEEE; 2005. p. 60–5.

55. Palmer MW. Fractal geometry: A tool for describing spatial patterns of plant communities. Vegetatio. 1988;75:91–102.

56. Critten DL. Fractal dimension relationships and values associated with certain plant canopies. J Agric Eng Res. 1997;67:61–72.

57. Alados CL, Escos J, Emlen JM, Freeman DC. Characterization of branch complexity by fractal analyses. Int J Plant Sci. 1999;160:147–55.

58. Morávek Z, Fiala J. Fractal dynamics in the growth of root. Chaos, Solitons & Fractals. 2004;19:31–4.

59. Alados CL, Pueyo Y, Navas D, Cabezudo B, Gonzalez A, Freeman DC. Fractal analysis of plant spatial patterns: A monitoring tool for vegetation transition shifts. Biodivers Conserv. 2005;14:1453–68.

60. Ruderman DL, Bialek W. Statistics of natural images: Scaling in the woods. Phys Rev Lett. 1994;73:814–7.

61. Ruderman DL. Origins of scaling in natural images. Vision Res. 1997;37(23): 3385–98.

62. Chauveau J, Rousseau D, Richard P, Chapeau-Blondeau F. Multifractal analysis of three-dimensional histogram from color images. Chaos, Solitons & Fractals. 2010;43(1):57–67.

63. Chauveau J, Rousseau D, Chapeau-Blondeau F. Fractal capacity dimension of three-dimensional histogram from color images. Multidimensional Syst Signal Process. 2010;21(2):197–211.

64. Chapeau-Blondeau F, Chauveau J, Rousseau D, Richard P. Fractal structure in the color distribution of natural images. Chaos, Solitons & Fractals. 2009;42(1):472–82.

65. Chéné Y, Belin E, Rousseau D, Chapeau-Blondeau F. Multiscale analysis of depth images from natural scenes: Scaling in the depth of the woods. Chaos, Solitons & Fractals. 2013;54:135–49.

66. Ferraro P, Godin C, Prusinkiewicz P. Toward a quantification of self-similarity in plants. Fractals. 2005;13:91–109.

67. Martinez Bruno O, de Oliveira Plotze R, Falvo M, de Castro M. Fractal dimension applied to plant identification. Inf Sci. 2008;178:2722–33.

68. Da Silva D, Boudon F, Godin C, Sinoquet H. Multiscale framework for modeling and analyzing light interception by trees. Multiscale Model Simul. 2008;7:910–33.

69. Chéné Y, Rousseau D, Lucidarme P, Bertheloot J, Caffier V, Morel P, et al. On the use of depth camera for 3D phenotyping of entire plants. Comput Electron Agric. 2012;82:122–7.

70. Chandra M, Rani M. Categorization of fractal plants. Chaos Solitons & Fractals. 2009;41(3):1442–7.

71. Scheuring I, Riedi RH. Application of multifractals to the analysis of vegetation pattern. J Vegetation Sci. 1994;5(4):489–496.

72. Bradshaw G, Spies TA. Characterizing canopy gap structure in forests using wavelet analysis. J Ecol. 1992;80:205–15.

73. Tastumi J, Yamauchi A, Kono Y. Fractal analysis of plant root systems. Ann Bot. 1989;64(5):499–503.

74. Izumi Y, Iijima M. Fractal and multifractal analysis of cassava root system grown by the root-box method. Plant Production Sci. 2002;5(2):146–51.

75. Ketipearachchi KW, Tatsumi J. Local fractal dimensions and multifractal analysis of the root system of legumes. Plant Production Sci. 2000;3(3):289–95.

13

Remote, aerial phenotyping of maize traits with a mobile multi-sensor approach

Frank Liebisch[1*], Norbert Kirchgessner[1], David Schneider[2], Achim Walter[1] and Andreas Hund[1]

Abstract

Background: Field-based high throughput phenotyping is a bottleneck for crop breeding research. We present a novel method for repeated remote phenotyping of maize genotypes using the Zeppelin NT aircraft as an experimental sensor platform. The system has the advantage of a low altitude and cruising speed compared to many drones or airplanes, thus enhancing image resolution while reducing blurring effects. Additionally there was no restriction in sensor weight. Using the platform, red, green and blue colour space (RGB), normalized difference vegetation index (NDVI) and thermal images were acquired throughout the growing season and compared with traits measured on the ground. Ground control points were used to co-register the images and to overlay them with a plot map.

Results: NDVI images were better suited than RGB images to segment plants from soil background leading to two separate traits: the canopy cover (CC) and its NDVI value ($NDVI_{Plant}$). Remotely sensed CC correlated well with plant density, early vigour, leaf size, and radiation interception. $NDVI_{Plant}$ was less well related to ground truth data. However, it related well to the vigour rating, leaf area index (LAI) and leaf biomass around flowering and to very late senescence rating. Unexpectedly, $NDVI_{Plant}$ correlated negatively with chlorophyll meter measurements. This could be explained, at least partially, by methodical differences between the used devices and effects imposed by the population structure. Thermal images revealed information about the combination of radiation interception, early vigour, biomass, plant height and LAI. Based on repeatability values, we consider two row plots as best choice to balance between precision and available field space. However, for thermography, more than two rows improve the precision.

Conclusions: We made important steps towards automated processing of remotely sensed data, and demonstrated the value of several procedural steps, facilitating the application in plant genetics and breeding. Important developments are: the ability to monitor throughout the season, robust image segmentation and the identification of individual plots in images from different sensor types at different dates. Remaining bottlenecks are: sufficient ground resolution, particularly for thermal imaging, as well as a deeper understanding of the relatedness of remotely sensed data and basic crop characteristics.

Keywords: Remote sensing, Aerial phenotyping, Near infrared imaging, Image analysis, NDVI, Thermal imaging, *Zea mays*

Background

Field-based high-throughput phenotyping methods are urgently needed by plant breeding research [1,2]. Whereas laboratory-based phenotyping platforms that monitor the performance of single plants of model species have advanced greatly in recent years (e.g. [3], for a

review see [4]), the development of field-based phenotyping approaches has lagged. For field-based methods, progress has been made mostly using camera-based approaches that are mounted on ground-based vehicles like tractors (e.g. [5,6]; for a review see [2,7]). Yet, there is little progress on methods and platforms that operate from the air [1] although currently drones are becoming increasingly popular for aerial photography. However, high quality camera systems often still exceed the payload of available drones. Automation of data processing,

* Correspondence: frank.liebisch@usys.ethz.ch
[1]Institute of Agricultural Sciences, ETH Zürich, Universitätstrasse 2, 8092 Zürich, Switzerland
Full list of author information is available at the end of the article

difficulties in extraction of meaningful parameters and blurry images taken from conventional carrier systems such as airplanes travelling at relatively high altitude are other reasons which presently restrict fast methodological advances. Nevertheless, the potential throughput of airborne phenotyping approaches is intrinsically higher than that of ground-based approaches, for several reasons: (1) wider viewing angle from the air, (2) potentially higher travelling speed, (3) absence of physical contact with and hence no mechanical distraction of the growing crop and (4) independence of wet soil conditions that prevent traffic on the ground.

Maize is one of the most important staple crops and has gained an enormous importance in tropical and temperate regions as a food, fodder, and energy crop. As a consequence there is a high need to develop high-throughput methods for hybrid breeding of maize in order to increase selection efficiency [8-11]. Relevant breeding approaches require field-based testing of their genotypes [12]. Often hundreds or thousands of genotypes need to be investigated for their performance in the field and hence need to be grown and assessed synchronously side by side. It is widely accepted that in such breeding programs, phenotyping of traits that are related to yield and quality is currently constituting a serious bottleneck [2,13], for which the development of technological possibilities has not kept pace with the genomic characterization of the germplasm.

Therefore, we aimed to develop a concept allowing for 1) continuous measurements of genotypes throughout the growing season using RGB and near infrared imaging and thermography, 2) develop protocols to automatically identify individual field plots in images derived by the different sensors at different dates and from slightly different angles, 3) identify suitable traits and optimal plot size based on the repeatability and 4) relate remotely sensed data to ground truth data.

More specifically, this study investigates the application of a camera combination consisting of (1) a standard RGB camera, (2) a camera to determine the normalized difference vegetation index (NDVI) and (3) a high-resolution thermal camera (Table 1). This sensor array was operated manually on a Zeppelin aircraft offering regular sight-seeing round trips. The maize experiment was placed on one of the flight tracks in order to ensure frequent monitoring during the growing season (Figure 1). The experimental field contained 16 different maize genotypes, arranged in a well-designed plot structure with plots of multiple sizes (i.e. different number of rows). Each genotype x plot size combination was replicated four times (Additional file 1). From the acquired images, parameters such as the canopy cover, leaf greenness and canopy temperature were detected, and a software routine was developed that allowed for (semi-)

automated identification of and data extraction from the field plot structure. The extracted parameter values were then correlated with ground measurements of relevant crop traits collected throughout the crop development. They comprise phenological traits (like the time needed to reach certain key developmental stages) and morphological characteristics (like plant height and leaf biomass) that contribute to the performance of a genotype in a given environment. We hypothesize that the elaborated methods of image capture and analysis can be used to identify genotypic differences and changes during development of maize throughout the season and that remotely sensed parameters can be related sufficiently well to ground measured plant properties and traits relevant for breeding.

Results
Processing of image-based signals
A semi-automated recognition of black metal markers on 2 m high poles worked well for all sensor outputs. The markers were distinguishable from the soil and plant signal by all three cameras (Figure 1). However, cleaning of the field markers before flight campaigns was necessary, especially during pollen shedding. White tarps that were put on the ground were less useful as markers since they were easily covered with soil, particularly after rains. Cleaning of the white tarps proved to be too laborious and time consuming. Moreover, with increasing plant size the white tarps were progressively obscured by the maize plants, changing the detectable marker shape and increasing the need for manual co-registration of the images. However, the tarps were very useful as landmarks for the pilots when the plants were small, and the plots were difficult to detect.

After identification in the images, the black metal markers were used to clip the images to the area of interest (AoI) and subsequently co-register all sensor output images and correct them for trapeze distortion (Figure 1). On top of the aligned and corrected image stacks, the prepared experimental plot mask was projected for subsequent plotwise data extraction.

The most basic information derived from the images of the vegetation camera, was the plot-based NDVI ($NDVI_{Plot}$) including both, plant and soil information. A segmentation process based on the NDVI or RGB information led to two additional traits: the canopy cover (CC) and the NDVI values of the area covered by plants ($NDVI_{Plant}$). CC was best calculated from NDVI images since throughout the season the same segmentation threshold could be used whereas for RGB images, it had to be adjusted for each measurement campaign (Figure 2). For NDVI images, only maximal signal intensity needed to be adjusted to comparable conditions for the different measurement campaigns. Using a NDVI threshold of 0.1 excluded non-plant material

Table 1 Size of ground images (length and width) and effective pixel dimensions (Instantaneous Field of View = IFoV) as affected by sensor resolution and measurement altitude

Camera	Lens (focal length)	Sensor resolution	Sensor dimensions	Image parameter	Altitude 290	300	310
	mm	pixel	mm		m	m	m
NIR	60	4282 × 2848	22.2 × 14.8	length	107.6	111.3	115.5
				width	71.5	74.0	76.5
				IFoV[a]	0.025	0.026	0.028
RGB	60	3898 × 2595	22.2 × 14.8	length	107.3	111.1	114.7
				width	71.4	73.9	76.4
				IFoV	0.0275	0.0285	0.0295
IR	75	640 × 480	14.9 × 11.2[b]	length	61.9	64	66.1
				width	46.4	48	49.6
				IFoV	0.096	0.1	0.103

[a]IFoV = (pixel dimension*distance)/focal length.
[b]IFoV = (sensor pixel size* distance to ground)/focal length, derived from companies IFoV calculator.

and included green and senescent plant material. A threshold of 0.2 excluded large parts of the senescent plant area as well (data not shown).

Seasonal development of canopy cover was most reliably evaluated in plots with more than two rows

The canopy cover increased until flowering (540°Cd) and decreased during the late senescence phase (after 892°Cd; Figure 3). The corresponding dates and growth stages can be found in Table 2. The canopies of nearly all investigated genotypes were closed at the onset of flowering indicated by CC values above 0.95. At this stage, only genotypes 6 and 15 showed CC below 0.9 (data not shown). A small reduction of CC was observed during flowering and shortly after flowering between 540 and 793°Cd (Figure 3) and likely was related to green leaf area overlayed by tassels and anthers.

To elucidate which plot size (row number) was sufficient to differentiate among genotypes we used the repeatability, i.e. the proportion of the genotypic variation compared to the overall phenotypic variation. The repeatability of CC was above 0.95 before onset of tassels for the three and four row plots decreasing slightly to 0.85 at maturity. As expected, the smaller plot size showed lower repeatability, especially for the one row plots. There, values ranged from 0.9 before flowering down to 0.7 close to maturity. Clearly, three- to four row plots were preferable for this type of aerial observations.

Seasonal development of NDVI was most reliably measured in plots with more than one row

$NDVI_{Plot}$ and $NDVI_{Plant}$ (Figure 4) showed similar seasonal trends, but $NDVI_{Plant}$ had less variance within each measurement point in time. Yet, differences were higher for $NDVI_{Plant}$ (indicated by lower HSD), demonstrating the value of image segmentation. In general, NDVI

increased until 892°Cd (Figure 4), whereas for some genotypes a plateau in NDVI values was observed from 727°Cd onwards (data not shown). At 940°Cd, NDVI dropped and subsequently decreased slowly to the lowest values observed at 1366°Cd.

The repeatability of NDVI values were high ($h^2 > 0.85$) with the highest values being observed during and after flowering (Figure 4). Repeatability was generally higher in plots with more than one row, but there was only a minor advantage of having more than two rows. The reduction of the repeatability of $NDVI_{Plot}$ towards the end of the growing season was not observed for $NDVI_{Plant}$ in the three and four row plots.

The skewness of NDVI (Additional file 2 section 6) showed a different seasonal pattern than NDVI being relatively constant with a small reduction at the beginning of flowering and a stronger increase at the end of the season. The repeatability of the skewness was generally lower than the one for the CC and NDVI. Plot size effects were more pronounced making it less suited to differentiate among genotypes.

Highest repeatability of canopy temperature was found on temperate days

Canopy temperatures (T_C) ranged from 22 to 27°C during flight campaigns with significant differences between day of measurement (Figure 5). Across the season, T_C was highly correlated to air temperature T_A, measured by the close-by weather station (Table 3). To exclude the temperature effect, T_C was normalized to T_A resulting in the temperature difference dT. As to be expected the relationship of dT to T_A was not significant but the effect of radiation parameters (actual PAR and sun hours) and evapotranspiration (ETo) were more pronounced (Table 3). Measurements of dT taken between 612 and 940°Cd were negative, indicating the canopy was cooler

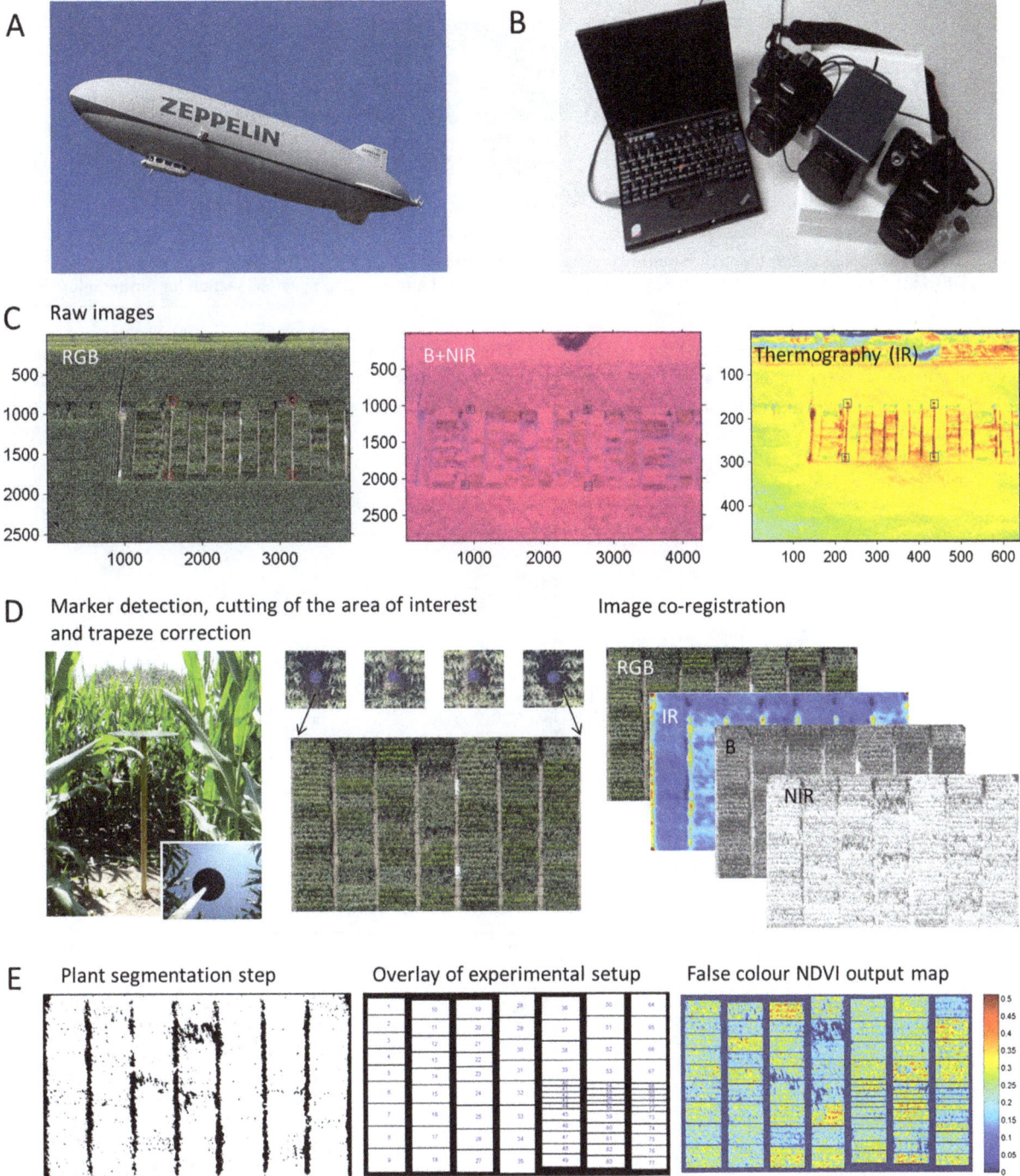

Figure 1 Illustration of the imaging and analysis pipeline. (A) Aerial platform Zeppelin NT, **(B)** handheld sensor array consisting of three cameras, **(C)** images derived by: a consumer camera (RGB), a modified consumer camera for vegetation detection (B + NIR) and an infrared camera (Thermography, IR). Squares indicate location of field markers. **(D)** Round black metal plates serve as field markers (left) for automatic detection and subsequent clipping of the area of interest (AoI) and trapeze correction of the raw images (middle). Co-registered tiff images for the RGB composite and the three data channels IR, B and NIR (right). **(E)** Three steps of image procession: mask to segment plant from soil pixel (left), mask to identify plots (middle) and a combined output map of NDVI$_{Plant}$ values within plots in false colour (right). The shown images represent one of three parts of the field separated for measurement purposes, details described in the material section. See Additional file 1 for a complete image of the experiment.

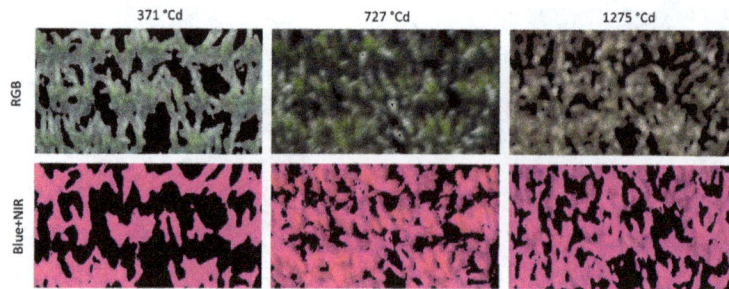

Figure 2 Comparison of HSB and NDVI based threshold segmentation masks laid over unprocessed RGB and B + NIR images, respectively, when the same colour or NDVI thresholds were applied throughout the season (see materials section for further info).

than ambient air measured by the weather station. Only at 1275°Cd when plants were in an advanced senescent stage dT was positive. The skewness of T_C showed a small but significant reduction over time but no genotype effect (data not shown).

The repeatability of T_C was highly affected by plot size and day of measurement (Figure 5) which differed in the prevailing climate conditions (Table 4). In one row plots, T_C showed a very low repeatability except at 727 and 940°Cd when high values were found in plots of all sizes

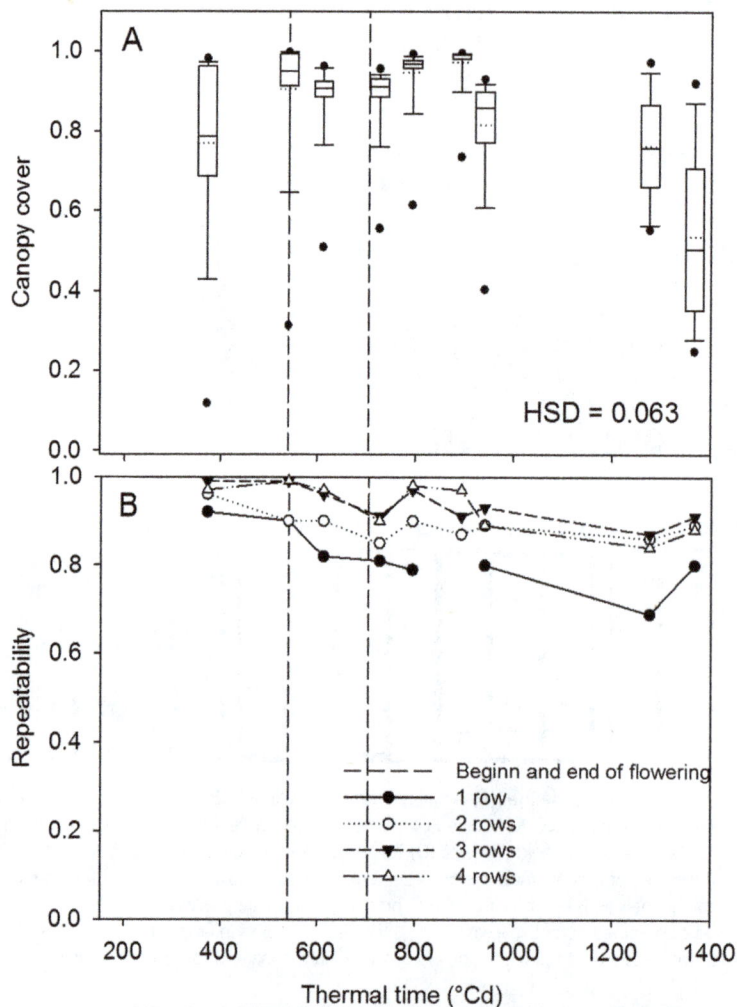

Figure 3 Canopy cover of four row plots (A) and repeatability of canopy cover affected by plot size (B). Boxplots are based on mean values of the 16 genotypes. The solid line in the box indicates the median and the dotted line the mean.

Table 2 Dates of aerial image acquisition

Flight	Date[a]	Time[b]	DAS[c]	TT[d]	Camera line up	Growth stage
1	16.06.2011	8:38	56	371	NIR, RGB	leaf 6 to 9 fully developed
2	05.07.2011	17:18	75	540	NIR, RGB	Begin of tasseling
3	11.07.2011	17:47	81	612	NIR, RGB, IR	Most genotypes tasseling (>50%)
4	26.07.2011	16:43	96	727	NIR, RGB, IR	All genotypes tasseling (100%)
5	02.08.2011	17:22	103	793	NIR, RGB, IR	Begin of corn filling
6	12.08.2011	17:22	113	893	NIR, RGB, IR	Begin of leaf senescence
7	16.08.2011	17:26	117	940	NIR, RGB, IR	
8	15.09.2011	14:30	147	1275	NIR, RGB, IR	Late senescence, upper leaf levels affected
9	29.09.2011	17:47	161	1366	NIR, RGB	Full maturity of most genotypes (black layer observed)

[a]day. month. year, [b]Central European time, [c]days after sowing, [d]thermal time (in °Cd).

(Figure 5). Interestingly, the highest repeatability values of 0.65 to 0.85 were observed in the three row plots. Surprisingly, the highest repeatability was not observed on the hot days, but on the two days with the lowest T_C (Additional file 1: Figure A10). At these days, we observed the strongest cooling effect of the canopy compared to ambient temperature reflected by dT.

Relationship of remotely sensed parameters to ground measured plant properties

The observed maize development can be divided into three phases that can be distinguished by NDVI measurements: 1) early development until canopy closure (up to 540°Cd), 2) flowering and early senescence (540–793°Cd) and 3) late senescence up to maturity (after 793°Cd). We used different ground truth measurements depending on the developmental phase to evaluate the value of remotely sensed parameters (Table 5). For canopy cover, we considered ground truth measurements related to canopy structure and architecture but evaluated also early vigour and stay green. During the early phase, i.e. at the single measuring campaign between 303 and 371°Cd, the remotely sensed CC was highly correlated to plant density (r = 0.67) early vigour rating (r = 0.77)

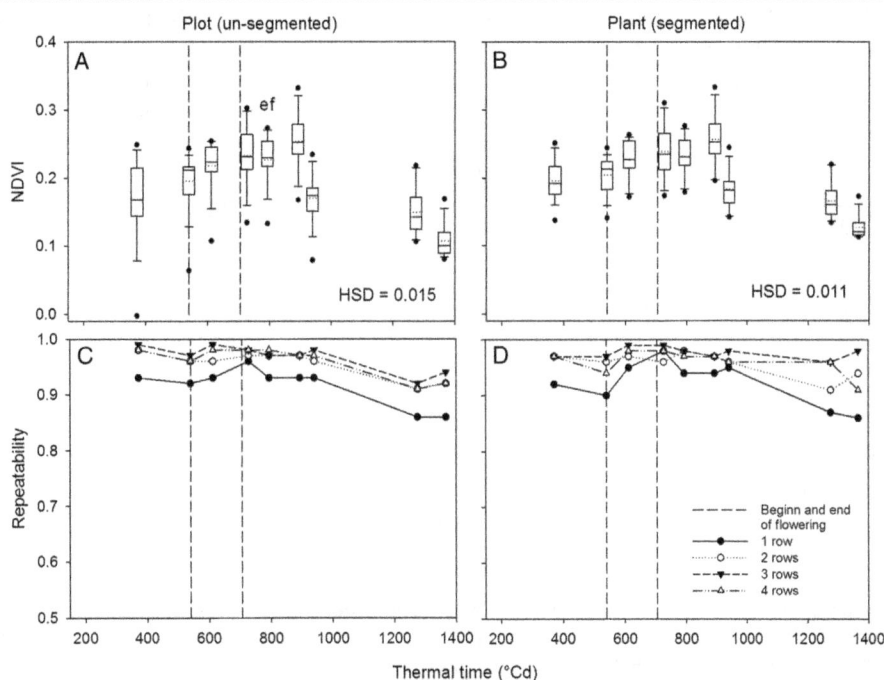

Figure 4 Plot and plant NDVI and their repeatability as affected by plot size. NDVI is shown in four row plots **(A, B)** and repeatability of NDVI in one, two, three and four row plots **(C, D)** of NDVI$_{Plot}$ **(A, C)** and NDVI$_{Plant}$ **(B, D)**. Boxplots are based on mean values of the 16 genotypes. The solid line in the box indicates the median and the dotted line the mean.

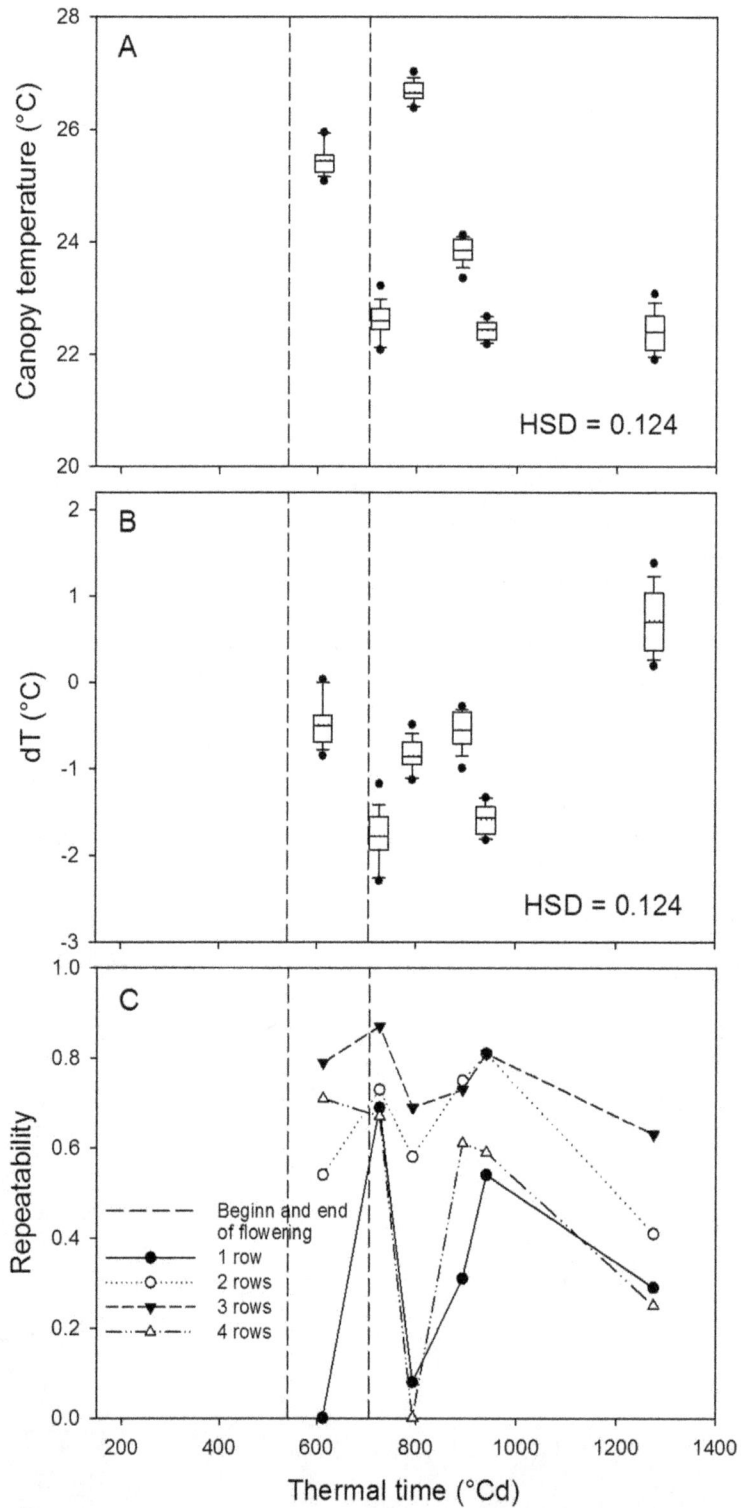

Figure 5 Canopy temperature (A), difference of canopy temperature to air temperature (B) and their repeatability (C). A + B shown for four row plots and the repeatability of one, two, three and four row plots **(C)**. Boxplots are based on mean values of the 16 genotypes. The solid line in the box indicates the median and the dotted line the mean.

Table 3 Correlation coefficients determined for canopy temperature and dT with climate conditions measured at the same time, as daily averages and as cumulated values for precipitation

Climate parameters		Canopy temperature	dT
		°C	°C
Canopy temperature (T_C)	°C	1 ***	0.72
Air temperature (T_A)	°C	0.96 **	0.51
Daily maximum temperature (T_M)	°C	0.93 *	0.7
Gust speed	m s^{-1}	−0.58	−0.14
Vapour pressure deficit (VPD)	kPa	0.23	−0.06
Actual radiation	µmol s^{-1} m^{-2}	0.68	0.89 *
Precipitation (7 days cumulated)	mm	0.26	0.44
Evapotranspiration (ETo)	mm day^{-1}	0.45	0.88 *
Sun hours	h day^{-1}	0.74	0.97 **

Thermal data from das 1275°Cd, was excluded because plants showed already advanced senescence.
Significance codes are: '***' p-value < 0.001, '**' p-value < 0.01 and '*'p-value < 0.05.

and leaf size (r = 0.67). During flowering and early senescence, CC was closely correlated to plant density (r = 0.73) and radiation interception (r = 0.75 to 0.86) but less to total plant biomass and leaf area index (r = 0.22 − 0.38). During senescence, CC was again closely correlated to radiation interception measured at the early senescence phase (893° Cd; r = 0.6 to 0.71) but less to late radiation interception (1275°Cd; r = 0.2 to 0.49). A correlation between stay green rating and CC was only detected during very late senescence (r = 0.36 to 0.53). No correlation was found for leaf biomass.

For NDVI$_{Plant}$ we considered ground truth data related to leaf greenness, senescence and canopy size (Table 5). During the early phase, i.e. at the single measuring campaign between 303 and 371°Cd, early vigour, was highly correlated to NDVI$_{Plant}$ (r = 0.64). At the flowering phase, correlations to leaf biomass, plant height and leaf area index were moderate to high (0.29 - 0.58). At advanced senescence, a positive correlation to stay-green rating was observed (r = 0.53). The most striking result was the moderate negative correlation with SPAD values throughout the season (r = −0.45 to −0.59), where a strong positive relationship would be expected. This strong discrepancy may be related to camera-based constraints or to the influence of plant architecture.

For the differences in canopy temperature, negative relations were observed with radiation interception, crop vigour rating, biomass, height and LAI (Table 6). Positive correlations were found for stay green rating and leaf temperature (LTMP) during late stages of development (>940°Cd). We found no correlations with leaf stomatal conductance (LSC) and LTMP before 940°Cd

most likely caused by the large time difference between ground and aerial measurements.

Genotypic differences
Clear differences between genotypes were detected for all parameters (both in remotely sensed parameters and ground-truth traits). A detailed discussion of these effects would exceed the scope of this manuscript and will be done in a future publication. Yet, in order to demonstrate the value of the presented method, it is important to point out a few cardinal differences detected for the remotely sensed parameters (Figure 6). For CC, differences between hybrids and inbred lines were pronounced because of the larger canopies of the hybrids. Their canopies were mostly closed in a 6 to 9 leaf stage (at 371°Cd), when the CC of the inbred lines was still below 0.9 (Figure 6). At the beginning of tasseling (540°Cd; data not shown), the majority of the genotypes had a CC higher than 0.9. Evaluating CC of the inbred lines, we observed genotype 9 to be significantly lower than the genotypes 8, 11, 12 and 14.

the genotypes differed for both, plant size and CC, NDVI$_{Plot}$ clearly distinguished among genotypes. However, NDVI$_{Plant}$ showed the difference of leaf greenness independent of differences in CC due to emergence rates or canopy architecture. This effect was clearly found for the three genotypes 6, 9 and 15 with plant densities below 70% of the original sowing rate (data not shown), where NDVI values markedly increased when measured on a plant basis instead of a plot basis. It is clearly visible that hybrids had generally higher NDVI$_{Plant}$ values compared to inbred lines.

Significant differences of T_C between genotypes were found at 612, 727, 893 and 940°Cd, but not on 793 and 1275°Cd (data not shown). On 793°Cd we measured the highest T_A of 27.5°C during IR image acquisition. However, it did not lead to a good separation among genotypes in contrast to dT. At flowering (727°Cd) when T_A was lower, T_C ranged between 22.0 and 24.0°C and were 1.4 to 2.3°C lower than T_A depending on genotype (Figure 6). The highest dT was found for genotype 3 and the lowest for genotype 13.

Discussion
Imaging platform, sensors and experimental field site
Many non-destructive measuring techniques are ground based or stationary using fixed, handheld or motorized systems (e.g. tractor mounted sensor platforms or crane systems). Thus, they are often limited to relatively small measurement areas and low numbers of replicates or genotypes. Furthermore, they are rather labour and time intensive, and do seldom cover temporal plant development [1,14,15]. For example, three tractors and several workers would be needed to measure a typical breeding

Table 4 Selected weather conditions prevailing at the time of thermal image capture (A) and some daily average values (B)

Date	year-month-day	2011-07-11	2011-07-26	2011-08-02	2011-08-12	2011-08-16	2011-09-15
Time	hh:mm:ss	17:45:38	16:45:38	17:25:38	17:25:38	17:25:38	14:25:38
Days after sowing	days	81	96	103	113	117	147
Thermal time[a]	°C days	612	727	793	893	940	1275
A at time of thermal capture							
Air temperature (T_A)	°C	25.95	24.40	27.52	24.40	24.01	21.71
Relative humidity (rH)	%	51.25	37.75	53.25	56.25	69.75	56.25
Dew point	°C	15.1	9.1	17.2	15.1	18.2	12.6
Wind speed	m s^{-1}	0.93	1.3	0.74	1.3	0.56	0.93
Gust speed	m s^{-1}	2.23	3.15	1.86	3.53	2.23	2.41
Soil temperature in 5 cm depth (T_S)	°C	20.63	17.34	19.20	18.70	20.15	17.32
Vapour pressure deficit (VPD_{air})[b]	kPa	0.62	0.76	0.62	0.54	0.37	0.50
Photosynthetically active radiation (PAR)[c]	µmol s^{-1} m^{-2}	1422	1082	nd[d]	1611	1043	nd
B based on daily data							
Maximum air temperature (T_{MAX})	°C	27.3	25.0	28.1	26.1	26.1	21.9
Evapotranspiration (ET_O- Penman)[e]	mm day^{-1}	4	3.1	3.7	4.5	3.1	2.1
Radiation[e]	kWh m^{-2}	7.0	5.4	6.5	5.5	5.4	4.6
Sun hours[e]	h	12	9.0	11.0	11.0	10	9.0
7 day cumulated precipitation	mm	40.6	37.0	30.4	36.0	22.8	24.8
Total cumulated precipitation	mm	230.2	329.8	360.2	415.4	434.4	508.0

[a]$TT = \Sigma if \geq 0((T_{max} + T_{min})/2)- T_{base}$, T_{base} of 8°C, [b]calculated according equation, [c]measured with a line quantum sensor, [d]not detected, [e]provided by the meteorological service (LTZ, Baden-Württemberg, Germany).

set of 20'000 plots in a few hours [1]. In contrast, aerial remote sensing offers the potential to cover large areas planted with many plots in relatively short time. In our study, the experimental field of 0.4 ha (30 × 132 m) was imaged from the air within 10 s. Accordingly, it would take around 6 minutes to monitor 20'000 two row plots of 1.5 × 4.75 m covering an area of 14.25 ha. Of course, additional time might be required depending on allocation, alignment and shape of the field.

The Zeppelin proved to be a valuable remote sensing platform due to the limited restrictions for sensor weight and its slow speed during image acquisition. Too high travel speeds can cause blurring effects and thus mix target and non-target information deteriorating the quality of the measurement. With a cruising altitude of 300 m at the highest speed of 20 km h^{-1}, the lowest image resolution (thermal camera) was 10 × 10 cm of ground cover per image pixel. During the opening of the shutter of 50 ms, the thermal camera moved 0.28 m along the row resulting in a blurring effect of 3 pixels. In order to keep the blurring effect within the row, it is preferable to fly in row direction rather than crossway of it. That blurring effect was less affecting the cameras with higher resolution and shutter speed and is reduced by lower cruising speed.

Costs of the Zeppelin operation were low compared to tractor-based operation, since a touristic route was used and only one man-hour plus ticket costs were required to take the images. Of course, this makes the choice of the test location extremely inflexible. Limits of the Zeppelin system are mainly the weather conditions and the mission area. Weather conditions restricting flights are mainly wind, rain and thunderstorms. Wind speeds above 25 m s^{-1}, and thunderstorms keep the Zeppelin on the ground or force the pilots to return to the air field (personal communication with Zeppelin NT). Rainy conditions do not interfere with the ability and allowance to operate the Zeppelin but affect data values of the image due to high water content in the light path [16]. In the beginning of the experiment the unpredicted occurrence of bad weather conditions prevented image acquisition during the early growth stages of maize.

The shown remote phenotyping approach may be adapted to other aerial platforms such as blimps, fixed wing or helicopter drones or even planes. A review of platforms and sensors would exceed the focus of this paper (for an overview of platforms see [16] for sensors [17]). However, drones combined with lower weight sensors and sensing technology as used in precision agriculture [18] seem to offer the temporal and spatial

Table 5 Selected coefficients of correlation between crop traits measured in the field and remote detected canopy cover and NDVI$_{Plant}$ shown for different times of measurement

Ground plant parameter	DAS[a]	Canopy cover							
		56	75	81	96	103	113	117	147
Plant density	29	0.67***	0.78***	0.73***					
Plant vigour rating	47	0.77***	0.58***	0.48***					
Leaf length	49	0.67***	0.51***	0.34**					
Radiation interception	75	0.77***	0.74***	0.76***					
Radiation interception	97			0.82***	0.75***	0.86***			
Total plant biomass	81-97			0.34*	0.22 ns	0.3*			
LAI[b]	81-97			0.32*	0.3*	0.38**			
Radiation interception	112					0.6***	0.63***	0.71***	
Stay green rating	117					-0.22 ns	-0.2 ns	-0.16 ns	
Stay green rating	147						-0.33**	-0.12 ns	0.34**
Radiation interception	147						0.2 ns	0.49***	0.41**
		NDVI$_{Plant}$							
Plant density	29	0.37***	0.52***	0.39***					
Plant vigour rating	47	0.64***	0.55***	0.34***					
Radiation interception	75	0.47***	0.62***	0.40***					
SPAD	96			-0.45***	-0.59***	-0.55***			
Radiation interception	97			0.49***	0.29*	0.50***			
Leaf biomass	81-97			0.39***	0.43***	0.35***			
Plant height	81-97			0.41***	0.18 ns	0.37**			
LAI	81-97			0.58***	0.44***	0.58***			
SPAD	103					-0.54***	-0.50***	-0.50***	
Radiation interception	112					0.43***	0.27***	0.38***	
Stay green rating	147						0.38**	0.36***	0.53***

[a]Days after sowing, see Table 2 for conversion in thermal time, [b]Leaf area index (m^2 m^{-2}).
Significance codes are: '***' p-value < 0.001, '**' p-value < 0.01 and '*'p-value < 0.05.

flexibility needed for phenotyping (for a review on drones see [19]). Because of pay load restrictions most drone approaches use a single sensor or sensors restricted in measurement capabilities. Remote sensing platforms capable of carrying a high payload and thus multiple or high weight sensors such as large drones and air planes are restricted in use by issues such as costs, law, region, manpower and training [19] similar to the Zeppelin platform. Additionally, the high flying altitude and speed resulting in high ground pixel sizes limits their use for crop phenotyping. In the future, light weight or micro drones combined with low weight sensor technology as currently investigated for precision farming applications might enable flexible high throughput crop phenotyping with multiple sensors and high temporal resolution as needed for breeding research.

The field markers that were used to semi-automatically match the images from the different sensors were an important feature of the experiment. Their identification was the prerequisite for the semi-automated registration of the area of interest and the ortho-correction process. Detection of the markers in RGB and NIR images may be improved by including white centres on the black plates, but it is not yet clear how that would interfere with thermal detection. Single plot labelling, as described by Jones et al. 2009 [20] for ground IR imaging would require too much investment for large, remotely sensed field setups. We consider markers to identify the corners and intermediate way points of the experimental area as sufficient to correct for image distortion and to allow for a correct positioning of a plot map.

Once the processing pipeline was established, data processing needed relatively little labour for input file conversion, check of the correct identification of the field markers, identifying and setting thresholds for segmentation and for creating the plot overlay. Time needed for conversion of proprietary input files (*cr2* from Canon and *irb* from InfraTec) into the open tagged image file format (*tiff*) can be minimized by using acquisition software saving in open formats in future

Table 6 Selected coefficients of correlation between crop traits measured in the field and remote detected temperature difference (dT) shown for different times of measurement

Ground plant trait		dT				
	DAS[a]	81	96	103	113	117
Plant vigour rating	47	−0.45***	−0.4**			
LSC[b]	81	0.09 ns	−0.08 ns			
LTMP[c]	81	−0.16 ns	−0.1 ns			
Radiation interception	81	−0.59***	−0.55***			
Radiation interception	97	−0.55***	−0.24 ns			
Total plant biomass	81-97	−0.38*	−0.34*			
Plant height	81-97	−0.46***	−0.53***			
LAI[d]	81-97	−0.52***	−0.49***			
LSC	103			−0.07 ns	−0.27*	0.17 ns
LTMP	103			−0.06 ns	−0.19 ns	0.09 ns
Stay green rating	105			0.12 ns	0.2 ns	0.34**
Stay green rating	112			0.1 ns	0.12 ns	0.36**
LSC	117			−0.17 ns	−0.21 ns	−0.13 ns
LTMP	117			0.45***	0.53***	0.28*
Radiation interception	118			−0.52***	−0.29*	−0.66***

[a]Days after sowing, see Table 2 for conversion in thermal time, [b]Leaf stomatal conductance (mmol m^{-2} s^{-1}), [c]Leaf temperature (°C), [d]Leaf area index (m^2 m^{-2}). Significance codes are: '***' p-value < 0.001, '**' p-value < 0.01 and '*'p-value < 0.05.

measurement campaigns. Nevertheless, a manual inspection of the correct identification of the field markers was necessary. It was facilitated by an automatically created overview and by a manual interface to identify the centre of the field marker, if necessary. Further development of reliable field markers in combination with suitable software will be necessary to improve automation for large areas monitored with higher throughput.

Image segmentation to distinguish into canopy cover and the NDVI value of the canopy itself was useful

The threshold settings for image segmentation for the seasonal imaging campaigns were similar for the NDVI but had to be adjusted for grey intensity (shades vs. nonshade) according to radiation condition prevailing during image capture. This procedure may be optimized by using the relationship of global radiation to grey intensity threshold in future campaigns. However, changing canopy properties might complicate that. Another option could be the placement of additional reference markers with different grey scales or colour to enable subsequent adjustment of the exposure values. This would also enhance the comparability of values between different flight campaigns.

Segmentation in images above mm scale resolution does always result in mixed pixels of either plant or soil features along edges or feature borders. A rule of thumb is, that for precise object identification a minimum object size of three times the instantaneous field of view

(IFoV), i.e. 3 ground pixels is needed [21]. Considering an IFoV of 2.5-3 cm for the vegetation camera, it is evident that three pixels were not available for object identification in regions of leaf tips, edges and where tassels cover the leaves. Such impurities were apparently marginal influential for determination of CC in both RGB and B-NIR images but should be minimized to a certain degree e.g. by using cameras with higher resolution. In our case the detection of CC appeared relatively robust, particularly CC was higher than 0.5. However, we were not able to detect young seedlings, which had approximately two fully developed leaves (first flight data not shown). Accordingly, an improvement is to be expected by using higher resolution cameras, especially to estimate germination rate, development during early growth stages, or if the detection of changes in tassel colour is anticipated in order to determine the time of flowering.

We used image segmentation to generate two independent parameters: CC to measure the canopy cover and $NDVI_{Plant}$ to measure leaf greenness independent of differences in canopy cover. The alternative, average NDVI signal of the whole plot without segmentation reflects a combination of temporal, spatial and genotypic variation of leaf greenness and CC. Thus, $NDVI_{Plot}$ should be interpreted carefully if used for plant phenotyping. Only when canopies are closed differences in leaf greenness measured on plot base can be considered reliable. Here the skewness of $NDVI_{Plot}$ can be used

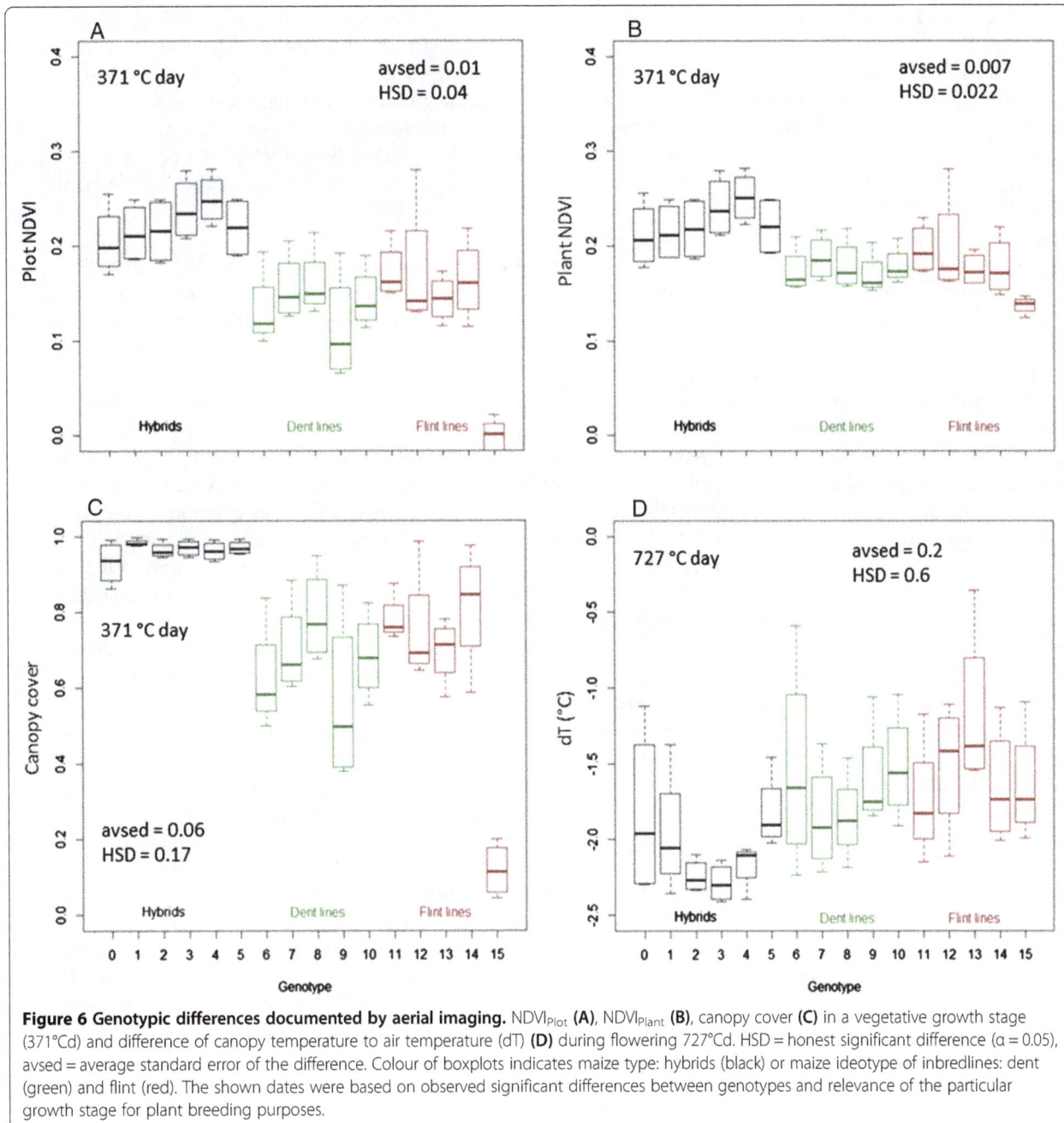

Figure 6 Genotypic differences documented by aerial imaging. NDVI$_{Plot}$ **(A)**, NDVI$_{Plant}$ **(B)**, canopy cover **(C)** in a vegetative growth stage (371°Cd) and difference of canopy temperature to air temperature (dT) **(D)** during flowering 727°Cd. HSD = honest significant difference ($\alpha = 0.05$), avsed = average standard error of the difference. Colour of boxplots indicates maize type: hybrids (black) or maize ideotype of inbredlines: dent (green) and flint (red). The shown dates were based on observed significant differences between genotypes and relevance of the particular growth stage for plant breeding purposes.

as a quality parameter with values higher than zero indicating plots with lower CC. At lower CC, NDVI-$_{Plot}$ is to a large extent influenced by soil pixel and, thus, information about leaf greenness is masked by differences in CC (see Additional file 2, section 6 for details). Thus, detection of leaf greenness reduction during senescence based on NDVI$_{Plot}$ must be hampered as soon as CC becomes sparse.

We also tested, whether the distribution parameter of the skewness of the NDVI$_{Plant}$ would be valuable as indicator for senescence. Senescence increases the

patchiness of green, yellow and brown leaf parts [22]. In our study, the explanatory power of skewness was limited because effective pixel size was too small to disentangle soil and plant signal sufficiently (see Additional file 2 section 6).

For the detection of NDVI$_{Plant}$, even one row plots were sufficient but clearly two and more rows improved repeatability further. We consider two row plots as a good balance between the precision to measure genotypes remotely and the necessity to screen large numbers of genotypes.

Thermal imaging had too little resolution for a suitable segmentation

For the thermal images no effective segmentation could be conducted because of the large IFoV of 0.3×0.1 m (discussed above). It was possible to detect maize rows, interspaces and larger patches of soils but no single leaves (Additional file 2). Therefore, the investigated whole plot signal reflects a plant and soil mixture. Accordingly, it is important to take canopy cover into consideration when comparing among genotypes. Similar observations were reported by Jones et al. [20] and Costa et al. [23]. Thermal measurements with the here reported ground resolution are applicable without restrictions in crops with closed canopies or in orchards, where plant area and unplanted inter row spaces are large and generally have a different temperature than the targeted plants as in [15]. Using the Zeppelin, an increased resolution can be achieved by reduced flight altitude (down to 80 m) and speed (down to 0 m s^{-1}). This would bring the ground resolution below 0.03 by 0.03 m instead of 0.1 by 0.3 m as in the present study. Alternatively, software solutions such as multiframe super resolution ([24]) or sensors with a higher resolution are an option.

Due to the mixed signal, the repeatability of T_C was higher in larger plots. The partially observed low repeatability of T_C in the four row plots is assumed to be caused by the intensive use of the four row plots for regular ground truth measurements. These frequent activities led to broken lower leaves and may have compacted the soil in the inter-row spaces leading to additional random noise. This indicates that entering plots for thermal measurement on a regular base should be avoided.

Seasonal development of remotely sensed traits

The observed development of CC and NDVI$_{Plant}$ as an indicator for leaf greenness in this study is in agreement with many studies, which recognized two to three phases of maize development depending on sensor type and parameter used [25-27]. In this study the initial growth phase was only represented by one flight campaign but the early vigour rating was still well related to the measurements at the end of this phase. This indicates the possibility to phenotype early growth stages of maize by remote sensing. Thereby the limit of early growth measurements is defined by the IFoV as discussed above and by plant size. In this study, plants were sufficiently large from the 4 to 6 leaf stage onwards. The following phase with relatively constant and high NDVI corresponds to a so-called plateau phase [26]. The duration of this phase and the height of the NDVI values depended on the genotype-specific flowering time and beginning of senescence. The last phase of maize development was well

identified by a decrease in NDVI$_{Plant}$ and its skewness following progressive senescence. The late start of the senescence detected by remote sensing as compared to rating on the ground, can be related to the fact that senescence starts at lower, older leaves [25] which are not detectable from above. Nevertheless, the NDVI parameters did reflect stay green rating as the inverse of the senescence after 900°Cd.

Correlation of image-based parameters and plant traits

We aimed to evaluate whether the remotely sensed parameters, NDVI, CC and T_C, reflect plant traits measured on the ground, like biomass, radiation interception, plant density and plant vigour.

Measurements of CC during early maize development and after beginning of corn filling, where genotypic variation was highest, seem to be promising indicators for early vigour and delayed senescence. Although very early measurements are missing in this experiment the derivation of CC from aerial images was successful and relationships to ground truth parameters were validated. The strongest correlation was found to radiation interception which itself is a strong non-destructive indicator for canopy size and leaf area traits. Clearly, the measurement of canopy coverage and density in the field is very time consuming compared to the aerial approach [28], justifying its application for large populations.

NDVI$_{Plant}$ appeared to measure the opposite of leaf greenness, as measured by the SPAD meter. This negative relationship was not expected because SPAD values indicate leaf greenness as a function of chlorophyll content (Additional file 2: Figure A4) and thus should be positively correlated to NDVI values [11,29]. Such a negative relationship of camera-based vegetation indices to SPAD was also observed by [30] who reported them to be in contrast to narrow or broad band indices. The camera channels B and NIR of the vegetation camera used in this study covered a range of 370 to 480 nm for the blue channel and 675 to 775 nm for the near infrared channel, respectively. The SPAD values are measured as transmission difference of two narrow bands (<10 nm range), with a high chlorophyll absorbance at 650 nm and a low absorbance of chlorophyll at 940 nm with a stable light emission by a red and a NIR LED (Konica Minolta Sensing Inc., Osaka, Japan). Accordingly, the NDVI camera uses a much broader range of the spectrum and different wavelengths than the SPAD measurement. Measurements with more precise narrow-band imaging sensors will likely improve the detection of leaf greenness as related to leaf chlorophyll content with spectral indicators such as NDVI$_{Plant}$.

Despite the failure to measure leaf greenness as a function of chlorophyll content as observable with the SPAD meter, we believe that the strong correlations of

$NDVI_{PLant}$ with other plant characteristic support the applicability of this method for breeding approaches. Its correlation to the leaf area index around flowering is very useful for remote-sensing applications, especially as canopy cover was little related to the leaf area index during this phase. Plant density and early vigour reflect germination rate and the genotypes' ability for fast establishment in the field, respectively. These two important traits which describe early development of maize are used for breeding purposes [31]. Stay green is a trait reflecting the plant's ability to maintain the photosynthesis functioning in the final growth stage. It is linked to increased yields as well as enhanced stress tolerance [11,32,33].

The effect of transpiration cooling of plants can be shown by the normalization to standard temperatures such as T_A or temperatures of certain standard surfaces [20]. We measured T_A with a weather station resulting in reasonable dT values: when maize was imaged in a green and transpiring stage, the cooling effect was −0.5 to −2°C; when it was imaged in the senescent non-transpiring stage, dT was lightly positive. Leaf temperatures can also be higher than T_A when radiation intensity is very high (e.g. at noon) and wind conditions are stagnant as shown in lab and field studies [9,34]. In this study, high radiation conditions during remote measurement campaigns were avoided due to late afternoon flights when radiation is lower. At the conditions presented here, the temperature normalization (dT) enabled a meaningful comparison between genotypes, measurements on different days with different climate conditions.

Highest repeatability, i.e. best differentiation among genotypes was achieved at days with moderate T_A when the largest cooling effect of the canopy was observed. This observation is in contrast to studies were T_C is measured mostly during hot days at the hottest time of the day around noon as an indicator for drought tolerance adaptation of genotypes [8,9] or crop water status [35]. Most of such studies were conducted in a different climate and with different research questions than this study and thus cannot fully be compared. Certainly, the optimal time of the day and temperature for IR measurements for plant breeding might still be a question to be answered. Due to unfavourable thermic conditions in the target area at midday, the company operating the Zeppelin did not allocate regular flights to the area where the experiment was placed. This made it impossible to test which time of the day would be optimal for thermal imaging.

The strong, negative correlation of dT with plant size and coverage information such as radiation interception, LAI and biomass confirms the applicability of the IR camera to measure T_C. In canopies with higher biomass,

coverage and plant density T_C is lower reflecting a higher transpiring area and cooling effect. Additionally, the correlation to radiation interception and LAI may be explained by the large pixel size and thus the mix of soil and plant information in the signal. The low correlation of T_C and dT to stomatal conductance and leaf temperature measured with the porometer might be explained by methodical differences as well as genotypic differences. The porometer measurements in the field reflect two point measurements per plot at the youngest fully developed leaf in a four row plot and thus only a marginal part of the IR plot image. This is supported by the observation of a better correlation of T_C to leaf temperature in the later growth stages when stomatal conductance is reduced due to advancing senescence and, thus, genotypic properties affecting T_C are less important. The positive correlation of Tc to the stay green rating may be an effect of differences in CC and canopy architecture properties. A lower CC results in lower leaf area as well as higher soil area in the image and thus a smaller dT of the canopy.

Conclusion

We developed a multi-channel remote-sensing pipeline with semi-automated image analysis.

The comparably low cruising altitude and cruising speed of the Zeppelin combined with high ground resolution enabled image segmentation. Accordingly we could distinguish into canopy cover (CC) and the normalized difference vegetation index of the segmented canopy ($NDVI_{Plant}$). Such segmentation was not possible for the thermal images with their comparably lower resolution. For CC and $NDVI_{Plant}$, two row plots enabled a sufficient differentiation among genotypes; for thermal imaging, more than two rows are preferable.

The NDVI camera could be used to measure different traits, depending on the time of the year. Early in the season, CC was related to early vigour, leaf length and plant density, later it was related to radiation interception. $NDVI_{Plant}$, was well related to the vigour rating and to very late senescence rating. More important, it was related to the leaf area index during flowering, when canopy cover did not correlate well with the trait. Most strikingly, $NDVI_{Plant}$ was negatively related to leaf chlorophyll content measured with the SPAD-meter. This discrepancy demands for an in-depth evaluation of this phenomenon.

For the thermography, highest repeatability of canopy temperature was observed on large plots on temperate days with strongest differences in canopy cooling.

The presented aerial phenotyping approach is applicable to other crops and larger field experiments and genotypic sets as well as other aerial carrier and sensor systems. Similar approaches might be realistic with light weight aerial carriers in the future when sensor technology evolves and sensor weight decreases, especially for thermal imaging. Such approaches can help to close the gap between phenotyping and genotyping and reduce the constraints currently limiting breeding advances.

Method

Experimental set up

The experimental field was placed below one of the frequently operated touristic routes of the Zeppelin (Zeppelin NT, Friedrichshafen, Germany) in the area of Lake Constance. It was embedded in a maize field near Salem in Germany (47° 46' 15.37" N, 9° 17' 15.16" E, 440 m.a.s.l.). The soil was a cambisol [36] classified according to soil texture as a sandy loam. The experimental setup was organised as a split plot design with four replications, the number of rows per plot (one to four) as the whole plot factor and a set of 16 genotypes as the split plot factor (Additional files 1 and 2). To avoid neighbour effects between hybrids (entries 0 to 5) and inbred lines (entries 6 to 15), the two groups were randomized in two separate blocks within the split plots. The plot length was 4 m and row spacing was 0.75 m. Additional single row plots of 10 m length were created at the end of each block for destructive samplings and measurements.

For the precise detection of the experimental plots in the aerial imagery we placed field markers within and around the experimental field (Figure 1). Two types of markers were used. Nine white plastic tarps (1 × 2 m) were placed on the ground in diagonal cross-like form just after sowing. Eight round, black metal plates (Ø 70 cm) were placed on top of 2 m poles along the edge of the experimental field after canopy closure.

Maize genotypes and cultivation

We selected 16 maize genotypes which reflect a large variability in plant development and morphology. The selection comprised six commercial hybrids (entries 0 to 5), five dent (entries 6 to 10) and five flint inbred lines (entries 11 to 15). The genotypes were Lapriora (entry 0, KWS SAAT AG, Einbeck, Germany), DKC2960 (entry 1, DeKalb Genetics Corp., Dekalb, IL, USA), Tiago, Pralinia, Bonfire, Swiss301, DSP1771, DSP5009S3, DSP5049A31, DSP5145X1, DSP5164A3, DSP2563E3, DSP2637A (entries 2 to 12, Delley seeds and plants, Delley, Switzerland), UH003 and UH008 (entries 13 and 14, University of Hohenheim, Germany) and SMxxx (entry 15, Freiherr von Moreau Saatzucht GmbH, Altburg, Germany).

The genotypes were planted on April 21, 2011 with a planting density of 9 plants per m^2 using a single-seed drilling machine (type TRM, Wintersteiger AG, Austria). The maize was cultivated according to best management

practices by the local farmer (for details see Additional file 2). For spraying of pesticides a tractor mounted wing sprayer with a wing length of 15 m was used (no crossing through the experiment).

Climate and weather conditions

Air temperature, relative humidity (2 m above ground), precipitation, wind- and gust speed (3 m above ground) and soil temperature (5 cm in the soil) were recorded with an on-site weather station (Onset Hobo, Pocasset, USA) installed at the edge of the field, at a distance of 150 m to the experiment. Thermal time (TT) was calculated as $TT = \Sigma_{if \geq 0}((T_{max} + T_{min})/2) - T_{base}$, with a base temperature (T_{base}) of 8°C [37] and is expressed in degree days (°Cd). Vapour pressure deficit (VPD) was calculated as $VPD = ((100 - rH)/100)*SVP$, with the saturation vapour pressure: $SVP (Pa) = 610.7*10^{7.5T/(237.3+T)}$.

Additionally, for days with thermal measurements evapotranspiration (ETO- Penman), radiation and number of sun hours were derived from a close by commercial weather station at Ailingen (47° 41' 30.49" N, 9° 28' 11.79" E, 440 m.a.s.l.) managed by the local meteorological service (LTZ, Baden-Württemberg, Germany).

Aerial imaging equipment

In this experiment we used a Zeppelin operated by Zeppelin NT (Deutsche Zeppelin-Reederei GmbH, Friedrichshafen, Germany) as remote sensing platform. In this proof of concept study, we decided to buy tourist tickets and to acquire images out of the open side window instead of a fixed on-board installation of our equipment. A fix installation would have demanded for an aviation certification and training for the pilots for using the imaging equipment. The sensor array was secured against falling off. During flight campaigns (Table 2) the Zeppelin was directed along the experimental field in south to north direction. Images were captured at approximate nadir position (view angle 90° to the soil surface) at an altitude of about 300 m and cruising speed between 0 and 20 km h^{-1} depending on wind situation.

For image capture we used a handheld camera system (Figure 1), which consisted of two consumer grade cameras and an optionally attached thermal camera. The consumer grade cameras were a 10.1 megapixel CMOS RGB camera (Canon EOS 400D, Canon, Tokyo, Japan) and a two-channel, 12.2 megapixel CMOS vegetation camera (Canon EOS 450D NDVI, modified by LDP LLD, Carlsted, USA) with a sensitivity range of 370 to 480 nm (blue channel, B) and 675 to 775 nm (near infrared channel, NIR). More information on sensor sensitivity and NIR photography can be found in Nijland [38]. The two cameras were equipped with Canon EF-S 60 mm f/2.8 Macro USM lenses and mounted on an aluminium frame with handles and interconnected with a remote trigger cable for simultaneous image capture. The aperture size was adjusted shortly before the field capture using a random maize field between the airport and the experimental field. The focus was centre weighed and set to AI servo mode, the ISO was set to 100 and all other settings were set to automatic. During the flight over the experimental field a series of images was taken.

The thermal camera was an industrial grade thermal infrared (IR) camera VarioCAM head 600 (Infratec GmbH, Dresden, Germany). It was attached only for selected missions on hot summer days (Table 2). The IR camera measures in the spectral range between 7.5 and 14 μm, a spatial resolution of 640×480 pixels and a thermal resolution of better than 0.03 K at 30°C. A 75 mm lens was attached and shutter speed was 50 ms. For mobile image acquisition the IR camera needed an additional battery and a laptop connected by firewire for camera control and data saving. The camera was attached to the handle bar between the two consumer grade cameras and was run in video mode recording five images per second during flight over the experimental field with the focus adjusted automatically (every 40 seconds) shortly before the field was reached.

Images were recorded in raw format (.cr2 for the Canon cameras and .irb for the IR camera). The total imaging setup had a weight of 7.2 kg, a detailed description (incl. information about the ground cover and spatial resolution for the three cameras at altitudes of 300 ± 10 m) can be found in Table 1.

Image processing and analysis

For analysis of the aerial imagery the macro array of field plots arranged in the experimental field was split into three sub arrays (rectangles), each one covered by a separate image scene (a detailed sketch of the field plot macro array can be found in the Additional file 1). For each camera images were selected manually and transformed from the respective raw file format to 16 bit .tiff images (Figure 1C). Selection criteria were that the target rectangle was well focused and central in the image to minimize vignette effects.

The image processing scripts were developed in Matlab (2011a Natick, MA, USA). The black field markers (Figure 1D) were used to automatically identify, match and co-register the sub arrays in the different images. For the RGB and the NDVI camera the blue channel was used to identify the markers in the images. In order to accurately transform these images into the same coordinate system, the marker positions were determined consistently by normalized cross correlation (NCC) [39]. Here, the marker regions of the NIR images

served as templates and their best position (most similar position) in the blue channel of the RGB image was determined in the region of the corresponding markers by NCC. In the thermal images, the black field markers emitted higher temperatures than plants and soil. However, the success of the automatic marker detection procedure was manually adjusted in a few cases. Subsequently, the marker positions were used to rectify the sub array images by applying a projective transformation using bi-cubic interpolation. For IR images the resolution was up-scaled to the resolution of the other sensors before applying the projective transformation. The result is a set of images from all sensors transformed to the same coordinates for each sub array (see Figure 1D).

For evaluation of plant features we differentiated soil from plant pixels, by means of segmentation [21]. This procedure was tested for the blue-near infrared images (B-NIR) and the RGB images separately. For B-NIR we calculated the normalized difference vegetation index (NDVI) based on the blue band instead of the red band: $NDVI = (NIR - B)/(NIR + B)$, where B is the blue channel and NIR is the near infrared channel. For the B-NIR images the segmentation of plants was performed using two separate threshold procedures. The first segmentation was based on NDVI with a threshold of 0.1 meaning, all pixels with a higher values were regarded as plant pixels. The threshold 0.1 was chosen for all images to allow detection of maize leaves with reduced greenness particularly during the late development stages (senescence). The second segmentation step was done after converting the images to monochrome images, which shows the reflection intensity, in order to remove highly shaded areas. It was directly affected by the actual radiation and thus was set individually for each flight campaign depending on radiation conditions (thresholds can be found in Additional file 2). The resulting masks were combined by multiplication. For the segmentation of the RGB images the images were transformed to the HSB colour space. Thresholds were set for hue, saturation and brightness, respectively and for each flight individually.

To identify the sampling plots and to exclude unwanted areas such as tracks around the sampling plots, a mask file was prepared (Figure 1). The mask was used as overlay to clip and save an area of interest (AoI) for each experimental plot. For each camera and field rectangle an image stack was generated for visual control of the output images.

The per plot extracted data comprised the median and skewness (distribution parameter) for the RGB channels, the NDVI and canopy temperature (T_C) with and without segmentation, respectively. The skewness of a distribution is a rating of the asymmetry of its histogram relatively to its distribution mean (Additional file 2: Figure A5) and is defined as:

$$s = \frac{\frac{1}{n}\sum_{i=1}^{n}(x_i - \bar{x})^3}{\left(\sqrt{\frac{1}{n}\sum_{i=1}^{n}(x_i - \bar{x})^2}\right)^3}$$

where n is the number of distribution elements, x_i is the i-th element and $\bar{x}$ is the mean. Negative values are encountered if the median of the distribution is greater than the distribution mean and positive values if it is smaller. If the distribution is symmetric to its mean, the skewness is zero.

The canopy cover (CC) was extracted as the fraction of plant pixels from the segmented NDVI images. From T_C we calculated the difference to air temperature (dT) using the actual air temperature (T_A) measured by the weather station on-site at the time of the image capture.

Maize development and ground truth measurements

Unless reported otherwise, all observations presented here, were taken from the four-row plots. Evaluations and measurements started 0.5 m behind the first plant to minimize edge effects. Emergence was evaluated on 4 m rows 155°Cd corresponding to 29 days after sowing (DAS). The corresponding dates, degree days, DAS and approximate growth stages can be found in Table 2. Tasseling was evaluated during the period from 540 to 727°Cd on ten adjacent plants in approximately 3-day intervals. The exact dates at which 50% of the plants were tasseling were determined by linear interpolation. Leaf and total above ground biomass, plant height and number of leaves were determined on five adjacent plants in the sampling plots when the respective genotype was considered fully tasseling. Fresh weight biomass was determined with an electric field balance and height was measured with a yard stick. Stay green (development or delay of senescence) was evaluated five times from 815 to 1275°Cd on ten plants per plot by counting green leaves below the ear [40].

The leaf area index (LAI), was calculated from leaf biomass taking advantage of the narrow relationship between leaf area and specific leaf fresh weight (SLW) determined on a subset of plants (n = 24, $r^2 = 0.98$): LAI (cm^2 cm^{-2}) = SLW (mg cm^{-2}) -3.96/27.4, with SLW cm^{-2}) = leaf biomass (g)/(SL (cm)*70 cm)*1000, where SL is the sampling length and the row distance is 70 cm. Details on the sub-experiment can be found in the Additional file 2 in section 5.

Leaf chlorophyll content, canopy radiation interception and stomatal conductance were determined throughout the season mostly parallel to the aerial imaging campaign. Leaf chlorophyll content was determined with a SPAD meter (Konica Minolta Sensing Inc., Osaka, Japan) on 10 leaves per plot. Before silking, measurements were done on the youngest fully developed leaf. After silking, SPAD was measured on the second leaf above the ear leaf. Photosynthetic active radiation (PAR) was measured with a 1 m line quantum sensor (LI-186-line, LI-COR, Lincoln, Nebraska, USA). Measurements were taken in the middle row on the ground (PAR transmitted) and above the canopy (incident PAR) at noon on clear days or days with stable cloud cover. The proportion of PAR radiation absorbed by the crop (radiation interception) was calculated as the ratio of the difference between incident and transmitted PAR to incident PAR [41]. Leaf stomatal conductance (LSC) and leaf temperature (LTMP) were measured with a steady state diffusion leaf porometer (SC-1, Decagon Devices, Pullman, WA, USA) on the same leaf as SPAD. Measurements were taken in early afternoon (12–14:00) with two measurements per genotype and block for time reasons.

Statistical analysis

The investigated dataset consisted of three levels of data: (1) genotype level: ground truth data collected in the destructive sampling plots, which were not the same as aerial survey plots, (2) plot level: ground truth measured in plots at the same day and on the same plots as the aerial survey (generally four row plots) and (3) the remote sensing level: data available for all plots and plot sizes measured at the same time. Data from different levels of measurement were combined by time of measurement (TT), genotype and block. Data consisting of more than one measurement per plot (SPAD, LSC, silking, stay-green rating, and biomass) were averaged before entering data analysis.

Statistics were calculated with R version 3.0.1 [42]. Boxplots show the 25 and 75% quantiles as the lower and upper limit of the box with the median as solid line in between (mean values shown as dotted line in some cases). The lower and upper whiskers represent the 5 and 95% percentile or the minimum and maximum value if no individual points (outliers) are plotted.

Comparisons between genotypes or measurement dates were done by means of a mixed model analysis using the package 'asreml' version 3.0 for R [43] followed by a HSD test. The variance components to estimate repeatability of the one to four rowed plots were determined by setting block as fixed and genotypes as random factor. Repeatability was calculated as $h^2 = \sigma_{gen}^2/(\sigma_{gen}^2 + \sigma_\varepsilon^2/4)$, where σ_{gen}^2 is the estimated genetic variance and σ_ε^2 is the residual error variance. We used the repeatability to elucidate which plot size (row number) was sufficient to differentiate among genotypes, depending on measurement time and traits. Coefficients of correlation (r) were calculated by the Pearson product moment correlation. The used significance codes are: '***' p-value < 0.001, '**' p-value < 0.01 and '*'p-value < 0.05.

Additional files

Additional file 1: Figure A1. Overview of the experimental field set up shown as a scheme (A), as aerial side view (B) and top down images of the three measurement arrays as used for data extraction (C). The columns shown in A represent the columns that can be seen in the field in B and C. The grey coloured plots in A are the experimental four row plots and the destructive sampling plots (numbers stand for maize genotypes), black areas are field markers and targets white areas represent edge rows, the one to three row plots or walkways.

Additional file 2: Word document containing additional information about: 1. Experimental field setup, 2. Field management, 3. Weather information, 4. Camera set up, 5. Ground measurements, 6. Observed skewness of NDVI$_{Plant}$ of three genotypes during the season and 7. Canopy temperature as related to air temperature and repeatability.

Abbreviations

NDVI: Blue band Normalized Difference vegetation Index; CC: Canopy cover; IR: Infrared; HSB: hue-saturation-brightness (color space); TT: Thermal time; DT: Difference between canopy and air temperature; LAI: Leaf area index; LWI: Leaf weight index; IFoV: Instantaneous Field of View; AoI: Area of interest; T$_C$: Canopy temperature; T$_A$: Air temperature; SCO: Leaf stomatal conductance; LTMP: Leaf temperature.

Competing interests

The authors declare that they have no competing interests.

Authors' contributions

FL (field and flight campaign, data analysis), DS (field and flight campaign, data handling), NK (image processing), AW (project planning), AH (project planning, field campaign, data analysis). All authors contributed to manuscript writing. All authors read and approved the final manuscript.

Acknowledgements

We would like to thank Karl-Heinz Camp (Delley seeds and plants, Ltd) for the composition of the set of test genotypes; Roland Lohr for support during sowing, Matthias Hagge (Farm manager Schloss Salem) for field management; and the Zeppelin NT crew, particularly Dietmar Blasius (manager research and special missions), Susanne Federle (tickets and organisation) Kate Board and Fritz Günter (pilots) for the flexible and professional support during flight campaigns. Thanks to Niclas Freitag, Cathrine Meyer, Chantal le Marié and Michael Mielewczik (members of ETH crop science) for support during the field campaigns.

Author details

[1]Institute of Agricultural Sciences, ETH Zürich, Universitätstrasse 2, 8092 Zürich, Switzerland. [2]Norddeutsche Pflanzenzucht, Hohenlieth Holtsee D-24363, Germany.

References

1. White JW, Andrade-Sanchez P, Gore MA, Bronson KF, Coffelt TA, Conley MM, et al. Field-based phenomics for plant genetics research. Field Crop Res. 2012;133:101–12.
2. Araus JL, Cairns JE. Field high-throughput phenotyping: the new crop breeding frontier. Trends Plant Sci. 2014;19:52–61.
3. Rajendran K, Tester M, Roy SJ. Quantifying the three main components of salinity tolerance in cereals. Plant Cell Environ. 2009;32:237–49.
4. Walter A, Studer B, Kölliker R. Advanced phenotyping offers opportunities for improved breeding of forage and turf species. Ann Bot. 2012;110:1271–9.
5. Montes JM, Technow F, Dhillon BS, Mauch F, Melchinger AE. High-throughput non-destructive biomass determination during early plant development in maize under field conditions. Field Crop Res. 2011;121:268–73.
6. Kipp S, Mistele B, Schmidhalter U. Identification of stay-green and early senescence phenotypes in high-yielding winter wheat, and their relationship to grain yield and grain protein concentration using high-throughput phenotyping techniques. Funct Plant Biol. 2013;41:227–35.
7. Deery D, Jimenez-Berni J, Jones H, Sirault X, Furbank R. Proximal remote sensing buggies and potential applications for field-based phenotyping. Agronomy. 2014;4:349–79.
8. Zia S, Romano G, Spreer W, Sanchez C, Cairns J, Araus JL, et al. Infrared thermal imaging as a rapid tool for identifying water-stress tolerant maize genotypes of different phenology. J Agron Crop Sci. 2013;199:75–84.
9. Romano G, Zia S, Spreer W, Sanchez C, Cairns J, Luis Araus J, et al. Use of thermography for high throughput phenotyping of tropical maize adaptation in water stress. Comput Electron Agric. 2011;79:67–74.
10. Cairns JE, Crossa J, Zaidi PH, Grudloyma P, Sanchez C, Araus JL, et al. Identification of drought, heat, and combined drought and heat tolerant donors in maize. Crop Sci. 2013;53:1335–46.
11. Cairns JE, Sanchez C, Vargas M, Ordonez R, Araus JL. Dissecting maize productivity: ideotypes associated with grain yield under drought stress and well-watered conditions. J Integr Plant Biol. 2012;54:1007–20.
12. Nelissen H, Moloney M, Inzé D. Translational research: from pot to plot. Plant Biotechnol J. 2014;12:277–85.
13. Furbank RT, Tester M. Phenomics - technologies relieve the phenotyping bottleneck. Trends Plant Sci. 2011;16:635–44.
14. Soudani K, Hmimina G, Delpierre N, Pontailler JY, Aubinet M, Bonal D, et al. Ground-based network of NDVI measurements for tracking temporal dynamics of canopy structure and vegetation phenology in different biomes. Remote Sens Environ. 2012;123:234–45.
15. Sakamoto T, Gitelson AA, Nguy-Robertson AL, Arkebauer TJ, Wardlow BD, Suyker AE, et al. An alternative method using digital cameras for continuous monitoring of crop status. Agr Forest Meteorol. 2012;154:113–26.
16. Jones HG, Vaughan RA. Remote Sensing of Vegetation, Principles, Techniques And Applications. Oxford: Oxford University Press; 2010.
17. Lee WS, Alchanatis V, Yang C, Hirafuji M, Moshou D, Li C. Sensing technologies for precision specialty crop production. Comput Electron Agric. 2010;74:2–33.
18. Zhang C, Kovacs JM. The application of small unmanned aerial systems for precision agriculture: a review. Precis Agric. 2012;13:693–712.
19. Watts AC, Ambrosia VG, Hinkley EA. Unmanned aircraft systems in remote sensing and scientific research: classification and considerations of use. Remote Sens. 2012;4:1671–92.
20. Jones HG, Serraj R, Loveys BR, Xiong L, Wheaton A, Price AH. Thermal infrared imaging of crop canopies for the remote diagnosis and quantification of plant responses to water stress in the field. Funct Plant Biol. 2009;36:978–89.
21. Jähne B. Digital Image Processing. 5th ed. Berlin: Springer; 2002.
22. Thomas H. Senescence, ageing and death of the whole plant. New Phytol. 2013;197:696–711.
23. Costa JM, Grant OM, Chaves MM. Thermography to explore plant–environment interactions. J Exp Bot. 2013;64:3937–49.
24. Farsiu S, Robinson MD, Elad M, Milanfar P. Fast and robust multiframe super resolution. IEEE Trans Image Process. 2004;13:1327–44.
25. Escobar-Gutierrez AJ, Combe L. Senescence in field-grown maize: from flowering to harvest. Field Crop Res. 2012;134:47–58.
26. Verhulst N, Govaerts B, Nelissen V, Sayre KD, Crossa J, Raes D, et al. The effect of tillage, crop rotation and residue management on maize and wheat growth and development evaluated with an optical sensor. Field Crop Res. 2011;120:58–67.
27. Colomb B, Kiniry JR, Debaeke P. Effect of soil phosphorus on leaf development and senescence dynamics of field-grown maize. Agron J. 2000;92:428–35.
28. Fletcher AL, Johnstone PR, Chakwizira E, Brown HE. Radiation capture and radiation use efficiency in response to N supply for crop species with contrasting canopies. Field Crop Res. 2013;150:126–34.
29. Liebisch F, Küng G, Damm A, Walter A. Characterization of Crop Vitality and Resource Use Efficiency By Means of Combining Imaging Spectroscopy Based Plant Traits. In: Workshop on Hyperspectral Image and Signal Processing, Evolution in Remote Sensing, vol. 6. Lausanne, Switzerland: IEEE International; 2014.
30. Hunt ER, Doraiswamy PC, McMurtrey JE, Daughtry CST, Perry EM, Akhmedov B. A visible band index for remote sensing leaf chlorophyll content at the canopy scale. Int J Appl Earth Observation Geoinform. 2013;21:103–12.
31. Maydup ML, Graciano C, Guiamet JJ, Tambussi EA. Analysis of early vigour in twenty modern cultivars of bread wheat (Triticum aestivum L.). Crop Pasture Sci. 2012;63:987–96.

32. Lopes MS, Araus JL, van Heerden PDR, Foyer CH. Enhancing drought tolerance in C4 crops. J Exp Bot. 2011;62:3135–53.

33. Zheng HJ, Wu AZ, Zheng CC, Wang YF, Cai R, Shen XF, et al. QTL mapping of maize (Zea mays) stay-green traits and their relationship to yield. Plant Breed. 2009;128:54–62.

34. Schymanski SJ, Or D, Zwieniecki MA. Stomatal control and leaf thermal and hydraulic capacitances under rapid environmental fluctuations. PLoS One. 2013;8:e54231.

35. Alchanatis V, Cohen Y, Cohen S, Moller M, Sprinstin M, Meron M, et al. Evaluation of different approaches for estimating and mapping crop water status in cotton with thermal imaging. Precis Agric. 2010;11:27–41.

36. FAO. World Reference Base for Soil Resources. Rome: FAO, ISRIC and ISSS; 1998.

37. Lizaso JI, Batchelor WD, Westgate ME. A leaf area model to simulate cultivar-specific expansion and senescence of maize leaves. Field Crop Res. 2003;80:1–17.

38. Basics of infrared photography, infrared light. [http://www.ir-photo.net/ir_imaging.html, access date 12.05.2013], 12.05.2013

39. Lewis J. Fast Normalized Cross-Correlation, Vision Interface. 1995. p. 120–3.

40. Fox RH, Piekielek WP, Macneal KE. Comparison of late-season diagnostic tests for predicting nitrogen status of corn. Agron J. 2001;93:590–7.

41. Calderini DF, Dreccer MF, Slafer GA. Consequences of breeding on biomass, radiation interception and radiation-use efficiency in wheat. Field Crop Res. 1997;52:271–81.

42. R-Development-Core-Team. R Foundation for Statistical Computing, Vienna, Austria ISBN 3-900051-07-0. 2008. URL http://www.r-project.org/.

43. Butler D. asreml: asreml() Fits the Linear Mixed Mode. R package version 3.00. R package version 3.00. 2006.

14

Automated characterization of flowering dynamics in rice using field-acquired time-series RGB images

Wei Guo[1], Tokihiro Fukatsu[2] and Seishi Ninomiya[1*]

Abstract

Background: Flowering (spikelet anthesis) is one of the most important phenotypic characteristics of paddy rice, and researchers expend efforts to observe flowering timing. Observing flowering is very time-consuming and labor-intensive, because it is still visually performed by humans. An image-based method that automatically detects the flowering of paddy rice is highly desirable. However, varying illumination, diversity of appearance of the flowering parts of the panicles, shape deformation, partial occlusion, and complex background make the development of such a method challenging.

Results: We developed a method for detecting flowering panicles of rice in RGB images using scale-invariant feature transform descriptors, bag of visual words, and a machine learning method, support vector machine. Applying the method to time-series images, we estimated the number of flowering panicles and the diurnal peak of flowering on each day. The method accurately detected the flowering parts of panicles during the flowering period and quantified the daily and diurnal flowering pattern.

Conclusions: A powerful method for automatically detecting flowering panicles of paddy rice in time-series RGB images taken under natural field conditions is described. The method can automatically count flowering panicles. In application to time-series images, the proposed method can well quantify the daily amount and the diurnal changes of flowering during the flowering period and identify daily peaks of flowering.

Keywords: Time-series RGB image, SIFT, BoVWs, SVM

Background

The dynamics of flowering is an important trait for paddy rice and affects the maturation timing of rice grain [1,2]. Great effort is invested in observing flowering time. Diurnal variance in flowering time is also important because heat reduces pollen fertility and pollination efficiency, reducing yield and degrading grain quality. Facing global warming, rice breeders are now trying to find early-morning flowering lines to avoid heat at the time of flowering [3,4]. The search for early-morning-flowering lines requires observers to remain in fields, for several hours daily, starting early morning.

Machine learning and digital image processing techniques are becoming readily available for field-based agronomic applications. For example, methods for measuring or estimating crop growth parameters such as canopy coverage, leaf area index, and plant height [5-12] and for monitoring crop growth status [13-15] have been recently proposed. In particular, methods for extracting the phenotypic characteristics of specific plant organs (leaf, fruit, flower, grain, etc.) have been helpful for researchers and breeders attempting to understand the performance of crop genetic resources [16-20]. In view of such innovative applications of image analysis for crops, an image-based method that automatically detects and quantifies the flowering behavior of paddy rice appears feasible.

Generally, flowering in paddy rice occurs by anther extrusion between the opening and closing of the spikelet. Active flowering generally lasts for 1–2.5 h daily during

* Correspondence: snino@isas.a.u-tokyo.ac.jp
[1]Institute for Sustainable Agro-ecosystem Services, Graduate School of Agricultural and Life Sciences, The University of Tokyo, 1-1-1. Midori-cho, Nishi-Tokyo, Tokyo 188-0002, Japan
Full list of author information is available at the end of the article

Figure 1 An example of the same panicles' appearance in one day. The daily active flowering time is short. In this example, active flowering starts around 11:00 and lasts until anthers begin shrinking around 13:00. The red elliptic circles indicate examples of actively flowering panicles.

the reproductive phase, and it is very sensitive to external environmental factors such as temperature, solar radiation, etc. [21,22]. For example in Figure 1 active flowering is observed only in the image acquired at around 12 PM. Moreover, because the crop grows under natural conditions, varying illumination, diverse orientations, various appearances of panicles, shape deformation by wind and rain, partial occlusion, and complex background make image-based methods challenging. Figure 2 shows examples of various appearances of flowering panicles of rice, and Figure 3 demonstrates how they change with growth and the external environment. Figure 3a shows physical size and shape changes due to growth in two panicles taken over three days. Figure 3b and c show images taken within a 5-min interval may be very different because of color changes under natural light conditions and shape changes due to leaf overlapping.

In this study, we combined a local feature descriptor, the scale-invariant feature transform (SIFT) [23], an image representation method, the bag of visual words (BoVWs) [24,25], and a machine learning model, the support vector machine (SVM) [26] to overcome these difficulties, and attempted to develop a model able to detect flowering panicles of paddy rice in normal RGB images taken under natural field conditions. The method is based on generic object-recognition technology, which is still challenging in machine vision. We evaluated the performance of the

proposed method by monitoring the diurnal/daily flowering pattern and the flowering extent of paddy rice during the flowering period. Although some methods such as the color based method for lesquerella [27] and the spectral reflectance based method for winter wheat [28] have been studied to identify flowers under natural condition, no digital image-based identification method of paddy rice flowering has been proposed to date.

Results

We acquired two independent time series images of two paddy rice varieties, *Kinmaze* and *Kamenoo* and provided three datasets, Dataset 1, Dataset 2 and Dataset 3 to verify the flowering identification capabilities of the proposed method. The images were taken every 5 minutes from 8:00 to 16:00 between days 84 and 91 after transplanting considering the flowering period of the varieties. Dataset 1 and Dataset 3 are composed of the original 645 and 768 full size images of *Kinmaze* and *Kamenoo* respectively whereas Dataset 2 is composed of the central parts of the images cropped from Dataset 1. A total of 700 image patches sampled from 21 images of Dataset 1 were used to train the support vector machine (SVM) model for detecting the flowering in the proposed method. The 21 images were removed from Dataset 1 and Dataset 2 when the datasets were used for the model verifications.

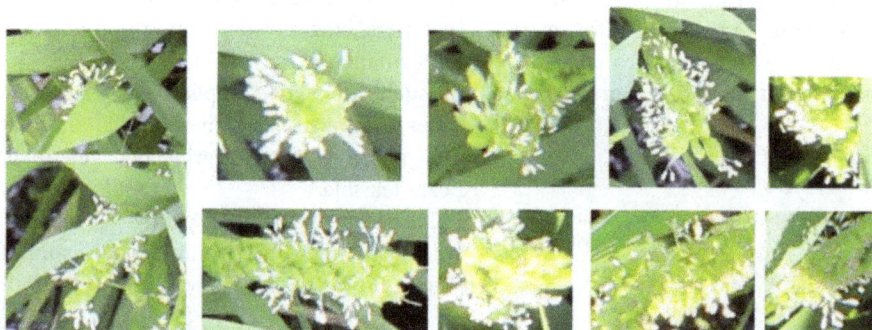

Figure 2 Various appearances of flowering panicles.

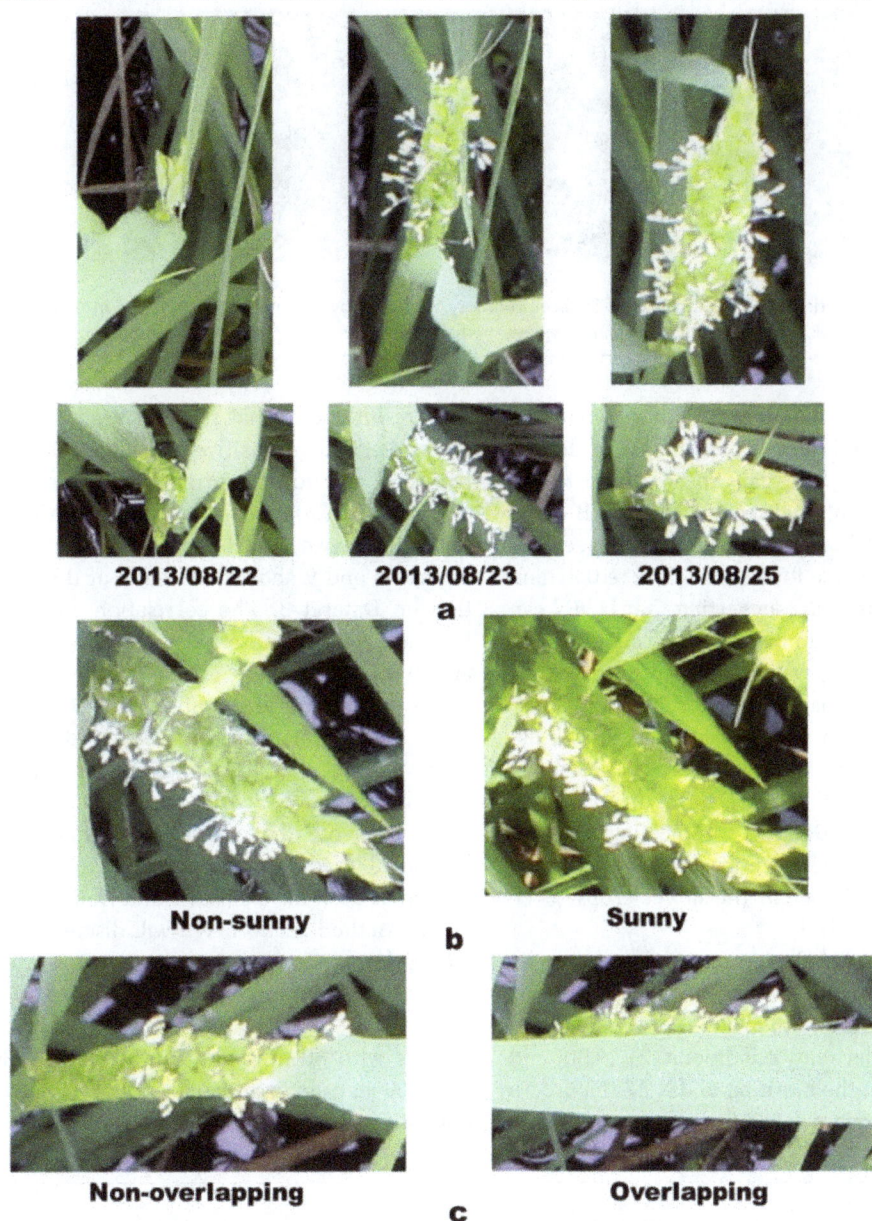

Figure 3 Changes in the appearance of identical flowering panicles. (a) Images of two identical flowering panicles taken over three consecutive days. Physical size and shape change owing to growth; **(b)** Images of an identical flowering panicle. The appearance changes under different light conditions; **(c)** Images of an identical flowering panicle. The appearance is changed by an overlapping leaf.

Figures 4 and 5 show examples of the flowering detections in Dataset 1 and Dataset 2. Each small block of violet red color shown in Figures 4b and 5b indicates a sliding window that was assessed as a flowering part (s). The red rectangles in Figure 5c show the regions which surround the connected violet red blocks in Figure 5b and they successfully detected most of the flowering panicles. In additional, a video was provided to demonstrate the detected result during whole experimental period (Additional file 1), the image Datasets and demo matlab Pcode used in this experiment also available on

our website[a]. Figure 6a and b show the results of flowering detection between days 84 and 91 after transplanting of Dataset 1 and Dataset 2. Because of transmission errors of the image acquisition system for *Kinmaze*, some of the images, particularly on day 86, are missing. Green, black, and blue circles indicate the number of blocks assigned as flowering parts of panicles (FBN), the number of regions of connected blocks (FCBN), and the number of visually counted flowering panicles (FPN), respectively. The daily flowering patterns shown by FBN and FCBN were similar to the actual number of

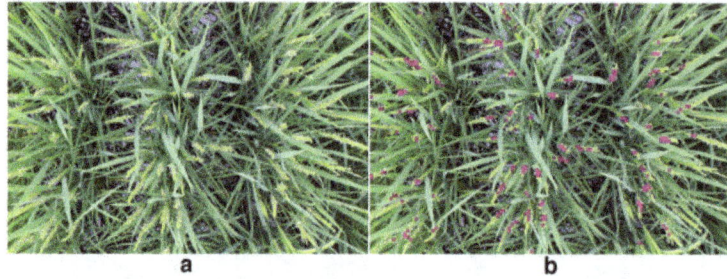

Figure 4 An example of flowering panicle detection of Dataset 1(variety, _Kinmaze_) by the method developed in this study. (a) Original image from Dataset 1; **(b)** Each violet block indicates a sliding window in which part of a flowering panicle was detected.

flowering panicles (FPN). Thus, the method quantified well the daily amount and the diurnal changes of flowering, including identifying the daily peak of flowering. The correlation coefficients between FPN and FBN and between FPN and FCBN were 0.80 and 0.82 respectively for Dataset_1 whereas those for Dataset 2 were 0.81 and 0.82. FCBN was close to FPN, suggesting that FCBN can be used to estimate the number of flowering panicles. Dataset 2 (cropped images) was used to evaluate the influence of the marginal image distortion by the 24 mm wide lens on the detection accuracy but the results did not indicate any influence on the accuracy. Moreover, the curves for FCBN and FBN for Dataset 1 were much smoother than those for Dataset 2, indicating that the larger images could provide more stable detections because of the larger number of the target crops to be detected in an image.

Figure 6 shows that the flowering number normally reached a maximum around 12:00 on all days except day 87, when it reached a maximum around 15:00, Rice does not start flowering under rainy conditions [21,29,30] and it was in fact raining on the morning of day 87 (Figure 7). We observed that the rain delayed flowering on this day.

This result shows that the proposed method can accurately detect such sensitive physiological responses of rice by identifying flowering timing and extent.

Dataset 3 (_Kamenoo_) was used to verify the applicability of the above model used for Dataset 1 and Dataset 2. Figures 8 and 9 show the results of the flowering detection on Dataset 3. The correlation coefficients between FPN and FBN and between FPN and FCBN were 0.64 and 0.66, respectively. Although the correlation coefficients were lower than those for Dataset 1 and Dataset 2, the detected patterns of daily and diurnal flowering of _Kamenoo_ were well quantified by the model which was trained only by the images of a different variety, _Kinmaze_. Note that the sliding window size used for Dataset 3 to detect the flowering blocks was different from that used for Dataset 1 and Dataset 2 as mentioned in the Method section. We will discuss this point in the Discussion section.

Using our computer system (Microsoft Windows 8 PC with a 4-core i7 CPU and 16 GB of memory), the learning process with 600 training image patches (300 flowering and 300 non-flowering) takes approximately 30s. Using only 60 training image patches (30 flowering and 30 non-

Figure 5 An example of flowering panicle detection of Dataset 2 by the method developed in this study. (a) Original image from Dataset 2; **(b)** Each violet block indicates a sliding window in which part of a flowering panicle was detected. **(c)** Each red-outlined rectangle indicates a region of connected blocks.

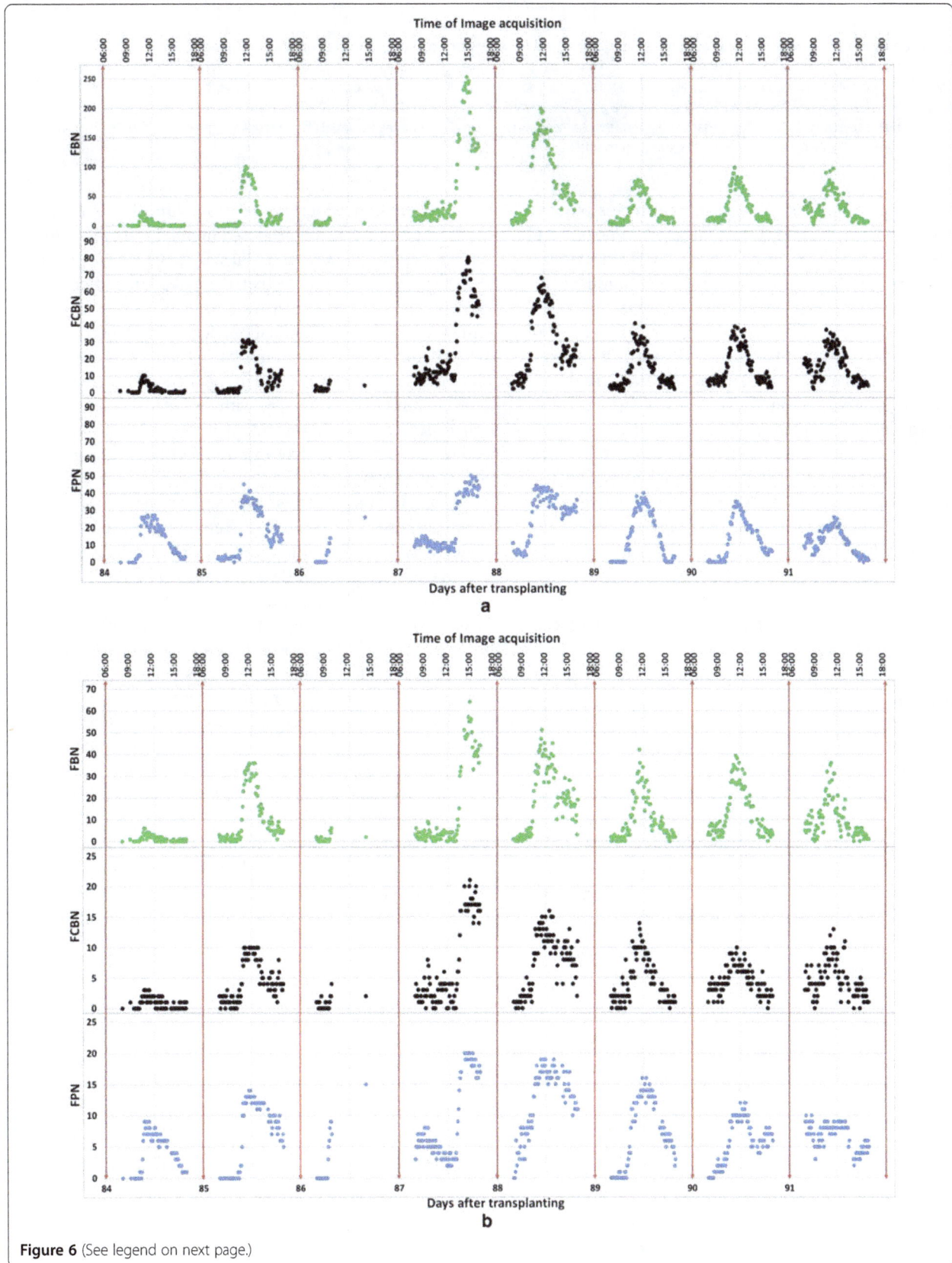

Figure 6 (See legend on next page.)

(See figure on previous page.)
Figure 6 Comparison of manually and automatically determined numbers of flowering panicles of Dataset 1 and Dataset 2. FBN: the number of the blocks which are judged to contain the flowering parts of panicles; FCBN: the number of the regions of connected blocks; FPN: the number of visually counted flowering panicles. **(a)** Dataset 1 for the original full size time series images of *Kinmaze*; **(b)** Dataset 2 for the cropped time series images of *Kinmaze*; The images were acquired every 5 minutes from 08:00 to 16:00 during the flowering period between days 84 and 91 after transplanting. Note that the system sometimes failed to acquire the images, which is particularly obvious on day 86. The failure was caused mainly by unstable network status in the field.

flowering) takes only 10s. The detection process requires approximately 480 s for each test image of Dataset 1 and Dataset 3 (5184 × 3456 pixels), and 70s for Dataset 2 (2001 × 1301 pixels). Although parallel computing helps us to process four images simultaneously, detection is still computationally expensive (22 ~ 30 h for Dataset 1 and Dataset 3, and 5 ~ 6 h for Dataset 2). We accordingly conducted a preliminary test on Dataset 2 to evaluate the effect of image resolution on the accuracy of the detection, aiming to reduce the computational cost of the method. The original images were resized to 75% and 50% of their original resolution and the accuracy of detection was evaluated (Figure 10). The 75% reduction did not affect accuracy (the correlation coefficient between FPN and FCBN was 0.83), whereas the 50% reduction clearly decreased accuracy (the correlation coefficient was 0.72). These results show that reduction of the test image resolution in an appropriate range reduced computing cost without loss of detection accuracy.

Discussion

The developed method accurately detected flowering rice panicles in time series of RGB images taken under natural field conditions. It was suggested to use the larger images to cover the larger number of crops, because the detections seemed to be more stable with more crops in a scene. The fact that the distortion of the images in the marginal parts did not influence the accuracy of the detections supported the suggestion. Although, the time series images in this study were acquired regardless

of light condition which varied from time to time, the results indicated that the proposed method was rather robust in detecting daily and diurnal flowering patterns. However, we also observed that the detection sometimes failed by specular reflection over panicles caused by extremely strong sunny illumination, degrading the accuracy of the detection. At this moment, we do not have any solution for the issue but it might be a good idea to automatically remove such images with specular reflections as outliers from frequently acquired images. To do so, we need to develop a new algorithm to identify such specular reflections in images.

The general versatility is required for the method to be widely used. As the first step, we examined the applicability of the model trained by the images of *Kinmaze* to a different variety *Kamenoo*. The result indicated that the model could quantify the daily and diurnal patterns of the flowering of the different variety but the correlation coefficients between FPN and FBN and between FPN and FCBN were worse than those for *Kinmaze*. We expect that many factors can cause such degradation. One possible cause of the degradation is the difference in the resolution of the panicle images between two varieties, because the proposed method detects the flowering depending on the spatial features of the images and the spatial features vary with image resolution. Actually, the observed plant heights of *Kinmaze* and *Kamenoo* at the flowering stage were around 107 cm and 145 cm respectively, so that the positions of the panicles of *Kamenoo* were much closer to the

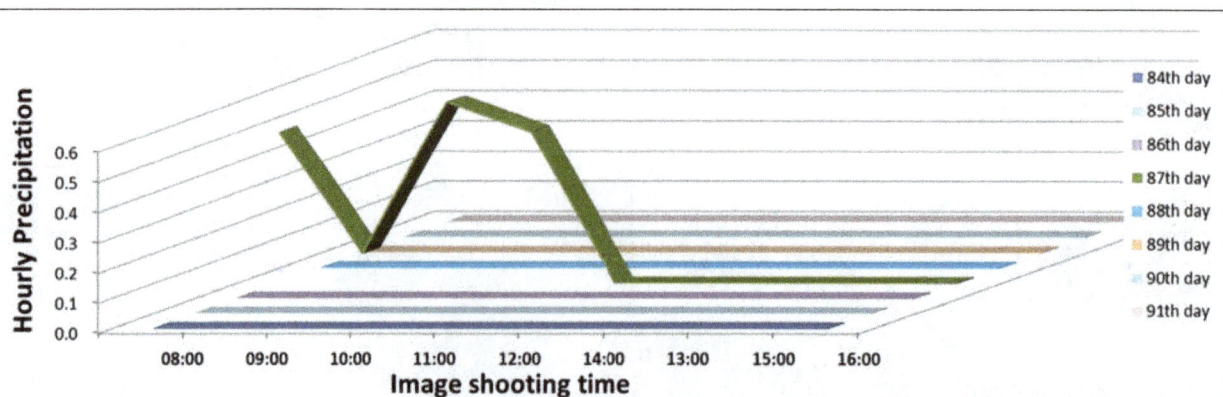

Figure 7 Hourly precipitation during seven consecutive flowering days from days 84 to 91 after transplanting. Each line indicates the hourly precipitation from 8:00 to 16:00. Note that it was raining on the morning of day 87 (green line).

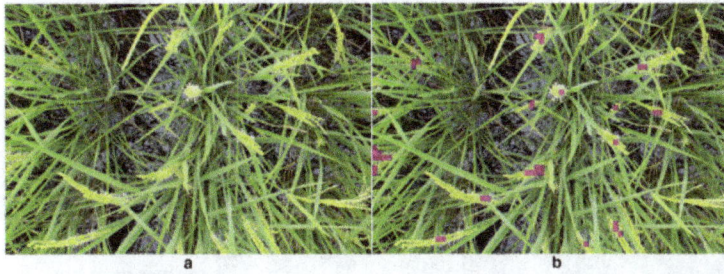

Figure 8 An example of flowering panicle detection of Dataset 3 (variety, *Kamenoo*) by the method developed in this study. (a) Original image from Dataset 3; **(b)** Each violet block indicates a window in which part of a flowering panicle was detected.

camera lens, making the resolution of the panicle images of *Kamenoo* higher. We tried to compensate this issue by adjusting the optimal size of the sliding window to detect the flowering for each variety in a preliminary test. Currently, the adjustment was done *ad hoc* through trial and error and we first need to develop an algorithm to conduct automatic adjustments of the sliding window size. In order to improve the proposed method for its general applicability in paddy rice, we also need to identify other causes of the degradation by using a wide range of varieties.

Generic object recognition is still an important target of pattern recognition studies and continues to be developed. For example, BoVWs count only the occurrences of visual words based on local image features, and ignores location and color information of each feature

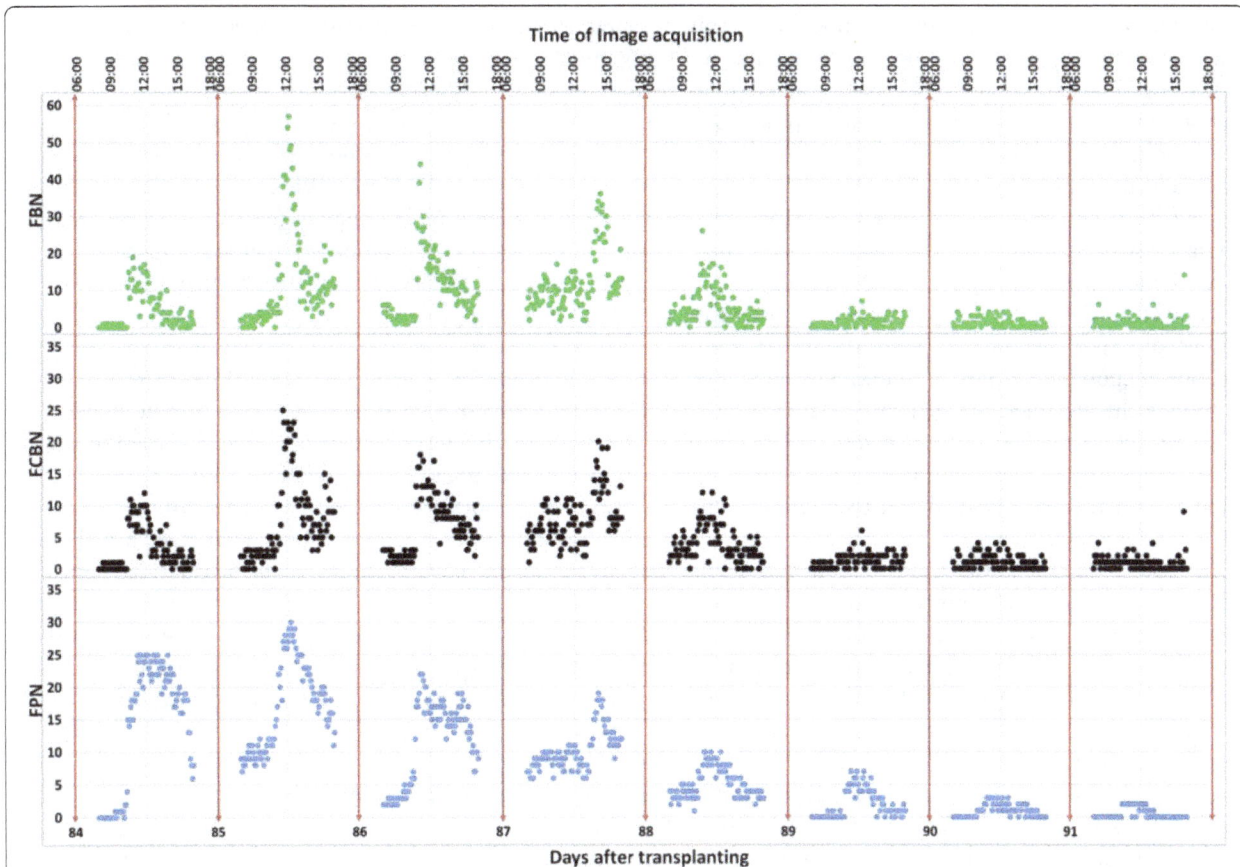

Figure 9 Comparison of manually and automatically determined numbers of flowering panicles of Dataset 3. FBN: the number of the blocks which are judged to contain the flowering parts of panicles; FCBN: the number of the regions of connected blocks; FPN: the number of visually counted flowering panicles. The images were acquired every 5 minutes from 08:00 to 16:00 during the flowering period between days 84 and 91 after transplanting.

2013/08/25, 14PM

(a) 100% (Original size)   **(b) 75%**   **(c) 50%**

Figure 10 An example of flowering detection at three different image resolutions. The resolution of the original image (2001 × 1301 pixels) was reduced by 75% (1501 × 976) and 50% (1001 × 651) and the efficiencies of detection were compared. The detection in the 75% reduction case **(b)** was almost the same as that in the original resolution **(a)** and the correlation coefficient between FPN and FCBN is 0.83, whereas the missed detection in the 50% case **(c)** was obvious and the correlation was 0.73.

that may improve the accuracy of the model. For this reason, studies are now focusing on increasing the dimensions of BoVWs by adding more statistical variables such as a vector of locally aggregated descriptors [31], super vector coding [32], a Fisher vector [33], and a vector of locally aggregated tensors [34]. These new concepts have been proposed to accurately recognize and classify large scale images in the real world. We expect that such concepts will contribute to the improvement of our flowering detection method as well as the development of other agricultural applications for high-throughput phenotyping in future studies. Our next step is to improve the accuracy and general versatility of the flowering detection method. To reach this goal, we will also need to identify the optimal quantity and quality of the training image patches in addition to improving the model.

In this study, a camera was fixed, targeting a single plot. However, providing a camera for each plot is impractical when a number of plots are to be observed. Therefore, we are now developing a movable camera system, which can cover several plots only with a single camera. We also expect to use an unmanned aerial vehicle (UAV) to cover a large numbers of plots.

Though we need further improvements of the method as discussed above, the overall results in this study showed a high performance in detecting the flowering panicles of rice. We expect that our method will contribute to practical rice farming management as well as to rice research. Although flowering timing is one of the most important indicators in optimal management and

characterization of rice, it is still judged visually, requiring much time. In particular, when a large number of small plots with different flowering timings are to be observed, our method can be especially useful. A typical example is rice breeding, where a large number of plots must be observed efficiently. We expect that the combination of a movable camera system/UAV and the improved version of the proposed method applicable to paddy rice in general will dramatically ease and accelerate the breeding process.

Notably, the diurnal flowering timing of rice is becoming important because of the trend of global warming. The pollination of rice occurs at the timing of spikelet anthesis and the fertility depends strongly on the air temperature at pollination. Therefore, rice varieties flowering early morning before the temperature rises are being sought [3]. In breeding for such varieties, breeders at present must observe many plots of candidate lines continuously for a few hours early morning every day during the expected flowering period. The proposed method, which can accurately detect diurnal flowering timing, is expected to be highly helpful in such cases.

Methods
Experimental materials and growth conditions
In this study, the japonica rice (*Oryza sativa* L.) varieties, *Kinmaze* and *Kamenoo*, were used. Seeds were sown on April 26 and transplanted on May 31, 2013 in the field at the Institute for Sustainable Agro-ecosystem Services, University of Tokyo (35°44′22″N, 139°32′34″E and 67 m above the sea level). The area of the experimental field

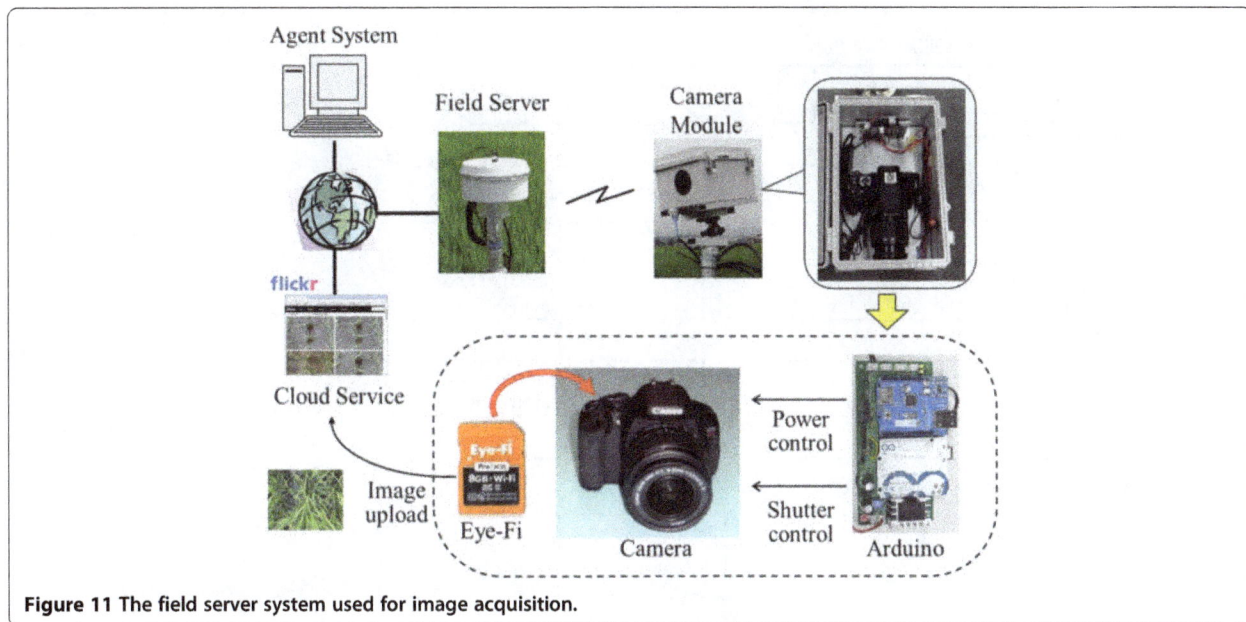

Figure 11 The field server system used for image acquisition.

was approximately 250 m², and the planting density was 28 plants/m². From June to September, the average temperature, the average humidity, total rainfall, and total solar radiation were 26.2°C, 62%, 653.0 mm, and 1980.5 MJ/m², respectively.

Image acquisition

A Field Server system [35,36] was used to acquire the experimental images (Figure 11). The camera module of the system is based on a digital single-lens reflex (DSLR) camera, the Canon EOS Kiss X5 camera, with an EF-S18-55 mm lens (Canon Inc., Tokyo) that provides high-quality and high-resolution (18 megapixels) image data. The power and shutter of the camera are controlled by a preprogrammed microcontroller board, the Arduino Uno (http://arduino.cc). The captured image data were sent to a free cloud service, Flickr (www.flickr.com) by a wireless uploading SD card, Eye-Fi (Eye-Fi, Inc., Mountain View) through WI-FI hotspots provided by the Field Servers at the field site. The Agent System [37] automatically grabs the images from the webpage of Flickr, arranges them, and saves them into a database at the National Agriculture and Food Research Organization using their EXIF data.

The cameras are set to view the rice canopy from 2 m above the ground. At this distance, the image resolution is approximately 43 pixels/cm at the ground level and the resolution of crop images increases according to the crop growth. Using the system, time-series images of two paddy varieties were acquired every 5 min from 08:00 to 16:00 between days 84 and 91 after transplanting. Some of the images of the variety *Kinmaze* are missing because the system failed to acquire them. The

failure was mainly due to the unstable network status in the field and was particularly obvious on day 86. Finally, a total of 645 images for *Kinmaze* (Dataset 1) and 768 images for *Kamenoo* (Dataset 3) were obtained. The images (5184 × 3456 pixels) corresponded to a field size of 138 cm × 98 cm and the number of the crops included in an image was around 30. Then, we cropped the original images of *Kinmaze* (Dataset 1) to the central regions in order to create a new time series image dataset named Dataset 2. The cropped image corresponded to a field size of 30 × 45 cm that contained three rice plants. Figure 12 shows the cropping, by which the original image of 5184 × 3456 pixels was cropped to a central region of 2001 × 1301 pixels. We used Dataset 2 to evaluate the influences of both the crop number included in an image and the distortion of the marginal area of the image caused by the camera lens on the accuracy of the flowering detection, comparing with the full size image dataset of *Kinmaze* (Dataset 1). To evaluate the flowering

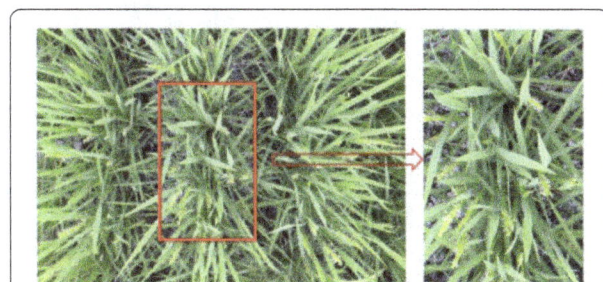

Figure 12 Cropping of the original image. The central region of each original image of the variety *Kinmaze* was cropped. The cropped region was corresponded to a field size of 30 × 45 cm that contained three rice plants.

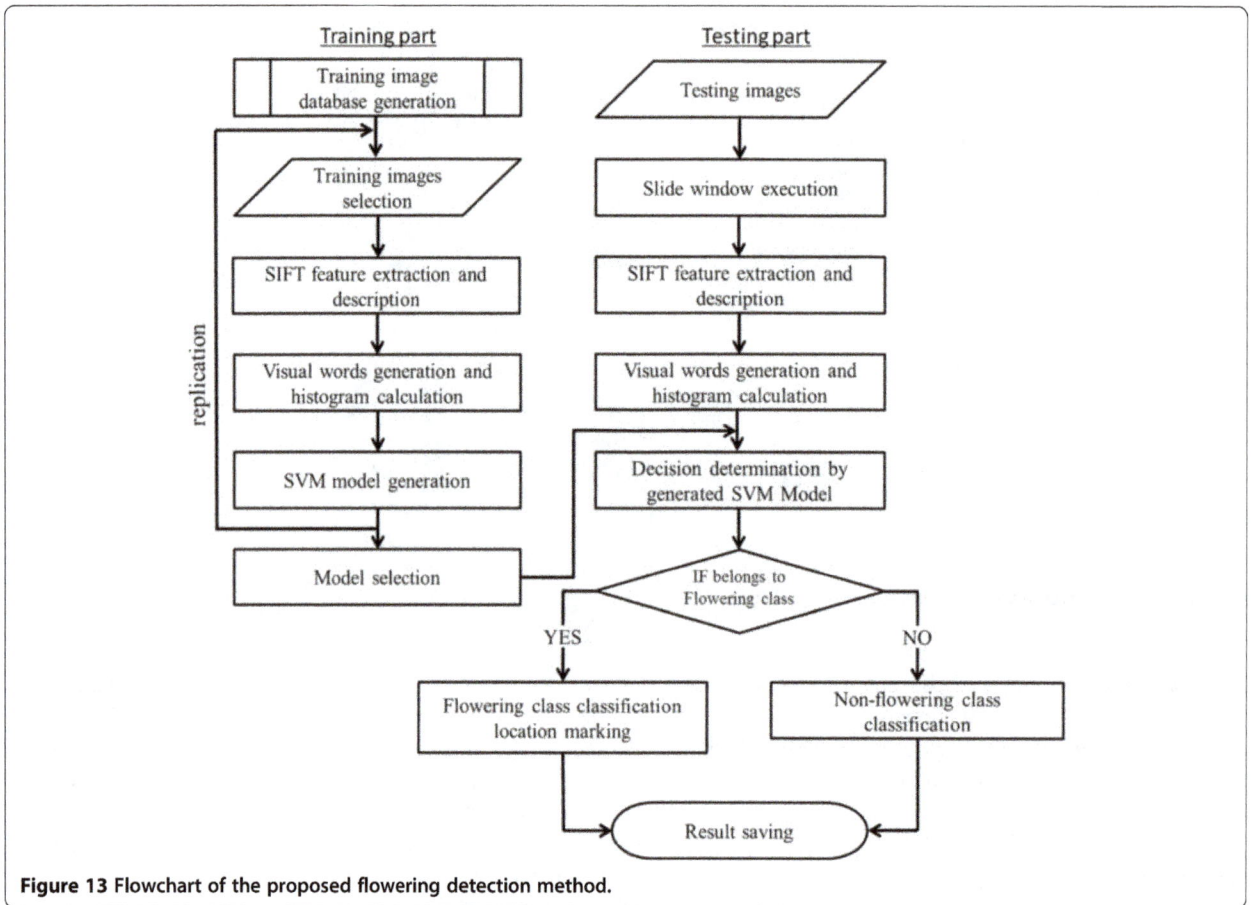

Figure 13 Flowchart of the proposed flowering detection method.

detection performance by the proposed method, the numbers of flowering panicles in all of the acquired images were counted visually.

Flowering panicle detection

The full process is illustrated in Figure 13 and can be separated into two parts: training and testing. The process comprises the following steps:

1. Creating the training database by manually cropping the experimental images to yield rectangular regions. We created a database of training image patches of two classes, the positive class (flowering panicles) and the negative class (the background). Twenty one images from Dataset 1 were selected to obtain training data, considering the variation of the weather conditions in photographing (sunny, rainy, and cloudy conditions), the growth stage during the flowering

Figure 14 Some examples of training image patches. (a) Positive data which contain flowering parts of panicle (s); **(b)** Negative data which does not contain flowering parts of panicle (s), the training image patches were sampled from 21 images of Dataset 1.

Figure 15 An example of dense sampling and SIFT feature point description. (a) SIFT descriptors are computed at regular grid points with a spacing of 15 pixels, as represented by the red circle; **(b - e)** At each point, SIFT descriptors are calculated on four different scales using four different radii: r = 4, 6, 8, and 10 pixels. The descriptor of each scale has 16 patches, represented by the red rectangles, which are rotated to the dominant orientation of the feature point. Each patch is described in gradient magnitudes of eight directions (red bins inside the red rectangles).

period (initial, middle, and final flowering stages), and the positions (with and without occlusions and overlaps by other panicles and leaves). Finally, we obtained 300 image patches that contained part (s) of rice flowering panicles and 400 image patches that did not contain any part (s) of flowering panicles. An example of those training image patches are shown in Figure 14. Note that the sizes of the training image patches are not necessarily the same.

2. Extracting local feature points and descriptors of those points from training image patches. In this study, we used SIFT descriptors [23] and dense sampling [38] to extract the points. In dense sampling, regular grid points with a space of M pixels are overlaid on an image and the SIFT descriptors are computed at each grid point of the image (Figure 15). In this study, we used M = 15 based on a preliminary test and used four circular support patches with radii r = 4, 6, 8, and 10 pixels to calculate scale-invariant SIFT descriptors. Consequently, each point was characterized by four SIFT descriptors, each of which comprised a 128-dimensional vector (Figure 15). The descriptor of each scale is based on a square with 16 patches [red squares in Figure 15 (b–e)]. The square is rotated to the dominant orientation of the feature point, and each patch in the square is described in the gradient magnitudes of eight different directions resulting in a total of 128 variables for each scale.

3. Generating visual words using the *k-means* method, which has been reported to perform well in object-recognition approaches [25,39]. The choice of the initial centroid position and the number of clusters (k) affects the resulting vocabulary in the *k-means*

clustering method. In this study, we predefined k = 600 (number of visual words). We then ran *k-means* several times with random initial assignments of points as cluster centers, and used the best result to select the best-performing vocabulary. Note that these visual words do not contain location information of points.

4. Training the SVM as a flowering detection model, using the visual words as training data. SVM is one of the most popular machine learning models for object generic recognition. We used the SVM with a χ^2 kernel, which is particularly powerful with data in histogram format [40,41]. A homogeneous kernel map was used to approximate the χ^2 kernel to accelerate the learning process. The map transforms the data into a compact linear representation that reproduces the desired kernel to a very good level of approximation. This representation enables very fast linear SVM solvers [42]. The source code is available from the VLFeat open source library [43].

5. Verifying the performance of the generated SVM model for detecting the flowering parts of panicles in the test images. We used a sliding-window approach to apply the SVM model to the test images. The concept of the sliding window is to scan a whole test image without any overlaps using a predefined window size and then decide whether or not each scan window contains flowering parts, with reference to the trained model. In each scan window, the distribution of the visual words by the *k-means* method based on the entire set of sampling grid points where SIFT descriptors were calculated was used as an input to the generated SVM model. The most appropriate

Table 1 Relationship between the number of training images and the performance of flowering detection

Training number	5	15	30	50	100	300
Accuracy[+]	0.74± 0.05	0.81± 0.04	0.83± 0.03	0.79± 0.03	0.73± 0.02	0.64
TP rate[+]	0.65± 0.13	0.61± 0.12	0.59± 0.09	0.49± 0.08	0.31± 0.04	0.09
TN rate[+]	0.8 ± 0.09	0.95± 0.03	0.99± 0.00	0.99± 0.00	1 ± 0.00	1.0

[+]Accuracy, TP rate, and TN rate, were defined as follows:
Accuracy $= \frac{TP + TN}{TP + FP + TN + FN}$, TPrate $= \frac{TP}{TP+FN}$, TN rate $= \frac{TP}{FP+TN}$
where TP, TN, FP, and FN represent the numbers of true positives, true negatives, false positives, and false negatives, respectively, of the confusion matrix.

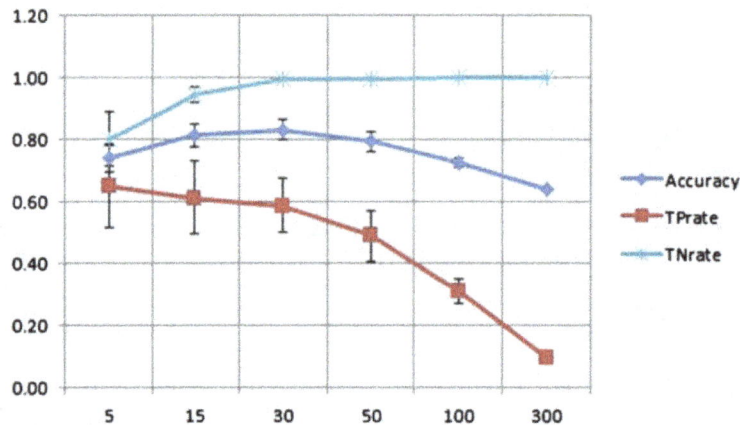

Figure 16 Performance of SVM models under different numbers of training image patches. Please see Table 1 for the definition of Accuracy, TPrate and TNrate. Considering Accuracy, TP rate and TN rate, the performance of the model is most well balanced when 30 training image patches were used.

sliding-window size was determined by a preliminary test as 140×140 pixels for Dataset_1 and Dataset_2, and 170×170 pixels for Dataset_3, given that the size strongly affects flowering detection by the method.

The whole process was implemented using the software package MATLAB (MathWorks Inc., Natick) on a Microsoft Windows 8 PC with a 4-core CPU and 16 GB memory. Correlation analysis was performed with the statistical software package R (R Development Core Team, 2012).

Training data selection

Because the training image patches were manually selected, there was no guarantee that all of them provided "good" training data sets for training the flowering detection model. In addition, our preliminary test showed that the full use of the 300 positive and 400 negative training image patches did not provide the best performance compared with the use of the smaller number. Therefore, in lieu of using all the training image patches, we sought to determine how to select the most appropriate training image patches. We evaluated the accuracy

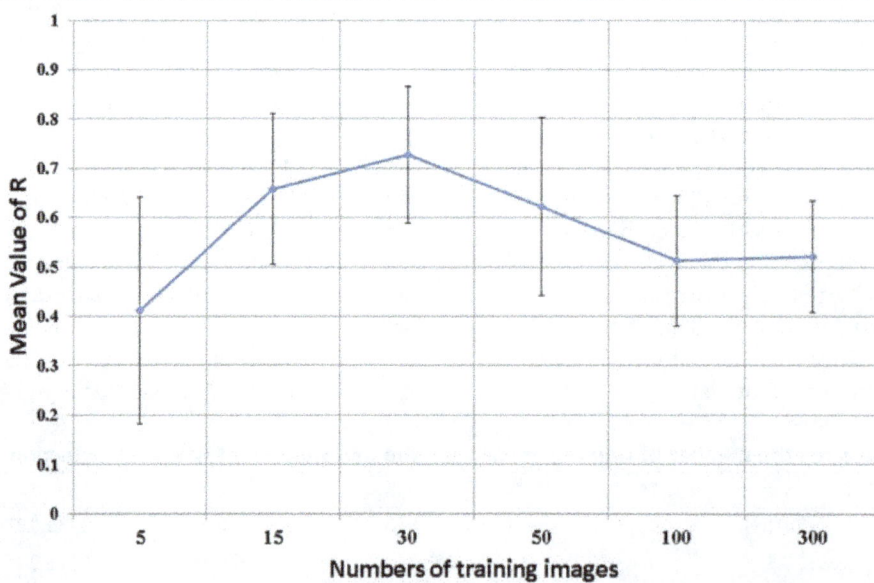

Figure 17 Relationship between numbers of training image patches and flowering detection performance. Performance is represented by the correlation coefficients between visually determined numbers of flowering panicles (FPN) and automatically detected numbers of flowering panicles (FCBN) in each case. The performance is best when 30 training image patches were used.

of flowering detection using a different number of training image patches, for both positive and negative data with the aim of determining the optimal number, as follows: 5, 15, 30, 50, 100, and 300 (full use). Each set of images was randomly selected from the training image database with 10 replications, except when all 300 images were used. Then, using each of the training data sets, the SVM model was trained and its accuracy for flowering detection in the training image patches was evaluated. To evaluate the performance of the detection, three indices, accuracy, TP rate, and TN rate, were used. They are defined as follows:

$$Accuracy = \frac{TP + TN}{TP + FP + TN + FN}$$

$$TP\ rate = \frac{TP}{TP + FN}$$

$$TN\ rate = \frac{TN}{FP + TN}$$

where TP, TN, FP, and FN represent the numbers of true positives, true negatives, false positives, and false negatives of the confusion matrix, respectively. Accuracy measures the model detection ability for both flowering and background classes over the whole test data. The true positive rate, TP rate, measures the proportion of detected flowering images in the flowering class, whereas the true negative rate, TN rate, measures the detected background images in the background class. The means and standard deviations of the values from the 10 replications under different training image numbers are shown in Table 1 and Figure 16. The result shows that the performance of the model as measured by accuracy, TP rate, and TN rate is most well balanced with the training image number 30.

To verify the performance of flowering panicle detection by each model, we calculated the correlation coefficient (R) between visually determined flowering panicle numbers and numbers of blocks detected that contain flowering panicles (Figure 17). The R values increased with the number of training image patches until it reached 30, and then declined rapidly as the number increased. Thus, we again concluded that the training image number of 30 was optimal for flowering detection and used the training data set of 30 images that performed best among the 10 replicates in this study.

We originally expected that the full set of training image patches would perform best, but a much smaller number actually demonstrated the best performance in flowering detection. We can expect that the complexity of the background class generates widely varying SIFT descriptors within the class, and the more the training data, the more variation will appear. Such a variation in the SIFT features within a class may affect accuracy, although further studies are needed to identify the reason.

Endnote

[a]http://park.itc.u-tokyo.ac.jp/nino-lab/labhome/PhenotypingTools/RiceFlower.html

Additional file

Additional file 1: A demo video that shows flowering panicle detection on Dataset_1.

Abbreviation

SIFT: Scale-Invariant feature transform; BoVWs: Bag of visual words; SVM: Support Vector Machine; DSLR: Digital Single-Lens Reflex; TP: True positive; TN: True negative; TPrate: True positive rate; TNrate: True negative rate; FBN: The number of the blocks which are judged to contain the flowering parts of panicles; FCBN: The number of the regions of connected blocks; FPN: The number of visually counted flowering panicles.

Competing interests

The authors declare that they have no competing interests.

Authors' contributions

WG performed the experimental work, constructed the flower detecting model, carried out the data analysis and interpretation, and wrote the manuscript. FT developed the Field Sever and image acquisition modules for the field monitoring system. SN supervised the study. All authors contributed in reading, editing and approving the final manuscript.

Acknowledgements

The authors would like to thank all the technical support staffs in the Institute of Sustainable Agro-ecosystem Services, The University of Tokyo for their helping the field experiments. This research was partially funded by "Research Program on Climate Change Adaptation" of Ministry of Education, Culture, Sports, Science and Technology, Japan.

Author details

[1]Institute for Sustainable Agro-ecosystem Services, Graduate School of Agricultural and Life Sciences, The University of Tokyo, 1-1-1. Midori-cho, Nishi-Tokyo, Tokyo 188-0002, Japan. [2]National Agriculture and Food Research Organization, 3-1-1 Kannondai, Tsukuba, Ibaraki 305-8666, Japan.

References

1. Wassmann R, Jagadish SVK, Heuer S, Ismail A, Redona E, Serraj R, Singh RK, Howell G, Pathak H, Sumfleth K: Climate Change Affecting Rice Production: The Physiological and Agronomic Basis for Possible Adaptation Strategies. In: Donald L Sparks, editor. Advances in Agronomy Volume 101; 2009. P.59-122
2. Jagadish SVK, Craufurd PQ, Wheeler TR. High temperature stress and spikelet fertility in rice (Oryza sativa L.). J Exp Bot. 2007;58:1627–35.
3. Ishimaru T, Hirabayashi H, Ida M, Takai T, San-Oh YA, Yoshinaga S, et al. A genetic resource for early-morning flowering trait of wild rice Oryza officinalis to mitigate high temperature-induced spikelet sterility at anthesis. Ann Bot. 2010;106:515–20.
4. Shah F, Huang J, Cui K, Nie L, Shah T, Chen C, et al. Impact of high-temperature stress on rice plant and its traits related to tolerance. J Agric Sci. 2011;149:545–56.
5. Confalonieri R, Foi M, Casa R, Aquaro S, Tona E, Peterle M, et al. Development of an app for estimating leaf area index using a smartphone. Trueness and precision determination and comparison with other indirect methods. Comput Electron Agric. 2013;96:67–74.
6. Liu J, Pattey E. Retrieval of leaf area index from top-of-canopy digital photography over agricultural crops. Agric For Meteorol. 2010;150:1485–90.
7. Liu J, Pattey E, Admiral S. Assessment of in situ crop LAI measurement using unidirectional view digital photography. Agric For Meteorol. 2013;169:25–34.
8. Royo C, Villegas D: Field Measurements of Canopy Spectra for Biomass Assessment of Small-Grain Cereals. In: Matovic D, editor. Biomass - Detect Prod Usage; 2011.27-52

9. Sakamoto T, Shibayama M, Kimura A, Takada E. Assessment of digital camera-derived vegetation indices in quantitative monitoring of seasonal rice growth. ISPRS J Photogramm Remote Sens. 2011;66:872–82.

10. Torres-Sánchez J, Peña JM, de Castro AI, López-Granados F. Multi-temporal mapping of the vegetation fraction in early-season wheat fields using images from UAV. Comput Electron Agric. 2014;103:104–13.

11. Guo W, Rage UK, Ninomiya S. Illumination invariant segmentation of vegetation for time series wheat images based on decision tree model. Comput Electron Agric. 2013;96:58–66.

12. Sritarapipat T, Rakwatin P, Kasetkasem T. Automatic rice crop height measurement using a field server and digital image processing. Sensors. 2014;14:900–26.

13. Yu Z, Cao Z, Wu X, Bai X, Qin Y, Zhuo W, et al. Automatic image-based detection technology for two critical growth stages of maize: Emergence and three-leaf stage. Agric For Meteorol. 2013;174–175:65–84.

14. Sakamoto T, Gitelson AA, Nguy-Robertson AL, Arkebauer TJ, Wardlow BD, Suyker AE, et al. An alternative method using digital cameras for continuous monitoring of crop status. Agric For Meteorol. 2012;154:113–26.

15. Nguy-Robertson A, Gitelson A, Peng Y, Walter-Shea E, Leavitt B, Arkebauer T. Continuous monitoring of crop reflectance, vegetation fraction, and identification of developmental stages using a four band radiometer. Agron J. 2013;105:1769.

16. Yoshioka Y, Iwata H, Ohsawa R, Ninomiya S. Quantitative evaluation of the petal shape variation in Primula sieboldii caused by breeding process in the last 300 years. Heredity (Edinb). 2005;94:657–63.

17. Iwata H, Ebana K, Uga Y, Hayashi T, Jannink J-L. Genome-wide association study of grain shape variation among Oryza sativa L. germplasms based on elliptic Fourier analysis. Mol Breed. 2009;25:203–15.

18. Yoshioka Y, Fukino N. Image-based phenotyping: use of colour signature in evaluation of melon fruit colour. Euphytica. 2009;171:409–16.

19. Remmler L, Rolland-Lagan A-G. Computational method for quantifying growth patterns at the adaxial leaf surface in three dimensions. Plant Physiol. 2012;159:27–39.

20. Mielewczik M, Friedli M, Kirchgessner N, Walter A. Diel leaf growth of soybean: a novel method to analyze two-dimensional leaf expansion in high temporal resolution based on a marker tracking approach (Martrack Leaf). Plant Methods. 2013;9:30.

21. Yoshida S: Fundamentals of Rice Crop Science. Los Banos; International Rice Research Institute; 1981. http://books.irri.org/9711040522_content.pdf

22. Kobayasi K. Effects of Solar Radiation on Fertility and the Flower Opening Time in Rice Under Heat Stress Conditions. In: Babatunde EB, editor. Solar Radiation. 2012. p. 245–66.

23. Lowe D. Distinctive image features from scale-invariant keypoints. Int J Comput Vis. 2004;60(2): 91–110. doi:10.1023/B:VISI.0000029664.99615.94

24. Gabriella C, Dance CR, Fan L, Willamowski J, Bray C. Visual categorization with bags of keypoints. In: Work Stat Learn Comput Vision, ECCV. 2004. p. 1–22.

25. Sivic J, Zisserman A. Video Google: a text retrieval approach to object matching in videos. In: Comput Vision, 2003 Proceedings Ninth IEEE Int Conf, vol. 2. 2003. p. 1470–7.

26. Vapnik VN. Statistical learning theory. 1st ed. New York: Wiley-Interscience; 1998. p. 768.

27. Thorp KR, Dierig DA. Color image segmentation approach to monitor flowering in lesquerella. Ind Crops Prod. 2011;34:1150–9.

28. Scotford IM, Miller PCH. Estimating tiller density and leaf area index of winter wheat using spectral reflectance and ultrasonic sensing techniques. Biosyst Eng. 2004;89:395–408.

29. Matsuo T, Hoshikawa K. Science of the rice plant (Volume One): phsiology. Tokyo: Food and Agriculture Policy Research Center; 1993. p. 686 [Science of the Rice Plant].

30. Kiyochika H. The growing rice plant (In Japanese). Tokyo: Rural Culture Association Japan; 1973. p. 317.

31. Jegou H, Douze M, Schmid C, Perez P. Aggregating local descriptors into a compact image representation. In: 2010 IEEE Comput Soc Conf Comput Vis Pattern Recognit. IEEE. 2010. p. 3304–11.

32. Zhou X, Yu K, Zhang T, Huang TT. Image classification using super-vector coding of local image descriptors. In: Daniilidis K, Maragos P, Paragios N, editors. Comput Vis – ECCV 2010 SE - 11, vol. 6315. Berlin Heidelberg: Springer; 2010. p. 141–54 [Lecture Notes in Computer Science].

33. Perronnin F, Liu Y, Sanchez J, Poirier H. Large-scale image retrieval with compressed Fisher vectors. In: Comput Vis Pattern Recognit (CVPR), 2010 IEEE Conf. 2010. p. 3384–91.

34. Picard D, Gosselin P-H. Improving image similarity with vectors of locally aggregated tensors. In: Image Process (ICIP), 2011 18th IEEE Int Conf. 2011. p. 669–72.

35. Fukatsu T, Kiura T, Hirafuji M. A web-based sensor network system with distributed data processing approach via web application. Comput Stand Interfaces. 2011;33:565–73.

36. Fukatsu T, Watanabe T, Hu H, Yoichi H, Hirafuji M. Field monitoring support system for the occurrence of Leptocorisa chinensis Dallas (Hemiptera: Alydidae) using synthetic attractants, Field Servers, and image analysis. Comput Electron Agric. 2012;80:8–16.

37. Fukatsu T, Hirafuji M, Kiura T. An agent system for operating web-based sensor nodes via the internet. J Robot Mechatronics. 2006;18:186–94.

38. Nowak E, Jurie F, Triggs B. Sampling strategies for bag-of-features image classification. Comput Vision–ECCV 2006. 2006;3954:490–503.

39. Lazebnik S, Schmid C, Ponce J. Beyond bags of features: spatial pyramid matching for recognizing natural scene categories. In: Comput Vis Pattern Recognition, 2006 IEEE Comput Soc Conf. 2006. p. 2169–78.

40. Jiang Y-G, Ngo C-W, Yang J. Towards optimal Bag-of-features for object categorization and semantic video retrieval. In: Proc 6th ACM Int Conf Image Video Retr. New York, NY, USA: ACM; 2007. p. 494–501 [CIVR '07].

41. Zhang J, Marszałek M, Lazebnik S, Schmid C. Local features and kernels for classification of texture and object categories: a comprehensive study. Int J Comput Vis. 2007;73:213–38.

42. Vedaldi A, Zisserman A. Efficient additive kernels via explicit feature maps. IEEE Trans Pattern Anal Mach Intell. 2012;34:480–92.

43. Vedaldi A, Fulkerson B. Vlfeat: an open and portable library of computer vision algorithms. In: Proc Int Conf Multimed. New York, NY, USA: ACM; 2010. p. 1469–72.

15

Two-dimensional multifractal detrended fluctuation analysis for plant identification

Fang Wang[1*], Deng-wen Liao[2], Jin-wei Li[3] and Gui-ping Liao[3]

Abstract

Background: In this paper, a novel method is proposed to identify plant species by using the two- dimensional multifractal detrended fluctuation analysis (2D MF-DFA). Our method involves calculating a set of multifractal parameters that characterize the texture features of each plant leaf image. An index, I_0, that characterizes the relation of the intra-species variances and inter-species variances is introduced. This index is used to select three multifractal parameters for the identification process. The procedure is applied to the Swedish leaf data set containing leaves from fifteen different tree species.

Results: The chosen three parameters form a three-dimensional space in which the samples from the same species can be clustered together and be separated from other species. Support vector machines and kernel methods are employed to assess the identification accuracy. The resulting averaged discriminant accuracy reaches 98.4% for every two species by the 10 − fold cross validation, while the accuracy reaches 93.96% for all fifteen species.

Conclusions: Our method, based on the 2D MF-DFA, provides a feasible and efficient procedure to identify plant species.

Keywords: Plant identification, Multifractal detrended fluctuation analysis, Support vector machines and kernel methods

Introduction

The increasing interest in biodiversity and biocomplexity, together with the growing availability of digital images and image analysis algorithms, makes plant species identification and classification a topic that has attracted many researchers' attention. In general, many parts of a plant such as flowers, seeds, roots, and leaves can be used to identify plant species [1-3]. In this paper, we focus on the usage of image of leaves as they are widely available. Leaf's shape, color, vein properties, texture and contours are important features for plant identification. For example, leaf shapes were used in [4-6]; complex veins and contours of leaves were used in [7] and leaf texture was used in [8-11] for plant species identification. For plant species identification using digital morphometrics, we refer the reader to [12-14] and the references therein.

Note that in [7], a monofractal method was used to extract plant leaf's features from leaf images. This method was then used in [15,16]. It's been recognized that the monofractal method cannot fully extract detailed information from the leaf image and therefore cannot be efficiently applied to process the images of the objects that are locally irregular [17]. To overcome this difficulty, several multifractal analysis (MFA) methods were proposed [18-22]. For example, Backes *et al.* [18,19] used multi-scale fractal dimensions to describe the texture property of leaf's surface to identify plants, which turned out to be very efficient. Note that the classical MFA is based on capacity measurement or probability measurement and thus describes only stationary measurements [17]. For a leaf image, the surface itself is hardly stationary. Therefore, the multifractal detrended fluctuation analysis (MF-DFA) method that can deal with non-stationary is a desirable method for leaf image analysis [23]. Though the MF-DFA method has been successfully applied in many fields for non-stationary series and surfaces [24-30], to the best of our knowledge, no work yet has applied the MF-DFA on leaf images for plant identification and classification. In this paper, we attempt to identify plant species *via* leaf images by using the MF-DFA. More precisely, we first adopt the MF-DFA to extract important texture features from leaf images and obtain several key multifractal parameters, and then we apply the support vector machines and kernel methods (SVMKM) to distinguish leaves from different plant species. The widely used Swedish leaf data set [31] containing leaves from fifteen

* Correspondence: popwang619@163.com
[1]College of Science, Hunan Agricultural University, Changsha 410128, China
Full list of author information is available at the end of the article

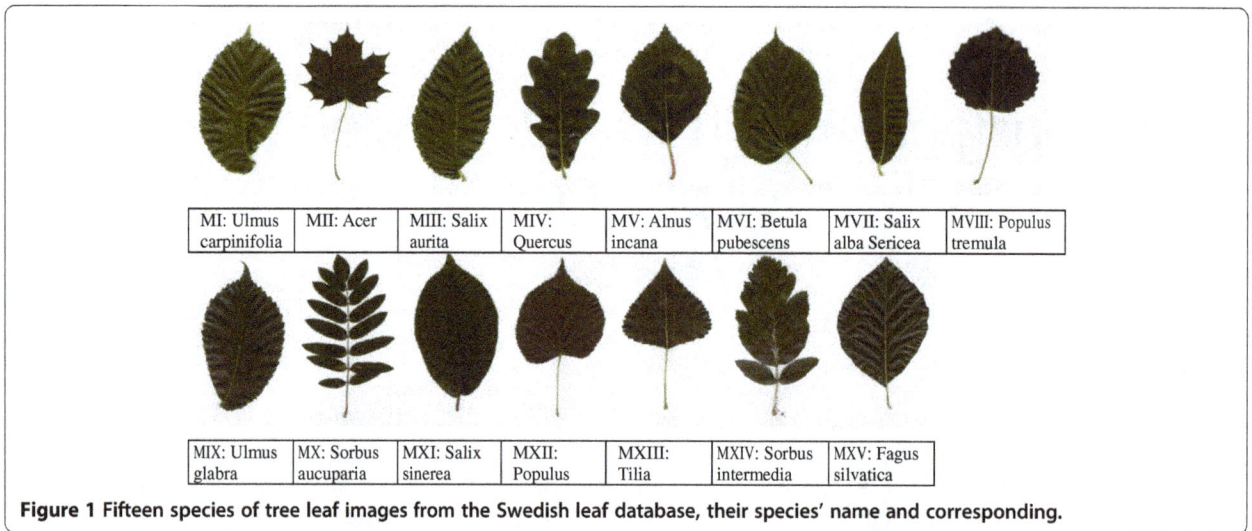

Figure 1 Fifteen species of tree leaf images from the Swedish leaf database, their species' name and corresponding.

different Swedish tree species are used for our experiments. Our results show that the average accuracy is 98.4% for every two species by the 10 – fold cross validation; for the over-all species, the average accuracy reaches 93.96% by the same validation criterion.

We organize the rest of this paper as follows: in Methods and materials we adopt the two-dimensional (2D) MF-DFA to calculate the multifractal parameters. In Results and discussion, we present and discuss our results. Our method is then further tested in Model test. A summary is provided in Conclusions.

Methods and materials
Multifractal detrended fluctuation analysis
We first adopt the 2D MF-DFA method proposed in [32] to our setting as follows:

Step 1: Regard a leaf image as a self-similar surface and represent it by an $M \times N$ matrix $X = (X(i, j))$, $i = 1, 2,...,$

M and $j = 1, 2,..., N$. Partition the surface into $M_s \times N_s$ non-overlapping square sub-surface of equal length s, where $M_s \equiv [M / s]$ and $N_s \equiv [N / s]$ are positive integers (Here $[u]$ stands for the largest integer that is less than or equal to u). Each sub-surface is denoted by $X_{m,n} = X_{m,n}(i, j)$ with $X_{m,n}(i, j) = X(r + i, t + j)$ for $1 \le i, j \le s$, where $r = (m-1)s$ and $t = (n-1)s$. Note that M and N are not necessarily multiples of the length s, therefore, the sub-surfaces in the upper-right and the bottom may not be taken into consideration. We can then repeat the partitioning procedure starting from the other three corners.

Step 2: For each sub-domain $X_{m,n}$, find its cumulative sum

$$G_{m,n}(i,j) = \sum_{k_1=1}^{i} \sum_{k_2=1}^{j} {}_{m,n} Xm(k_1, k_2),$$ (1)

where $1 \le i, j \le s$, $m = 1, 2, ..., M_s$ and $n = 1, 2, ..., N_s$. Then $G_{m,n} = G_{m,n}(i, j)$ $(i, j = 1, 2, \cdots, s)$ itself is a surface.

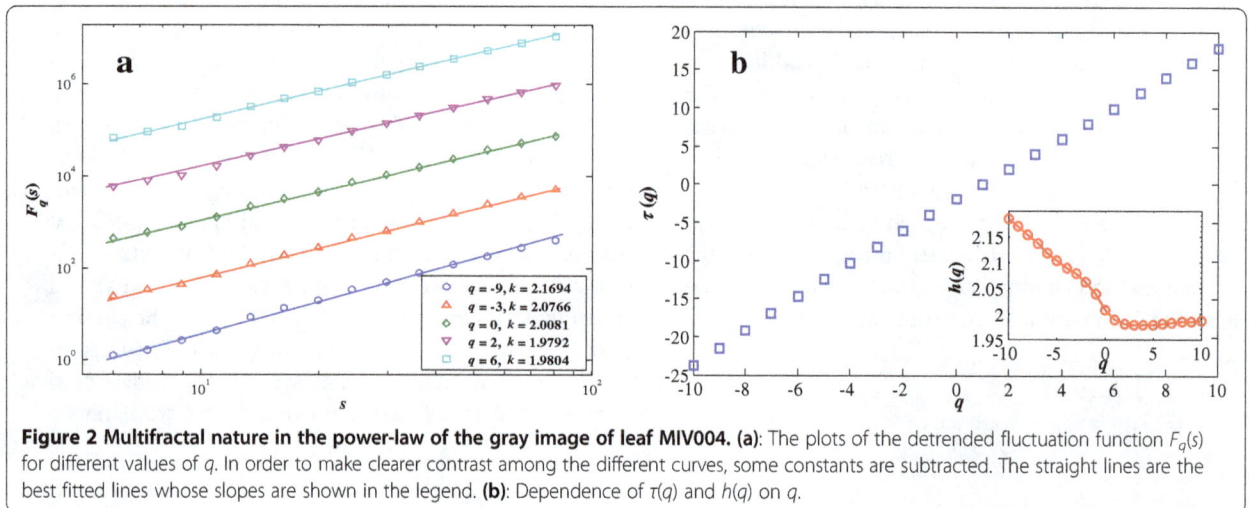

Figure 2 Multifractal nature in the power-law of the gray image of leaf MIV004. (a): The plots of the detrended fluctuation function $F_q(s)$ for different values of q. In order to make clearer contrast among the different curves, some constants are subtracted. The straight lines are the best fitted lines whose slopes are shown in the legend. **(b)**: Dependence of $\tau(q)$ and $h(q)$ on q.

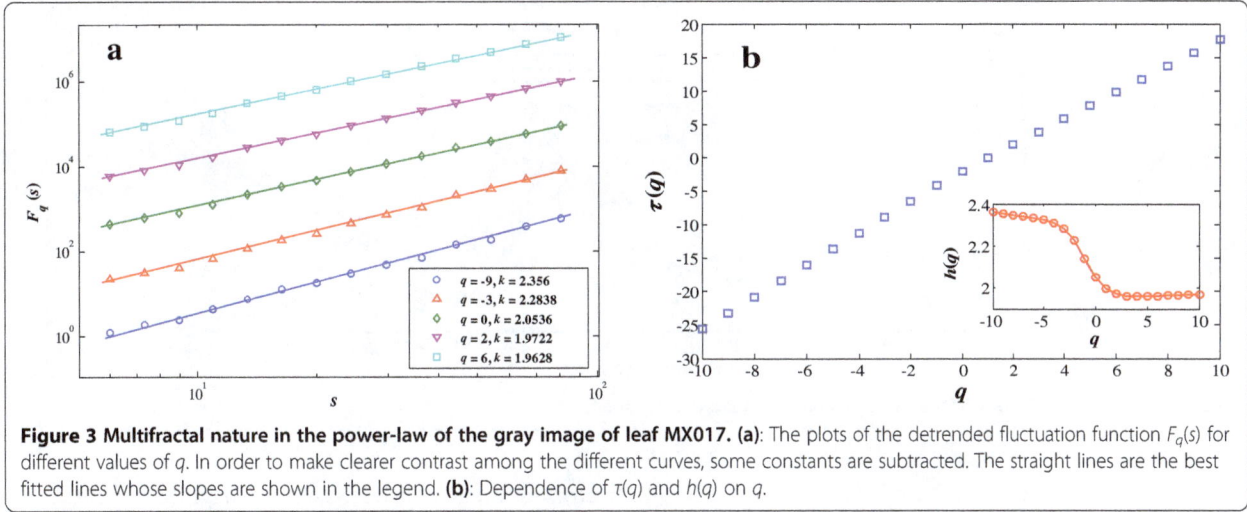

Figure 3 Multifractal nature in the power-law of the gray image of leaf MX017. (a): The plots of the detrended fluctuation function $F_q(s)$ for different values of q. In order to make clearer contrast among the different curves, some constants are subtracted. The straight lines are the best fitted lines whose slopes are shown in the legend. **(b)**: Dependence of $\tau(q)$ and $h(q)$ on q.

Step 3: For each surface $G_{m,n}$, obtain a local trend $\tilde{G}_{m,n}$ by fitting it with a pre-chosen bivariate polynomial function. In this paper, we choose the trending function as

$$\tilde{G}_{m,n}(i,j) = ai + bj + c, \tag{2}$$

a.where $1 \leq i, j \leq s$ and a, b and c are free parameters to be determined by the least-squares method. The residual matrix is then given by $y_{m,n} = y_{m,n}(i,j)$ with

$$y_{m,n}(i,j) = G_{m,n}(i,j) - \tilde{G}_{m,n}(i,j). \tag{3}$$

Step 4: Define the detrended fluctuation function $F(m, n, s)$ for the segment $X_{m,n}$ as follows:

$$F^2(m,n,s) = \frac{1}{s^2} \sum_{i=1}^{s} \sum_{j=1}^{s} y_{m,n}(i,j)^2 \tag{4}$$

and the qth-order fluctuation function

$$F_q(s) = \left[\frac{1}{M_s N_s} \sum_{m=1}^{M_s} \sum_{n=1}^{N_s} [F(m,n,s)]^q \right]^{1/q}, q \neq 0. \tag{5}$$

$$F_q(s) = \exp\left\{ \frac{1}{M_s N_s} \sum_{m=1}^{M_s} \sum_{n=1}^{N_s} \ln[F(m,n,s)] \right\}, q = 0. \tag{6}$$

Step 5: Vary the value of s ranging from 6 to $\min(M,N)/4$. If there is long-range power-law correlation for large values of s, then

$$F_q(s) \propto s^{h(q)}.$$

This allows us to obtain the scaling exponent $h(q)$ via linearly regressing $\ln F_q(s)$ on $\ln s$. Note that $h(2)$ is the so called Hurst index of the surface, we then call $h(q)$ the generalized Hurst index of the surface. For each q, the corresponding classical multifractal scaling exponent $\tau(q)$ is given by:

$$\tau(q) = qh(q) - D_f = qh(q) - 2, \tag{7}$$

where D_f is the fractal dimension of the geometric support of the multifractal measure, and takes the value of $D_f = 2$ in our work. The generalized multifractal dimension D_q is then given by

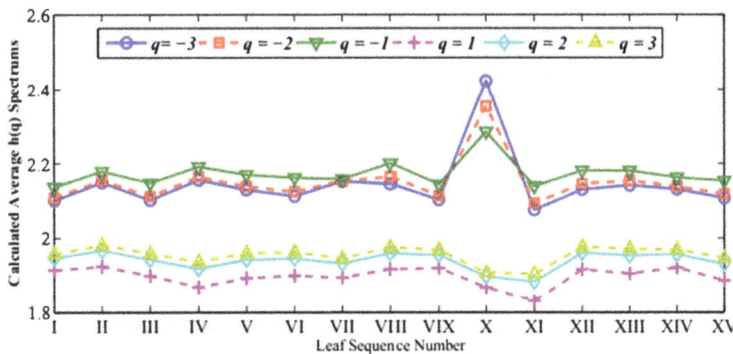

Figure 4 The generalized Hurst exponents $h(q)$ for each species.

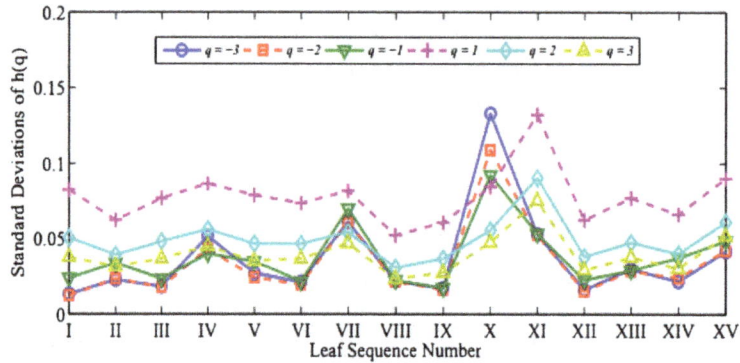

Figure 5 The standard deviations of the averaged $h(q)$ calculated in Figure 4.

$$D_q = \frac{\tau(q)}{q-1} = \frac{qh(q)-2}{q-1}, \quad q \neq 1. \tag{8}$$

In the case where $q = 1$, D_1 can be obtained *via* a linear regression of $\sum_{m=1}^{M_s} \sum_{n=1}^{N_s} P_{m,n} \ln P_{m;n}$ against $\ln s$, where

$$P_{m,n} = \frac{\sum_{1 \leq i,j \leq s} X_{m,n}(i,j)}{\sum_{1 \leq i \leq M} \sum_{1 \leq j \leq N} X(i,j)}.$$

The other two indicators characterizing the singularity strength of the multifractal surface are the *Hölder* exponent $\alpha(q)$ and the singularity spectrum $f(\alpha)$, which are given by

$$\alpha(q) = \tau'(q) = h(q) + qh'(q), \quad f(\alpha)$$
$$= q\alpha(q) - \tau(q) = q[\alpha - h(q)] + 2. \tag{9}$$

Here $\alpha(q)$ characterizes the local singularity of an image texture, and $f(\alpha)$ measures the global singularity of an image texture. Varying the value of q in the range from -15 to 15 determines $\Delta\alpha$ and Δf as follows:

$$\Delta\alpha = \alpha_{\max} - \alpha_{\min}, \Delta f = f(\alpha_{\max}) - f(\alpha_{\min}), \tag{10}$$

$\alpha_{\max} = \max\{\alpha(q), \quad q \in [-15,15]\}$ and $\alpha_{\min} = \min\{\alpha(q), q \in [-15,15]\}$. Note that the index $\Delta\alpha$ is considered as an

indicator to measure the absolute magnitude of the gray scale volatility. The larger value of $\Delta\alpha$, the smaller even distribution of probability measure and the more roughness image surface will be expected. The index Δf is the *Hausdorff* dimension of the measure object, which measures the degree of confusion. Therefore both $\Delta\alpha$ and Δf are important multifractal parameters in describing the characteristics of an image in our study.

Experiment materials

To demonstrate our method of identifying plant species by using the leaf texture, we use the Swedish leaf data set [31] for our experiment, which is widely employed in computer vision and pattern recognition fields [4,33,34], plant taxon fields [1] and image processing fields [6,35]. This leaf data set has images of 15 species of leaves with 75 sample images per species. We label the fifteen species by MI, MII, ⋯, MXV (See Figure 1).

We first transform the color image to gray scale so that each image can be viewed as a three- dimensional surface with the first two coordinates (i, j) denoting the 2D position and the third coordinate z denoting the gray level of the corresponding pixel.

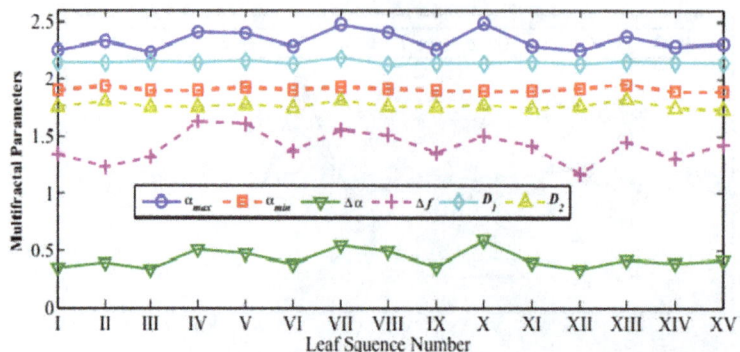

Figure 6 The averaged values for six related multifractal parameters based the MF-DFA estimation for each species.

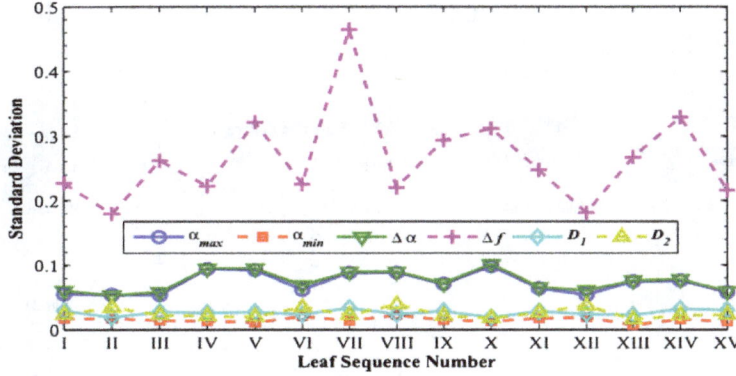

Figure 7 The standard deviations of the averaged parameter values calculated in Figure 6.

Multifractal nature of image surfaces

Each image is stored as a 2D matrix in 256 grey levels. This allows us to follow the procedure introduced in Multifractal detrended fluctuation analysis to calculate the associated $h(q)$ and $\tau(q)$. If $\tau(q)$ is nonlinear in q, that is $h(q)$ is not independent of q, then the image possesses the multifractal nature.

For the Swedish leaf data set, we find that the leaf images all possess the multifractal nature. Figure 2 and Figure 3 demonstrate the multifractal nature of two randomly chosen leaf images, namely, image MIV004 and image MX017, the former has 1793 × 979 pixels and the latter has 2934 × 1771 pixels. In each the left panel illustrates the dependence of the detrended fluctuation function $F_q(s)$ as a function of the scale s for different q. The well fitted straight lines indicate the evident power law scaling of $F_q(s)$ versus s. The right panel shows that $\tau(q)$ is nonlinear in q, indicated by the fact that $h(q)$ depends on q.

Results and discussion

For each image, we can calculate the generalized Hurst exponents $h(q)$ and six other multifractal parameters including α_{max}, α_{min}, Δf, D_1 and D_2. For each tree species, we take the averaged value over the 75 samples and report our calculated values in Figures 4 and 5. Their standard deviations are given in Figures 6 and 7, respectively.

As seen in Figure 4, comparing with $h(2)$ and $h(3)$, the estimations of $h(-3)$, $h(-2)$, $h(-1)$ and $h(1)$ vary in relatively wider dynamic ranges and thus demonstrate better abilities to distinguish textures among different species.

Yet, one notes that there are relatively large variations in the standard deviations among the 75 samples for the $h(q)$ exponents in Figure 5. This suggests that this indicator alone may not be adequate to identify the fifteen tree species. Also as seen in Figure 6 that the three parameters, α_{max}, $\Delta\alpha$, and Δf admit wider dynamic ranges than the other three parameters do. The variations among the 75 samples in the same tree species are notably large as shown in Figure 7.

For species i (i = I, II, $\cdots$, XV), with respect to each calculated multifractal parameter, we denote the standard deviation of the 75 samples by $\sigma_{in}(i)$ and define σ_{in} as

$$\sigma_{in} = \frac{1}{15} \sum_{i=1}^{XV} \sigma_{in}(i), \tag{11}$$

which represents the intra-species variance. Note also that for each indicator, we can calculate its value corresponding to each species and there are 15 values in total for those 15 species. We define $\sigma_{bet.}$ as the standard deviation of these 15 calculated values. Then the term $\sigma_{bet.}$ represents the inter-species variance for each multifractal indicator. We now define an index, I_0, as

$$I_0 = \frac{\sigma_{bet.}}{\sigma_{in}}. \tag{12}$$

From the definition, we note that the multifractal parameter with larger I_0 serves better as an indicator to distinguish species. We present the calculated values of I_0 in Table 1.

Table 1 The calculated $\sigma_{bet.}$, σ_{in} and I_0 for the 12 multifractal parameters

Parameters	$h(-3)$	$h(-2)$	$h(-1)$	$h(1)$	$h(2)$	$h(3)$	α_{max}	α_{min}	$\Delta\alpha$	Δf	D_1	D_2
$\sigma_{bet.}$	0.0605	0.0403	0.0363	0.0255	0.0237	0.0233	0.0845	0.0183	0.0806	0.1327	0.0140	0.0259
σ_{in}	0.0368	0.0342	0.0381	0.0779	0.0496	0.0395	0.0722	0.0151	0.0646	0.2645	0.0264	0.0252
I_0	**1.6459**	1.1810	0.9548	0.3280	0.4777	0.5891	1.1705	**1.2132**	**1.2469**	0.5015	0.5316	1.0284

Tip: the symbol bold numbers mean the best choice yielding the top three I_0 indices.

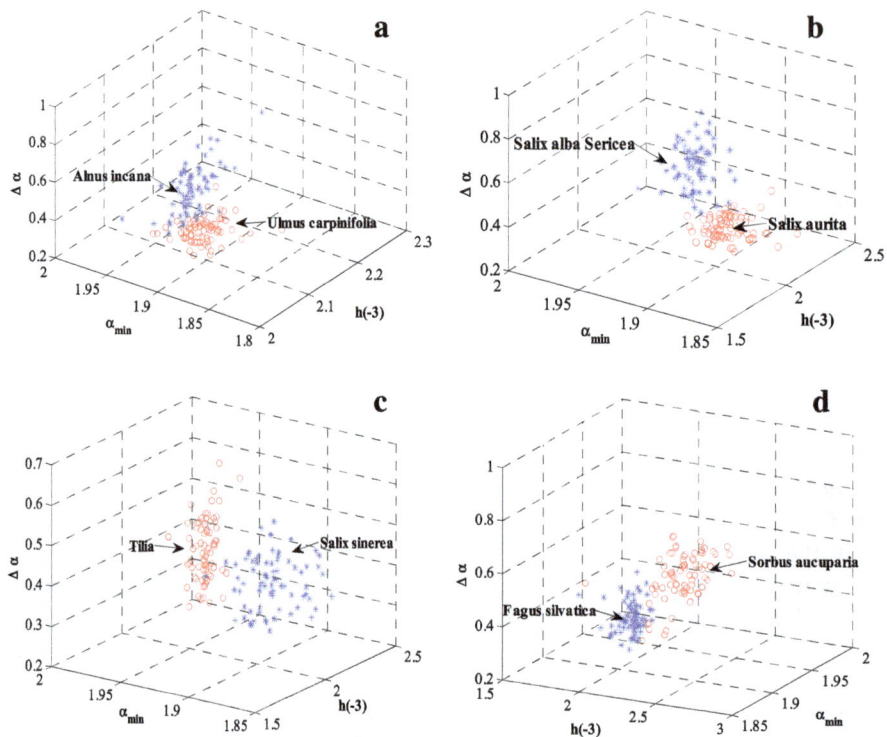

Figure 8 Visualization of two tree species in the $\{h(-3), \alpha_{min}, \Delta\alpha\}$ space. (a): *Ulmuscarpinifolia* versus *Alnusincana*; (b): *Salixaurita* versus *SalixalbaSericea*; (c): *Salixsinerea* versus *Tilia*; (d): *Sorbusaucuparia* versus *Fagussilvatica*.

We choose the combination of three multifractal parameters with larger I_0 values, namely, $\{h(-3), \alpha_{min}, \Delta\alpha\}$, as the feature descriptors for our classification purpose and apply the support vector machines and kernel methods (SVMKM) with the heavy-tailed radial basis function-'*htrfb*' as the kernel [36]. It is worth mentioning that the combination of 4 or more parameters does not lead to significant higher accuracies, but at a cost with much longer computational time and with no visual advantages. In this sense, the combination of the above three parameters is optimal. For the total sample set containing $75 \times 15 = 1125$ samples, we use the K – fold cross validation to evaluate the learning performance. This means that $100\,(K-1)/K\%$ samples are randomly chosen as a training set and the remaining $100/K\%$ samples are considered as a test set. The calculation process is then repeated 10 times to eliminate the impact of randomness.

In our first identification experiment, we test the proposed method through examining the distinguishing

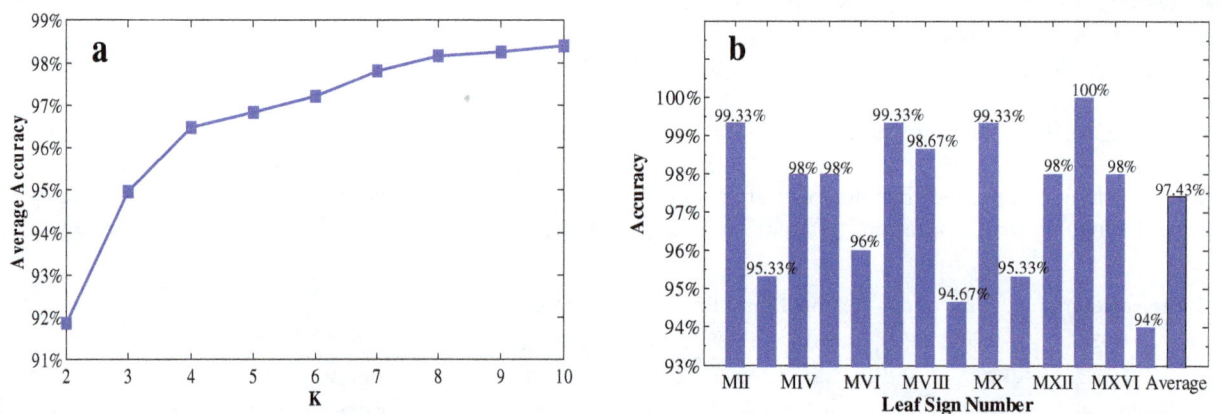

Figure 9 Average identification accuracies. (a): the average accuracies of every two species using different values of K; (b): The accuracies of identifying species *Ulmus carpinifolia* versus the other 14 species using K = 10.

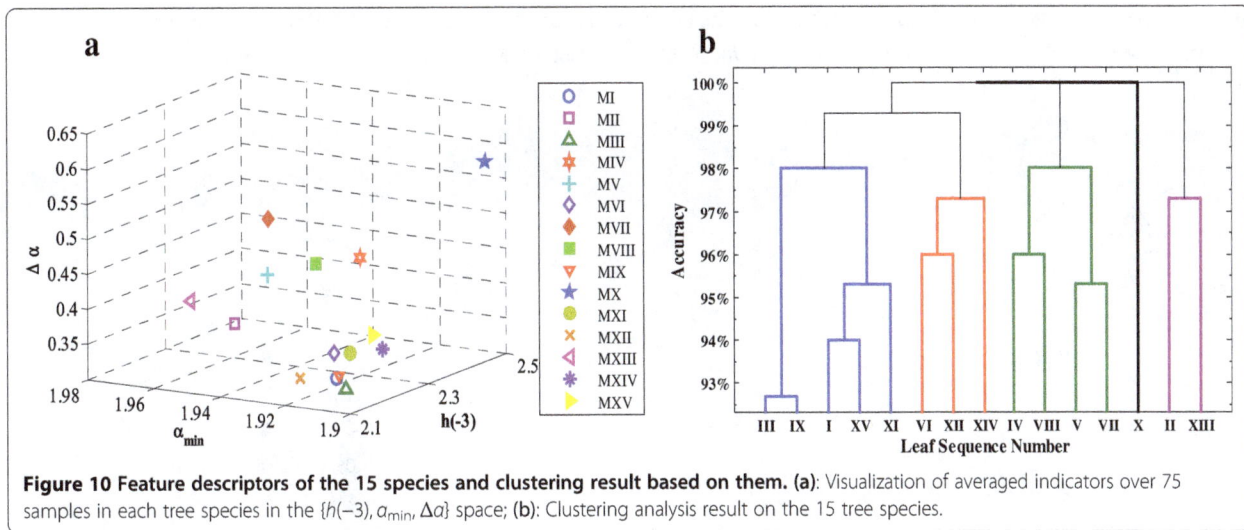

Figure 10 Feature descriptors of the 15 species and clustering result based on them. (a): Visualization of averaged indicators over 75 samples in each tree species in the $\{h(-3), \alpha_{min}, \Delta\alpha\}$ space; **(b)**: Clustering analysis result on the 15 tree species.

effect for every two species. To this end, we form a three-dimensional parameter space with components given by the above chosen feature descriptors $\{h(-3), \alpha_{min}, \Delta\alpha\}$. In this space, one point represents a leaf sample image. In Figure 8(a)-(d), we plot the corresponding points for *Ulmus carpinifolia* versus *Alnus incana*, *Salix aurita* versus *Salix alba Sericea*, *Salix sinerea* versus *Tilia* and *Sorbus aucuparia* versus *Fagus silvatica*, respectively. As shown in these plots, the samples from the same tree species are clustered together reasonably well.

In addition, we calculate the discriminant accuracies of every two tree species by SVMKM using the K–fold cross validation with different K values. The average accuracies of 10 trials are shown in Figure 9(a). To display the applicability of identifying different tree species by our proposed method, as an example, we plot the accuracy of identifying species MI (*Ulmus carpinifolia*) versus other 14 species with K = 10 in Figure 9(b). As expected,

the average accuracy of every two species is increasing with respect to K. The obtained best accuracy is 98.40%, higher than 96.82% reported in [35], which requires a very complex pre-processing process for leaf images. It is seen from Figure 9(b) that there are accuracy variations between species *Ulmus carpinifolia* and the other 14 species. Five species, namely, *Salix aurita*, *Betula pubescens*, *Ulmus glabra*, *Salix sinerea* and *Fagus silvatica*, have accuracies below the average accuracy. This suggests that species *Ulmus carpinifolia* has high similarity with the above mentioned five species, which agrees with the observation from Figure 1.

For each species, the averaged $\{h(-3), \alpha_{min}, \Delta\alpha\}$ of the 75 samples is represented by a single point in the three-dimensional parameter space (see Figure 10) in which different points representing different species may be clustered into several groups. We use the calculated discriminant accuracy of every two species as the distance between these two points (species). This allows us to

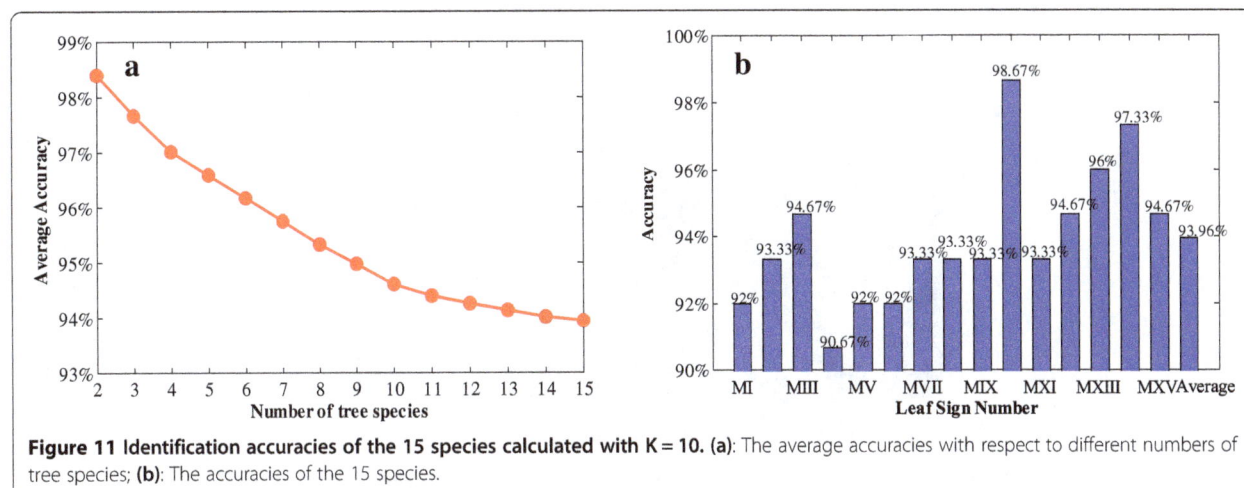

Figure 11 Identification accuracies of the 15 species calculated with K = 10. (a): The average accuracies with respect to different numbers of tree species; **(b)**: The accuracies of the 15 species.

Table 2 The results of identification for the fifteen species of tree leaves by the method of SVMKM with K = 10

	MI	MII	MIII	MIV	MV	MVI	MVII	MVIII	MIX	MX	MXI	MXII	MXIII	MXIV	MXV
MI	69	0	2	0	1	0	0	0	3	0	0	0	0	0	0
MII	0	70	0	0	1	0	1	0	1	1	0	1	0	0	0
MIII	1	0	71	0	0	0	0	0	0	0	2	1	0	0	0
MIV	1	0	1	68	1	0	0	3	1	0	0	0	0	0	0
MV	0	2	0	0	69	0	0	2	1	0	0	1	0	0	0
MVI	1	0	1	1	0	69	0	0	1	0	1	1	0	0	0
MVII	0	2	0	1	2	0	70	0	0	0	0	0	0	0	0
MVIII	0	0	0	1	2	0	1	70	0	0	1	0	0	0	0
MIX	1	0	0	0	1	1	0	0	70	0	1	1	0	0	0
MX	0	0	0	0	0	0	0	0	0	74	1	0	0	0	0
MXI	1	0	1	0	0	0	0	0	3	0	70	0	0	0	0
MXII	0	0	1	1	0	1	1	0	0	0	0	71	0	0	0
MXIII	0	0	0	0	0	0	0	0	0	0	0	0	72	0	3
MXIV	0	0	0	0	0	0	0	0	0	0	0	0	0	73	2
MXV	0	0	0	0	0	0	0	0	0	0	0	0	2	2	71

conduct a cluster analysis for all samples of the 15 species by the method of hierarchical clustering [37]. The result is given in Figure 10(b), which suggests that the 15 tree species' leaf samples can be clustered into five groups: (i) {*Ulmus carpinifolia, Salix aurita, Ulmus glabra, Salix sinerea, Fagus silvatica*}; (ii) {*Betula pubescens, Populus, Sorbus intermedia*}; (iii) {*Quercus, Alnus incana, Salix alba Sericea, Populus tremula*}; (iv) {*Acer, Tilia*} and (v) {*Sorbus aucuparia*}. This is consistent with visualizing the images directly from Figure 1 showing our proposed approach is applicable.

As another important aspect of identification experiment, we next test our method through calculating the identification accuracies for different numbers of species. The averaged accuracy result calculated when K = 10 is shown in Figure 11(a). Note that the average accuracy is decreasing as the number of tree species increases. This is due to the increasing probability of incorrect classification. However, under the worst situation, all 75 × 15 = 1125 sample leaf images are well mixed together, which gives the lowest average accuracy: 93.96%. This is still very convincing that our approach is feasible. We calculate the identification accuracy also when K = 10 for each species and report the result in Figure 11(b), while the identification result for each species is displayed in Table 2. The best three accuracies reach 98.67%, 97.33% and 96%, and the corresponding species are *Sorbus aucuparia, Sorbus intermedia* and *Tilia*. As is seen in Figure 1, these three species are clearly distinct from the other species in leaf shapes and textures. This again shows that our method is effective and feasible.

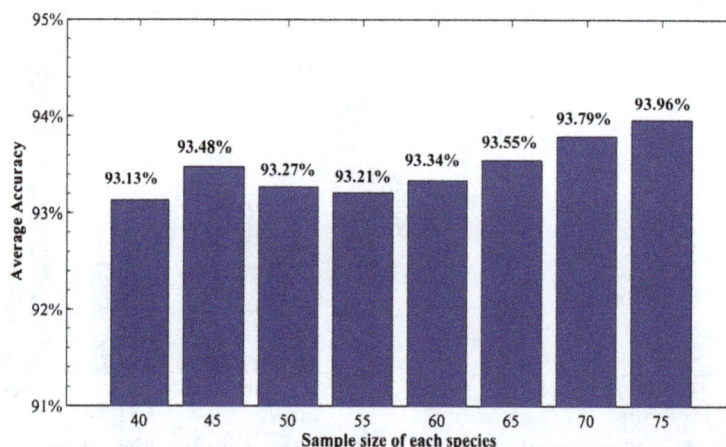

Figure 12 Average identification accuracies of the 15 species calculated with K = 10 with different species sample sizes.

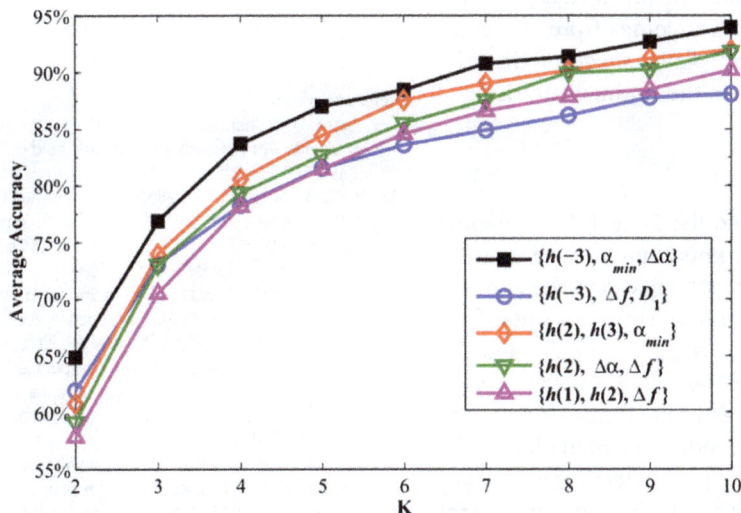

Figure 13 The average accuracies of the 15 species for the selected combinations with increasing K.

We remark that the sample size of each species has little effect on the average discriminant accuracy. To justify this, we randomly choose n ($n \leq 75$) leaf samples for each species and run the procedure. Then repeat the process 10 times and take the average accuracy, which is reported in Figure 12. It can be seen from Figure 12 that as the number of samples changes from 40 to 75, the accuracy changes only 0.73%.

Model test

In this section, we test our proposed method to demonstrate its efficiency. More precisely, we test the validity of the optimal multifractal parameter combination $\{h(-3), \alpha_{\min}, \Delta\alpha\}$. To this end, we choose other four combinations composed by three multifractal parameters to construct four three-dimensional spaces from Table 1. These four choices are $\{h(-3), \Delta f, D_1\}$, $\{h(2), h(3), \alpha_{\min}\}$, $\{h(2), \Delta\alpha, \Delta f\}$ and $\{h(1), h(2), \Delta f\}$. One notes that each of the first three combinations contains one multifractal parameter from $\{h(-3), \alpha_{\min}, \Delta\alpha\}$ and the fourth combination consists of the three parameters that produce the three smallest I_0 values. As in the procedure proposed in the previous subsection, we place the 1125 leaf samples into the four new three-dimensional spaces and also use the SVMKM to distinguish them. Under the K–fold cross validation, the discriminant accuracies with increasing K are shown in Figure 13. Obviously, the highest accuracy

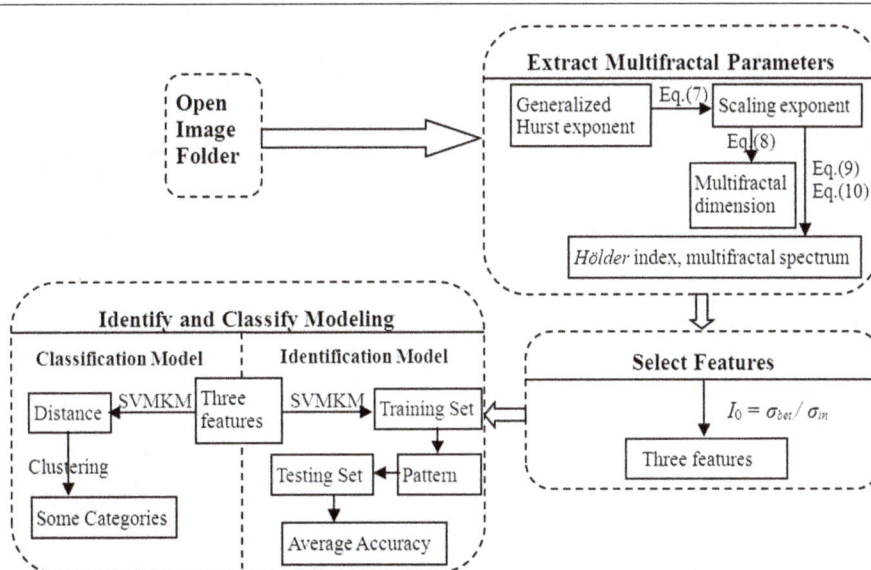

Figure 14 The flow chart of software programing base on our model is as follows. Detailed codes are available upon request.

still comes from the combination $\{h(-3), \alpha_{\min}, \Delta\alpha\}$ for each K and the lowest accuracy comes from the combination $\{h(1), h(2), \Delta f\}$. This again suggests that the index I_0 successfully indicates the optimal multifractal parameter combination.

Conclusions

In this paper we have adopted the 2D MF-DFA method proposed in [32] to extract important texture features from leaf images. This allow us to calculate the generalized Hurst exponents, $h(q)$, and several other multifractal parameters including $\alpha_{\max}$, $\alpha_{\min}$, $\Delta\alpha$, Δf, D_1 and D_2. By defining an index, I_0, which examines the variation of the inter-species variances and the intra-species variances, we are able to find an optimal combination of the multifractal parameters that best characterizes the key features of plant species allowing high accuracy in plant species identification. For the Swedish leaf data set which contains 15 species and $75 \times 15 = 1125$ samples in total [31], the combination of $\{h(-3), \alpha_{\min}, \Delta\alpha\}$ turns out to be optimal compared to other combinations of parameters. We have obtained 98.4% of averaged discriminant accuracy for every two species by SVMKM with the 10 – fold cross validation, while the accuracy reaches 93.96% for the over-all 15 species. Software based on our work can be designed and coded, for that purpose, we provided the corresponding flow chart in the Figure 14.

We should point out that most of the existing work on texture image recognition focuses mainly on the standard multifractal analysis. Our work has shown that the MF-DFA is of particular practice for plant leaf identification as the MF-DFA multifractal parameters can be combined to distinguish similar but different leaf textures.

Competing interests
The authors declare that they have no competing interests.

Authors' contributions
FW initiated the study, carried out modeling, programming and drafted the manuscript. DL participated in the design of the study and performed the statistical analysis. JL participated in programming. GL participated in revising the final version. All authors read and approved the final manuscript.

Acknowledgments
The authors wish to thank the two reviewers and the editor-in-Chief Prof. Brian G. Forde for their comments and suggestions, which led to a great improvement to the presentation of this work.
FW was supported by the Young Scholar Project Funds of the Education Department of Hunan Province (14B087). JL and GL were supported by the Major project funds of Science and technology program of Hunan Province (2013FJ1006-4). DL was partially supported by a grant from the Forestry Department of Hunan Province, China.

Author details
[1]College of Science, Hunan Agricultural University, Changsha 410128, China. [2]Forestry Department of Hunan Province, Quality Testing and Inspection Centre of Forest Products, Changsha 410007, China. [3]Agricultural Information Institute, Hunan Agricultural University, Changsha 410128, China.

References
1. Agarwal G, Belhumeur P, Feiner S, Jacobs D, Kress WJ, Ramamoorthi R, et al. First steps toward an electronic field guide for plants. Taxon. 2006;55(3):597–610.
2. Judd W, Campbell CS, Kellog EA, Stevens PF. Plant Systematics: A Phylogenetic Approach. Massachusetts: Sinauer Associates; 1999.
3. Kumannn MH, Hemsley AR. The Evolution of Plant Architecture. Kew, London: Royal Botanic Gardents; 1999.
4. Ling H, Jacobs DW. Shape classification using the inner-distance. IEEE trans Pattern Analysis and Machine Intelligence. 2007;29(2):286–99.
5. Neto JC, Meyer GE, Jones DD, Samal AK. Plant species identification using elliptic Fourier leaf shape analysis. Comput Electron Agr. 2006;50:121–34.
6. Wang B, Gao Y. Hierarchical string cuts: a translation, rotation, scale, and mirror invariant descriptor for fast shape retrieval. IEEE Trans Image Process. 2014;23(9):4101–11.
7. Bruno OM, de Oliverira Plotze R, Falvo M, de Castro M. Fractal dimension applied to plant identification. Inform Sciences. 2008;178:2722–33.
8. Backes AR, Bruno OM. Plant Leaf Identification Using Multi-Scale Fractal Dimension, in International Conference On Image Analysis and Processing. Berlin/ Heidelberg: Springer; 2009. p. 143150.
9. Backes AR, Goncalves WN, Martinez AS, Bruno OM. Texture analysis and classification using deterministic tourist walk. Pattern Recogn. 2010;43:685–94.
10. Cope JS, Remagnino P, Barman S, Wilkin P. Plant Texture Classification Using Gabor co-Occurrences, in International Symposium on Visual Computing. Las Vegas Nevada, USA: Springer-Verlag; 2010.
11. Liu J, Zhang S, Deng S. A method of plant classification based on wavelet transforms and support vector machines. Emerging Intelligent Computing Technology and Applications. 2009;5754:253–60.
12. Cope JS, Corney D, Clark JY, Remagnino P, Wilkin P. Plant species identification using digital morphometrics: a review. Expert Syst Appl. 2012;39:7562–73.
13. Du JX, Wang XF, Zhang GJ. Leaf shape based plant species recognition. Appl Math Comput. 2007;185:883–93.
14. Zhang SW, Lei YK, Dong TB, Zhang XP. Label propagation based supervised locality projection analysis for plant leaf classification. Pattern Recogn. 2013;46:1891–7.
15. Kong Y, Wang SH, Ma CW, Li BM, Yao YC. Effect of image processing of a leaf photograph on the calculated fractal dimension of leaf veins. Computer and computing technologies in agriculture. 2008;259:755–60.
16. Rossatto DR, Casanova D, Kolb RM. Fractal analysis of leaf-texture properties as a tool for taxonomic and identification purposes: a case study with species from Neotropical Melastomataceae (Miconieae tribe). Plant Syst Evol. 2011;29:103–16.
17. Chaudhuri BB, Sarkar N. Texture segmentation using fractal dimension. IEEE Trans Pattern Anal Mach Intell. 1995;17:72–7.
18. Backes AR, Bruno OM. Plant leaf identification using color and multi-scale fractal. Dimension Image and Signal Processing. 2010;6134:463–70.
19. Backes AR, Bruno OM. Plant leaf identification using multi-scale fractal dimension. Image Analysis and Processing. 2009;5716:143–50.
20. Lopes R, Dubois P, Bhouri I, Bedoui MH, Maouche S, Betrouni N. Local fractal and multifractal features for volumic texture characterization. Pattern Recogn. 2011;44:1690–7.
21. Wang F, Liao GP, Wang XQ, Li JH, Li JW, Shi W. Feature description for nutrient deficiency rape leaves based on multifractal theory. Transactions of the Chinese Society of Agricultural Engineering (Transactions of the CSAE). 2013;29(24):181–9 (in Chinese with English abstract).
22. Xia Y, Feng DG, Zhao RC. Morpholgy-based multifractal estimation for texture segmentation. IEEE trans image processing. 2006;15:614–22.
23. Kantelhardt JW, Zschiegner SA, Bunde EK, Havlin S, Bunde A, Stanley HE. Multifractal detrended fluctuation analysis of nonstationary time series. Phys A. 2002;316:87–114.
24. Du G, Ning X. Multifractal properties of Chinese stock market in Shanghai. Physica A. 2008;387:261–9.
25. Hedayatifar L, Vahabi M, Jafari GR. Coupling detrended fluctuation analysis for analyzing coupled nonstationary signals. Phys Rev E. 2011;84:021138.

26. Wang F, Liao GP, Li JH, Zhou TJ, Li XC. Multifractal detrended fluctuation analysis for clustering structures of electricity price periods. Physica A. 2013;392(22):5723–34.

27. Wang F, Liao GP, Zhou XY, Shi W. Multifractal detrended cross-correlation analysis for power markets. Nonlinear Dynamics. 2013;72(1–2):353–63.

28. Wang F, Li JW, Shi W, Liao GP. Leaf image segmentation method based on multifractal detrended fluctuation analysis. J Appl Phys. 2013;114(21):214905.

29. Wang F, Li ZS, Liao GP. Multifractal detrended fluctuation analysis for image texture feature representation. Int J Pattern Recognit Artif Intell. 2014;28(3):1455005.

30. Wang F, Li ZS, Li JW. Local multifractal detrended fluctuation Analysis for Non-stationary Image's Texture Segmentation. Appl Surf Sci. 2014;233:116–25.

31. Söderkvist OJO. Computer Vision Classification of Leaf from Swedish Trees, Master Thesis, Linköping University, 2001

32. Gu GF, Zhou WX. Detrended fluctuation analysis for fractals and multifractals in higher dimensions. Phys Rev E. 2006;74:061104.

33. Lei YK, Zou JW, Dong T, You ZH, Yuan Y. Orthogonal locally discriminant spline embedding for plant leaf recognition. Comput Vis Image Underst. 2014;119:116–26.

34. Wu JX, Rehg JM. A visual descriptor for scene categorization. IEEE trans Pattern Analysis and Machine Intelligence. 2010;33(8):1489–501.

35. Zhang SW, Zhang CL, Cheng L. Plant leaf image classification based on supervised orthogonal locality preserving projections. Transactions of the Chinese Society of Agricultural Engineering. 2013;29(5):125–31.

36. Canu S, Grandvalet Y, Guigue V, Rakotomamonjy A. SVM and kernel methods matlab toolbox, perception systemes et information. Rouen, France: INSA de Rouen; 2005.

37. Székely GJ, Rizzo ML. Hierarchical clustering via joint between within distances: extending ward's winimum variance method. J Classif. 2005;22:151–83.

16

Imaging of polysaccharides in the tomato cell wall with Raman microspectroscopy

Monika Chylińska, Monika Szymańska-Chargot and Artur Zdunek[*]

Abstract

Background: The primary cell wall of fruits and vegetables is a structure mainly composed of polysaccharides (pectins, hemicelluloses, cellulose). Polysaccharides are assembled into a network and linked together. It is thought that the percentage of components and of plant cell wall has an important influence on mechanical properties of fruits and vegetables.

Results: In this study the Raman microspectroscopy technique was introduced to the visualization of the distribution of polysaccharides in cell wall of fruit. The methodology of the sample preparation, the measurement using Raman microscope and multivariate image analysis are discussed. Single band imaging (for preliminary analysis) and multivariate image analysis methods (principal component analysis and multivariate curve resolution) were used for the identification and localization of the components in the primary cell wall.

Conclusions: Raman microspectroscopy supported by multivariate image analysis methods is useful in distinguishing cellulose and pectins in the cell wall in tomatoes. It presents how the localization of biopolymers was possible with minimally prepared samples.

Keywords: Plant cell wall, Polysaccharides, Raman imaging, PCA, MCR

Background

The primary cell wall of fruits and vegetables is a heterogonous structure mainly composed of polysaccharides, including a wide variety of pectins and hemicelluloses as well as cellulose. Most plant cell wall components are important elements of human nourishment, known as dietary fiber [1]. The plant cell wall is a kind of cellular skeleton that controls cell shape and determines the relationship between turgor pressure and cell volume. The composition of plant cell walls is also important from the point of view of macroscopic mechanical properties and water transport [2-4].

The cell wall is a composite of cellulose, hemicellulose and pectin, with the addition of other, non-polysaccharide components like proteins, lipids, enzymes and aromatic compounds. Generally, the primary cell wall is composed of approximately 25% cellulose, 25% hemicellulose and 35% pectins, with up to 8% structural proteins (on a dry-weight basis). However, large deviations from these values may be found [5].

Cellulose is a highly stable and linear polymer containing between 1,000 and one million D-glucose residues linked by β-1,4 glycosidic bonds. Pectins and hemicelluloses are represented by wide variety of compounds [6,7]. Xyloglucan is the most common hemicellulose present in primary cell wall of higher plants [5,8]. Among pectins the main are homogalacturonan, rhamnogalacturonan I and rhamnogalacturonan II. According to the common plant cell wall model, the cellulose is organized in microfibrils, which are tethered and connected via hemicelluloses. The cellulose-hemicellulose network is embedded in a highly hydrated matrix composed of pectin polysaccharides [9]. It is suggested that the key role in the biomechanics of plant cell walls is played by the xyloglucan-cellulose network, which has loadbearing properties [10]. In addition, pectins are responsible for controlling cell wall porosity and inducing the binding of neighboring cells [11]. It is thought that the percentage of the components of the plant cell wall has an important influence on the mechanical properties of fruits and vegetables [12,13].

Up to now, many analytical [14,15] and microscopic (e.g., optical [16] and electron microscopy [17]) methods have been used for the evaluation of the plant cell wall's

* Correspondence: a.zdunek@ipan.lublin.pl
Institute of Agrophysics, Polish Academy of Sciences, Doswiadczalna 4, 20-290 Lublin, Poland

structure and composition. So far, antibody probes have served as the most common and - probably - the only effective method to provide micro-scale insight into the fruit cell wall's biochemical structure. The labelling of polysaccharides with antibodies was applied in several studies. Immunocytochemistry was used to depict the spatial deployment of pectins in potato cell walls [18] and in ripening tomatoes [19]. Also, the spatial distribution of hemicelluloses in fruit cell walls was investigated [20,21]. However, despite its many advantages, this method is very selective, quite expensive and time-consuming.

Raman imaging has been suggested as being a complementary or even alternative method to antibody labeling in studying the spatial distribution of fruit cell wall components [22]. In brief, Raman microscope is a connection of microscope and Raman spectroscope. Raman spectra are collected while scanning over a sample, providing the spatial and chemical labeling of various components within the sample simultaneously. In this way a map of spatial distribution of sample's components can be obtained. However, the resolution is limited by the diffraction limit, which depends on the wavelength of the excitation light and the objective used. Typically it cannot be lower than 200 nm. This is insufficient to depict single polysaccharide in cell wall which have diameter of about few to several dozen nanometers however it could be used to visualize the spatial distribution at the micro-scale level. The Raman spectroscopy based on "Raman effect" was described for the first time in 1928 [23]. When a sample is illuminated by monochromatic laser light, mainly elastic scattering ("Rayleigh scattering") of photons is observed . It means that the energy of incident radiations is equal to the energy of scattered light. However, a small amount of incident light (about one in a million photons) is scattered with different frequency. In case of such inelastic scattering, the energy of photons of scattered light can be lower or higher than the energy of incident light. In Raman spectrum lines corresponding to lower frequencies are presented as a plot of intensity versus frequency shift [24].

Raman microspectroscopy has been shown to be a useful method in the study of polymers and biopolymers [25]. One of the first applications of Raman imaging was in the evaluation of the distribution of chemical components in flex (*Linum usitatissimum* L.) stem tissue [25,26]. The potential of Raman microspectroscopy was frequently demonstrated on wood cell walls, like the distribution of lignin and cellulose in black spruce wood (*Picea mariana*) [27] and the lignin spatial content in the cell wall of poplar wood [22]. Furthermore, single carotenoid crystals were detected directly in carrot cells, without any compound extraction [28]. Vibrational spectroscopic imaging techniques - especially Raman microscopy - have seen numerous applications in the medical sciences (e.g., in cancer diagnosis) [29].

A literature review revealed a lack of an application of Raman imaging technique to the evaluation of the chemical composition and distribution of polysaccharides in the cell wall in fruits and vegetables. Taking into account the simplicity of Raman imaging and the relatively easier sample preparation (which does not require extensive chemical treatment prior to imaging), this method would benefit in more effective characterization of the composition of cell wall with particular emphasis on dynamic changes which occur during development and maturation of fruits and vegetables. However, at present an application of this method requires developing an approach to handle of a large set of data which combines both spatial and spectral information. Beside of traditional single band imaging, multivariate techniques could be used to visualize in a more complex manner the spatial distribution of cell wall constituents due to fact that each polysaccharide is represented in Raman spectra by more than one band. Therefore, the general aim of the present work is to depict the spatial distributions of the main cell wall compounds in tomato tissue using Raman imaging. The tomato fruit was chosen due to the fact that it has become a well-known model system in the study of fleshy fruits [30]. In order to evaluate the spatial and spectral data, three methods for the visualization of the distribution of polysaccharides in cell walls were used: single Raman band imaging for preliminary analysis, principal component analysis (PCA) and multivariate curve resolution (MCR).

Results and discussion
Assignment of Raman spectra to cell wall polysaccharides
Table 1 presents assignment of bands in the Raman bands characteristic for the most abundant plant cell wall polysaccharides based on the literature [31-33]. As a consequence of similar chemical compositions, many Raman bands originating from the vibrations of the same functional groups were overlapping for most of the polysaccharides [34]. For example, the band of $v(C = O)$'s vibration at 1,742 cm^{-1} could be assigned to both hemicellulose (xyloglucan, glucomannan) and pectin [31,35]. Theoretically, this ester band allows us to distinguish between hemicellulose and cellulose. However, in practice, in plant cell wall samples this is almost impossible due to significant pectin interference. Nevertheless, it may be that there are some particular Raman shifts which are unique to each polysaccharide and which could be used for the identification of these compounds in the cell wall material. In the CH stretching region from 2,700 cm^{-1} to 3,060 cm^{-1} characteristic for carbohydrate, intense bands assigned to pectin, hemicellulose and cellulose were observed at 2,952, 2,930, 2,897 cm^{-1}, respectively. The peaks at 1,121 cm^{-1} and 1,098 cm^{-1} are assigned to the symmetric and asymmetric stretching mode of COC in the glycosidic bond in

Table 1 Assignment of bands in the Raman spectra of cell wall polisaccharides based on the literature [31-33]: C – cellulose; P- pectins; H – hemicelluloses

Raman wavenumber [cm⁻¹] (literature)	Assignment	Origin
2952	ν(CH)	P
2930	ν(CH)	H
2897	ν(CH)	C
1742	ν(C = O) ester	P, H
1378	δ(HCC), δ(HCO), δ(HOC)	C
1256	δ(CH), δ(COH)	H
1121	ν(COC) glycosidic, symetric	C
1098	ν(COC) glycosidic, assymetric	C
971	ρ(CH$_2$)	C
854	(COC) skeletal mode of α-anomers	P
817	ν(COH) ring	P
478	ν(COC) glycosidic	P
380	δ(CCC) ring	C

cellulose, respectively, and it should be noted that these bands are highly sensitive to the orientation of microfibrils along the fiber. Bands centered at 1,378 and 971 cm⁻¹ Raman shift might also indicate the presence of cellulose in the sample. Additionally, the band at 380 cm⁻¹ related to the ring deformational modes is also characteristic for the cellulose. Moreover, the characteristic sharp band at 854 cm⁻¹ can be considered to be a marker band for α-glycosidic bonds in pectin. In particular, this band is very sensitive to O-acetylation [36]. Therefore, it could take values from 850 to 862 cm⁻¹ according to the increasing degree of pectin acetylation. On the average spectrum of the tomato cell wall, other bands denoting pectin at 817 and 478 cm⁻¹ are also present.

Figure 1 presents the reference Raman spectra of isolated cell wall components (pectins, cellulose and hemicellulose) and the spectrum of material isolated from the tomato cell wall. For the pectins, bands at 2,948, 1,750, 850, 817 and 479 cm⁻¹ were assigned (Figure 1A). Meanwhile, for the hemicellulose (xyloglucan), bands at 2,897, 1,459 and 1,256 cm⁻¹ (Figure 1B) are shown, while for the cellulose, bands at 2,894, 1,378, 1,121, 1,098, 971, 897 and 380 cm⁻¹ (Figure 1C) are presented. In the averaged Raman spectrum of the tomato cell wall material (Figure 1D), many bands characteristic for pure substances were broadened and overlapped and compared to the spectra of the pure polysaccharides (Figures 1A-C). Nevertheless, it was possible to identify and link bands from the tomato's spectrum to bands characteristic for functional groups of pure polysaccharides. The vibrations assigned for individual cell wall polysaccharides presented in Table 1 were slightly shifted in

comparison with the experiment (Figure 1). For example, the bands corresponding to the CH stretching mode are shifted - in the pectins spectrum from 2,952 to 2,948 cm⁻¹, and in the cellulose spectrum from 2,897 to 2,894 cm⁻¹. Nevertheless, the characteristic bands obtained for isolated cell wall components have been used to visualize the spatial distribution of these polysaccharides from Raman images.

Single Raman band imaging

Figure 2A presents an optical image of size 300 × 400 μm where three cell walls of tomato pericarp tissue are visible. For this experiment, a place on a tomato slice was chosen as an example to analyze the spatial distribution of polysaccharides in tomato cell walls. The recorded map had the dimensions 46 × 54 μm (23 × 27 pixels) and was subjected to a smoothing operation. The map focused on the area of the cell wall corner, since this location is supposed to be rich in pectin. The chemical image (Figure 2B) was obtained by integrating over 2,940 cm⁻¹, since this band reflecting the CH-stretching vibrations is characteristic for all primary cell wall components (i.e., cellulose, hemicellulose and pectin) [37].

The chemical maps calculated by integrating over wavenumber ranges corresponding to the strong Raman bands characteristic for individual cell wall polysaccharides are presented in Figures 2C and 2D. The peak at 854 cm⁻¹ (Figure 2C) corresponded to the band for the (COC) skeletal mode of α-anomers in pectin [32]. Integration over the 854 cm⁻¹ band confirmed higher concentration of pectin inwards of the cell wall. The highest intensity (red color in Figure 2C) was found close to the cell corner of tomato parenchyma. The cell corner and middle lamella are particularly enriched in pectin, especially homogalacturonan. The compounds of this part of the tissue are responsible for cell-cell adhesion [38]. The cellulose distribution is presented at the 1,090 cm⁻¹ band (Figure 2D), assigned to the glycosidic bond in cellulose. This band is characteristic for cellulose oriented in parallel to excitation light [32]. Integration over this wavelength displays the largest content of cellulose on the right side of the studied area (red color in Figure 2D). Due to the fact that the sample's components are homogeneously localized in the studied cell wall, high concentrations of both pectin and cellulose were found in similar locations.

Unfortunately, it was impossible to localize the hemicellulose in this tomato pericarp tissue. The chemical structure of hemicellulose is very similar to cellulose and therefore the bands from the Raman spectrum characteristic for hemicellulose were overlapped by the bands characteristic for cellulose. Moreover, it seems that this group of compounds was distributed homogeneously in the cell wall. The low percentage of hemicellulose in the cell wall material (less than 10%) could be the other

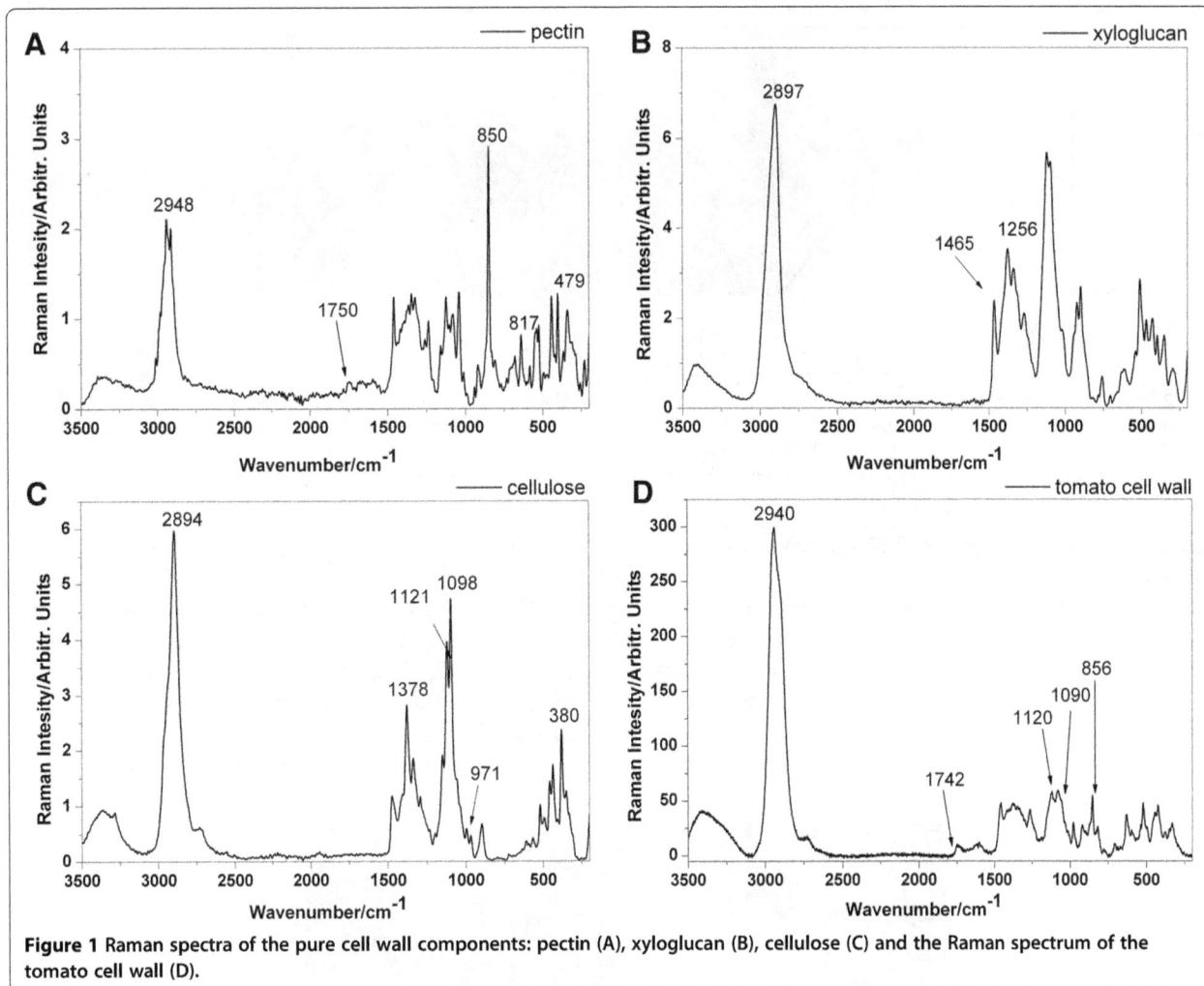

Figure 1 Raman spectra of the pure cell wall components: pectin (A), xyloglucan (B), cellulose (C) and the Raman spectrum of the tomato cell wall (D).

reason why the spatial distribution of those components could not be visualized.

Principal component analysis (PCA)

Since each polysaccharide is represented by multiple bands in the Raman spectra, single band imaging cannot accurately represent the distribution of polysaccharides in cell walls. The PCA method considers loading from the entire spectra and, therefore, can be used for the imaging of the spatial distribution. For this purpose, loadings for each PC are analyzed to identify which bands had the most influence on the component, such that it is possible to determine which polysaccharide influenced it the most. Under this method, a spatial distribution of the principal components is drawn [39]. The first principal component (PC1) describes the combination of spectral locations with the greatest spectral variance in the map. PC1 typically explains the majority of variability - therefore, PC1 only varies the scores within each group, indicating that it might reflect the standard deviation of the recorded spectra and thus might not be useful for the visualization

of the polysaccharides' distribution [40]. PC1 corresponds to all polysaccharide composites in the sample, without distinguishing between the different compounds.

Figure 3 shows the score images and loadings spectra for PC2 and PC3 obtained in this experiment for the studied fragment of the tomato cell walls. Some relationship between the PCs loadings and the pure components (Figures 1A-C) could be established.

PC2 brought information mainly concerning pectin. The large positive loading was observed for the band at 2,958 cm^{-1}, assigned to the CH–stretching modes for pectin, while there was a large negative loading at 2,897 cm^{-1} attributed to the CH–stretching modes for cellulose. The second-largest values of the positive loadings was for a pectin marker band at 854 cm^{-1}, which might be associated with the skeletal mode of the α-anomers in the pectin molecules. In addition, a band at 820 cm^{-1} (ν(COH) ring) which positively influenced the loading spectrum was observed. Also, bands at 1,115 and 1,094 cm^{-1} which negatively influenced the loadings might correspond with the vibration of the cellulose molecules'

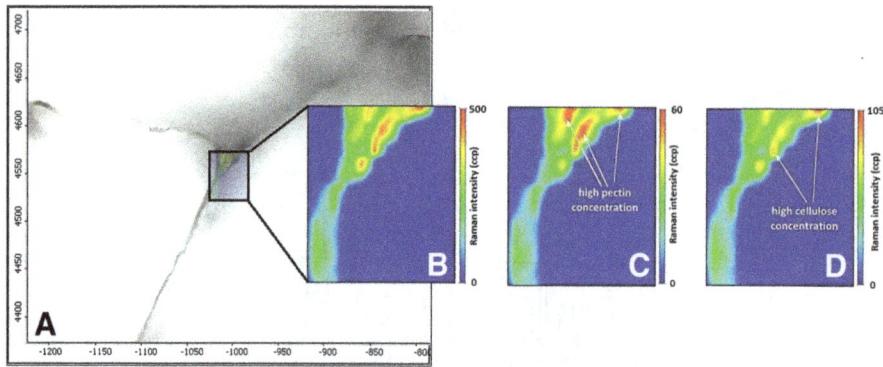

Figure 2 Images of the tomato cell wall. A) microscopic image (300 × 400 μm); **B)** Raman image of all primary cell wall polysaccharides at 2, 940 cm^{-1}, γ(CH); **C)** Raman image of pectin at 854 cm^{-1}, the (COC) skeletal mode of α-anomers; **D)** Raman image of cellulose at 1, 090 cm^{-1}, γ(COC) glycosidic.

modes, and were present in the PC2 spectrum loadings (Figure 3D).

PC3 predominantly provides information about cellulose. Analysis of PC3's loading spectrum (Figure 3E) showed that the majority of the bands positively influenced the loadings that were characteristic for cellulose vibration (1,122, 1,093, 380 cm^{-1}), and large negative loading was observed for the band at 2,949 cm^{-1} that

can be assigned to the CH–stretching modes for pectin.

Figure 3C shows the imposed score images for pectin (PC2) and cellulose (PC3). It can be observed that, in some parts, both the PC2 and PC3 scores overlap, but in some places - especially in the middle of the investigated cell wall - the PC2 scores occurred separately. It might be considered that the PC2 scores (red pixels) correspond to

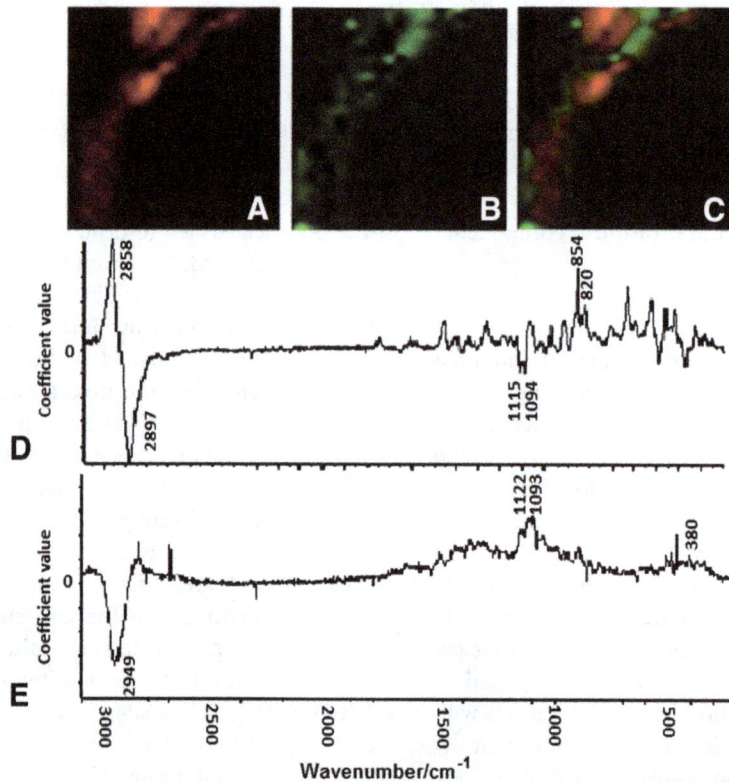

Figure 3 PCA score images and loading spectra. Based on comparison with the reference spectra, it was concluded that: **A)** PC2 mainly represents pectins; **B)** PC3 mainly represents cellulose; **C)** PC2 and PC3 are depicted as revealing the distribution of pectins and cellulose; **D)** PC2 loadings; and **E)** PC3 loadings.

the middle lamella – the area between the primary walls, enriched in pectins and responsible for cell-cell adhesion.

Unfortunately, the identification of the hemicellulose's localization was not possible by PCA analysis. PC4's loading spectrum (not shown) could not be linked with any polysaccharides from the plant cell wall.

Multivariate curve resolution (MCR)

The MCR method allows for the recovery of the response profiles of more than one component in unresolved and unknown mixtures, and therefore provides information about the nature and composition of these mixtures. Here, the goal in using the MCR technique was to estimate which pure components are present in the active area map, as well as the locations and concentrations of those 'new' MCR components. The three main components were extracted and the their

Raman spectra and concentration maps could be found (Figure 4).

Based on a comparison of the pure-components' spectra from the MCR method (Figures 4A-C) and the pure reference spectra of the polysaccharides (Figures 1A-C), the sample constituents could be identified. The similarity between the pectin spectrum (Figure 1A) and the component 1 spectrum (Figure 4A) is noticeable. The spectrum of component 1 contains Raman bands centered at 2,949, 1,751, 855 and 816 cm^{-1} characteristic of pectin polysaccharide (Table 2). Components 2 and 3 could not be clearly distinguished due to the presence in the spectra bands characteristic for cellulose and hemicelluloses. However, it is more likely that component 2 corresponds to cellulose, since its spectrum included more bands characteristic for cellulose molecules' vibration modes. In the component 2 spectrum (Figure 4B), the

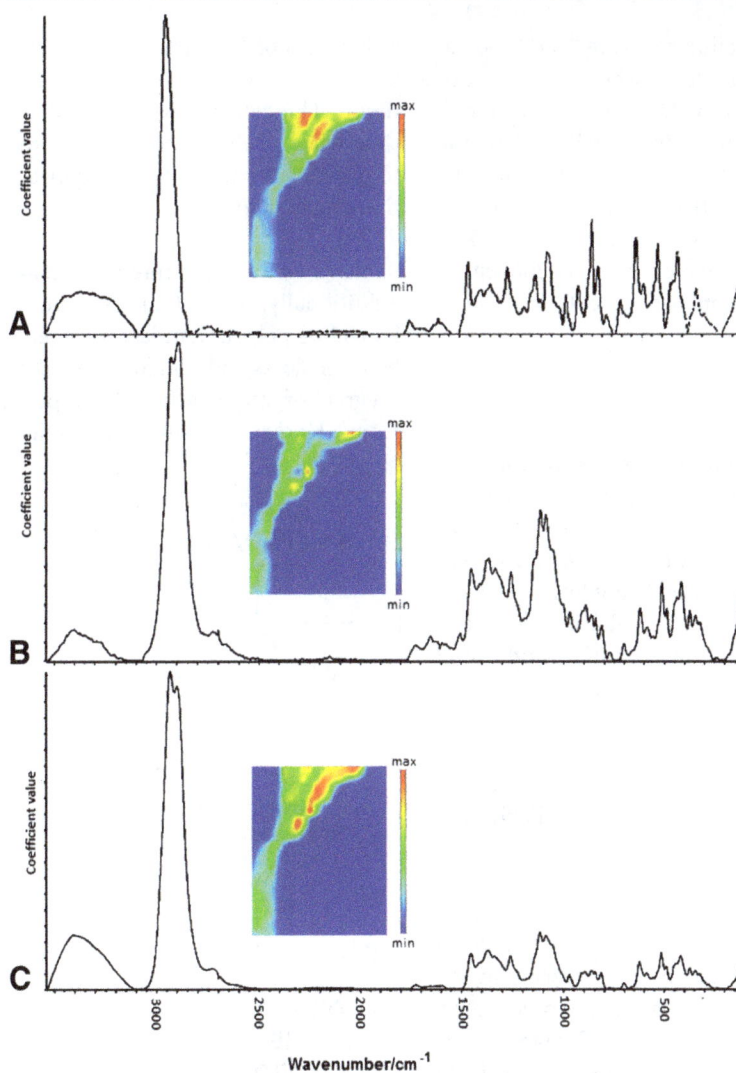

Figure 4 MCR concentration images and the spectra of the sampled pure-components: A) component 1; B) component 2; C) component 3.

Table 2 The pure components' bands (components 1–3) which were detected in the experiments with assignments from the literature [31-33]: C – cellulose; P – pectins; H – hemicelluloses

Component 1		Component 2		Component 3	
Raman wavenumber [cm^{-1}]	Assignment, Origin	Raman wavenumber [cm^{-1}]	Assignment, Origin	Raman wavenumber [cm^{-1}]	Assignment, origin
2949	2948 ν(CH), P	2932	2932 ν(CH), H	2935	2932 ν(CH), H
1751	ν(C = O) ester, P,H	2895	2897 ν(CH), C	2895	2897 ν(CH), C
855	855 (COC)skeletal mode of α-anomers, P	1378	1376 δ(HCC), δ(HCO), δ(HOC), C	1459	1460 δ(CH$_2$), δ(COH), C,H
816	817 ν(COH)ring, P	1257	1256 δ(CH) δ(COH), H	1256	1256 δ(CH), δ(COH) ,H
		1122	1121 ν(COC) glycosidic, symetric, C		
		1093	1098 ν(COC) glycosidic, assymetric, C		
		971	ρ(CH$_2$), C		
		381	380 δ(CCC)ring, C		

bands characteristic for cellulose were centered at 1,378, 1,122, 1,093 and 971 cm^{-1}, whereas for hemicellulose they were centered at 2,933 and 1,257 cm^{-1}. Additionally, the band at 2,895 cm^{-1} attributed to the CH–stretching modes for cellulose had greater intensity in the component 2 spectrum than in the component 3 spectrum. In the component 3 spectrum (Figure 4C), only bands characteristic of hemicellulose (1,256 cm^{-1}) and for both hemicellulose and cellulose (1,459 cm^{-1}) were observed.

Conclusions

This study showed that Raman microspectroscopy supported by multivariate image analysis methods is useful in the chemical imaging of polysaccharides' distributions in the cell wall of the tomato. It was shown that distinguishing between pectin and cellulose was possible from minimally-prepared samples. However, the imaging of hemicellulose's distribution was not possible due to overlapping characteristic bands with cellulose and pectins.

Methods

The main steps of the experimental procedure are presented on the scheme below (Figure 5). Each step is described in detail in what follows.

Sample preparation for Raman imaging

The fully ripe fruit of tomato (*Lycopersicon esculentum* Mill. Cv Conqueror) was used in the experiment. In preliminary experiments, a method for sample preparation was developed to study fruit tissue with Raman microspectroscopy. Prior to cutting, tomato parenchyma tissue was frozen and then lyophilized. The lyophilized tissue was cut using a vibratome (Leica VT 1000S) into slices of

a thickness of 130 μm. The cut slices were then soaked in acetone in order to remove pigments from the tomato tissue. The samples were placed on a microscope glass slide which, beforehand, was covered with aluminum foil to avoid interference from the glass Raman spectrum on the sample map.

Raman spectra of the reference materials

Additionally, to obtain exemplary Raman spectra of individual polysaccharides, commercially available pectins, hemicelluloses and cellulose were studied. High methylated (degree of methylation 80%) pectins were purchased from Herbstreit and Fox (Neuenbürg, Germany),

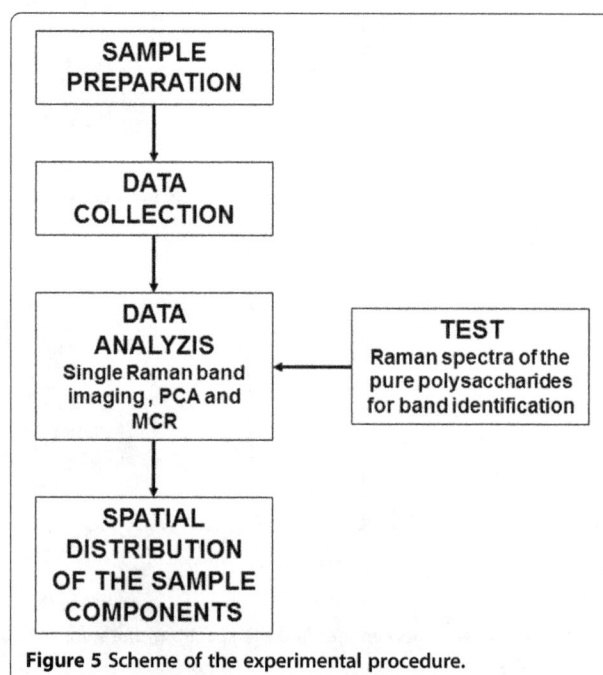

Figure 5 Scheme of the experimental procedure.

microcrystalline cellulose powder (ca. ~ 20 μm) from Sigma Aldrich and xyloglucan (tamarind, purity >95%) from Megazyme (Bray, Ireland). All the chemicals were used without further purification. Also, the Raman spectrum from the tomato cell wall was collected for comparison. Characteristic bands of the individual polysaccharides from the reference materials were used for the calibration of the Raman imaging method.

Raman microscope

The imaging system used in this study was a DXR Raman Microscope (Thermo Scientific, Waltham, USA), equipped with a diode-pumped, solid state (DPSS) green laser ($\lambda = 532$ nm) with a maximum power of 10.0 mW, a diffraction grating of 900 lines per mm and a pinhole confocal aperture of 25 μm. The Raman light was detected with an air-cooled CCD detector with a spectral resolution of 4 cm^{-1}. The 20x/0,40NA objective was used.

The map was recorded with a spatial resolution of 2 μm in both directions, x, y. The vertical z displacement was fixed. The integration time (8 s) was fixed for each scan. A single spectrum at each point was recorded within the range of 3,500–150 cm^{-1} of Raman shift for an average of 12 scans. Each pixel corresponds to one average spectrum. The spectra were not normalized.

Also, the Raman spectra of the reference materials were collected on a DXR Raman Microscope (Thermo Scientific, Waltham, MA, USA), with a green laser ($\lambda = 532$ nm) and a maximum power of 10.0 mW. The spectra were recorded within the range of 3,500–150 cm^{-1}.

Visualization methods

The Omnic Atlμs program (Thermo Scientific, Omnic 8.1, USA) was used in collecting the Raman spectra and performing the data analysis in order to obtain chemical images. Moreover, PCA and MCR were performed using the Omnic program.

Single Raman band imaging allows the generation of two–dimensional images based on the integral of the different Raman bands that are characteristic for sample components. It is used for the preliminary analysis and initial identification and localization of the biopolymers in the sample.

PCA is mathematical technique used for reducing the dimensionality of data from hundreds of spectral data points into a few orthogonal PCs. Each PC explains a part of the total information contained in the original data but not always corresponding to one specific chemical component (especially when several pure components' spectra are overlapping) [41]. The first basis spectrum - or principal component - accounts for the maximum variance in the data if the data is mean-centered prior to analysis. The second basis spectrum accounts for the next most variance, and so

on. These spectra bases are created such that they are orthogonal to each other and, therefore, contain no overlapping spectral information. These principal components are fitted to the imaging data set and are used to create a two-dimensional image, which will provide a map of how the spectral features represented by the principal components are distributed in the sample. This map can be correlated with the Raman spectra of known chemicals [42].

Another exploratory analysis of the chemical data (especially the spectroscopic and chromatographic data) is given by MCR [43]. This technique allows for the estimation of which pure components (and, therefore, chemical species) are present in the active area map and shows the locations of those components. The aim of MCR is to mathematically decompose an instrumental response for a mixture into the pure contributions of each component involved in the system studied [44]. MCR maximizes the explained variance in the data - as PCA would - while also delivers physical or chemical information about the system rather than the mathematical or statistical constraints as for PCA. The profiles of the pure-components are given a chemical meaning and can be straightforwardly interpreted as concentration profiles and spectra [43]. In the results of the analysis, the pure-components' spectra are obtained. Another important feature of MCR is that although the method itself does not require prior knowledge concerning the sample, additional information can always be incorporated to facilitate the analysis when available. For example, the reference spectra of known existing components can be used as an initial estimate. However, the MCR method requires an estimation of the number of pure components in the system [41].

Single Raman band imaging allows the generation of images of the studied component but only based on the one Raman band from the spectrum. Therefore it is used for the preliminary analysis. Whereas PCA and MCR methods relate to the data from the entire spectrum or from the range selected from the spectrum. Those methods deliver more details and are much more reliable than imaging based on one band (even characteristic for studied component).

For PCA, two regions of the spectra were selected (3,500–2,500 and 1,800–200 cm^{-1}) and four principal components were analyzed. MCR analysis was performed using the whole spectra and three pure components were provided.

The graphical presentation of the individual spectra was prepared on the OriginPro program (Origin Lab v8.5 Pro, Northampton, USA).

Abbreviations

MCR: Multivariate curve resolution; PCA: Principal component analysis.

Competing interests
The authors declare that they have no competing interests.

Authors' contributions
MC and MS performed the experiment. MC wrote the manuscript. MS and AZ participated in the preparation of the manuscript. MS and AZ supervised the experiment and preparation of the manuscript. All authors read and approved the final manuscript.

Acknowledgement
The study was partially supported under project no. 2011/01/D/NZ9/02494 under the Polish National Science Centre (NCN).

References
1. Brownlee IA: **The physiological roles of dietary fibre.** *Food Hydrocoll* 2011, **25**:238–250.
2. Fanta SW, Abera MK, Tri Ho Q, Verboven P, Carmeliet J, Nicola BM: **Microscale modeling of water transport in fruit tissue.** *J Food Eng* 2013, **2**:229–237.
3. Cybulska J, Zdunek A, Psonka-Antonczyk KM, Stokke BT: **The relation of apple texture with cell wall nanostructure studied using an atomic force microscope.** *Carbohyd Polym* 2013, **1**:128–137.
4. Zdunek A, Kurenda A: **Determination of the Elastic Properties of Tomato Fruit Cells with an Atomic Force Microscope.** *Sensors* 2013, **13**:12175–12191. doi:10.3390/s130912175.
5. Taiz L, Zeiger E: *Plant Physiology.* 3rd edition. Sunderland: Sinauer Associates; 2002.
6. Szymańska-Chargot M, Cybulska J, Zdunek A: **Sensing the Structural Differences in Cellulose from Apple and Bacterial Cell Wall Materials by Raman and FT-IR Spectroscopy.** *Sensors* 2011, **11**:5543–5560. doi:10.3390/s110605543.
7. Zugenmeier P: *Crystalline Cellulose and Derivatives. Characterization and Structure.* Berlin/Heidelberg: Springer-Verlag; 2008.
8. Fry SC: *The growing plant cell wall: chemical and metabolic analysis.* UK: The Blackburn Press. Longman Scientific & Technical; 1988.
9. Carpita NC, Gibaut DM: **Structural models of primary cell walls in flowering plants: consistency of molecular structure with the physical properties of the walls during growth.** *Plant J* 1993, **3**:1–30.
10. Park YB, Cosgrove DJ: **A revised architecture of primary cell walls based on biomechanical changes induced by substrate-specific endoglucanases.** *Plant Physiol* 2012, **158**:1933–1943.
11. Lacayo CI, Malkin AJ, Holman H-YN, Chen L, Ding S-Y, Hwang MS, Thelen MP: **Imaging Cell Wall Architecture in Single *Zinnia elegans* Tracheary Elements.** *Plant Physiol* 2010, **154**:121–133.
12. Agoda-Tandjawa G, Durand S, Gaillard C, Gernier C, Doublier JL: **Properties of cellulose/pectins composites: implication for structural and mechanical properties of cell wall.** *Carbohyd Polym* 2012, **90**:1081–1091.
13. Cybulska J, Konstankiewicz K, Zdunek A, Skrzypiec K: **Nanostructure of natural and model cell wall materials.** *Int Agrophysics* 2010, **24**:107–114.
14. Van Soest PJ: **Use of detergents n the analysis of fibrous feeds. A rapid method for determination of fiber and lignin.** *J Association of Official Analytical Chemists* 1963, **46**:155–160.
15. Redgwell RJ, Curti D, Gehin-Delval C: **Physicochemical properties of cell wall materials from apple, kiwifruit and tomato cell wall material.** *Eur Food Res Technol* 2008, **227**:607–618.
16. Peng F, Westermark U: **Distribution of conifer alcohol and coniferaldehyde groups in the cell wall of spruce fibers.** *Holzforschung* 1997, **51**:531–536.
17. Fujino T, Sone Y, Mitsuishi Y, Itoh T: **Characterization of Cross-Links between Cellulose Microfibrils, and Their Occurrence during Elongation Growth in Pea Epicotyl.** *Plant Cell Physiol* 2000, **41**:486–494.
18. Bush MS, McCann MC: **Pectic epitopes are differentially distributed in the cell walls of potato (*Solanum tuberosum*) tubers.** *Physiol Plantarum* 1999, **107**:201–213.
19. Steele NM, McCann MC, Roberts K: **Pectin Modification in Cell Walls Ripening Tomatoes Occurs in Distinct Domains.** *Plant Physiol* 1997, **114**:373–381.
20. Terao A, Hyodo H, Satoh S, Iwai T: **Changes in the distribution of cell wall polysaccharides in early fruit pericarp and ovule, from fruit set to early fruit development, in tomato (*Solanum lycopersicum*).** *J Plant Res* 2013, **126**:719–728.
21. McCartney L, Marcus SE, Knox JP: **Monoclonal Antibodies to Plant Cell Wall Xylans and Arabinoxylans.** *J Histochem & Cytochem* 2005, **4**:543–546.
22. Gierlinger N, Schwanninger M: **Chemical imaging of poplar wood cell walls by confocal Raman microscopy.** *Plant Physiol* 2006, **4**:1246–1254.
23. Raman CV, Krishnan KS: **A New Type of Secondary Radiation.** *Nature* 1928, **121**:501–502.
24. Sun D: *Modern Techniques for Food Authentication, First Edition.* Elsevier Inc; 2008.
25. Gierlinger N: **Imaging of Plant Cell Walls.** In *Confocal Raman Microscopy.* Edited by Dieing T, Hollricher O, Toporski J. Berlin/Heidelberg: Springer-Verlag; 2010:225–235.
26. Himmelsbach DS, Khahili S, Akin DE: **Near-infrared–Fourier-transform–Raman microspectroscopic imaging of flax stems.** *Vib Spectrosc* 1999, **19**:361–367.
27. Agarwal UP: **Raman imaging to investigate ultrastructure and composition of plant cell walls: distribution of lignin and cellulose in black spruce wood (*Picea mariana*).** *Planta* 2006, **224**:1141–1153.
28. Barańska M, Barański R, Grzebelus E, Roman M: **In situ detection a single carotenoid crystal In a plant cell using Raman microspectroscopy.** *Vib Spectrosc* 2011, **56**:166–169.
29. Brożek-Płuska B, Placek I, Kurczewski K, Morawiec Z, Tazbir M, Abramczyk H: **Breast cancer diagnostics by Raman spectroscopy.** *J Mol Liq* 2008, **3**:145–148.
30. Carrari F, Fernie AR: **Metabolic regulation underlying tomato fruit development.** *J Exp Bot* 2006, **57**:1883–1897.
31. Schulz H, Barańska M: **Identification and quantification of valuable plant substances by IR and Raman spectroscopy.** *Vib Spectrosc* 2007, **1**:13–25.
32. Gierlinger N, Sapei L, Paris O: **Insights into chemical composition of Equisetum hyamale by high resolution Raman imaging.** *Planta* 2008, **227**:969–980.
33. Séné CFB, McCann MC, Wilson RH, Crinter R: **Fourier-Transform Raman and Fourier-Transform Infrared Spectroscopy.** *Plant Physiol* 1994, **106**:1623–1631.
34. Gierlinger N, Keplinger T, Harrington M: **Imaging of plan cell walls by confocal Raman microscopy.** *Nat Protoc* 2012, **7**:1694–1708.
35. Lacayo CI, Malkin AJ, Holman HY, Chen L, Ding S-Y, Hwang MS, Thelen MP: **Imaging cell wall architecture in single Zinnia elegans tracheary elements.** *Plant Physiol* 2010, **154**:121–133.
36. Synytsya A, Čopikova J, Matějka P, Machovič V: **Fourier transform Raman and infrared spectroscopy of pectins.** *Carbohydr Pol* 2003, **54**:97–106.
37. Richter S, Mussig J, Gierlinger N: **Functional plant cell wall design revealed by the Raman imaging approach.** *Planta* 2011, **233**:768–772.
38. Marry M, Roberts K, Jopson SJ, Huxham IM, Jarvis MC, Corsar J, Robertson E, McCann MC: **Cell-cell adhesion in fresh sugar beet root parenchyma requires both pectin esters and calcium cross –links.** *Physiol Plantarum* 2006, **126**:243–256.
39. Hedegaard M, Matthäus C, Hassing S, Krafft C, Diem M, Popp J: **Spectral unmixing and clustering algorithms for assessment of single cells by Raman microscopic imaging.** *Theor Chem Acc* 2011, **130**:1249–1260.
40. Szymańska-Chargot M, Zdunek A: **Use of FT-IR Spectra and PCA to the Bulk Characterization of Cell Wall Residues of Fruits and Vegetables Along a Fraction Process.** *Food Biophysics* 2013, **8**:29–42.
41. Zhang L, Henson MJ, Sekulic SS: **Multivariate data analysis for Raman imaging of a model pharmaceutical tablet.** *Anal Chim Acta* 2005, **545**:262–278.
42. Shafer-Peltier KE, Haka AS, Motz JT, Fitzmaurice M, Dasari RR, Feld MS: **Model-Based Biological Raman Spectral Imaging.** *J Cell Biochem* 2002, **39**:125–137.
43. Ruckebusch C, Blanchet L: **Multivariate curve resolution: A review of advanced and tailored applications and challenges.** *Anal Chim Acta* 2013, **765**:28–36.
44. Garrido M, Rius FX, Larrechi MS: **Multivariate curve resolution–alternating least squares (MCR-ALS) applied to spectroscopic data from monitoring chemical reactions processes.** *Anal Bioanal Chem* 2008, **390**:2059–2066.

A rapid and sensitive method for determination of carotenoids in plant tissues by high performance liquid chromatography

Prateek Gupta, Yellamaraju Sreelakshmi* and Rameshwar Sharma*

Abstract

Background: The dietary carotenoids serve as precursor for vitamin A and prevent several chronic-degenerative diseases. The carotenoid profiling is necessary to understand their importance on human health. However, the available high-performance liquid chromatography (HPLC) methods to resolve the major carotenoids require longer analysis times and do not adequately resolve the violaxanthin and neoxanthin.

Results: A fast and sensitive HPLC method was developed using a C30 column at 20°C with a gradient consisting of methanol, methyl-*tert*-butyl ether and water. A total of 15 major carotenoids, including 14 all-trans forms and one cis form were resolved within 20 min. The method also distinctly resolved violaxanthin and neoxanthin present in green tissues. Additionally this method also resolved geometrical isomers of the carotenoids.

Conclusion: The HPLC coupled with C30 column efficiently resolved fifteen carotenoids and their isomers in shorter runtime of 20 min. Application of this method to diverse matrices such as tomato fruits and leaves, Arabidopsis leaves and green pepper fruits showed the versatility and robustness of the method. The method would be useful for high throughput analysis of large number of samples.

Keywords: Carotenoids screening, HPLC, C30 column, cis-Isomers, Tomato

Background

Carotenoids have been extensively studied in different matrices to analyze their distribution and levels, as diet rich in carotenoids imparts health benefit properties. They are the most widely distributed pigments in nature [1] and more than 700 different carotenoids have been identified so far [2]. In plants, they act as accessory pigments for photosynthesis and precursor to plant hormone ABA and strigolactones [3]. Though the forms of carotenoids found in foods are not many, however, their composition is very complex and varies both qualitatively and quantitatively. Moreover, biological matrices often contain hundreds to thousands of other plant metabolites that interfere with detection of carotenoids [4].

Carotenoids are made up of polyene hydrocarbon chain consisting of eight isoprene units and are classified into two groups: hydrocarbons (carotenes) and their

oxygenated derivatives (xanthophylls). The presence of conjugated double bonds and cyclic groups at ends leads to the formation of variety of stereoisomers. In nature, carotenoids exist as cis/trans isomer, however, they exist primarily in the more stable all-trans isomeric form, but cis isomers do occur. Different carotenoids vary significantly in their absorption maxima as well as in their fine structure. Absorption spectra of carotenoids are unique, mostly showed three proximately distinct peaks [5]. The ratio of absorption peak heights from the trough between peak II and III is used for distinguishing carotenoids and their isomers. The isomers can also often be tentatively identified by the presence of a "cis peak".

Animals are unable to synthesize carotenoids and acquire them from plants through diet. The carotenoid composition of plants is affected by several factors such as cultivar or variety; part of the plant; stage of maturity; climate or geographic site of production; harvesting and postharvest handling; processing and storage [6]. During food processing, the levels of cis-isomers increase due to the isomerization of the trans-isomers. Consequently,

---

* Correspondence: syellamaraju@gmail.com; rameshwar.sharma@gmail.com
Repository of Tomato Genomics Resources, Department of Plant Sciences, School of Life Sciences, University of Hyderabad, Hyderabad 500046, India

effective quantitation of carotenoids in both foods and biological samples is necessary to understand their importance in body metabolism and health.

A variety of methods have been employed to detect the carotenoids in food samples ranging from thin layer chromatography, to high pressure liquid chromatography (HPLC) and combination of HPLC with mass spectrometry including MALDI-TOF. The most commonly used method for identification and quantification of carotenoids utilizes HPLC combined with UV–vis absorption detection. Though carotenoid separation can be carried out using both normal phase and reverse phase HPLC, however, normal phase HPLC is not suitable for carotenoid separation due to poor separation of nonpolar carotenoids. In contrast, reverse phase HPLC enables a significant increase in the interaction between analyte and non-polar stationary phase leading to enhanced resolution of carotenoids [7]. Among the columns, C18 columns with isocratic or gradient mode are preferred for carotenoid separation [8]. However, C18 column do not resolve geometrical isomers and inefficiently resolves positional isomers, particularly lutein and zeaxanthin. To maximize chromatographic resolution and selectivity, Sander et al. [9] developed a non-endcapped RP-HPLC column with triacontyl (C30) ligands to resolve carotenoids and its isomers. The polymeric C30 columns also possess abilities to resolve cis/ trans-carotenoids [10]. Nevertheless, the efficiency of C30 column to resolve geometrical isomers of carotenoids is offset by requirement of longer run times needing 60 minutes or more for complete separation of carotenoids resulting in low throughput.

Recently, UHPLC technology has been used to analyze carotenoids in various matrices [11]. UHPLC offers several advantages over HPLC such as higher peak capacities, smaller peak widths, gain in sensitivity and higher chromatographic resolution. The shorter analysis times also considerably save mobile phase solvents. Since C30 stationary phase columns are not commercially available for UHPLC, C18 columns have been used for carotenoid separation despite the limitation that C18 columns poorly resolve carotenoid isomers.

In a recent study, C18 UHPLC columns were compared with a C30 HPLC column. Though usage of C30 HPLC column resulted in better resolution of carotenoids, particularly of geometrical isomers compared to C18 UHPLC columns, this advantage was negated by longer run time needed for C30 HPLC column (100 min versus 23 min for C18 UHPLC column) [12]. In another study, Maurer et al. [13], demonstrated the separation of 11 major carotenoids in 13.50 minutes using C18 column on UHPLC-UV for faster analysis. However, using this method they could detect only few geometrical isomers. Though currently available UHPLC methods allow

shorter run times, these methods are poor in resolving cis isomers of different carotenoids.

Supercritical fluid chromatography (SFC) and comprehensive two-dimensional LC (LC × LC) has also been applied for the separation of carotenoids, to enhance the chromatographic separations of complex carotenoid mixtures [2]. Although both techniques have better potential to carry out complex carotenoid separations, nonetheless require more specialized and expensive instruments and longer analysis times nearly double than HPLC. The longer analysis times also necessitate special precautions to avoid carotenoid degradation.

The availability of a rapid HPLC method using a polymeric C30 column would greatly aid separation, identification and quantitation of various carotenoid isomers with better sensitivity and selectivity. In this study, we developed a rapid HPLC method using C30 column to determine the carotenoid profiles along with identification of carotenoid isomers from different plant organs. The method is also capable of separation and quantitation of the major carotenoids present in fruit and green leaves.

Results and discussion
HPLC analysis of carotenoids standards

It is well known that the separation of carotenoids is strongly influenced by the properties of stationary phase. Among the stationary phases used, polymeric, non-endcapped stationary phases with C30 ligands give optimal separation of carotenoids and their isomers. In contrast, C18 stationary phases are of insufficient thickness to allow full penetration of carotenoid molecules leading to poor isomer separation due to weak solute-bonded phase interactions. It is by virtue of these properties a C30 column provides better resolution of carotenoids and their geometrical isomers than a C18 column [14]. Using a C30 column in HPLC, Fraser et al. [15] separated carotenes, xanthophylls, ubiquinones, tocopherols, and plastoquinones in a single run, however, their analysis time was 42 minutes. Since a longer run time is a major drawback for high throughput analysis, we improved the analytical method by modifying solvent percentages of mobile phase to reduce the run time.

Nelis and de Leenheer [14] advocated the use of non-aqueous reversed-phase liquid chromatography for the separation of complex carotenoid mixtures, citing optimal sample solubility. This results in minimum risk of sample precipitation on the column, increased sample capacity, excellent chromatographic efficiency, and prolonged column life. Though carotenoids are sparingly soluble in water, most studies employ solvent mixtures containing a small fraction of water. In this study, a small fraction of water was used to resolve the early eluting free xanthophylls, such as those present in leaves and mature green fruits. In initial optimization phase, when we used binary solvents containing (B) methanol/water (95:5, v/v) and (C) tert-methyl butyl ether, lutein coeluted with chlorophyll b marring the chromatographic resolution and analyte identification. In successive trials by introducing new steps in the gradient, the relative separation was considerably increased. Of all the gradients tested, the best separation was achieved with (A) methanol/water (98:2) and (C) MTBE for initial 2 minutes, which clearly resolved all-trans-violaxanthin and all-trans-neoxanthin; lutein and Chl b peaks, followed by next 10 minutes run with solvent (B) methanol/water (95:5, v/v) and (C) tert-methyl butyl ether. The separation of chlorophylls from the carotenoids also eliminated the need for saponification of samples for removal of chlorophylls.

The optimization of carotenoid separation needs compatibility between injection solvent and mobile phase. Ideally the injection solvent should be either compatible with the mobile phase or more polar than the reverse phase to provide on-column concentration of samples. In case carotenoids are more soluble in the injection solvent than in the mobile phase, and especially when solution is nearly saturated, the carotenoids will precipitate on injection, leading to peak tailing. Alternatively, they will remain in the injection solvent while passing through the column, resulting in broad bands and doubled peaks [16]. On the other hand, the sample will not dissolve fully in mobile phase if the injection solvent is too weak.

In order to improve resolution, we tried different injection solvents like dichloromethane, tetrahydrofuran, ethyl acetate, acetone, and different ratios of mobile phase solvents (MeOH and MTBE). We found that the mixture of mobile phase solvents (MTBE and MeOH) is most ideal for injections of standards and samples in the column. Our analysis revealed 2:3 ratio of MTBE:MeOH is ideal for green tissues whereas 3:1 ratio of MTBE:MeOH is best for red ripe tomato fruits (Additional file 1: Figure S1). The usage of two different ratios was related to difference in composition of carotenoids in green tissues and tomato fruits respectively. As green tissues are rich in xanthophylls, the high content of methanol in injection solvent improves the resolution of early eluting xanthophylls specifically all-trans-violaxanthin and all-trans-neoxanthin. The red ripe tomato fruits are enriched in lycopene, therefore higher content of MTBE is required to completely dissolve lycopene.

For optimal resolution of carotenoids, it is established that column temperature is an important factor in improving separations and reproducibility of retention times, which is critical for correct identification of peaks in complex mixtures. The lower temperatures (ca. 13°C) maximize selectivity for a set of cis/trans isomers, whereas high temperatures (i.e. 38°C) efficiently resolve different carotenoids [17]. At lower temperature, the carotenoids like lutein, zeaxanthin, β-carotene and lycopene were better resolved, while separation of echinenone and α-carotene improved as the temperature increased [16]. Bohm [18] reported that a column temperature of $23 \pm 1°C$ seems to be the best compromise for the separation of most prominent carotenoids, including their cis isomers. Considering the above, we examined carotenoid separation at 12°C, 18°C, 20°C and 25°C. Amongst them, the maximum selectivity was obtained at 20°C column temperature that was used for all other optimization.

After careful evaluation of different mobile phases/conditions, a gradient mobile phase consisting of methanol, water and methyl-tert butyl ether, as described in methods section was selected for the analysis of carotenoids and their isomers. Figure 1 shows the chromatogram of carotenoids standards mix. In a run time of 20 minutes, fifteen major carotenoids were separated and identified. The elution profile indicates that good separation efficiency along with shorter separation time was achieved for carotenoid analysis. Table 1 presents the chromatographic and the quantification data for the carotenoids. To assess the solvent strength of the mobile phase the k value (retention factor) is used. The k values of all peaks ranged between 1.66 and 8.92, indicating that a proper solvent strength of the mobile phase was maintained. Generally, it is accepted that for optimum separation, the k value should range from 2 to 10, however, when complicated mixture of compounds are to be separated it can range between 0.5 and 20 [19]. The separation or selectivity factor (α) values describes the separation of two species on a column. The selectivity factor is always greater than one. When α is close to unity, k is optimized first and then α is increased by changing the mobile phases, column temperature or composition of stationary phase. In our results, for all the peaks, the α were greater than 1.0, implying that a good selectivity of mobile phase to sample components was achieved.

For extraction of carotenoids from plant samples, mixture of different extraction solvents comprising diethyl ether: chloroform (1:2), methanol: chloroform: dichloromethane (1:2:1), methanol: chloroform: acetone (1:2:1) and dichloromethane: chloroform (1:2) were evaluated. Of these, a better resolution and recovery was obtained with dichloromethane: chloroform (1:2), as it does not contain methanol, therefore it also eliminates the metabolites interfering with the detection of carotenoids.

Photoisomerization of standards
For identification of cis-carotenoids, the fifteen carotenoid standards were illuminated to accelerate cis isomers

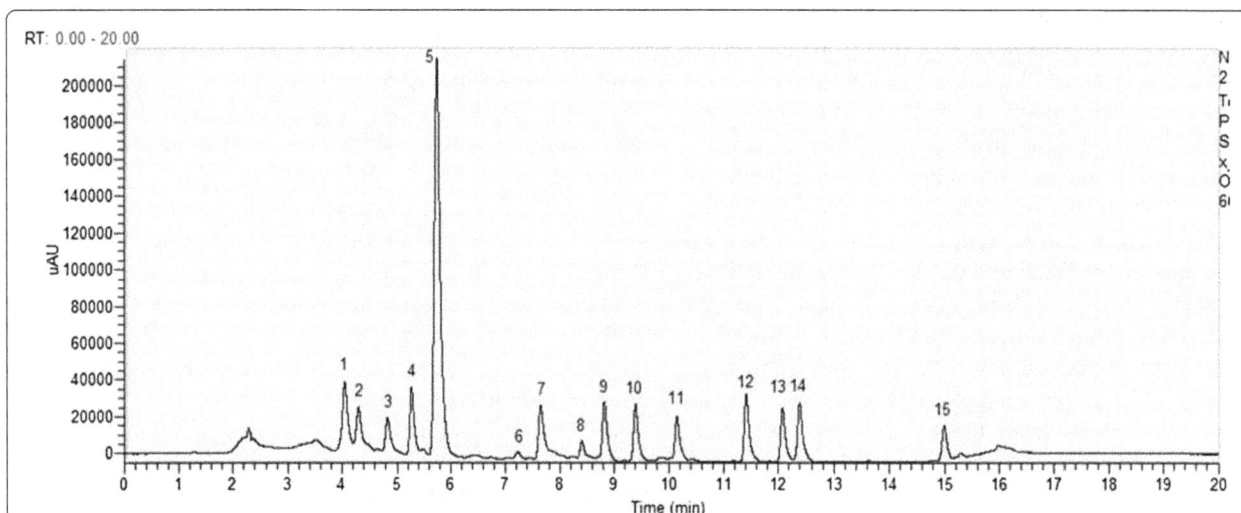

Figure 1 HPLC profile of carotenoid standards recorded in the range of 250.00-700.00 nm. The compounds are (1) violaxanthin; (2) neoxanthin; (3) anthraxanthin; (4) lutein; (5) zeaxanthin; (6) phytoene; (7) β-cryptoxanthin; (8) phytofluene; (9) α-carotene; (10) β-carotene; (11) ζ-carotene; (12) δ-carotene; (13) γ-carotene; (14) neurosporene; (15) lycopene.

formation by using a procedure described in the methods section. The retention time and absorption spectral characteristics of carotenoid isomers were used in identifying the unknown peaks in the samples. Figure 2 shows the HPLC chromatograms of photoisomerized carotenoid standards, including α-carotene (Figure 2A), antheraxanthin (Figure 2B), β-carotene (Figure 2C), β-cryptoxanthin (Figure 2D), γ-carotene (Figure 2E), lutein (Figure 2F), neoxanthin (Figure 2G), neurosporene (Figure 2H), violaxanthin (Figure 2I), zeaxanthin (Figure 2J), δ-carotene (Figure 2K), lycopene (Figure 2L), phytofluene (Figure 2M),

phytoene (Figure 2N) and ζ-carotene (Figure 2O). Table 2 represents the chromatographic and quantification data for the cis isomers of carotenoids. The identification of the cis isomers was based on the cis peak, wavelength spectrum and Q ratio's (ratio of the height of the cis-peak to the main absorption peak) with those in the literature. The retention time and absorption spectral characteristics of carotenoid isomers were used for identifying the unknown peaks in the samples. The on-line PDA spectra of all-trans and isomerized standards are presented in Additional file 2: Figure S2.

Table 1 Identification and chromatographic data of carotenoid standards

Peak no.	Compound	RT (min)	λ (nm) found	%III/II found	Regression equation	R^2	LOD	LOQ	k	α	%CV
1	All-trans-violaxanthin	4.05	416.0, 439.0, 469.0	88.8	$y = 1.011x - 0.270$	0.991	0.075	0.250	1.66	1.10	5.45
2	All-trans-neoxanthin	4.31	415.0, 437.0, 465.0	78.2	$y = 1.042x - 1.842$	0.996	0.075	0.250	1.83	1.17	5.30
3	All-trans-antheraxanthin	4.80	(424.0), 446.0, 474.0	60.0	$y = 1.057x - 1.919$	0.994	0.075	0.250	2.15	1.13	4.94
4	All-trans lutein	5.25	(424.0), 445.0, 474.0	62.5	$y = 1.000x - 0.040$	0.997	0.075	0.250	2.45	1.12	4.43
5	All-trans-zeaxanthin	5.73	(428.0), 451.0, 478.0	40.0	$y = 1.046x - 1.725$	0.998	0.075	0.250	2.76	1.37	5.66
6	15-cis-phytoene	7.29	(277.0), 286.0, (298.0)	n.c.	$y = 1.066x - 1.081$	0.989	0.120	0.400	3.79	1.06	7.74
7	All-trans- β-cryptoxanthin	7.66	(424.0), 452.0, 479.0	35.0	$y = 1.028x - 1.282$	0.991	0.075	0.250	4.03	1.12	6.84
8	All-trans-phytofluene	8.43	332.0, 348.0, 367.0	90.9	$y = 1.001x - 0.118$	0.998	0.120	0.400	4.54	1.06	6.60
9	All-trans- α-carotene	8.84	424.0, 446.0, 475.0	63.3	$y = 1.012x - 0.892$	0.991	0.075	0.250	4.81	1.07	7.94
10	All-trans- β-carotene	9.42	(425.0), 452.0, 479.0	30.0	$y = 1.016x - 1.176$	0.990	0.075	0.250	5.19	1.09	6.36
11	All-trans- ζ-carotene	10.18	380.0, 403.0, 426.0	115.4	$y = 1.015x - 1.092$	0.991	0.075	0.250	5.69	1.14	7.69
12	All-trans- δ-carotene	11.43	432.0, 457.0, 488.0	71.4	$y = 1.014x - 1.015$	0.989	0.075	0.250	6.51	1.06	5.22
13	All-trans- γ-carotene	12.09	439.0, 462.0, 492.0	53.3	$y = 1.008x - 0.396$	0.983	0.075	0.250	6.95	1.02	8.40
14	All-trans-neurosporene	12.40	416.0, 440.0, 469.0	92.3	$y = 1.010x - 1.041$	0.991	0.075	0.250	7.15	1.25	4.86
15	All-trans-lycopene	15.09	446.0, 472.0, 503.0	73.9	$y = 1.027x - 0.630$	0.993	0.075	0.250	8.92	1.37	10.67

Parentheses indicate a shoulder.
%III/II represents the ratio of peak heights from the trough between peak II and III.

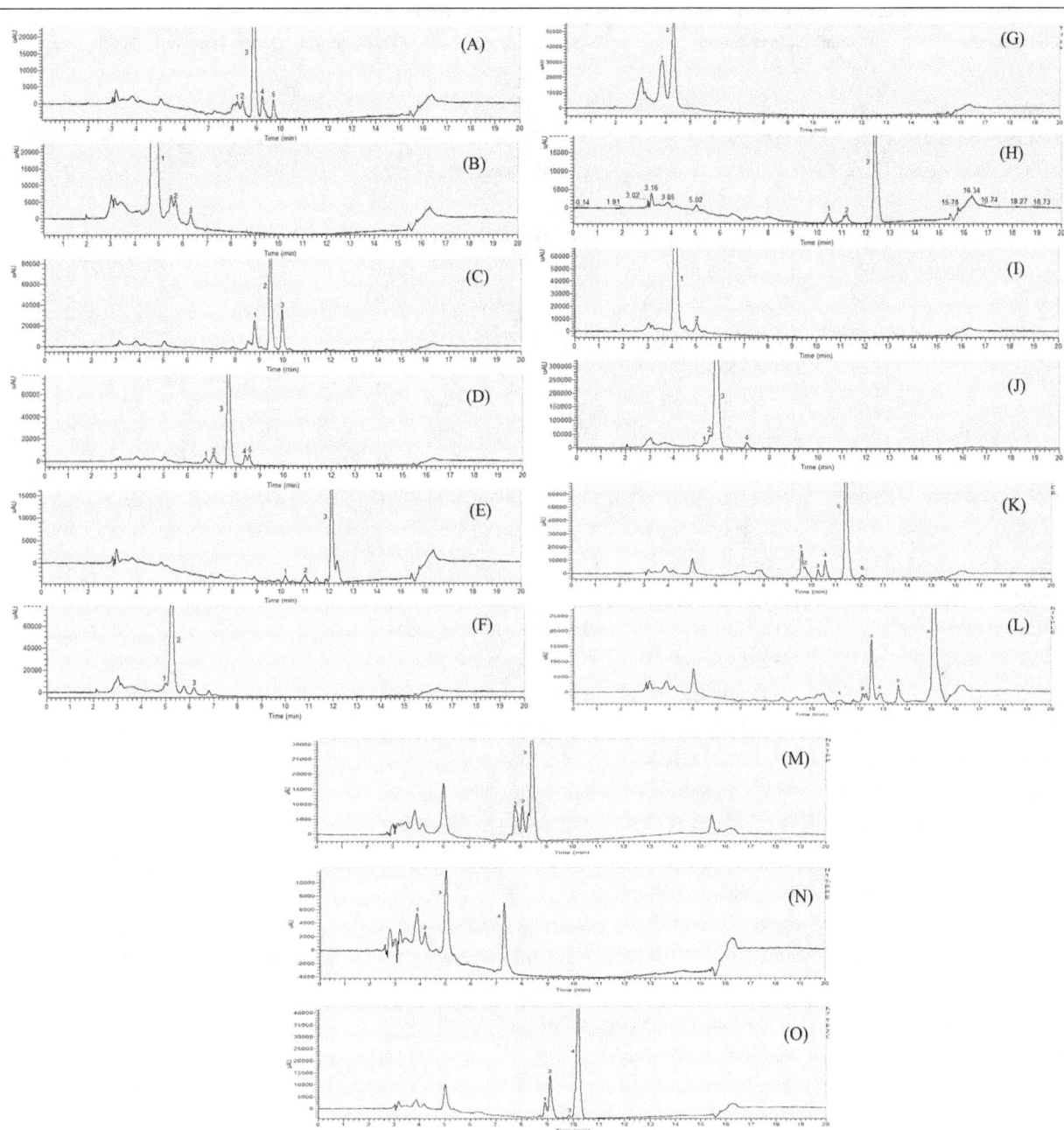

Figure 2 HPLC chromatogram of different carotenoids and their isomers after photoisomerization. (A) α-carotene, (B) antheraxanthin, (C) β-carotene, (D) β-cryptoxanthin, (E) γ-carotene, (F) lutein, (G) neoxanthin, (H) neurosporene, (I) violaxanthin, (J) zeaxanthin, (K) δ-carotene, (L) lycopene, (M) phytofluene, (N) phytoene and (O) ζ-carotene. The Arabic numeral in bold preceding carotenoid isomer refers to its peak on chromatogram. (A): 1; 13-cis-α-carotene; 2; 13′-cis-α-carotene; 3; all-trans-α-carotene; 4; 9-cis-α-carotene; 5; 9′-cis-α-carotene. (B): 1; all-trans-antheraxanthin; 2; 9-cis-antheraxanthin; 3; 9′-cis-antheraxanthin. (C): 1; 13-cis-β-carotene; 2; all-trans-β-carotene; 3; 9-cis-β-carotene. (D): 1; 13-cis-β-cryptoxanthin; 2; 13′-cis-β-cryptoxanthin; 3; all-trans-β-cryptoxanthin; 4; 9-cis-β-cryptoxanthin; 5; 9′-cis-β-cryptoxanthin. (E): 1; cis-γ-carotene 2; cis-γ-carotene 3; all-trans-γ-carotene. Lutein (F): 1; 13- or 13′-cis-lutein; 2; all-trans-lutein; 3; 9- or 9′-cis-lutein. Neoxanthin (G): 1; 9-cis-neoxanthin; 2; all-trans-neoxanthin. Neurosporene (H): 1; 15-cis-neurosporene; 2; 13-cis-neurosporene; 3; all-trans-neurosporene. Violaxanthin (I): 1; all-trans-violaxanthin; 2; 9-cis-violaxanthin. Zeaxanthin (J): 1; 15-cis-zeaxanthin; 2; 13-cis-zeaxanthin; 3; all-trans-zeaxanthin; 4; 9-cis-zeaxanthin. δ-carotene (K): 1; δ-carotene isomer 1, 2; δ-carotene isomer 2, 3; δ-carotene isomer 3, 4; δ-carotene isomer 4, 5; all-trans-δ-carotene, 6; δ-carotene isomer 5. Lycopene (L): 1; di-cis-lycopene 1, 2; di-cis-lycopene 2, 3; 15-cis-lycopene, 4; 13-cis-lycopene, 5; 9-cis-lycopene, 6; all-trans-lycopene, 7; 5-cis-lycopene. Phytofluene (M): 1; 15,9′-cis-phytofluene, 2; phytofluene isomer, 3; all-trans-phytofluene. Phytoene (N): 1; phytoene isomer 1, 2; phytoene isomer 2, 3; phytoene isomer 3, 4; all-trans-phytoene. ζ-carotene (O): 1; 9.15,9′ cis- ζ-carotene, 2; ζ-carotene isomer 1, 3; ζ-carotene 2, 4; all-trans-ζ-carotene. The HPLC profiles were recorded in the range of 250.00-700.00 nm.

Table 2 Tentative identification of photoisomerised isomers after illumination of all-trans standards

S. no.	Compound	RT (min)	λ (nm) found	λ (nm) reported	Q ratio found	Q ratio reported	%III/II	Ref.
1	9-cis-neoxanthin	3.78	328.0,(418.0),439.0, 468.0	327.0,416.0, 439.0, 468.0	0.13	0.16	55.55	[20]
2	Phytoene isomer 1	3.89	277.0					
3	Phytoene isomer 2	4.18	265.0, 272.0				19.08	
4	13-cis or 13'-cis-Lutein	5.00	330.0,(420.0),441.0, 466.0	332.0,416.0, 440.0, 468.0	0.39	0.39	23.07	[21]
5	Phytoene isomer 3	5.02	270.0					
6	9-cis-violaxanthin	5.03	327.0, 412.0, 435.0, 464.0	328.0,412.0, 436.0, 464.0	0.12	0.11	87.5	[22]
7	15-cis zeaxanthin	5.25	335.0, (422.0), 447.0, 470.0	338.0,422.0, 446.0, 470.0	0.48	0.45		[23]
8	13-cis zeaxanthin	5.49	337.0, (422.0), 444.0, 469.0	338.0,424.0, 446.0, 472.0	0.43	0.37	8.69	[23]
9	9-cis-antheraxanthin	5.68	330.0, (419.0), 442.0, 469.0	332.0,440.0, 468.0	0.13	0.09	5.33	[24]
10	Phytoene isomer	5.88	(276.0), 286.0, (298.0)					
11	9-cis or 9'-cis-Lutein	6.18	333.0, (419.0), 440.0, 468.0	332.0,416.0, 440.0, 470.0	0.08	0.13	64.70	[21]
12	9'-cis-antheraxanthin	6.29	330.0, (419.0), 441.0, 468.0	332.0,440.0, 468.0	0.07	0.08	53.57	[24]
13	13-cis-β-cryptoxanthin	6.73	336.0, (421.0), 449.0, 473.0		0.25			
14	9-cis-zeaxanthin	7.02	341.0, (422.0), 446.0, 473.0	338.0,422.0, 446.0, 474.0	0.11	0.12	36.36	[23]
15	13'-cis β-cryptoxanthin	7.08	336.0, (418.0), 444.0, 469.0	336.0,415.0, 443.0, 470.0	0.43	0.47	10.43	[25]
16	15,9'-cis-phytofluene	7.81	332.0, 348.0, 367.0	330.0, 347.0, 366.0			71.42	[26]
17	Phytofluene isomer 1	8.08	332.0, 348.0, 367.0				82.92	
18	13-cis-α-carotene	8.19	331.0, 418.0, 440.0, 466.0	330.0,417.0, 438.0, 466.0	0.43	0.43	33.33	[26,27]
19	9-cis-β-cryptoxanthin	8.38	336.0, (420.0), 446.0, 473.0	339.0,418.0, 445.0, 472.0	0.10	0.16	33.33	[25]
20	13'-cis α-Carotene	8.41	329.0, 416.0, 439.0, 466.0	332.0,416.0, 438.0, 465.0	0.35	0.41	33.33	[27]
21	9'-cis β-cryptoxanthin	8.55	338.0, (421.0), 447.0, 473.0	339.0,420.0, 445.0, 472.0	0.08	0.13	36.36	[25]
22	15-cis β-carotene	8.80	338.0, (420.0), 444.0, 468.0	337.0,420.0, 444.0, 470.0	0.42	0.41		[10]
23	9,15,9' cis-ζ-carotene	8.90	295.0, 378.0, 397.0, 423.0	296.0, 377.0, 399.0, 424.0	0.34	0.24	76.92	[28]
24	ζ-carotene isomer 1	9.10	296.0, 375.0, 395.0, 420.0	276.0, 375.0, 395.0, 420.0	0.25	0.23	86.30	[28]
25	9-cis α-carotene	9.24	330.0, 419.0, 441.0, 469.0	330.0,418.0, 441.0, 467.0	0.08	0.1	90.90	[26,27]
26	δ-carotene isomer 1	9.60	345.0, 425.0, 450.0, 480.0		0.48		53.12	
27	9'-cis α-Carotene	9.71	330.0, 421.0, 442.0, 469.0	330.0,421.0, 441.0, 469.0	0.12	0.08	50.00	[27]
28	δ-carotene isomer 2	9.75	345.0, 428.0, 453.0, 483.0	349.0, 430.0, 453.0, 482.0	0.54	0.42	42.85	[20]
29	ζ-carotene isomer 2	9.84	379.0, 401.0, 426.0	380.0, 401.0, 426.0			112.90	[28]
30	9-cis-β-carotene	9.97	346.0, (420.0), 447.0, 473.0	335.0, 421.0, 447.0, 472.0	0.10	0.09	31.80	[20,29]
31	Cis γ-carotene 1	10.18	348.0, 432.0, 454.0, 481.0		0.44		15.38	
32	δ-carotene isomer 3	10.27	426.0, 451.0, 481.0				71.42	
33	15-cis-neurosporene	10.46	330.0, 412.0, 435.0, 463.0	464.0	0.47	0.48	59.09	[30]
34	δ-carotene isomer 4	10.58	346.0, 425.0, 450.0, 480.0		0.39		64.51	
35	Cis γ-carotene 2	11.00	348.0, 433.0, 455.0, 486.0		0.20		40.00	
36	13-cis-neurosporene	11.20	330.0, 412.0, 434.0, 462.0	461.0	0.18	0.11	79.16	[30]
37	di-cis-lycopene 1	11.20	434.0, 457.0, 488.0	350.0, 458.0			57.0	[19]
38	di-cis-lycopene 2	12.10	460.0, 488.0	350.0, 464.0, 488.0	0.25		12.5	[19]
39	δ-carotene-isomer 5	12.14	428.0, 453.0, 484.0				73.07	
40	15-cis-lycopene	12.51	361.0, 440.0, 467.0, 495.0	360.0, 437.0, 466.0, 494.0	0.57	0.75	18.75	[19]
41	13-cis-lycopene	12.83	361.0, 442.0, 465.0, 495.0	360.0, 437.0, 463.0, 494.0	0.58	0.55	35.71	[31]

Table 2 Tentative identification of photoisomerised isomers after illumination of all-trans standards *(Continued)*

42	9-cis-lycopene	13.64	440.0, 467.0, 497.0	360.0, 438.0, 464.0, 494.0	0.27	0.13	66.66	[31]
43	di-cis-lycopene 3	14.19	446.0, 472.0, 503.0	446.0, 472.0, 503.0	0.15	0.08	70.58	[31]
44	5-cis-lycopene	15.08	446.0, 472.0, 503.0	362.0, 442.0, 470.0, 502.0		0.06	74.28	[31]

Parentheses indicate a shoulder.
Q-ratio is the height ratio of the cis-peak to the main absorption peak.
%III/II represents the ratio of peak heights from the trough between peak II and III.

Method validation

As mentioned in the above section, carotenoid standard curves were prepared for subsequent quantitation by HPLC–PDA. The amounts of carotenoids were calculated from the regression equations presented in Table 1. The limit of detection (LOD) and limit of quantification (LOQ) were calculated for each standard (Table 1). The limit of detection (LOD) was defined as the amount that resulted in a peak with a height three times that of the baseline noise respectively and the limit of quantification (LOQ) was determined as lowest injected amount which could be quantifiable reproducibly (RSD ≤ 5%). The precision was evaluated by the relative CV (%CV) which ranges from 4.43-10.67 (Table 1). The intra-day relative standard deviations (R.S.D.) were 0.008–0.02% for retention times of individual carotenoid and 0.54–2.13% for standard concentrations, whereas the inter-day R.S.D. were 0.04–0.08% for retention times and 1.13–3.97% for standard concentrations, demonstrating that a high reproducibility was achieved by using this method.

The accuracy of the extraction method was assessed by determining recovery of all-trans violaxanthin, neoxanthin, antheraxanthin, lutein, zeaxanthin, β-cryptoxanthin, α-carotene, β-carotene, ζ-carotene, δ-carotene, γ-carotene, neurosporene, lycopene, 15-cis-phytoene and all-trans phytofluene, with a mean value of 82.1, 93.3, 81.0, 86.5, 92.4, 83.2, 98.0, 80.0, 92.1, 82.2, 88.7, 98.0, 93.6, 94.4 and 94.6% being attained, respectively.

Application of the method to various plant tissues

To check the versatility of the method, carotenoid content was estimated from leaf and red ripe fruit tissue of field grown Indian tomato cultivar Arka Vikas. Carotenoids were also estimated from Arabidopsis leaf, and green capsicum fruits purchased from the local market. The extracts were analyzed using the C30 column on HPLC. In tomato leaves, three xanthophylls, lutein (26.82 μg/g FW), violaxanthin (14.27 μg/g FW) and neoxanthin (17.58 μg/g FW) and β-carotene (27.02 μg/g FW) as the principal carotene were present. In tomato leaves, other carotenoids were 9′-cis-α-carotene (1.08 μg/g FW) and 9-cis-β-carotene (4.36 μg/g FW). Tomato fruit tissue contains lutein (3.05 μg/g FW) as xanthophylls and all-trans-lycopene (66.2 μg/g FW) is the major carotenoid present along with β-carotene

(5.10 μg/g FW), carotenoid pathway precursors phytoene (8.46 μg/g FW) and phytofluene (1.23 μg/g FW). Using our chromatographic conditions, several cis isomers of lycopene were separated which cannot be resolved by C18 columns. The isomers of lycopene are of interest with respect to their dietary absorption and health beneficial effects [32]. This was also illustrated by the analysis of extracts of the yellow fruited mutant of tomato (PI114490), which contains negligible amounts of carotenoids [33]. The chromatogram of Arabidopsis leaf and green capsicum was similar to the tomato leaf with respect to carotenoid compounds, however the carotenoid content differed. In Arabidopsis leaf 15-cis-β-carotene (3.68 μg/g FW) was also detected. The chromatogram and carotenoid content are presented in Figure 3 and Table 3. We also checked the carotenoid content of tomato tangerine mutant fruits and the major carotenoid compounds 7,9,7′,9′-tetra-cis-lycopene, 9,9′-di-cis-zeta-carotene and 7,9,9′-cis-neurosporene (data not shown) were separated.

The literature is unanimous in reporting the efficiency of a polymeric C30 column over C18 column in resolving cis-isomers of carotenoids in fruits/vegetables and biological fluids/tissues [10,19,21,34]. However, there are still many studies using a C18 column for determination of major carotenoids like lutein, zeaxanthin, β-cryptoxanthin, β-carotene, γ-carotene and lycopene [35]. Commonly encountered drawbacks in these studies include poor resolution of lutein and zeaxanthin, which are often quantified together, and simultaneous elution of both all-trans- and cis-isomers of lycopene. In the current chromatographic conditions, lutein and zeaxanthin were resolved. Additionally, all trans-lycopene and its 7 cis-isomers were also separated and quantified. Most importantly all xanthophylls present in green photosynthetic tissues including violaxanthin and neoxanthin were distinctly resolved, indicating wide applicability of the developed method. In essence, the present method using C30 column offers a much improved resolution of carotenoids from different plant samples in shorter run times.

The earlier studies using C30 columns reported prolonged run times for adequate resolution of the carotenoids. The main advantage of our method is the improved resolution of carotenoids with significant

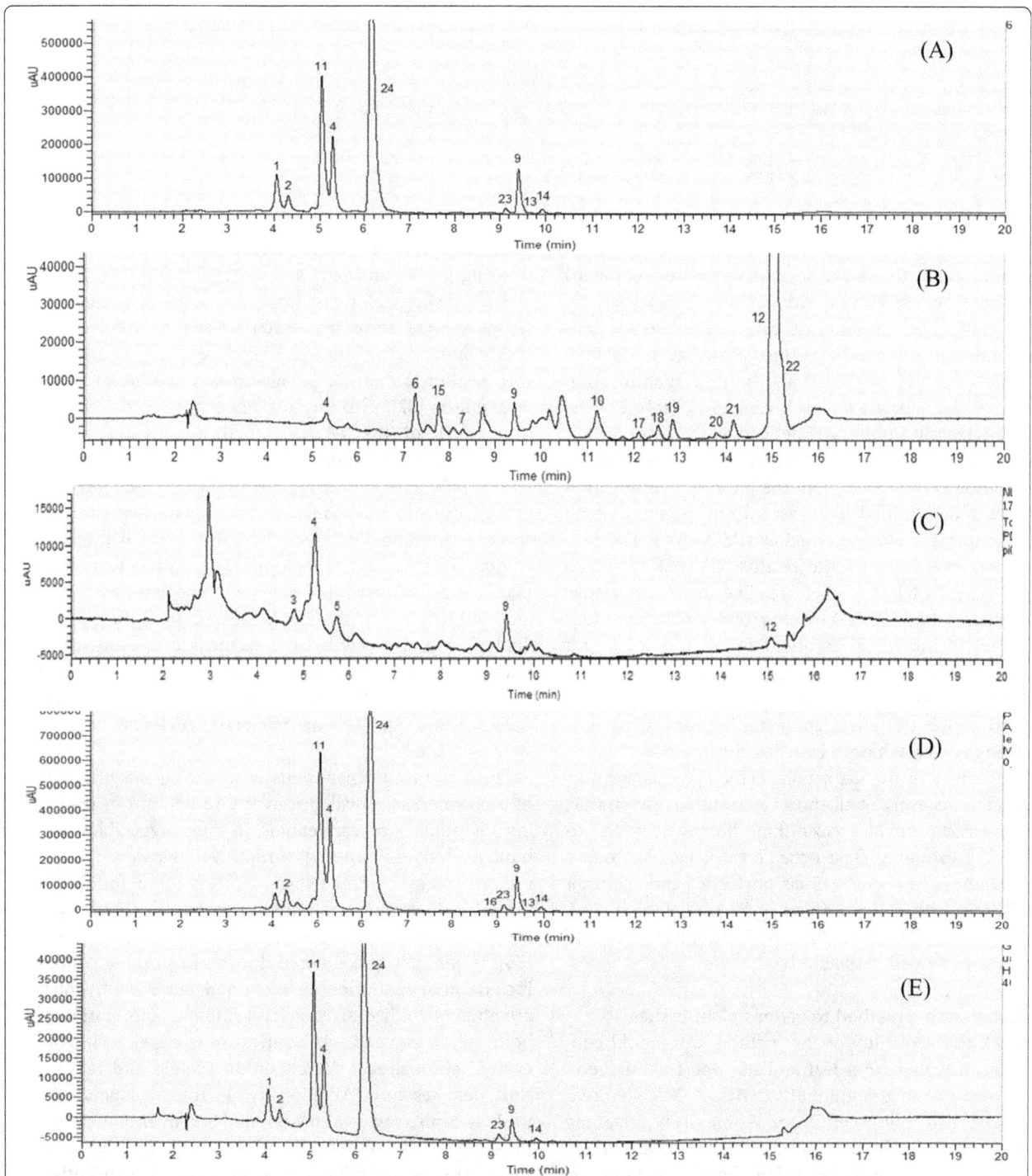

Figure 3 HPLC profiles of the major carotenoids in tomato leaf (A), tomato red ripe fruits (B), yellow fruited ripe tomato (PI114490) (C), **Arabidopsis leaf (D) and green capsicum (E).** Peak identification: (1) trans-violaxanthin, (2) trans-neoxanthin, (3) trans-antheraxanthin, (4) trans-lutein, (5) trans-zeaxanthin, (6) 15-cis-phytoene, (7) 15,9'-cis-phytofluene, (9) all-trans-β-carotene, (10) di-cis-lycopene 1 (11) chlorophyll b, (12) all-trans-lycopene, (13) 9'-cis-α-carotene, (14) 9-cis-β-carotene, (15) all-trans-phytofluene, (16) 15-cis-β-carotene, (17) di-cis-lycopene 2, (18) 15-cis-lycopene, (19) 13-cis-lycopene, (20) 9-cis-lycopene, (21) di-cis-lycopene 3, (22) 5-cis-lycopene, (23) chlorophyll a and (24) pheophytin b. The HPLC profiles were recorded in the absorbance range of 250.00-700.00 nm.

Table 3 Concentration (µg/g FW) of all-trans and cis- carotenoids in tomato cultivar Arka Vikas (AV) leaf and red ripe (RR) fruit, Arabidopsis leaf, tomato mutant line-PI 114490 ripe fruit and green capsicum

S. no.	Compound	AV leaf	AV RR fruit	PI 114490	Arabidopsis leaf	Green capsicum
1	All-trans-violaxanthin	14.27 ± 0.78	-	-	13.48 ± 1.08	2.75 ± 0.24
2	All-trans-neoxanthin	17.58 ± 1.66	-	-	32.77 ± 5.78	2.82 ± 0.22
3	All-trans-antheraxanthin	-	-	0.91 ± 0.08	-	-
4	All-trans-lutein	26.82 ± 3.87	3.05 ± 0.6	1.25 ± 0.17	67.40 ± 7.23	3.90 ± 0.54
5	All-trans-zeaxanthin	-	-	0.28 ± 0.02	-	-
6	15-cis-phytoene	-	8.46 ± 1.15	-	-	-
7	15,9'-cis-phytofluene	-	1.23 ± 0.31	-	-	-
8	All-trans-β-carotene	27.02 ± 3.48	5.10 ± 1.08	0.64 ± 0.2	62.05 ± 9.80	3.60 ± 0.43
9	All-trans-γ-carotene	-	1.47 ± 0.30	-	-	-
10	All-trans-lycopene	-	66.20 ± 4.00	0.37 ± 0.02	-	-
11	9'-cis-α-carotene	1.08 ± 0.28	-	-	1.97 ± 0.63	-
12	9-cis-β-carotene	4.36 ± 0.80	-	-	3.45 ± 0.96	-
13	Phytofluene isomer	-	2.42 ± 0.54	-	-	-
14	15-cis-β-carotene	-	-	-	3.68 ± 0.86	-
15	di-cis-lycopene 1	-	2.39 ± 0.37	-	-	-
16	di-cis-lycopene 2	-	1.47 ± 0.54	-	-	-
17	15-cis-lycopene	-	2.22 ± 0.45	-	-	-
18	13-cis-lycopene	-	1.54 ± 0.60	-	-	-
19	9-cis-lycopene	-	1.67 ± 0.37	-	-	-
20	di-cis-lycopene 3	-	1.76 ± 0.42	-	-	-
21	5-cis-lycopene	-	8.75 ± 1.08	-	-	-

reduction in chromatographic run time. Rajendran et al. [10] using polymeric C30 column coupled with HPLC separated all-trans- plus cis-carotenoids within 51 minutes. However their separation did not include precursors like phytoene and phytofluene. In a recent study, Hsu et al. [36] identified 30 carotenoids separated within 45 min in human serum. Using plant tissues, Fraser et al. [15] reported the separation of 25 carotenoids including their isomers in 42 min using C30 column. Fantini et al. [37] identified 45 carotenoids including isomers in 72 minutes in tomato fruits.

Considering that reverse phase HPLC takes longer analysis times, a recent study [12], compared UHPLC verses HPLC stationary phases for carotenoid separation. They concluded that while UHPLC analysis is more suited for rapid screening of carotenoids, for analysis of complex mixture, HPLC coupled with C30 columns gives better resolution. The advantage of C30 column was offset by the fact that analysis on this column took about four times longer than UHPLC (100 min versus 23 min, respectively). In another study, Rivera et al. [38] analyzed a mixture of 16 carotenoids by UHPLC–MS within 15 min but the method was not validated using biological samples. In recent study, Maurer et al. [13],

separated 11 major carotenoids in 13.5 min, but it did not include δ-carotene and γ-carotene.

Conclusions
HPLC coupled with a C30 column and a gradient used in this study resolved carotenoids rapidly and efficiently at a scale comparable to that reported for UHPLC with C18 column. In this method 15 carotenoids including their precursors were separated in shorter run time of 20 min and this method was validated using tomato and other plant samples. This method can be applied to determine the levels of cis/trans-carotenoids and their precursors phytoene and phytofluene in complex biological sample matrices. The method also efficiently resolved the carotenoids and xanthophylls from green tissues and red ripe fruits therefore is suitable for carotenoid analysis from wide range of plant samples. In future this method can be adapted to UHPLC on availability of C30 UPLC column with enhanced resolutions.

Materials and methods
Standards and solvents
Violaxanthin, neoxanthin, antheraxanthin, lutein, zeaxanthin, phytoene, β-cryptoxanthin, phytofluene, α-carotene,

β-carotene, ζ-carotene, δ-carotene, γ-carotene, neurosporene and lycopene were purchased from CaroteNature (Lupsingen, Switzerland). Methanol was purchased from Fisher Scientific (Waltham, MA, USA). Tert-methyl butyl ether (MTBE), chloroform and dichloromethane were purchased from Avantor Performance Materials (Panoli, Gujrat, India), hexane from Sigma chemical Co. (St. Louis, MO, USA) and ethanol from Hayman Ltd. (Essex, USA). All reagents were HPLC grade or higher. A Millipore Milli-Q water purification system was used to obtain high purity of water.

Standard preparation

Stock solutions of carotenes and xanthophylls were prepared in hexane and ethanol respectively of 0.1 mg/mL. The exact concentration of each stock solution was determined by spectrophotometry using the absorption coefficients A (1%, 1 cm) of the respective carotenoid (Additional file 3: Table S1). After determination of concentration, the standards were evaporated under nitrogen, and solubilized in methanol/MTBE (60/40, v/v) to obtain a final concentration of 5 μg/mL, that was used for HPLC analysis. Individual working solution of each standard was injected in the HPLC system.

Extraction procedure

Freeze-dried plant sample (~150 mg) was homogenized using a mortar and pestle or an IKA A11 basic grinder (IKA, Staufen, Germany) and to the homogenate 1.5 mL of chloroform:dichloromethane (2:1, v/v) was added. The resultant suspension was mixed for 20 min using a thermomixer at 1000 rpm at 4°C. Thereafter for phase separation, 0.5 mL of 1 M sodium chloride solution was added and contents were mixed by inversion. After centrifugation at $5000\,g$ for 10 min the organic phase was collected. The aqueous phase was re-extracted with 0.75 mL of chloroform:dichloromethane (2:1, v/v), centrifuged and again organic phase was collected. Both organic phases were pooled, dried by centrifugal evaporation, re-dissolved in 1 mL of methanol/MTBE (25/75, v/v) for red ripe fruit tissues and re-dissolved in 1 mL and 200 μL of methanol/MTBE (60/40, v/v) for leaf and mature green fruit respectively prior to analysis. A final volume of 20 μL was used for injection into HPLC.

Isomerization of carotenoid standards

For generation of *cis*-isomers of carotenoids, 1 mL solution (1 μg/mL) each of all-trans forms of violaxanthin, neoxanthin, antheraxanthin, lutein, zeaxanthin, β-cryptoxanthin, α-carotene, β-carotene, γ-carotene and neurosporene was subjected to photoisomerization as described by Rajendran et al. [10]. The tubes containing standards were illuminated with three 30 W fluorescent light tubes (Anchor B22-6500 K, Eurolite International,

Kowloon, Hong Kong) for 24 h at 25°C at a distance of 30 cm and light intensity of 2500–3500 lx. Stereomutation of δ-carotene, ζ-carotene, lycopene, phytoene and phytofluene was carried out by heating at 80°C for 60 minutes. Thereafter, the above standards were evaporated to dryness, dissolved in 100 μL MTBE/MeOH (75/25, v/v) and a 20 μL was injected for determination of retention time, absorption spectra and Q-ratio's.

HPLC-PDA analysis of carotenoids

Carotenoids were analyzed by reversed phase HPLC, using Thermo ACCELA U-HPLC (Thermo Fisher Scientific, Bremen, Germany) consisting of quaternary pump, an online degasser, a column oven controller and a photodiode array detector (PDA). Carotenoids were separated on a reverse-phase C30, 3 μm column (250 × 4.6 mm) coupled to a 20 × 4.6 mm C30 guard column (YMC Co., Kyoto, Japan) using mobile phases consisting of (A) methanol/ water (98:2, v/v), (B) methanol/ water (95:5, v/v) and (C) tert-methyl butyl ether. The gradient elution used with this column was 80% A, 20% C at 0 min, followed by linear gradient to 60% A, 40% C to 2.00 min at a flow rate of 1.4 mL/min, at 2.01 minute flow rate was changed to 1.00 mL/min with gradient changing to 60% B, 40% C followed by a linear gradient to 0% B, 100% C by 12 min and return to initial conditions by 13.00 min. A re-equilibration (7.00 min) was carried out at initial concentrations of 80% A, 20% C. The column temperature was maintained at 20°C. The eluting peaks were monitored at a range of 250 to 700 nm using PDA. Quantification was performed using Xcalibur software (version 2.2) comparing peak area with standard reference curves.

Identification and quantification of carotenoids in samples

Peaks were identified by comparing the retention times and UV–Vis spectral data with those of the corresponding standards. In addition, the *cis*-isomers of carotenoids were tentatively identified based on the absorption at near 330 or 360 nm (cis peak), wavelength spectrum and Q ratio's with reference to photoisomerized carotenoid standards and reported values in the literature.

Concentration of each analyte was calculated from the calibration curve of the corresponding standard. All standard solutions were prepared as described above in standard preparation section. Five-point external standard curves (ranging from 10, 25, 50, 75 and 100 ng) were constructed for the standard mix. Carotenoid concentrations were then calculated using a linear regression $y = ax + b$, where y = concentration and x = area of the five-point standard curve. The regression equation and correlation coefficient (R^2) were obtained using Microsoft® Excel 2013.

The cis-isomers of carotenoids were quantified using the standard curves of all-trans carotenoids because of similarity in extinction coefficient [21].

Validation Procedure

The developed method was validated in terms of separation, linearity, recovery and reproducibility. The retention factor (k) calculated by using the formula k = $(t_R - t_0)/t_0$, where t_R and t_0 denote retention time of sample components and sample solvent, respectively. Based on the retention factor of two neighboring peaks (k1 and k2), separation factor (α) was determined by using the formula, α = k2/k1 [39].

For recovery and reproducibility studies, plant sample was spiked with 0.5 μg/mL and 1.25 μg/mL concentration of each standard respectively. The spiked sample was then extracted adopting the method described in section 2.4. After performing HPLC analysis, the recovery of each carotenoid was calculated by R(%) = [(Cs – Cp)/Ca] × 100, where R(%) is percent recovery, Cs is total carotenoid content in the spiked sample, Cp is endogenous carotenoid content in the sample, and Ca is the amount of carotenoid standard added to the sample. The cis isomers of carotenoids were quantified using the recovery of their corresponding all-trans forms. For every sample analyses were performed in triplicates and the mean value was calculated. The reproducibility of this method was ascertained by taking mean of the two spiked concentrations in three replicates on the same day and on three different days. The %CV (percent coefficient of variation) for each carotenoid was calculated by %CV = $(SD/\bar{X}) * 100$ where SD is standard deviation and $\bar{X}$ is the mean. The limit of detection (LOD) and the limit of quantification (LOQ) were calculated as described by International Conference on Harmonization [39].

Additional files

Additional file 1: Figure S1. HPLC profile of carotenoid standards recorded in the range of 250.00-700.00 nm in different injection solvents. (A) Standard mix dissolved in MeOH:MTBE (25:75) and (B) Standard mix dissolved in MeOH:MTBE (60:40)   The compounds are (1) violaxanthin; (2) neoxanthin; (3) anthraxanthin; (4) lutein; (5) zeaxanthin; (6) phytoene; (7) β-cryptoxanthin; (8) phytofluene; (9) α-carotene; (10) β-carotene; (11) ζ-carotene; (12) δ-carotene; (13) γ-carotene; (14) neurosporene; (15) lycopene.

Additional file 2: Figure S2. On-line PDA spectra of all-trans and photoisomerised standards. (1) 9'-cis β-cryptoxanthin , (2) 9'-cis α-carotene, (3) 9'-cis antheraxanthin, (4) 9-cis β-cryptoxanthin, (5) 9-cis lutein, (6) 9-cis zeaxanthin, (7) 9-cis α-carotene, (8) 9-cis antheraxanthin, (9) 9-cis β-carotene, (10) 13-cis α-carotene, (11) 13'-cis α-carotene, (12) 13-cis β-cryptoxanthin, (13) 13'-cis β-cryptoxanthin, (14) 13-cis neurosporene, (15) 13 or 13'-cis lutein, (16) 13-cis zeaxanthin, (17) 15-cis neurosporene, (18) 15-cis β-carotene, (19) 15-cis zeaxanthin, (20) all-trans α-carotene, (21) all-trans antheraxanthin, (22) all-trans β-carotene, (23) all-trans β-cryptoxanthin, (24) all-trans δ-carotene, (25) all-trans γ-carotene, (26) all-trans lutein, (27) all-trans lycopene, (28) all-trans neoxanthin, (29) all-trans neurosporene, (30) all-trans violaxanthin, (31) all-trans zeaxanthin, (32) all-trans ζ-carotene, (33)

cis- γ-carotene 2, (34) cis- γ-carotene 1, (35) cis-neoxanthin, (36) cis-violaxanthin, (37) phytoene isomer, (38) phytoene, (39) phytofluene isomer (40) phytofluene (41) di-cis lycopene, (42) di-cis lycopene, (43) 15-cis lycopene, (44) 13-cis lycopene, (45) 9-cis lycopene, (46) di-cis lycopene and (47) 5-cis lycopene.

Additional file 3: Table S1. Absorption Coefficients of carotenoid standards.

Abbreviations

AV: Arka Vikas cultivar of tomato; HPLC: High-performance liquid chromatography; LOD: Limit of detection; LOQ: The limit of quantification; MTBE: tert-methyl butyl ether; PDA: Photodiode array; RP-HPLC: Reverse phase-HPLC; RSD: Relative standard deviations; UHPLC: Ultra-HPLC.

Competing interests

The authors declare that they have no competing interests.

Authors' contributions

The carotenoids extraction and analysis was done by PG, YS, PG, and RS were involved in planning of experiments and writing of manuscript. All authors read and approved the manuscript.

Acknowledgements

This work was supported by the Department of Biotechnology (grant no. BT/PR11671/PBD/16/828/2008 to R.S. and Y.S.), the Council of Scientific and Industrial Research (research fellowship to P.G.).

References

1. Schwartz SJ, von Elbe J, Giusti M. Colorants. In: Damodaran S, Parkin KL, Fennema OR, editors. *Fennema's Food Chem.* 4th ed. Boca Raton, FL: CRC Press/Taylor & Francis; 2008. p. 571–638.
2. Dugo P, Herrero M, Giuffrida D, Ragonese C, Dugo G, Mondello L. Analysis of native carotenoid composition in orange juice using C30 columns in tandem. J Sep Sci. 2008;31:2151–60.
3. Matthews PD, Luo R, Wurtzel ET. Maize phytene desaturase and zeta-carotene desaturase catalyse a poly-Z desaturation pathway: implications for genetic engineering of carotenoid content among cereal crops. J Exp Bot. 2003;54:2215–30.
4. Su Q, Rowley KG, Balazs NDH. Carotenoids: separation methods applicable to biological samples. J Chromatogr B. 2002;781:393–418.
5. Kohler BE. Electronic structure of carotenoids. In: Britton G, Liaaen-Jensen S, Pfander H, editors. Carotenoids, Volume 1B: Spectroscopy. Basel, Switzerland: Birkhauser Verlag; 1995.
6. Rodriguez-Amaya DB. Nature and distribution of carotenoids in foods. In: Charalambous G, editor. Shelf life studies of foods and beverages: Chemical, biological, physical and nutritional aspects. Amsterdam: Elsevier Science Publishers; 1993. p. 547.
7. Sander LC, Sharpless KE, Pursch M. C30 stationary phases for the analysis of food by liquid chromatography. J of Chromatogr A. 2000;880:189–202.
8. Khachik F, Spangler CJ, Smith JC, Canfield LM, Steck A, Pfander H. Identification, quantification, and relative concentrations of carotenoids and their metabolites in human milk and serum. Anal Chem. 1997;69:1873–81.
9. Sander LC, Sharpless KE, Craft NE, Wise SA. Development of engineered stationary phases for the separation of carotenoid isomers. Anal Chem. 1994;666:1667–74.
10. Rajendran V, Pu YS, Chen BH. An improved HPLC method for determination of carotenoids in human serum. J Chromatogr B. 2005;824:99–106.
11. Li H, Deng Z, Wu T, Liu R, Loewen S, Tsao R. Microwave-assisted extraction of phenolics with maximal antioxidant activities in tomatoes. Food Chem. 2012;130:928–36.
12. Bijttebier S, D'Hondt E, Noten B, Hermans N, Apers S, Voorspoels S. Ultra high performance liquid chromatography versus high performance liquid chromatography: stationary phase selectivity for generic carotenoid screening. J of Chromatogr A. 2014;1332:46–56.

13. Maurer MM, Mein JR, Chaudhuri SK, Constant HL. An improved UHPLC-UV method for separation and quantification of carotenoids in vegetable crops. Food Chem. 2014;165:475–82.

14. Nelis HJCF, de Leenheer AP. Isocratic nonaqueous reversed-phase liquid chromatography of carotenoids. Anal Chem. 1983;55:270–5.

15. Fraser PD, Pinto MES, Holloway DE, Bramley PM. Application of high-performance liquid chromatography with photodiode array detection to the metabolic profiling of plant isoprenoids. Plant J. 2000;24:551–8.

16. Craft NE. Carotenoid reversed-phase high-performance liquid chromatography methods: reference compendium. Methods Enzymol. 1992;213:185–205.

17. Bell CM, Sander LC, Wise SA. Temperature dependence of carotenoids on C_{18}, C_{30} and C_{34} bonded stationary phases. J of Chromatogr A. 1997;757:29–39.

18. Bohm V. Use of column temperature to optimize carotenoid isomer separation by C_{30} high performance liquid chromatography. J Sep Sci. 2001;24:955–9.

19. Lee MT, Chen BH. Separation of lycopene and its cis Isomers by liquid chromatography. Chromatographia. 2001;54:613–7.

20. de Faria AF, de Rosso VV, Merchandante AZ. Carotenoid composition of jackfruit (Artocarpus heterophyllus), determined by HPLC-PDA-MS/MS. Plant Food Hum Nutr. 2009;64:108–15.

21. Lin CH, Chen BH. Determination of carotenoids in tomato juice by liquid chromatography. J Chromatogr A. 2003;1012:103–9.

22. Vera De Rosso V, Mercadante AZ. Identification and Quantification of Carotenoids, By HPLC-PDA-MS/MS, from Amazonian Fruits. J Agric Food Chem. 2007;55:5062–72.

23. Inbaraj BS, Lu H, Hung CF, Wu WB, Lin CL, Chen BH. Determination of carotenoids and their esters in fruits of Lycium barbarum Linnaeus by HPLC-DAD-APCI-MS. J Pharm Biomed Anal. 2008;47:812–8.

24. Melendez-Martinez AJ, Britton G, Vicario IM, Heredia FJ. Identification of isolutein (lutein epoxide) as cis-antheraxanthin in orange juice. J Agric Food Chem. 2005;53:9369–973.

25. de Rosso VV, Mercadante AZ. HPLC–PDA–MS/MS of Anthocyanins and Carotenoids from Dovyalis and Tamarillo Fruits. J Agric Food Chem. 2007;55:9135–41.

26. Rodriguez-Amaya DB. A Guide to Carotenoid Analysis in Foods, International Life Sciences Institute, OMNI (Project). Washington: ILSI Press; 2001.

27. Emenhiser C, Englertb G, Sander LC, Ludwig B, Schwartz SJ. Isolation and structural elucidation of the predominant geometrical isomers of alpha-carotene. J Chromatogr A. 1996;719:333–43.

28. Breitenbach J. Sandmann G: zeta-Carotene cis isomers as products and substrates in the plant poly-cis carotenoid biosynthetic pathway to lycopene. Planta. 2005;220:785–93.

29. Nyambaka H, Ryley J. An isocratic reversed-phase HPLC separation of the stereoisomers of the provitamin A carotenoids (α- and β-carotene) in dark green vegetables. Food Chem. 1996;55:63–72.

30. Katayama N, Hashimoto H, Koyama Y, Shimamura T. High-performance liquid chromatography of cis trans isomers of neurosporene: discrimination of cis and trans configurations at the end of an open conjugated chain. J Chromatogr. 1990;519:221–7.

31. Schierle J, Bretzel W, Buhler I, Faccin N, Hess D, Steiner K, et al. Content and someric ratio of lycopene in food and human blood plasma. Food Chem. 1997;59:459–65.

32. Holloway DE, Yang M, Paganga G, Rice-Evans CA, Bramley PM. Isomerization of dietary lycopene during assimilation and transport in plasma. Free Radical Res. 2000;32:93–102.

33. Kang B, Gu Q, Tian P, Xiao L, Cao H, Yang W. A chimeric transcript containing Psy1 and a potential mRNA is associated with yellow flesh color in tomato accession PI 114490. Planta. 2014;240:1011–21.

34. Inbaraj BS, Chien JT, Chen BH. Improved high performance liquid chromatographic method for determination of carotenoids in the microalga Chlorella pyrenoidosa. J of Chromatogr A. 2006;1102:193–9.

35. Johnson E. The Role of Carotenoids in Human Health. J Nutri Clinic Care. 2002;5:56–65.

36. Hsu BY, Pu YS, Inbaraj BS, Chen BH. An improved high performance liquid chromatography-diode array detection-mass spectrometry method for determination of carotenoids and their precursors phytoene and phytofluene in human serum. J Chromatogr B. 2012;899:36–45.

37. Fantini E, Falcone G, Frusciante S, Giliberto L, Giuliano G. Dissection of tomato lycopene biosynthesis through virus-induced gene silencing. Plant Physiol. 2013;163:986–98.

38. Rivera S, Vilaró F, Canela R. Determination of carotenoids by liquid chromatography/mass spectrometry: effect of several dopants. Anal Bioanal Chem. 2011;400:1339–46.

39. International Conference on Harmonization (ICH). Guideline on the Validation of Analytical Procedures: Methodology Q2B, Geneva;1996. http://www.fda.gov/downloads/drugs/guidancecomplianceregulatoryinformation/guidances/ucm073384.pdf

18

Monitoring leaf water content with THz and sub-THz waves

Ralf Gente[*] and Martin Koch

Abstract

Terahertz technology is still an evolving research field that attracts scientists with very different backgrounds working on a wide range of subjects. In the past two decades, it has been demonstrated that terahertz technology can provide a non-invasive tool for measuring and monitoring the water content of leaves and plants. In this paper we intend to review the different possibilities to perform in-vivo water status measurements on plants with the help of THz and sub-THz waves. The common basis of the different methods is the strong absorption of THz and sub-THz waves by liquid water. In contrast to simpler, yet destructive, methods THz and sub-THz waves allow for the continuous monitoring of plant water status over several days on the same sample. The technologies, which we take into focus, are THz time domain spectroscopy, THz continuous wave setups, THz quasi time domain spectroscopy and sub-THz continuous wave setups. These methods differ with respect to the generation and detection schemes, the covered frequency range, the processing and evaluation of the experimental data, and the mechanical handling of the measurements. Consequently, we explain which method fits best in which situation. Finally, we discuss recent and future technological developments towards more compact and budget-priced measurement systems for use in the field.

Keywords: THz, Sub-THz, Water-status, Water-content, Drought-stress

Introduction

In comparison to other methods like measuring the water potential of the leaves or comparing their fresh and dry weight THz and sub-THz measurements have the advantage of being a non-invasive technique. This means that repeated measurements on the same sample over a long period of time are possible. With invasive techniques such measurements are problematic, because besides the obvious wastage of sample material with each measurement the extraction of tissue from a living organism always causes additional stress. This can obviously affect the result of the experiment.

Electromagnetic radiation in the THz and sub-THz frequency range is strongly absorbed by liquid water [1,2]. Various approaches for biological and medical applications of THz waves exist [3-10]. But the strong attenuation by water often tends to be a problem as samples in this field usually have a rather high water content [11,12]. As a result, this often makes these samples completely opaque for THz and sub-THz waves. Yet, for measuring the water content of a thin sample like a plant's leaf, the strong absorption turns out to be very convenient [13-21]. Also radiation in the neighboring microwave and infrared range has been used as a tool for water status measurements [22,23]. Often this is done via remote sensing taking several plants at once under observation [24], sometimes even with airborne or spaceborne sensors [25-28]. In contrast to this, the techniques in the sub-THz range (i.e. the upper microwave range) and the THz range, which we will take into focus here, are designed to be used locally on individual plants. Here, we review these new approaches and hope that they will get accepted and widely used by plant physiologists to monitor the water status of plants.

While the dry tissue of the leaf has little influence on the transmitted signal, the attenuation of the signal can be used directly for a qualitative observation of the leaf's water content. The high contrast between dry biomass and liquid water is caused by the polarity of the water molecules, which results in a high absorption coefficient in the THz frequency range. The capability of this approach for water status measurements was firstly demonstrated by the pioneering work of Hu et al. [13] and Mittleman et al. [14]. In their experiments Hu et al.

*Correspondence: ralf.gente@physik.uni-marburg.de
Faculty of Physics and Material Sciences Center, Philipps-Universität Marburg, Renthof 5, 35032 Marburg, Germany

recorded an image of a freshly cut leaf using THz time domain spectroscopy. In this image the veins of the leaf are clearly visible due to their higher water content and their higher thickness. After two days, the measurements were repeated and an increase of the overall transmission through the leaf was observed showing a decrease of the leaf water content. By performing similar measurements on a leaf of a living plant Mittleman et al. visualized the water uptake of the plant, which was previously subjected to drought stress, after rewatering. In such measurements a sample holder may be needed to keep a leaf in a defined position. Yet, the actual measurement is contact-free, which helps to keep the mechanical stress on the sample to a minimum [3,16,19]. THz and sub THz measurement systems, which are specially adapted for water status detection, are still subject to active research and development [17-19] and not commercially available so far. Yet, several different technical realizations of the underlying idea to use THz or sub-THz waves for water status measurements have been implemented and evaluated. Among these are THz time domain spectroscopy [29-33], THz continuous wave setups [34-38], and sub-THz continuous wave setups [39]. These techniques have different advantages and drawbacks, which make them suitable in different experimental situations. In the following, we will discuss the capabilities of the different approaches and the typical experimental configurations in which they can be used. For each of the different approaches we present experimental data, which demonstrates the capabilities of the technology and might serve as an inspiration for further experiments.

Review

Terahertz time domain spectroscopy

A typical THz time domain spectrometer consists of several components which serve to generate and detect a short electromagnetic pulse [29-33] and record its time trace. As this pulse typically consists of frequency components from a few hundred GHz to several THz, they are located in the electromagnetic spectrum between microwaves and infrared light. As shown in Figure 1 a central component of such a setup is a laser, which emits short pulses of light with a pulse duration of about 100 fs. A common technique for emitting and detecting THz pulses are photoconductive antennas [40-43]. The light pulses from the laser are used to excite both the emitter and the detector antenna by generating free carriers in the substrate material. At the emitter antenna a bias voltage is applied to accelerate the free carriers. This mechanism generates one terahertz pulse for each incoming light pulse. At the detector, the free carriers, which are generated by the light pulses, allow the terahertz pulses to induce a photocurrent. This photocurrent is measured using a lock-in amplifier or a transimpedance amplifier. In the optical path, which guides the light from the laser to the detector antenna, a delay unit is used to manipulate the time of arrival of the optical light pulses. By using this delay unit it is possible to scan across the THz pulse and record its shape in the time domain. For each measurement with a sample, a reference measurement is performed without the sample to record the characteristics of the measurement setup. Figure 2 shows an example for a THz time domain pulse trace and its representation in the frequency domain. The properties of the sample can be calculated by comparing the results of the sample measurement to the reference measurement. The figure shows how the THz pulse is attenuated and retarded by the sample. Frequency dependent data evaluation is possible by applying a Fourier transform to the time domain data.

A stationary laboratory setup is usually built upon an optical table, which holds the optical components in place. A more flexible and compact alternative are fiber-coupled measurement systems where glass fibers are used to guide the light from the laser to the photoconductive antennas. Moreover, combinations of both techniques are possible. In the following sections we will present different measurement systems, which can be used for different purposes and in different locations.

Automated long-term experiments

The THz time domain setup shown in Figure 3 is designed for long-term measurement series on a number of plants

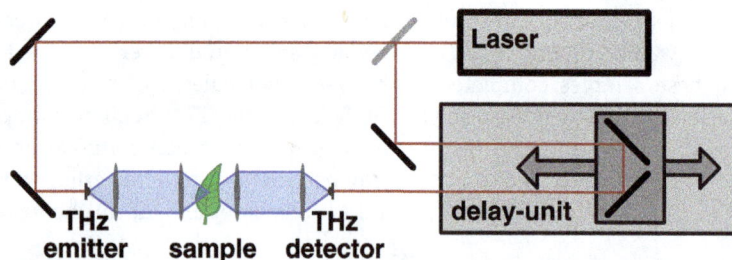

Figure 1 Schematic of a typical THz time domain setup. A laser emits short pulses of light, which are used for generation and detection of THz pulses. The delay unit enables scanning across the THz waveform (see also [29-33]).

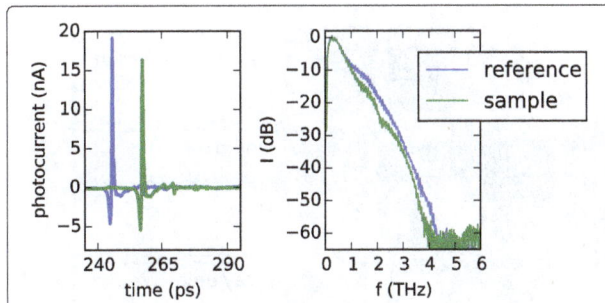

Figure 2 Example for a THz time domain measurement. The comparison of reference and sample measurement shows that the signal is attenuated and delayed by the sample. The data was recorded using a laboratory free space setup. To magnify the effect of the sample on the THz pulse, a 7mm thick block of polypropylene was used as a sample in this measurement (see also [29-33]).

over a course of several days or even weeks [16]. Once an experiment is started, all steps, which are necessary to carry out the measurements, are controlled by a computer. This way, measurement data can be taken continuously without any need for manual intervention.

The delay unit, which is used here, is built on an optical table as a conventional free space delay line. The remaining optical path to the antennas is fiber-coupled allowing to move the THz antennas and THz optics around on the table. These components are mounted on a measurement head, which can be moved around on a circular route by the motorized arm of a goniometer. Corresponding to that, the plants are arranged on the table in a circular

shape and a leaf from each plant is kept in a fixed position by a sample holder. The sample holders are adjusted to be exactly on the circular path of the measurement head. Thus, each leaf can be reached by the measurement head and a fully automated operation is possible. When 15 plants are placed in the setup one roundtrip of the measurement head takes about one hour. By using faster mechanics and data acquisition [44] the speed of the measurements could be increased at least by a factor of 20. Additionally, each pot sits on a computer controlled digital scale to keep track of the pot's weight as a measure of the amount of water, which is available to the plants. Above the table a high pressure sodium lamp is mounted as a light source for the plants, which is controlled by a time switch.

Figure 4 shows an example for long-term measurement data, which was recorded using this setup. Over a course of several weeks the water status of rye (*Secale cereale*) plants was observed while they were put under drought stress and finally rewatered. Besides the terahertz transmission the weight of the pots was recorded, too. Measurements were performed approximately once per hour on each of the 15 plants. The plot in Figure 4 shows the results from one of these plants. In the first days of the experiment, the water available to the plants was kept on a constant level. The plants were irrigated daily, which reflects in the sawtooth-like shape of the weight plot in these days. After 6 days, the plants were deprived from water. Comparison of the plot for the terahertz transmission and pot weight shows that it takes several days until the available amount of water is low enough to induce drought stress. The two small peaks in the THz transmission around the 7th and the 13th day of the experiment cannot be attributed to any particular event. The drought

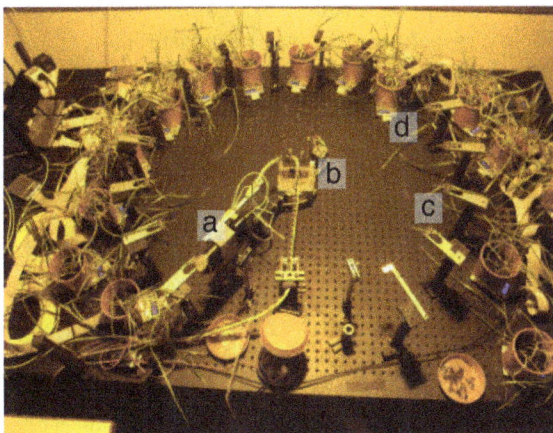

Figure 3 Photograph of a setup for automated long-term experiments. The plants (in this potograph: oat, lat. *Avena*) are placed on the table in a circular arrangement similar to the one used by Born et al. [16] to make them accessible to the measurement head **(a)** on the motorized arm of the goniometer **(b)**. One leaf from each plant is placed in a sample holder **(c)**. Below each pot a digital scale **(d)** is placed to keep track of the weight of the pots.

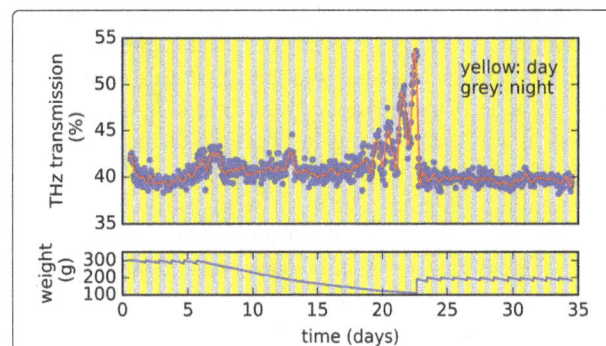

Figure 4 Result of a measurement series with the automated setup. The THz transmission through the leaf of a rye plant is plotted together with the weight of the pot. In agreement with Born et al. [16] higher transmission values stand for smaller water content in the leaf. Comparison of the two graphs shows how drought stress builds up while the amount of water available to the plant is decreased. Also the immediate reaction of the plant to rewatering is visible.

stress response starts to become visible on the 19th day. From the 20th day on the drought stress response during daytime is larger than twice the standard deviation of the transmission values before deprivation started ($\sigma = 0.87\%$). This applies to the difference between day and nighttime, too. We attribute the higher transmission values during daytime to the higher usage of water by the plant during this time. Additionally the opening and closing of the leaves' stomata might cause a slightly different scattering behavior of the radiation on the leaves' surface. But this would imply a frequency dependent effect, which has not been observed so far. The usage of water by the plant is constituted by the amount of water which is used for photosynthesis, and the amount which is lost due to physical drying by the incident light. During nighttime the water uptake by the plant from the soil can at least partly compensate for the water loss during the day.

Also the end of the drought stress on the 23rd day is clearly visible. The plot shows that the plant recovers immediately after rewatering and THz transmission comes back to its initial level.

An automated measurement setup like the one described above allows for a variety of experiments, where the development of leaf water content over time is under observation. Possible experiments for the future are the comparison of the behavior of different species of plants under drought stress and the comparison of leaves at different locations within one plant.

Mobile measurement systems for hand operation

Another approach for performing measurements on plants is to bring the measurement system to the plant rather than the plant to the measurement system. To make this possible, the measurement system needs to be a compact, self-contained unit. Figure 5 shows a fiber-coupled THz time domain system, which was designed to be used in a greenhouse. To fit the components of the spectrometer into one 19" rack case a solution for the delay unit had to be found. One possibility is to put a free space delay line in a sealed housing in order to address laser safety regulations and to guard it from dust and other environmental influences. But it is also possible to completely avoid free space optics by using a fiber-stretcher, which periodically stretches and releases several meters of optical fiber and thus generates the optical delay, which is needed for scanning over the THz time domain signal. While a free space delay line can be realized more cost-efficiently a fiber-stretcher allows for higher measurement speed. In either case, the bigger components of the spectrometer, like the laser and the delay unit, are located in the 19" rack case and the THz emission and detection take place in a handheld measurement head, which can be moved to the plant.

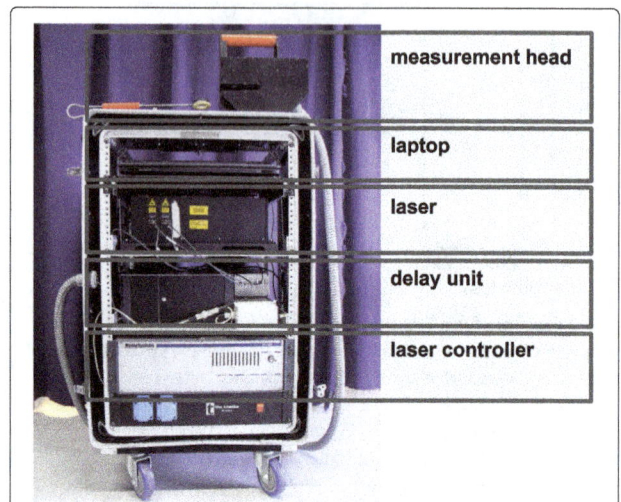

Figure 5 Mobile THz time domain setup for use in a greenhouse. All the components of the spectrometer are integrated in a 19" rack case which can be moved around on wheels [45,46].

This kind of setup allows for a flexible design of biological experiments as plants of different sizes in different locations can be easily reached as long as there is a way to bring the rack case with the spectrometer into a range of about 2 m from the plants. Though, the downside of this concept is that measurements need to be carried out by hand, which can be a time-consuming task depending on the number of plants in the experiment.

Figure 6 shows a measurement series, which was recorded to compare THz measurements using a mobile THz time domain system and conventional gravimetric measurements. These measurements were performed on ten leaves, which were detached from a barley (*Hordeum vulgare*) plant to make the gravimetric measurements possible. The leaves were dried in an oven. During the drying process they were taken out of the oven every ten minutes and THz measurements and gravimetric measurements were performed. As the THz measurements are non-invasive they are feasible in the same way also on leaves which are still alive and attached to a plant. How the water content is calculated from the THz data is explained in the next section.

Modeling a leaf as an effective medium

In Figure 4 we have shown that raw transmission values of the THz signal can already act as an indicator for the water status of a plant. But sometimes it's desirable to know the actual percentage of water in a leaf. Obtaining this value with gravimetric measurements is trivial, but also a destructive method. With sophisticated data analysis it is possible to calculate these values from non-destructive THz time domain measurements. One technique, which aims at this goal, is to use linear transforms like principal

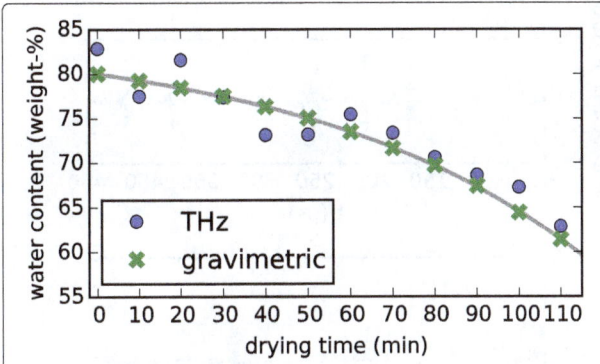

Figure 6 Comparison of THz measurements and gravimetric measurements of the water content of barley leaves [49]. For this experiment 10 leaves were detached from a barley plant and their water content was measured repeatedly. Detaching the leaves was only necessary for the gravimetric measurements, which were performed for comparison. The pearson correlation coefficient of the THz measurements and the gravimetric measurements is $r = 0.94$. A possible reason for the remaining deviations between the two methods is that a gravimetric measurement gives an average value for the whole leaf while a THz measurement is performed on a small spot. The THz measurements are non-invasive and can also be performed on living plants. The water content was calculated from the THz data using an effective medium model of the leaves [49]. The grey line is a 'guide to the eye'.

component analysis on the measured data [47]. Here, we will have a closer look at another concept, which is based on a physical model describing the transmission of a THz pulse through a leaf [48]. In this model a leaf consists of a mixture of water, dry biomass and air. Using an effective medium theory the dielectric properties of such a mixture can be calculated incorporating the properties of the components and their volumetric fractions. For building such a model the dielectric properties of the components need to be characterized separately. While the values for water and air can be taken from literature [1], for the dry biomass measurements in a laboratory setup are necessary. The absorption coefficient of dry biomass is small compared to the absorption coefficient of water, and thus the effect of the dry biomass on the results of the measurements is small, too. Still, for a reliable model the dry biomass should be characterized separately for each plant species. Jördens et al. [48] use the effective medium theory of Landau, Lifshitz and Looyenga for calculating the permittivity ϵ_L of the leaf material as follows:

$$\sqrt[3]{\epsilon_L} = a_W \sqrt[3]{\epsilon_W} + a_S \sqrt[3]{\epsilon_S} + a_A \sqrt[3]{\epsilon_A} \ , \quad a_W + a_S + a_A = 1$$

In this central formula of the model, the subindices W, S, and A stand for the three components water, solid matter, and air. One important property of this theory is that it does not make any assumptions about the inner structure of the mixture. When simulating the transmission of THz radiation through a leaf, also its thickness and surface

roughness need to be taken into account. For modeling the surface roughness Jördens et al. use a Raleigh roughness factor, which is based on the standard deviation of the height profile of the surface [48]:

$$\alpha = \alpha_{abs} + \left(\left(\sqrt{\epsilon_L} - 1 \right) \cdot \frac{4\pi \tau \cos(\theta)}{\lambda} \right)^2 \times \frac{1}{T}$$

In this formulation α is a combined expression for the attenuation of the signal by the leaf, which is caused by surface scattering and absorption. τ is the the standard deviation of the leaf's height profile, θ the angle of incidence ($\theta = 0$ for normal incidence), λ the free space wavelength, and T the thickness of the leaf. Based on this model for the sample material the so called transfer function of the sample is calculated. The transfer function is a frequency dependent representation of how electromagnetic radiation is delayed and attenuated when it is transmitted through the sample. The next step after building such a model is to reverse the problem and extract the model's parameters from real measured data. This can be done using an optimization algorithm, which fits the model to the measured transfer function [49]. The data in Figure 6, which has been mentioned before, was evaluated using this method.

Continuous wave THz setups

Instead of short pulses continuous THz radiation can also be used for water status measurements [17,18]. Continuous wave THz setups often use photomixing for generation and detection of the THz waves [38]. The concept is based on overlaying the light from two lasers, which are slightly detuned against each other. When this optical signal is focused onto a photoconductive antenna THz radiation at the difference frequency of the two lasers is generated. Similar to a THz time domain setup a part of the laser light is guided through a delay unit and onto the detector antenna. Similar to the former mentioned methods, the amplitude of the THz signal, which is transmitted through a leaf, is an indicator for the water status of the plant. An example for a measurement series, which was recorded with such a setup is shown in Figure 7. In the experiment shown in the figure, Kinder et al. [18] deprived a coffee plant (*Coffea arabica*) from water and performed measurements with a terahertz continuous wave setup over a course of 20 days. With evolving drought stress the transmission of the THz signal through the leaf increases accordingly.

Lasers for continuous wave generation are usually more compact and cost-efficient than a femtosecond laser. Though, as a usual continuous wave THz setup works at one single frequency, it is not possible to apply the effective medium method here. One possibility to overcome this limitation is to repeat the measurements with different detuning between the two lasers to obtain data for

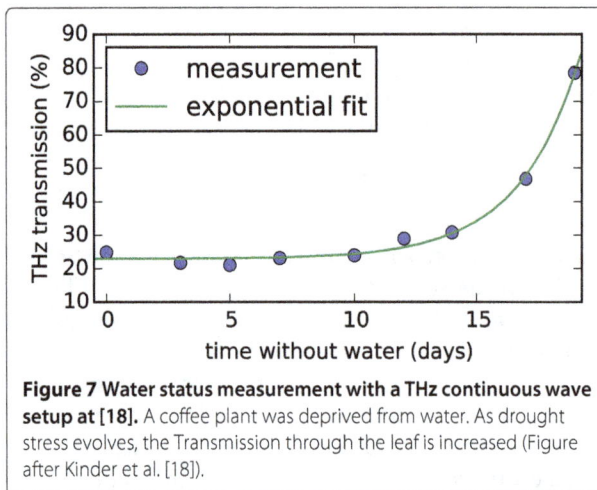

Figure 7 Water status measurement with a THz continuous wave setup at [18]. A coffee plant was deprived from water. As drought stress evolves, the Transmission through the leaf is increased (Figure after Kinder et al. [18]).

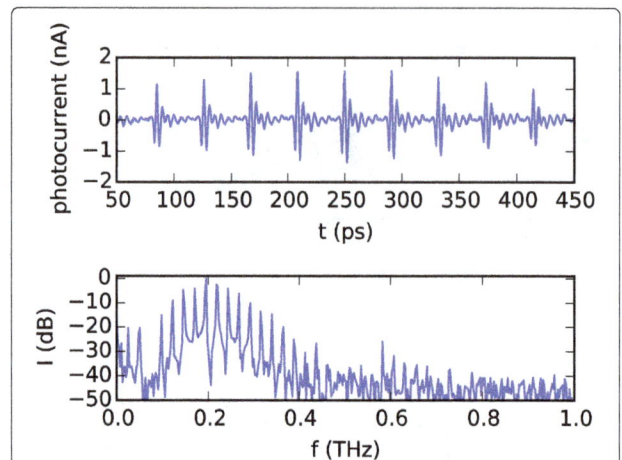

Figure 8 Signal from a THz quasi time domain setup. The upper graph shows the signal as it was recorded in the time domain. The Fourier transform of the time domain signal in lower graph reveals that the signal is constructed as the sum of many discrete frequency components [51,52].

different frequencies. A continuous wave setup can also be operated at several frequencies at once, either by overlaying more than two lasers or by using multimode laser diodes, which emit light at several wavelengths at once [50].

THz quasi time domain spectroscopy (QTDS)

A comparably new concept for generating THz radiation is quasi time domain spectroscopy. The setup is almost identical to a THz time domain spectrometer, but the femtosecond laser is replaced by an inexpensive multimode laser diode [51,52]. As such a laser diode generates light at many different wavelengths at once, there is a large number of difference frequencies, which are emitted by the photoconductive emitter antenna. In contrast to a continuous wave setup, which can also be operated with two multimode laser diodes, here the difference frequencies between the different modes of only one multimode laser diode are used. This results in a signal which looks like a train of THz pulses. These quasi pulses are where the name quasi time domain spectroscopy comes from. As shown in Figure 8 the difference frequencies show up in the spectrum of the recorded time domain signal when it is transformed into the frequency domain. THz QTDS setups for water status measurements are still under development, but first measurements, which are shown in Figure 9, show the feasibility of this approach. In these measurements a leaf of corn salad (*Valerianella locusta*) was periodically put in a QTDS setup for measuring the transmission of the THz signal through the leaf. Each QTDS measurement was accompanied by weighing the sample, so the actual loss of water was known, while the leaf was slowly drying. The plot shows a good correspondence of water loss and increase of transmission.

Replacing the expensive femtosecond laser with a cheap multimode laser diode enables QTDS setups to be built much more compact and cost-effectively. Using the QTDS

technology, compact, lightweight, battery-powered measurement systems for the use in the field get within reach. New compact, robust, and inexpensive delay concepts like the one proposed by Probst et al. [53] are an important step in this direction.

sub-THz-measurements

If we move to the upper part of the microwave spectrum, which is also called the sub-THz regime, more powerful emitters are available [39]. The longer wavelengths in the sub-THz regime make it impossible to focus the radiation onto a small spot, but instead the higher power can be

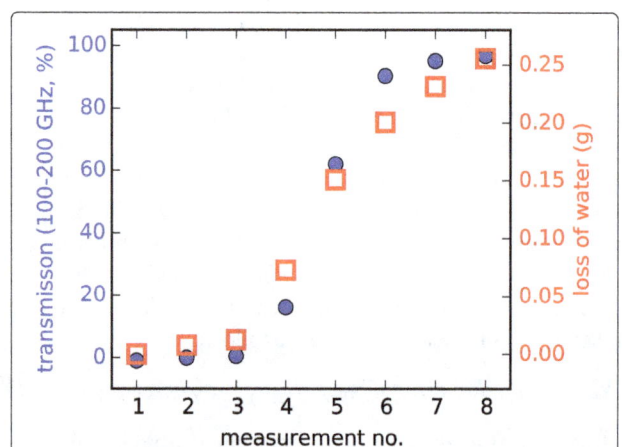

Figure 9 This experiment shows the feasibility of THz quasi time domain spectroscopy for water status measurements. A leaf of corn salad was periodically weighed and its THz transmission was measured using a QTDS setup. The results show a good correspondence of water loss and increase of transmission.

used to perform measurements on a bigger part of a plant. For this bigger part an average value is obtained instead of a measurement on one single leaf. When a plant is illuminated by a sub-THz beam with a diameter of several centimeters, a part of the radiation is transmitted straight through the plant, a part of it is absorbed and another part is scattered away in random directions. The scattered part of the radiation is one of the reasons why such measurements are not trivial. The amount and direction of the scattered radiation are determined by the random orientation of the leaves in the illuminated part of the plant. One possibility to take the scattered part of the radiation into account is to capture it by scanning around the plant with the detector [54]. In a setup like the one shown in Figure 10 the emitter and the plant are kept in a fixed position, while the detector is moved on a circular path around the plant. The biggest part of the radiation is transmitted straight through the plant and can be measured when the detector is directly facing the emitter. But still a significant amount of radiation can be detected at other angles besides the direct forward direction. How much of the radiation is scattered also depends on the water content of the plant. Because of this, the relationship between the directly transmitted signal and the scattered signal is nonlinear and the angular scan cannot be replaced by a simple linear proportionality (see Figure 2b in [54]). After integrating over the recorded data, the result of such a measurement is the sum of transmitted and scattered radiation. These values can directly be used as an indicator for the water status of a plant. When additional information about the size and geometry of the plant is available, it is also possible to calculate the actual water content of a plant in percent. Figure 11 shows the results of a measurement series, where one group of 11 barley plants

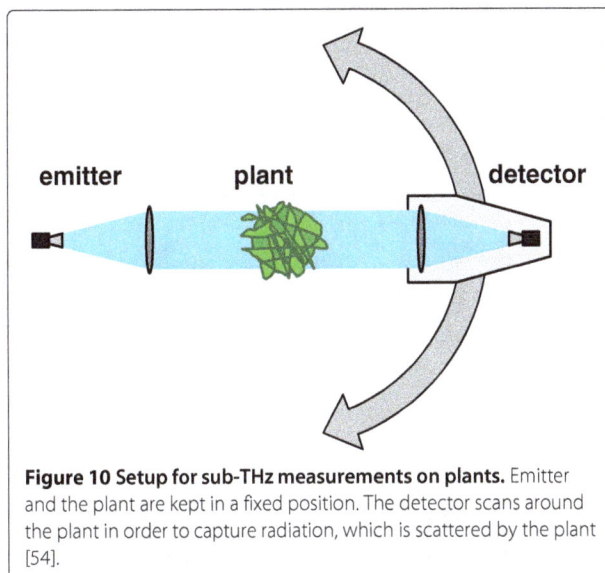

Figure 11 Result of a sub-THz measurement series on barley plants [54]. The red curve represents a group of 11 barley plants under drought stress. The green curve shows the corresponding results for the irrigated control group of the same size.

was sufficiently watered throughout the experiment, while another group of 11 plants was deprived from watering. Over a course of several weeks, sub-THz measurements were carried out regularly. The diverging water content of the two groups of plants is clearly visible in the plot. While the water content of the control group stays basically on a constant level, the water content of the stressed group starts to decrease some days after the last irrigation of the plants. The calculation of the water content was done using an effective medium model similar to the one described above. But as measurement values are available for only one frequency, additional information about the size of the plant is required [54]. Sub-THz setups still need to be developed further. Yet, they are good candidates for integration into automated high throughput phenotyping facilities, because no physical contact with the plant is needed for the measurements. The speed of the measurements is limited by the mechanical movement of the detector around the plant and not by the sub-THz technology. Additionally, in such phenotyping facilities additional sensors like cameras or laser scanners, which can be used to determine the size and geometry of the plants, often already exist. While a qualitative assessment of the development of a plant's water status over time is possible with the raw measurement data the accuracy of the calculated water content strongly depends on the information on the plants size and geometry. Compared to measurements in the THz regime no information on single leaves can be gained. If, for example, a plant gives up only some leaves when drought stress emerges while other leaves are still maintained, this behavior can not be detected using sub-THz measurements. On the other hand, averaging over a bigger part of a plant can also be an advantage, because then the results of the measurements do not depend on the individual variations of only a single or a few leaves, which are picked for the measurements.

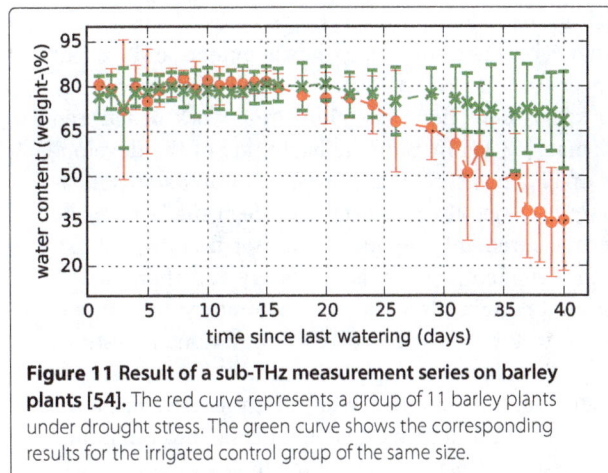

Figure 10 Setup for sub-THz measurements on plants. Emitter and the plant are kept in a fixed position. The detector scans around the plant in order to capture radiation, which is scattered by the plant [54].

Conclusions

We have outlined the early development and recent advances in the use of THz and sub-THz waves for water status detection. The pioneering work of Hu and Mittleman [13,14] marked the beginning of the development of THz measurement systems for water status measurements. Since then, efforts have been made to make these measurement systems more user-friendly, robust, and cost-efficient, which is necessary for them to be useful in a certain place, e.g. in a laboratory, in a greenhouse or in the field. The current measurement systems already allow for meaningful experiments to investigate the water status dynamics of plants. Further developments will have the aim to bring this technology into the hands of biologists and plant-breeders for their everyday work. In this context, especially the QTDS technology [51,52] is a promising candidate for building compact devices at competitive costs. But depending on the intended use, fiber-coupled TDS systems for automated measurements or sub-THz setups for high throughput facilities can also be a good choice. One important aim for the approaches discussed here is to reveal differences e.g. in the drought stress resistance between different genotypes. In general, this appears feasible. Yet, it is needless to say that the outcome of such experiments will depend on how strongly the drought stress tolerance differs between the genotypes. Variations below 5% in any parameter investigated will be hard to detect. With a broader use of these technologies in plant sciences more experience will be gained concerning measurements on different species as well as the detection of more subtle differences within one species.

Competing interests

The authors declare that they have no competing interests.

Authors' contributions

RG performed the measurements with the automated TDS setup and the QTDS setup, carried out the data evaluation with the effective medium model and drafted the manuscript. RG and MK did the literature review. All authors read and approved the final manuscript.

Acknowledgements

The authors wish to thank everyone, who was involved in carrying out the measurements, handling the plants, and taking care of the experiments from the biological point of view: W. Sannemann, M. Noschinski, A. Ballvora, J. Léon, M. A. Hohmann, and R. Snowdon for taking care of the experiments with the mobile THz setups. D. Behringer, S. Liepelt, U. Mattern, S. Beyer, and B. Ziegenhagen for fruitful discussions about the automated THz setup. M. Schwerdtfeger, W. Friedt, D. Selmar, and B. Breitenstein for the measurements with the THz continuous wave setup. A. Rehn for the sub-THz measurements. T. Probst, and M. Gerhard for their experienced help and advice with the QTDS technology. N. Born, M. Schwerdtfeger, M. Stecher, A. Rehn, K. Altmann, N. Voß, M. Vaupel, and S. Busch for countless small and big help and advice. Parts of this work are supported by the Johannes Hübner Foundation Giessen. The authors acknowledge support from the German Federal Ministry of Education and Research (BMBF, FKZ 0315532D), from the Federal Ministry of Food and Agriculture (BLE, FKZ 511-06.01-28-1-45.044-10, FKZ 28-1-45.015-10).

References

1. Liebe HJ, Hufford GA, Manabe T. A model for the complex permittivity of water at frequencies below 1 THz. Int J Infrared Millimeter Waves. 1991;12(7):659–75. doi:10.1007/BF01008897.
2. Rønne C, Åstrand P-O, Keiding SR. THz spectroscopy of liquid H_2O and D_2O. Phys Rev Lett. 1999;82:2888–91. doi:10.1103/PhysRevLett.82.2888.
3. Koch M. THz imaging: fundamentals and biological applications. In: Proc. SPIE 3828, Terahertz Spectroscopy and Applications II, vol. 3828. Munich, Germany: SPIE; 1999. p. 202–8. doi:10.1117/12.361035. http://dx.doi.org/10.1117/12.361035.
4. Whitmire SE, Wolpert D, Markelz AG, Hillebrecht JR, Galan J, Birge RR. Protein flexibility and conformational state: A comparison of collective vibrational modes of wild-type and {D96N} bacteriorhodopsin. Bioph J. 2003;85(2):1269–77. doi:10.1016/S0006-3495(03)74562-7.
5. He Y, Ku PI, Knab JR, Chen JY, Markelz AG. Protein dynamical transition does not require protein structure. Phys Rev Lett. 2008;101:178103. doi:10.1103/PhysRevLett.101.178103.
6. George DK, Knab JR, He Y, Kumauchi M, Birge RR, Hoff WD, et al. Photoactive yellow protein terahertz response: hydration, heating and intermediate states. Terahertz Sci Technol IEEE Trans. 2013;3(3):288–94.
7. Castro-Camus E, Johnston M. Conformational changes of photoactive yellow protein monitored by terahertz spectroscopy. Chem Phys Lett. 2008;455(4):289–92.
8. Siegel PH. Terahertz technology in biology and medicine. Microwave Theory Tech, IEEE Trans. 2004;52(10):2438–47. doi:10.1109/TMTT.2004.835916.
9. Pickwell E, Wallace VP. Biomedical applications of terahertz technology. J Phys D: Appl Phys. 2006;39(17):301.
10. Nagel M, Först M, Kurz H. THz biosensing devices: fundamentals and technology. J Phys: Condens Matter. 2006;18(18):601.
11. Oh SJ, Kim S-H, Jeong K, Park Y, Huh Y-M, Son J-H, et al. Measurement depth enhancement in terahertz imaging of biological tissues. Opt Express. 2013;21(18):21299–305. doi:10.1364/OE.21.021299.
12. Png GM, Choi JW, Ng BW-H, Mickan SP, Abbott D, Zhang X-C. The impact of hydration changes in fresh bio-tissue on THz spectroscopic measurements. Phys Med Biol. 2008;53(13):3501.
13. Hu BB, Nuss MC. Imaging with terahertz waves. Opt Lett. 1995;20(16):1716–8. doi:10.1364/OL.20.001716.
14. Mittleman DM, Jacobsen RH, Nuss MC. T-ray imaging. Sel Top Quantum Electron, IEEE J. 1996;2(3):679–92. doi:10.1109/2944.571768.
15. Hadjiloucas S, Karatzas LS, Bowen JW. Measurements of leaf water content using terahertz radiation. Microwave Theory Tech, IEEE Trans. 1999;47(2):142–9. doi:10.1109/22.744288.
16. Born N, Behringer D, Liepelt S, Beyer S, Schwerdtfeger M, Ziegenhagen B, et al. Monitoring plant drought stress response using terahertz time-domain spectroscopy. Plant Physiol. 2013;164(4):1571–7.
17. Breitenstein B, Scheller M, Shakfa MK, Kinder T, Müller-Wirts T, Koch M, et al. Introducing terahertz technology into plant biology: A novel method to monitor changes in leaf water status. J Appl Bot Food Qual. 2012;84(2):158–61.
18. Kinder T, Müller-Wirts T, Breitenstein B, Selmar D, Schwerdtfeger M, Scheller M, et al. In-vivo-messung des blattwassergehalts mit terahertz-strahlung. BioPhotonik. 2012;1:40–2.
19. Castro-Camus E, Palomar M, Covarrubias A. Leaf water dynamics of arabidopsis thaliana monitored in-vivo using terahertz time-domain spectroscopy. Sci R. 2013;3:1–5.
20. Federici J. Review of moisture and liquid detection and mapping using terahertz imaging. J Infrared, Millimeter, and Terahertz Waves. 2012;33(2):97–126. doi:10.1007/s10762-011-9865-7.
21. Parasoglou P, Parrott EPJ, Zeitler JA, Rasburn J, Powell H, Gladden LF, et al. Quantitative water content measurements in food wafers using terahertz radiation. Terahertz Sci Technol. 2010;3(4):176–82.
22. Sancho-Knapik D, Gismero J, Asensio A, Peguero-Pina JJ, Álvarez-Arenas VFTG, Gil-Pelegrín E. Microwave I-band (1730 mhz) accurately estimates the relative water content in poplar leaves. a comparison with a near infrared water index (R_{1300}/R_{1450}). Agric Forest Meteorology. (7):827–32. doi:10.1016/j.agrformet.2011.01.016.
23. Matzler C. Microwave (1-100 ghz) dielectric model of leaves. Geoscience Remote Sensing, IEEE Trans. 1994;32(4):947–9.
24. Claudio HC, Cheng Y, Fuentes DA, Gamon JA, Luo H, Oechel W, et al. Monitoring drought effects on vegetation water content and fluxes in

chaparral with the 970 nm water band index. Remote Sensing of Environ. 2006;103(3):304–11.

25. Tucker CJ. Remote sensing of leaf water content in the near infrared. Remote Sensing of Environ. 1980;10(1):23–32. doi:10.1016/0034-4257(80)90096-6.

26. Calvet J-C, Wigneron J-P, Mougin E, Kerr YH, Brito JL. Plant water content and temperature of the amazon forest from satellite microwave radiometry. Geoscience Remote Sensing, IEEE Trans. 1994;32(2):397–408.

27. Ferrazzoli P, Guerriero L. Passive microwave remote sensing of forests: A model investigation. Geoscience and Remote Sensing, IEEE Transactions on. 1996;34(2):433–43.

28. Hunt Jr ER, Li L, Yilmaz MT, Jackson TJ. Comparison of vegetation water contents derived from shortwave-infrared and passive-microwave sensors over central iowa. Remote Sensing Environ. 2011;115(9):2376–83.

29. Van Exter M, Fattinger C, Grischkowsky D. Terahertz time-domain spectroscopy of water vapor. Opt Lett. 1989;14(20):1128–30.

30. Jepsen PU, Cooke DG, Koch M. Terahertz spectroscopy and imaging - modern techniques and applications. Laser Photonic Rev. 2011;5(1):124–66. doi:10.1002/lpor.201000011.

31. Tonouchi M. Cutting-edge terahertz technology. Nat Photonics. 2007;1(2):97–105.

32. Hunsche S, Mittleman DM, Koch M, Nuss MC. New dimensions in T-ray imaging. IEICE Trans Electron. 1998;81(2):269–76.

33. Redo-Sanchez A, Laman N, Schulkin B, Tongue T. Review of terahertz technology readiness assessment and applications. J Infrared, Millimeter, and Terahertz Waves. 2013;34(9):500–18. doi:10.1007/s10762-013-9998-y.

34. McIntosh K, Brown E, Nichols K, McMahon O, DiNatale W, Lyszczarz T. Terahertz photomixing with diode lasers in low-temperature-grown gaas. Appl Phys Lett. 1995;67(26):3844–6.

35. Verghese S, McIntosh K, Calawa S, DiNatale WF, Duerr EK, Mahoney L. Photomixer transceiver. In: Optoelectronics' 99-Integrated Optoelectronic Devices. International Society for Optics and Photonics. San Jose, CA: SPIE; 1999. p. 7–13.

36. Gregory I, Tribe W, Baker C, Cole B, Evans M, Spencer L, et al. Continuous-wave terahertz system with a 60 db dynamic range. Appl Phys Lett. 2005;86(20):204104.

37. Siebert KJ, Quast H, Leonhardt R, Löffler T, Thomson M, Bauer T, et al. Continuous-wave all-optoelectronic terahertz imaging. Appl Phys Lett. 2002;80(16):3003–5.

38. Wilk R, Breitfeld F, Mikulics M, Koch M. Continuous wave terahertz spectrometer as a noncontact thickness measuring device. Appl Opt. 2008;47(16):3023–6. doi:10.1364/AO.47.003023.

39. Karpowicz N, Zhong H, Zhang C, Lin K-I, Hwang J-S, Xu J, et al. Compact continuous-wave subterahertz system for inspection applications. Appl Phys Lett. 2005;86(5):054105.

40. Jepsen PU, Jacobsen RH, Keiding S. Generation and detection of terahertz pulses from biased semiconductor antennas. JOSA B. 1996;13(11):2424–36.

41. Tani M, Matsuura S, Sakai K, Nakashima S-I. Emission characteristics of photoconductive antennas based on low-temperature-grown gaas and semi-insulating gaas. Appl Opt. 1997;36(30):7853–9.

42. Ezdi K, Heinen B, Jördens C, Vieweg N, Krumbholz N, Wilk R, et al. A hybrid time-domain model for pulsed terahertz dipole antennas. J Eur Opt Soc-Rapid Publications. 2009;4:09001-1–7.

43. Dietz RJB, Gerhard M, Stanze D, Koch M, Sartorius B, Schell M. THz generation at 1.55 μm excitation: six-fold increase in THz conversion efficiency by separated photoconductive and trapping regions. Opt Express. 2011;19(27):25911–7. doi:10.1364/OE.19.025911.

44. Bartels A, Cerna R, Kistner C, Thoma A, Hudert F, Janke C, et al. Ultrafast time-domain spectroscopy based on high-speed asynchronous optical sampling. Rev Sci Instrum. 2007;78(3):035107-1–8. doi:10.1063/1.2714048.

45. Gente R, Born N, Koch M. Water status monitoring of plants using thz systems. In: 2nd Annual Conference of COST Action MP1204 & International Conference on Semiconductor Mid-IR Materials and Optics SMMO2014. Germany: Marburg; 2014.

46. Gente R, Born MN. ansSchwerdtfeger, Rehn A, Koch M. Methods for water status detection with thz waves. In: 4th EOS Topical Meeting on Terahertz Science & Technology (TST 2014). Italy: Amogli; 2014.

47. Hadjiloucas S, Galvão RKH, Bowen JW. Analysis of spectroscopic measurements of leaf water content at terahertz frequencies using linear

transforms. J Opt Soc Am A. 2002;19(12):2495–509. doi:10.1364/JOSAA.19.002495.

48. Jördens C, Scheller M, Breitenstein B, Selmar D, Koch M. Evaluation of leaf water status by means of permittivity at terahertz frequencies. J Biol Phy. 2009;35(3):255–64. doi:10.1007/s10867-009-9161-0.

49. Gente R, Born N, Voß N, Sannemann W, Léon J, Koch M, et al. Determination of leaf water content from terahertz time-domain spectroscopic data. J Infrared, Millimeter, and Terahertz Waves. 2013;34(3-4):316–23. doi:10.1007/s10762-013-9972-8.

50. Scheller M, Baaske K, Koch M. Multifrequency continuous wave terahertz spectroscopy for absolute thickness determination. Appl Phys Lett. 2010;96(15):151112. doi:10.1063/1.3402767.

51. Scheller M, Koch M. Terahertz quasi time domain spectroscopy. Opt Express. 2009;17(20):17723–33. doi:10.1364/OE.17.017723.

52. Scheller M, Dürrschmidt SF, Stecher M, Koch M. Terahertz quasi-time-domain spectroscopy imaging. Appl Opt. 2011;50(13):1884–8.

53. Probst T, Rehn A, Busch SF, Chatterjee S, Koch M, Scheller M. Cost-efficient delay generator for fast terahertz imaging. Opt Lett. 2014;39(16):4863–6.

54. Gente R, Rehn A, Koch M. Contactless water status measurements on plants at 35 GHz. J Infrared, Millimeter, and Terahertz Waves. 2015;36(3):312–7. doi:10.1007/s10762-014-0127-3.

19

High-throughput phenotyping of seminal root traits in wheat

Cecile AI Richard[1*], Lee T Hickey[1], Susan Fletcher[2], Raeleen Jennings[2], Karine Chenu[3] and Jack T Christopher[4]

Abstract

Background: Water availability is a major limiting factor for wheat (*Triticum aestivum* L.) production in rain-fed agricultural systems worldwide. Root system architecture has important functional implications for the timing and extent of soil water extraction, yet selection for root architectural traits in breeding programs has been limited by a lack of suitable phenotyping methods. The aim of this research was to develop low-cost high-throughput phenotyping methods to facilitate selection for desirable root architectural traits. Here, we report two methods, one using clear pots and the other using growth pouches, to assess the angle and the number of seminal roots in wheat seedlings– two proxy traits associated with the root architecture of mature wheat plants.

Results: Both methods revealed genetic variation for seminal root angle and number in the panel of 24 wheat cultivars. The clear pot method provided higher heritability and higher genetic correlations across experiments compared to the growth pouch method. In addition, the clear pot method was more efficient – requiring less time, space, and labour compared to the growth pouch method. Therefore the clear pot method was considered the most suitable for large-scale and high-throughput screening of seedling root characteristics in crop improvement programs.

Conclusions: The clear-pot method could be easily integrated in breeding programs targeting drought tolerance to rapidly enrich breeding populations with desirable alleles. For instance, selection for narrow root angle and high number of seminal roots could lead to deeper root systems with higher branching at depth. Such root characteristics are highly desirable in wheat to cope with anticipated future climate conditions, particularly where crops rely heavily on stored soil moisture at depth, including some Australian, Indian, South American, and African cropping regions.

Keywords: Wheat breeding, Root angle, Root number, Adaptation, Drought

Background

Drought is a major limiting factor of wheat (*Triticum aestivum* L.) production world-wide [1]. Water deficit during critical periods of crop development such as grain filling, can greatly impact yield stability and productivity in rain-fed agricultural systems. Traditional wheat breeding relies heavily on selection for yield per se and has contributed to significant increases in yield. However, the rate of genetic progress has slowed in recent years [2]. Yield is a quantitative trait under complex genetic control, characterized by low heritability and high genotype by environment (G × E) interactions, particularly in drought environments [3]. Physiological approaches based on proxy traits, can offer higher heritability and lower G × E interactions than selection for yield itself and complement traditional breeding approaches to accelerate improvement in drought-prone environments.

Drought-adaptive traits related to root physiology and morphology have been identified in maize (*Zea mays*), sorghum (*Sorghum bicolor*), rice (*Oryza sativa*) and wheat [4-11]. Modelling studies performed using historical climate data for wheat grown throughout the Australian cropping region indicated that root architecture had significant functional implications for the timing and amount of subsoil water extraction [4,7,12]. Wheat cultivars with narrower lateral root distribution and higher proportion of roots at depth can access more soil moisture deep in the soil profile, particularly late in the season when marginal water-use efficiency for grain production is high [4,13-17]. Such root characteristics that facilitate improved access to soil moisture late in the season are highly desirable in rain-fed systems, particularly where crops rely

* Correspondence: c.richard@uq.edu.au
[1]The University of Queensland, QAAFI, St Lucia, QLD 4072, Australia
Full list of author information is available at the end of the article

heavily on stored soil moisture at depth, as in parts of some Australian, Indian, South American, and African cropping regions.

Two types of roots occur in wheat, the seminal roots coming directly from the embryo and the later, nodal roots emerging at the lower tiller nodes [18]. A more vertical angle of the seminal roots and a higher number of seminal roots in wheat seedlings have been linked to a more compact root system with more roots at depth in wheat [14,19-21]. Narrow root angle and a higher number of seminal roots are considered proxy traits for selection at early growth stages in wheat breeding programs [14,15,22,23]. The association between root angle and deeper rooting systems has been demonstrated in sorghum, maize and rice, and a number of quantitative trait loci (QTL) showing homology across species have been reported recently [9,24,25].

Despite rapid advances in genomic approaches to tackle complex traits [26,27], the lack of high-throughput and large-scale phenotyping methods for root traits remains a major bottleneck to elucidate the genetic control and enable selection for such traits in breeding programs. Both field- and laboratory-based methods for phenotyping root traits have been developed [28], including soil sampling [22,29,30], thermography [6,31], X-ray computed tomography [32-36], mini-rhizotrons [37-39], rhizotrons [40,41], and non-soil techniques [14,21,42-44]. However, most of these approaches are low-throughput. Laboratory-based methods can be limited in their ability to reproduce field-like conditions [45-47]. For example, soil-environment × genotype interactions significantly affect the root length of wheat cultivars grown in sandy soil compared to agar plates [48]. Yet, root studies performed in the laboratory are generally less laborious and less time-consuming than in the field, and can be conducted out-of-season. In addition, root measurements tend to be more precise and more reproducible because the plants are grown in a more homogeneous environment compared to the field.

In this study, we used a panel of 24 spring wheat cultivars to design and evaluate two high-throughput methods for measuring seminal root angle and number in controlled environment growth facilities, one based on clear pots and the other based on growth pouches. We discuss the advantages and disadvantages of these root trait phenotyping methods, along with the opportunity to exploit high-throughput phenotypic screening in breeding populations.

Results

Genetic variation for seminal root angle and number

In the clear pots, seedling roots grew along the wall and were clearly distinguished from the dark soil. At the time of imaging for seminal root angle (i.e. five days after sowing), the first pair of seminal roots had

elongated on each side of the radicle, with an average seminal root angle of 75.5° for the two clear pot experiments. By contrast, in the growth pouches, seedling roots grew freely in the air space between the moistened paper and the plastic. At the time of scanning (i.e. 20 days after sowing), first and often second pairs of seminal roots had elongated on each side of the radicle, however, only the angle between the first pair was considered here. The average seminal root angle across the two pouch experiments was 109.7°. The observed range in seminal root angle phenotypes varied between methods, the clear pot method provided a range in seminal root angle from 60.1 to 84.0°, while the growth pouch method produced a wider seminal root angle with a range from 100.8 to 117.4° (Figure 1A).

Seminal root number was measured six days later than seminal root angle in the clear pot experiments (i.e. at 11 days after sowing). In both clear pot experiments, the root number estimated non-destructively from the images was significantly lower (p-value < 0.001) compared to measures obtained by extracting the seedlings from the soil; average across the two experiments was 3.6 for imaged and 4.2 for extracted, respectively. In the pouch experiments, seminal root number was measured at the same time as root angle (i.e. at 20 days after sowing) and seedlings exhibited 3.9 roots on average across the experiments. The genotypic range in seminal root number phenotypes varied between methods, with the clear pot method providing the widest range in seminal root number (3.2–4.0 for imaged and 3.5–4.8 for extracted) compared to the growth pouch method (3.6–4.2) (Figure 1B).

Comparison of methods

The heritability for seminal root angle was higher for the clear pot method ($h^2 = 0.65$) compared to the growth pouch method ($h^2 = 0.52$) (Table 1). However, the heritability for each individual experiment displayed some variability within methods, with higher values for Clear_1 and Pouch_2 ($h^2 = 0.79$ and $h^2 = 0.63$, respectively) compared to Clear_2 and Pouch_1 ($h^2 = 0.51$ and $h^2 = 0.42$, respectively) (Table 1). For seminal root number, the heritability was the highest for the clear pot method, with higher heritability obtained for extracted root number ($h^2 = 0.80$) compared to imaged root number ($h^2 = 0.50$) (Table 1). The heritability for seminal root number was the lowest for the growth pouch method ($h^2 = 0.37$) (Table 1). Overall, the heritability for each individual experiment was quite consistent within methods (Table 1).

The error variance was higher than the genetic variance for all experiments (Table 1), indicating that there were more differences in the seminal root angle and number within cultivar individuals than across cultivar averages. Almost all variation was explained by the genetic and

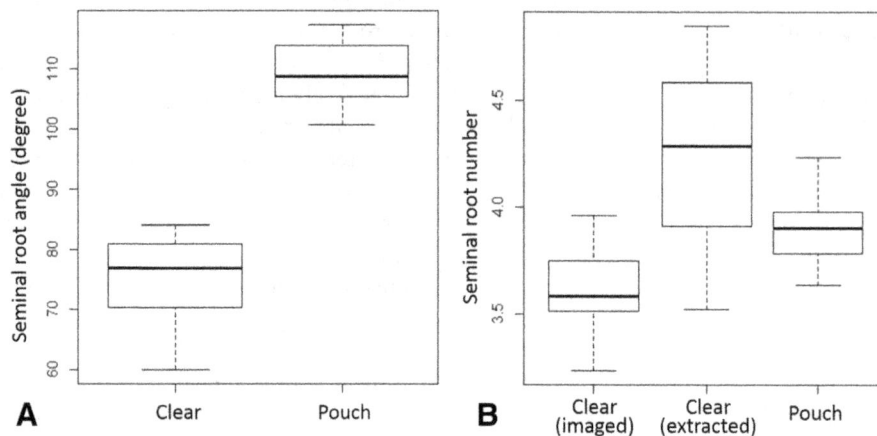

Figure 1 Genetic variation for seminal root angle and number. Box and whisker plots of **(A)** seminal root angle and **(B)** seminal root number, for the panel of 24 wheat cultivars evaluated using the clear pot and growth pouch methods. The values correspond to the average BLUPs per cultivar of the two clear pot experiments Clear_1 and Clear_2 (Clear) and the two growth pouch experiments Pouch_1 and Pouch_2 (Pouch). The seminal root number for the clear pot method was measured either via image analysis (imaged) or by counting roots after removing seedlings from soil (extracted). The bottom and the top of the boxes display the first and third quartile values for each experiment, respectively. The band inside the box displays the median and the ends of the whiskers display the minimum and maximum values.

error variance in the clear pot experiments. However, the random factors "Pouch" and "Box" had a significant effect in the growth pouch experiments.

The clear pot experiments (Clear_1 and Clear_2) used 10 reps per cultivar (i.e. 240 seeds in total per experiment), while the growth pouch experiments (Pouch_1 and Pouch_2) used only 6 reps per cultivar (i.e. 144 seeds in total per experiment). The number of observations for each experiment varied between experiments,

as in both methods some seeds didn't germinate and some roots were too short (<3 cm) to measure the seminal root angle. Using the clear pot method, some roots were also hidden by the soil on the images, making measurement impossible. Roots were sometimes hidden by the soil close to the surface, but visible deeper down, making the root angle measurement impossible but the imaged root number possible. In contrast, in the growth pouch method roots were always visible when present.

Table 1 Statistics for the seminal root angle and number

			Heritability (h^2)	Genetic variance	Error variance	Observations per cultivar
Seminal root angle	Clear	Clear_1	0.79	39%	61%	6.2/10
		Clear_2	0.51	16%	84%	5.7/10
		Clear average	**0.65**	**28%**	**72%**	**6.0/10**
	Pouch	Pouch_1	0.42	6%	55%	4.5/6
		Pouch_2	0.63	14%	78%	5.3/6
		Pouch average	**0.52**	**10%**	**67%**	**4.9/6**
Seminal root number	Clear (imaged)	Clear_1	0.45	9%	91%	8.2/10
		Clear_2	0.54	12%	86%	8.8/10
		Clear average	**0.50**	**10%**	**90%**	**8.5/10**
	Clear (extracted)	Clear_1	0.80	33%	66%	8.2/10
		Clear_2	0.79	30%	69%	8.8/10
		Clear average	**0.80**	**32%**	**68%**	**8.5/10**
	Pouch	Pouch_1	0.37	9%	88%	5.2/6
		Pouch_2	0.36	8%	70%	5.7/6
		Pouch average	**0.37**	**9%**	**79%**	**5.5/6**

Heritability h^2, genetic variance, error variance and average number of observations for seminal root angle and number for the panel of 24 wheat cultivars evaluated using different methods based on clear pots and growth pouches. The values correspond to the individual experiments. The values in bold correspond to the average of the two clear pot experiments Clear_1 and Clear_2 ('Clear average') and the two growth pouch experiments Pouch_1 and Pouch_2 ('Pouch average'). The seminal root number for the clear pot method was measured in two different ways: based on images (imaged) and after extracting the seedlings (extracted).

The average number of observations per cultivar for seminal root angle was 6.0 (out of 10) for the clear pot experiments and 4.9 (out of 6) for the growth pouch experiments (Table 1). For seminal root number, the average number of observations per cultivar were 8.5 (out of 10) for the clear pot method for both imaged and extracted seminal root number, while for the pouch method, observations were obtained for 5.5 (out of 6) plants per cultivar (Table 1).

The genetic correlations for the seminal root angle were the highest between the two clear pot experiments Clear_1 and Clear_2 ($r^2 = 0.82$) and the lowest between the two growth pouch experiments Pouch_1 and Pouch_2 ($r^2 = 0.11$) (Figure 2). The ranking of cultivars for root angle was almost the same across the two clear pot experiments, but differed markedly between the two growth pouch experiments. For instance, the cultivar Chara was the narrowest in Pouch_1, but one of the widest in Pouch_2 (data not shown). The genetic correlation between the two methods, clear pot and growth pouch, were medium (r^2 ranging 0.37–0.48) (Figure 2).

Genetic correlations between imaged and extracted seminal root number were high for both Clear_1 and Clear_2 experiments ($r^2 = 0.85$ and 0.75, respectively; Figure 3). The genetic correlations were high between the two clear pot experiments (Clear_1 and Clear_2) for the extracted seminal root number ($r^2 = 0.63$), but low for the imaged root number ($r^2 = 0.28$) (Figure 3). For the growth pouch method, the genetic correlation between the two experiments (Pouch_1 and Pouch_2) was medium ($r^2 = 0.53$), as well as the genetic correlations between clear pot (extracted) and growth pouch methods (r^2 ranging 0.37–0.64) (Figure 3). There was no significant genetic correlation between the seminal root angle and number for both the clear pot and growth pouch experiments (data not shown).

Diversity for root angle in Australian wheat cultivars

The cultivar ranking for seminal root angle was almost the same across the two clear pot experiments (Figure 4). Some trends based on genetic backgrounds could be observed, with all the Cook-type cultivars (EGA Wentworth, Giles, Janz, Lang, Sunco, and Sunvale) having narrower roots than all the Pavon-type cultivars (Diamonbird, Hartog, and Leichhardt) (Figure 4). The Cook/Pavon-type cultivars (Chara, EGA Edgetail, Silverstar, and Ventura) displayed a mixture of narrow and wide seminal root angle phenotypes as might be anticipated. Cultivars belonging to other genetic backgrounds did not show a consistent pattern of seminal root angle.

Discussion

The two phenotypic methods for seminal root traits evaluated in this study permitted differentiation of seminal root angle and number in the panel of 24 wheat cultivars. The clear pot method showed consistency across experiments and is considered the most suitable for large-scale and high-throughput screening of seedling root characteristics in crop improvement programs.

In this study, we examined the seminal root angle and number for a panel of 24 wheat cultivars measured using two methods; one based on clear pots and the other using growth pouches. The clear pot method provided a higher degree of variation for both seminal root traits with a range of 23.9° for root angle and 1.3 for extracted root number. This compared to the growth pouch method with a range of 16.6° and 0.6 roots per plant. It should be noted that these ranges may not represent the full extent of genetic variation in wheat germplasm, as this panel represents a limited set of genotypes and many share similar pedigrees and/or genetic backgrounds. Higher levels of variation for these traits were observed for the same 24 wheat cultivars in a previous

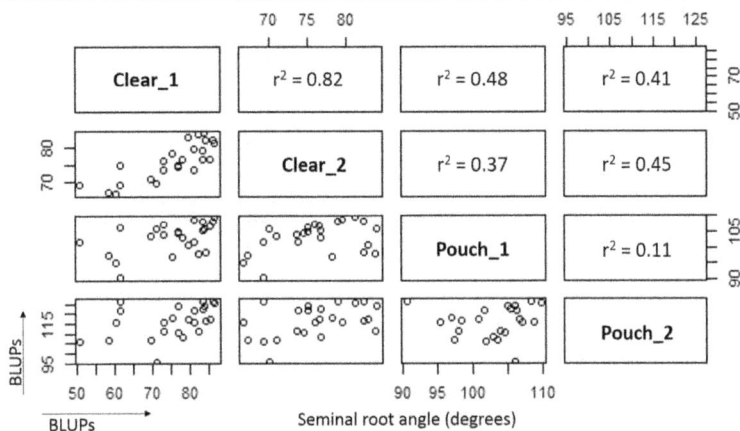

Figure 2 Genetic correlations of seminal root angle using clear pot and growth pouch methods. Genetic correlations (upper panels) and scatter plots (lower panels) of the BLUPs for seminal root angle (in degree) between the clear pot (i.e. Clear_1 and Clear_2) and the growth pouch (i.e. Pouch_1 and Pouch_2) experiments. Data represents average BLUPs of the 24 wheat cultivars.

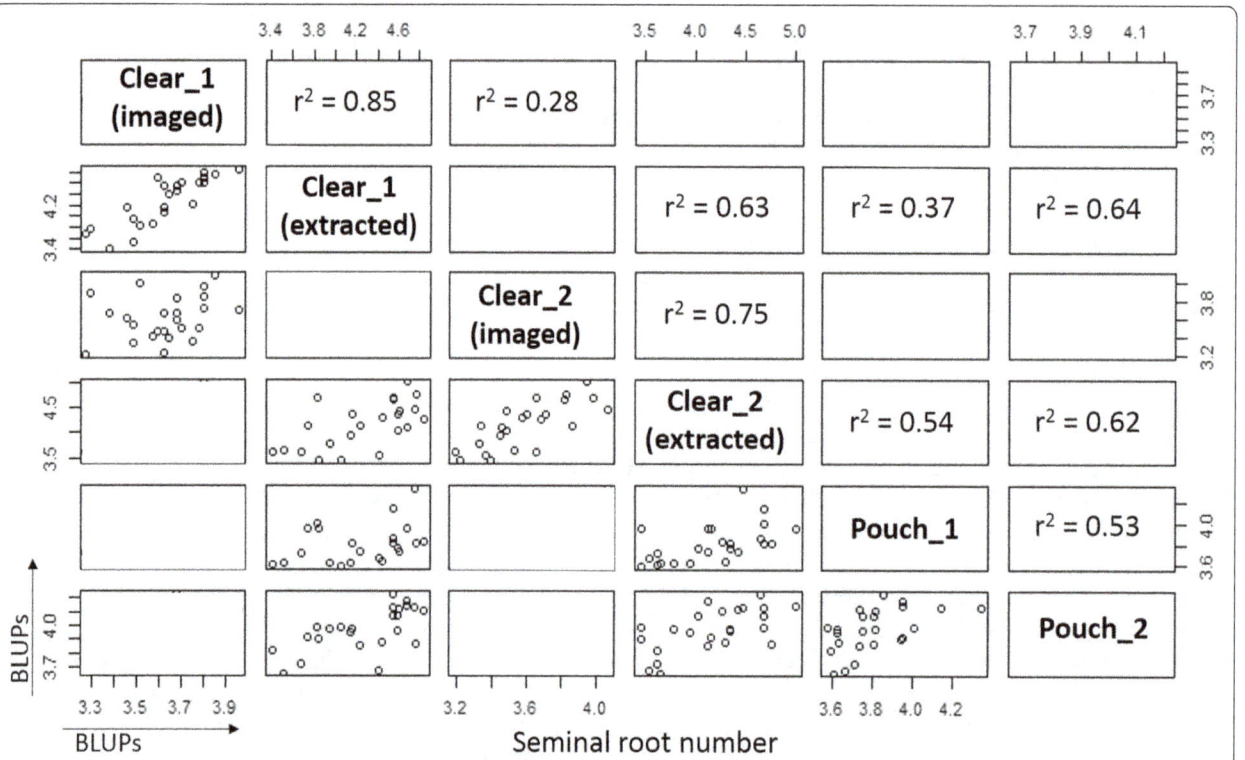

Figure 3 Genetic correlations of seminal root number using clear pot and growth pouch methods. Genetic correlations (upper panels) and scatter plots (lower panels) of the BLUPs for seminal root number counted based on images (imaged) and after extracting the seedlings (extracted) for each of the clear pot experiments (i.e. Clear_1 and Clear_2), and for the seminal root number with the growth pouch experiments (Pouch_1, and Pouch_2). Data represents average BLUPs of the 24 wheat cultivars.

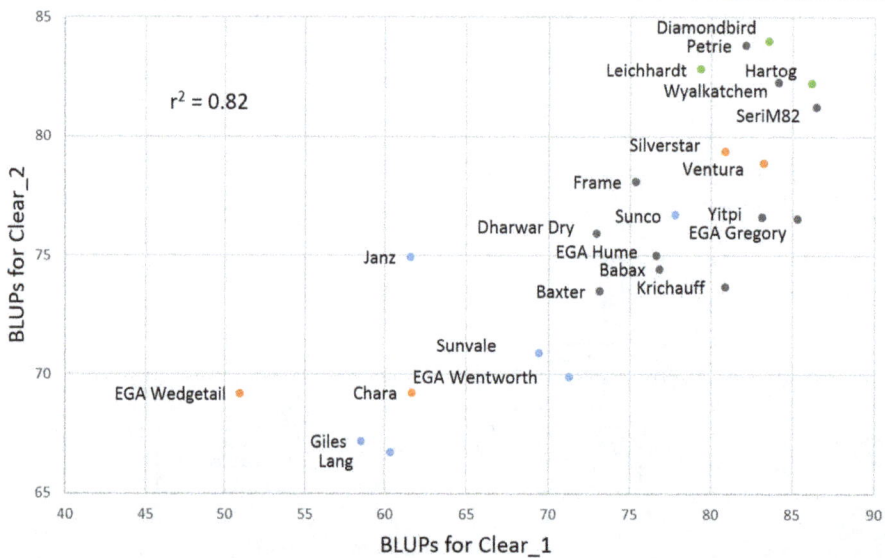

Figure 4 Seminal root angle of the panel of 24 wheat cultivars. Scatter plot of BLUPs for seminal root angle (in degrees) between the two clear pot experiments (i.e. Clear_1 and Clear_2) for 24 wheat cultivars. Blue dots = Cook-type, green dots = Pavon-type, orange dots = Cook/Pavon type, grey dots = other backgrounds.

study (72.4–112.6° i.e. 40.2° for root angle and for root number 3.2–5.0 i.e. 1.8 roots per plant) using a gel chamber method [14]. However, this method is labour intensive and not suitable for evaluation of large numbers of entries.

Despite variations within experiments, the heritability was higher using the clear pot method for both seminal root traits (i.e. $h^2 = 0.65$ for root angle and $h^2 = 0.80$ for extracted root number) compared to the growth pouch method ($h^2 = 0.52$ for root angle and $h^2 = 0.37$ for root number). If implemented in breeding programs, the relatively high heritability should enable genetic gain for this trait. The achieved number of observations for seminal root angle using the clear pot method was lower than the potential 10 observations due to the fact that some roots were hidden by soil in the images. As a consequence, this method requires a high number of repetitions (i.e. ~10) to ensure high heritability. The position of the seed at sowing (i.e. embryo pointed downwards and slightly towards the wall) is critical to ensure roots grow along the wall and are visible. The achieved number of observations for seminal root traits using the growth pouch method was close to the potential 6 observations due to the fact that roots were always visible when present. The heritability could be improved by increasing the number of reps, for example 10 reps instead of 6. The error variance was higher than the genetic variance for all experiments, which is not surprising considering that traits were measured for single plants. Results from the two clear pot experiments were more strongly correlated ($r^2 = 0.82$ for root angle and $r^2 = 0.63$ for extracted root number), when compared to results from the two growth pouch experiments ($r^2 = 0.11$ for root angle and $r^2 = 0.53$ for root number). The rank of the cultivars based on the seminal root angle and number was quite consistent across the two clear pot experiments, suggesting that the method is repeatable and has power to detect differences in root phenotypes (i.e. narrow/wide seminal root angle, low/high number of seminal roots). The wider range of root phenotypes obtained using the clear pot method enabled better differentiation among cultivars with more repeatable results, and thus appears superior to the growth pouch method for implementation in breeding programs.

Seminal root number was measured with the clear pot method in two different ways: by counting based on images and after seedlings were extracted from the soil. Roots were underestimated using the images because some roots were hidden by soil, resulting in a significantly lower average number of seminal roots for the imaged root number compared to the extracted root number. As expected, the extracted root number was more accurate than the imaged root number. For instance, the genetic variation, the heritability and the genetic correlations were higher for the extracted seminal root number than the imaged values. However, imaged and extracted seminal root number were strongly correlated ($r^2 > 0.75$) and ranking of cultivars using both techniques was also very similar. Despite a lower level of precision, estimation of seminal root number using the imaging technique is preferred for breeding purposes because this method doesn't require a labour intensive transplanting of the selected plants. For instance, the imaging method can be used to differentiate extreme phenotypes (i.e. low versus high root number), in order to enrich segregating populations with desirable genes via repetitive cycles of selection. However, to precisely phenotype or characterize fixed lines, counting the roots after pulling out the plants may be preferred.

The paper growth media in growth pouches and the agar gel of the gel-filled chamber method from Manschadi et al. [14] both provide conditions less representative of natural soils than the soil-based growth medium used in the clear pot system. Consequently the soil-based clear pot method may result in phenotypes more similar to those expressed in the field [48]. In addition, the growth pouch and gel-filled methods are very time-consuming and labour intensive to set up, thus, are better suited for evaluation of smaller numbers of cultivars compared to the clear pot method. For these reasons, we propose that the clear pot method is preferred for high-throughput and large-scale screening of seminal root angle and number.

The rank between cultivars based on the seminal root angle calculated with the clear pot method was almost identical across the two experiments and ranking seemed to correspond with the genetic background of the wheat cultivars. For instance, most of the cook-type cultivars displayed a narrow seminal root angle, while all the Pavon-type cultivars displayed wider seminal root angles, which is similar to previous studies [14,23]. The Cook-type cultivars tend to have a longer season maturity compared to the Pavon-type cultivars used in this study. Cultivars with a longer cycle are more likely to encounter terminal moisture stress in the season, particularly if grown in a summer dominant rainfall environment. Deeper rooting could be an adaptation for late cultivars to ensure photosynthetic and remobilization activities during grain filling in rainfed wheat production systems relying heavily on deep stored soil moisture. There was little consistency between the preferred growing region for the Australian wheat cultivars evaluated in this study and cluster analysis based on root angle phenotypes also failed to detect any obvious trends other than those associated with genetic background (data not presented). Although wheat breeders have likely indirectly selected for desirable root architecture where environmental pressure is frequent, this is not the only trait affecting drought adaptation. In fact, while drought types highly differ depending on the season and region [49-51], drought adaptation typically involves the interaction of a number of traits related to water utilization as well as other physiological processes [2,52]. As a result, breeders and pre-breeders are targeting other traits such as adapted phenology [53], transpiration efficiency [54], cooler canopy temperature [16,55,56] and reduce tillering [57]. While deep root architecture is likely important for adaptation in rainfed wheat production systems relying heavily on stored soil moisture (particularly at depth [4]), this trait may be less advantageous in other environments, for example where rainfall is more frequent through the growing season, where soils are compacted [58] or for late sown conditions [59].

Selection for combinations of physiological traits that underpin yield may be a more effective way to achieve genetic gain for yield in specific environment types, rather than direct selection for yield per se [3,51,60,61]. The clear pot method allows high-throughput and cost-effective screening of breeding populations at a rate of 600 plants/m^2 in a controlled environment within only five days for the seminal root angle and 11 days for the seminal root number. The technique is suitable for characterising both fixed lines and for screening large segregating populations (e.g. F2 and F3). As the system permits growing-on of the selected plants, repeated cycles of selection can be performed across consecutive generations to rapidly enrich breeding populations with desirable alleles for root traits. Alternatively, the method could be used to select parental lines with desired root traits for crossing. Therefore, the clear pot method has the potential to accelerate genetic gain for drought tolerance in breeding programs.

The technique is also well adapted for use in the "speed breeding" system developed and refined at The University of Queensland that achieves rapid plant growth by incorporating controlled temperature and constant light [62]. By combining speed-breeding growth conditions and the root trait phenotypic screening

method, it is possible to achieve up to 30 phenotypic screens within 12 months if plants are not grown to maturity. Alternatively, under optimised growth conditions, up to 6 consecutive cycles of selection could be achieved in 12 months with selections grown through to maturity producing seed in each generation. Thus, within a 12 month timeframe, it would be possible to make crosses, screen and produce seeds for F1 to F4 generations for desirable root traits, and produce F5:F6 lines with improved root traits. Also, seminal root trait screening can be easily integrated with other phenotypic screening methods adapted to the speed breeding system, such as adult plant resistance to rust pathogens [63] and grain dormancy for tolerance to pre-harvest sprouting [64]. We anticipate this methodology will accelerate identification of genetic diversity for root traits in wheat and propose that it could be applied to other crops, such as barley and chickpea.

Conclusions

Phenotyping root traits in wheat has been limited by the availability of suitable methods. In this study, we reported a new high-throughput method using clear pots to phenotype seminal root angle in 5-day-old wheat seedlings and seminal root number in 11-day-old wheat seedlings. This method has clear advantages over other previously reported techniques and could be easily integrated into wheat breeding programs targeting drought tolerance via improved access to deep soil water.

Methods

A panel of wheat cultivars differing for their region of adaptation and drought tolerance were assayed in clear pots and growth pouches for seminal root angle and number. In total, four experiments were conducted in this study – two based on clear pots (i.e. Clear_1 and Clear_2) and two based on growth pouches (i.e. Pouch_1 and Pouch_2) to assess the robustness and repeatability of each method.

Figure 5 Wheat seedlings phenotyped for seminal root traits in a high-throughput system using clear pots. (A) Wheat seedlings grown in clear pots under controlled environment conditions (picture taken five days after sowing). **(B)** The clear pots placed inside black pots to exclude light (picture taken at 11 days after sowing). **(C)** Images recorded for each plant of each pot using a camera fixed on a tripod, a black box with anti-reflection walls and a revolving stand.

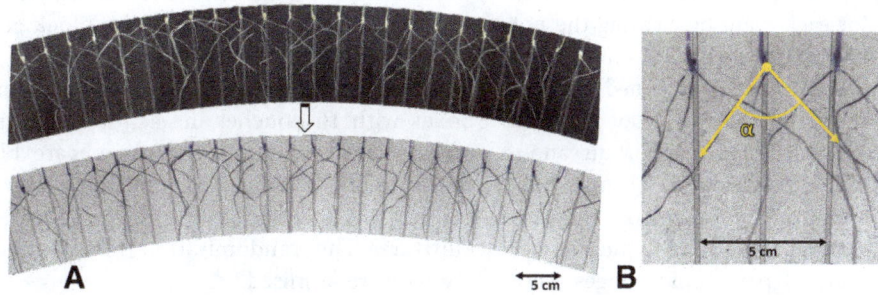

Figure 6 Measuring seminal root angle with the clear pot method. (A) Panoramic image of wheat seedling grown in the clear pot system obtained by stitching images of individual plants using software (*PhotoStitch*) and colours inverted to facilitate root identification. **(B)** For each plant, the angle (α) between the first pair of seminal roots was measured at approximately 3 cm distance from the seed using software (*ImageJ*).

Clear pot method

Two experiments using the clear pot method ('Clear_1', and 'Clear_2') were conducted successively under the same conditions to evaluate the panel of 24 wheat cultivars for seminal root angle and seminal root number.

Wheat seedlings were cultured in 4 L clear pots (ANOVApot®, 200 mm diameter, 190 mm height, http://www.anovapot.com/php/anovapot.php). The clear pots were filled with a pine bark potting media (70% composted pine bark 0–5 mm, 30% coco peat, pH 6.35, EC = 650 ppm, nitrate = 0, ammonia < 6 ppm and phosphorus = 50 ppm). Seeds were sown at a depth of 2 cm every 2.5 cm along the pot wall, providing a density of 24 seeds per pot (600 plants/m^2). The seeds were carefully placed vertically, embryo downwards and facing the wall to facilitate root growth along the transparent wall (Figure 5A). After sowing, the clear pots were placed inside 4 L black

pots (ANOVApot®, 200 mm diameter, 190 mm height) to exclude light from the developing roots (Figure 5B). The pots were watered after sowing and no additional water or nutrients were supplied thereafter.

The two experiments used randomised complete block designs where 24 cultivars were randomised across 10 pots, ensuring cultivars were present only once in each pot. Each pot represented one replicate block and one plant of each cultivar in each pot represented the experimental unit.

The two experiments were conducted in a walk-in, temperature-controlled growth facility. Constant temperature (17°C ± 2 C) was adopted over 24 hours with diurnal (12 hour) natural light.

Five days after sowing, images of the seminal roots visible through the clear wall were recorded using a camera (Canon PowerShot SX600 HS 16MP Ultra-Zoom Digital

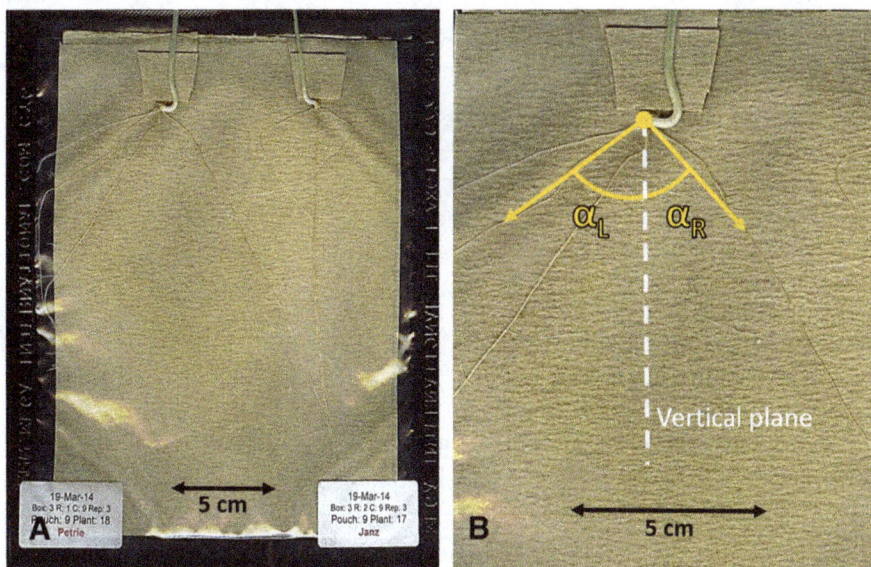

Figure 7 Illustration of a growth pouch. (A) Wheat seedlings were phenotyped for seminal root angle and number using growth pouches (picture taken 20 days after sowing). **(B)** For each plant, the left (α$_L$) and the right (α$_R$) angle between each of the first pair of seminal roots and the vertical plane was measured at approximately 3 cm distance from the seed using software (*Opengelphoto*).

Camera) fixed on a tripod (Slik F153 Tripod) (Figure 5C). Images were recorded for each plant by rotating the pot 15° in a clockwise direction. The images captured from each pot displayed some overlap and were joined together to create a panoramic image for the whole pot with the stitching software *PhotoStitch* (http://support-au.canon. com.au/contents/AU/EN/0200246607.html) (Figure 6A). This step reduced the picture file storage size and also improved image analysis speed by using 1 picture per pot instead of 24. Colours of panoramic images were inverted to enhance the contrast between roots and soil, facilitating root-trait measurements with the software *imageJ* (http://imagej.nih.gov/ij/) [65] (Figure 6A).

For each plant, the growth angle between the first pair of seminal roots was measured at approximately 3 cm distance from the seed (Figure 6B).

In this study, we tested two different ways to measure seminal root number at 11 days after sowing. The "imaged" number of seminal roots was measured based on the photographic images by counting the number of roots emerging from the seed. The "extracted" number of seminal roots was measured after pulling out the wheat seedlings and counting the number of roots.

Growth pouch method

Two experiments ('Pouch_1' and 'Pouch_2') were conducted successively under the same conditions to evaluate the panel of 24 wheat cultivars for seminal root angle and seminal root number using the growth pouch method.

The experiments were performed using Cyg germination growth pouches (Mega International, http://www. mega-international.com/index.htm). Measuring 18 cm × 16.5 cm, the plastic pouches contained perforated germination paper that has been folded to form a continuous trough along the top of the pouch, in which seeds are supported (Figure 7A). To avoid roots spatially interfering with each other during the initial growth period, each pouch contained only two seeds (Figure 7A). Pouches were pre-prepared by removing excess paper from the seed trough, leaving two individual troughs (Figure 7A). Tap water (15 mL) was added to each pouch and allowed to evenly distribute over the germination paper. Dry seeds were placed vertically into the troughs, with the embryo end pointing down, and the embryo facing out towards the plastic. Pouches were then placed vertically into containers, sandwiched between foam to maintain even pressure on the seeds and to reduce air spaces. Containers were covered in cling wrap to prevent moisture loss.

Pouches were placed into a plant growth cabinet at a constant temperature of 15°C with no light. After 12 days, lights were turned on using a 12 h photoperiod. Seedlings were grown for 20 days in total.

The pouch experiments used a resolvable block design where pouches constituted a block size of 2. This ensured pairs of cultivars were not in the same pouch together more than once. Each experiment had 6 boxes with 16 pouches in each box set out in a 2 × 8 array. Each box comprised a replicate block, with 1 replicate of the panel of 24 cultivars, 1 extra replicate for Hartog and SeriM82, and 1 replicate of 6 other cultivars. The randomisation for the pouch experiments were latinised.

Seminal root angle and number were measured using a scanner (Epson Perfection 4990 Photo) at 20 days after sowing. The images were analysed using a specifically-designed software program *Opengelphoto*, which enables measurement of angle of individual roots from a vertical plane. For each seedling the growth angle between each of the first pair of seminal roots (i.e. left and right first

Table 2 Name, origin and genetic background of the 24 wheat cultivars used in this study

Cultivar	Breeding program[1]	Genetic background
Babax	CIMMYT	Veery
Baxter	QDPI	CIMMYT/Cook
Chara	DPI Vic	Cook/Pavon
Dharwar Dry	Central India	CIMMYT
Diamondbird	NSW DPI	Pavon
EGA Gregory	EGA	Pelsart/Batavia
EGA Hume	EGA	Pelsart/Batavia
EGA Wedgetail	EGA	Cook/Pavon
EGA Wentworth	EGA	Cook
Frame	AGT	Condor/Gabo
Giles	QDPI	Cook
Hartog	QDPI	Pavon
Janz	QDPI	Cook
Krichauff	AGT	Condor/Gabo
Lang	QDPI	Cook
Leichhardt	QDPI	Pavon
Petrie	QDPI	Pelsart/Batavia
SeriM82	CIMMYT	CIMMYT/Veery
Silverstar	NSW DPI	Cook/Pavon
Sunco	Uni Syd	Cook
Sunvale	Uni Syd	Cook
Ventura	NSW DPI	Cook/Pavon
Wyalkatchem	AgWA	Condor/Gabo
Yitpi	AGT	Condor/Gabo

[1]Breeding program abbreviations: Queensland Department of Primary Industries (QDPI), Department of Primary Industries Victoria (DPI Vic), Australian Grain Technologies (AGT), New South Wales Department of Primary Industry (NSW DPI), International Maize and Wheat Improvement Center (CIMMYT), Enterprise Grains Australia (EGA), Western Australia Department of Agriculture (AgWA), University of Sydney (Uni Syd).

pair of seminal roots) and the vertical plane was measured at approximately 3 cm distance from the seed (Figure 7B). The root number was measured by counting the number of roots based on the scanned images.

Statistical analysis

A linear mixed model framework was used to analyse genotype-environment interactions across experiments based on clear pots (Clear_1 and Clear_2) and growth pouches (Pouch_1 and Pouch_2). The mixed model contained random components that identified the structure of the experimental design for each experiment: (i) Pot for the clear pot experiments, and (ii) Pouch and Box for the growth pouch experiment. Given the importance of genotype ranking across experiments, the random model formula also included Genotype as a random effect. The random model formula allows for estimation of variance heterogeneity for each of the random terms for each experiment. The residual maximum likelihood (REML) algorithm [66] was used to provide estimates of the variance components and the best linear unbiased predictions (BLUPs). Data were analysed with ASReml-R [67] using R software Version 3.0.0 (R Core team 2013).

For seminal root angle measured using the growth pouch method, each plant had two values corresponding to the angle between the left or right seminal roots and the vertical plane. Therefore, the dataset for seminal root angle measured using the growth pouch method had an additional factor Side (left and right). After the analysis, the BLUPs were multiplied by two to allow comparison with the seminal angle measured using the clear pot method. For seminal root number, a Student test was performed to compare the means between imaged and extracted root number using R software Version 3.0.0.

Plant material

The study was conducted using a panel of 24 spring wheat cultivars (Table 2), that was previously characterized for seminal root angle and root number using a gel-filled chamber method reported by Manschadi et al. [14]. In their study, Manschadi et al. [14] obtained seminal root angles ranging from 36.2° to 56.3° and number of seminal roots ranging from 3.2 to 5.0. These seminal root angle values corresponded to the angle between each of the seminal roots and the vertical plane and were multiplied by two to allow comparison with the seminal angle measured in this study.

The panel comprised 21 Australian spring wheat cultivars, including some of the most widely grown throughout Australia in recent years, two elite cultivars (Babax and SeriM82) from the International Maize and Wheat Improvement Center (CIMMYT) in Mexico and one wheat cultivar from India (Dharwar dry).

Competing interests

The authors declare that they have no competing interests.

Acknowledgements

This work was supported by the University of Queensland, Queensland Alliance for Agriculture and Food Innovation and was partially funded by the Grains Research and Development Corporation of Australia, including a PhD scholarship for Cecile Richard.

Author details

[1]The University of Queensland, QAAFI, St Lucia, QLD 4072, Australia. [2]Department of Agriculture, Fisheries and Forestry, Leslie Research Facility, Toowoomba, QLD 4350, Australia. [3]The University of Queensland, QAAFI, 203 Tor Street, Toowoomba, QLD 4350, Australia. [4]The University of Queensland, QAAFI, Leslie Research Facility, Toowoomba, QLD 4350, Australia.

References

1. Araus JL, Slafer GA, Royo C, Serret MD. Breeding for yield potential and stress adaptation in cereals. Crit Rev Plant Sci. 2008;27:377–412.
2. Fischer RA, Edmeades GO. Breeding and cereal yield progress. Crop Sci. 2010;50:S85–98.
3. Jackson P, Robertson M, Cooper M, Hammer G. The role of physiological understanding in plant breeding; from a breeding perspective. Field Crop Res. 1996;49:11–37.
4. Manschadi AM, Christopher JT, DeVoil P, Hammer GL. The role of root architectural traits in adaptation of wheat to water-limited environments. Funct Plant Biol. 2006;33:823–37.
5. Hammer GL, Dong Z, McLean G, Doherty A, Messina C, Schussler J, et al. Can changes in canopy and/or root system architecture explain historical maize yield trends in the U.S. Corn belt? Crop Sci. 2009;49:299–312.
6. Lopes MS, Reynolds MP. Partitioning of assimilates to deeper roots is associated with cooler canopies and increased yield under drought in wheat. Funct Plant Biol. 2010;37:147–56.
7. Lilley JM, Kirkegaard JA. Benefits of increased soil exploration by wheat roots. Field Crop Res. 2011;122:118–30.
8. Singh V, van Oosterom EJ, Jordan DR, Hunt CH, Hammer GL. Genetic variability and control of nodal root angle in sorghum. Crop Sci. 2011;51:2011–20.
9. Mace ES, Singh V, Van Oosterom EJ, Hammer GL, Hunt CH, Jordan DR. QTL for nodal root angle in sorghum (Sorghum bicolor L. Moench) co-locate with QTL for traits associated with drought adaptation. Theor Appl Genet. 2012;124:97–109.
10. Uga Y, Sugimoto K, Ogawa S, Rane J, Ishitani M, Hara N, et al. Control of root system architecture by DEEPER ROOTING 1 increases rice yield under drought conditions. Nat Genet. 2013;45:1097–102.
11. Borrell AK, Mullet JE, George-Jaeggli B, van Oosterom EJ, Hammer GL, Klein PE, et al. Drought adaptation of stay-green in sorghum associated with canopy development, leaf anatomy, root growth and water uptake. Journal of Experimental Botany. 2014;65(21):6251–63. doi:10.1093/jxb/eru232.
12. Veyradier M, Christopher J, Chenu K. 2013. Quantifying the potential yield benefit of root traits. In: Sievänen R, Nikinmaa E, Godin C, Lintunen A, Nygren P, editors. 7th International Conference on Functional-Structural Plant Models. Saariselkä, Finland. p 317-319.
13. Kirkegaard JA, Lilley JM, Howe GN, Graham JM. Impact of subsoil water use on wheat yield. Aust J Agric Res. 2007;58:303–15.
14. Manschadi A, Hammer G, Christopher J, DeVoil P. Genotypic variation in seedling root architectural traits and implications for drought adaptation in wheat (Triticum aestivum L.). Plant Soil. 2008;303:115–29.
15. Manschadi AM, Christopher JT, Hammer GL, Devoil P. Experimental and modelling studies of drought-adaptive root architectural traits in wheat (Triticum aestivum L). Plant Biosyst Int J Deal Aspects Plant Biol. 2010;144:458–62.
16. Olivares-Villegas JJ, Reynolds MP, McDonald GK. Drought-adaptive attributes in the Seri/Babax hexaploid wheat population. Funct Plant Biol. 2007;34:189–203.
17. Christopher JT, Manschadi AM, Hammer GL, Borrell AK. Developmental and physiological traits associated with high yield and stay-green phenotype in wheat. Aust J Agric Res. 2008;59:354–64.

18. Manske GGB, Vlek PLG. Root architecture – wheat as a model plant. In: Waisel Y, Eshel A, Kafkafi U, editors. Plant Roots: The Hidden Half, Third Edition. New York: Marcel Dekker; 2002. p. 249–59.

19. Nakamoto T, Oyanagi A. The direction of growth of seminal roots of triticum aestivum L. and experimental modification thereof. Ann Bot. 1994;73:363–7.

20. Araki H, Iijima M. Deep rooting in winter wheat: rooting nodes of deep roots in two cultivars with deep and shallow root systems. Plant Product Sci. 2001;4:215–9.

21. Bengough AG, Gordon DC, Al-Menaie H, Ellis RP, Allan D, Keith R, et al. Gel observation chamber for rapid screening of root traits in cereal seedlings. Plant Soil. 2004;262:63–70.

22. Wasson AP, Richards RA, Chatrath R, Misra SC, Prasad SVS, Rebetzke GJ, et al. Traits and selection strategies to improve root systems and water uptake in water-limited wheat crops. J Exp Bot. 2012;63:3485–98.

23. Christopher J, Christopher M, Jennings R, Jones S, Fletcher S, Borrell A, et al. QTL for root angle and number in a population developed from bread wheats (*Triticum aestivum*) with contrasting adaptation to water-limited environments. Theor Appl Genet. 2013;126:1563–74.

24. Omori F, Mano Y. QTL mapping of root angle in F2 populations from maize "B73" × teosinte "*Zea luxurians.*". Plant Root. 2007;1:57–65.

25. Uga Y, Okuno K, Yano M. Dro1, a major QTL involved in deep rooting of rice under upland field conditions. J Exp Bot. 2011;62:2485–94.

26. Yu J, Holland JB, McMullen MD, Buckler ES. Genetic design and statistical power of nested association mapping in maize. Genetics. 2008;178:539–51.

27. Buckler ES, Holland JB, Bradbury PJ, Acharya CB, Brown PJ, Browne C, et al. The genetic architecture of maize flowering time. Science. 2009;325:714–8.

28. Neumann G, George T, Plassard C. Strategies and methods for studying the rhizosphere—the plant science toolbox. Plant Soil. 2009;321:431–56.

29. Neill C. Comparison of soil coring and ingrowth methods for measuring belowground production. Ecology. 1992;73:1918–21.

30. Trachsel S, Kaeppler S, Brown K, Lynch J. Shovelomics: high throughput phenotyping of maize (Zea mays L.) root architecture in the field. Plant Soil. 2011;341:75–87.

31. Furbank RT, Tester M. Phenomics – technologies to relieve the phenotyping bottleneck. Trends Plant Sci. 2011;16:635–44.

32. Gregory PJ, Hutchison DJ, Read DB, Jenneson PM, Gilboy WB, Morton EJ. Non-invasive imaging of roots with high resolution X-ray micro-tomography. Plant Soil. 2003;255:351–9.

33. Lontoc-Roy M, Dutilleul P, Prasher S, Liwen H, Brouillet T, Smith D. Advances in the acquisition and analysis of ct scan data to isolate a crop root system from the soil medium and quantify root system complexity in 3-d space. Geoderma. 2006;137:231–41.

34. Hargreaves C, Gregory P, Bengough AG. Measuring root traits in barley (*Hordeum vulgare* ssp. *vulgare* and ssp. *spontaneum*) seedlings using gel chambers, soil sacs and X-ray microtomography. Plant Soil. 2009;316:285–97.

35. Mooney SJ, Pridmore TP, Helliwell J, Bennett MJ. Developing X-ray computed tomography to non-invasively image 3-D root systems architecture in soil. Plant Soil. 2012;352:1–22.

36. Mairhofer S, Zappala S, Tracy S, Sturrock C, Bennett MJ, Mooney SJ, et al. Recovering complete plant root system architectures from soil via X-ray mu-computed tomography. Plant Methods. 2013;9:8.

37. Hendrick R, Pregitzer K. Spatial variation in tree root distribution and growth associated with minirhizotrons. Plant Soil. 1992;143:283–8.

38. Ao J, Fu J, Tian J, Yan X, Liao H. Genetic variability for root morph-architecture traits and root growth dynamics as related to phosphorus efficiency in soybean. Funct Plant Biol. 2010;37:304–12.

39. Vamerali T, Bandiera M, Mosca G. Minirhizotrons in modern root studies. In: Mancuso S, editor. Measuring Roots. Berlin Heidelberg, Berlin: Springer-Verlag; 2012. p. 341–62.

40. Nagel KA, Putz A, Gilmer F, Heinz K, Fischbach A, Pfeifer J, et al. GROWSCREEN-Rhizo is a novel phenotyping robot enabling simultaneous measurements of root and shoot growth for plants grown in soil-filled rhizotrons. Funct Plant Biol. 2012;39:891–904.

41. Lobet G, Draye X. Novel scanning procedure enabling the vectorization of entire rhizotron-grown root systems. Plant Methods. 2013;9:1.

42. Miyamoto N, Steudle E, Hirasawa T, Lafitte R. Hydraulic conductivity of rice roots. J Exp Bot. 2001;52:1835–46.

43. Hund A, Trachsel S, Stamp P. Growth of axile and lateral roots of maize: I development of a phenotying platform. Plant Soil. 2009;325:335–49.

44. Iyer-Pascuzzi AS, Symonova O, Mileyko Y, Hao Y, Belcher H, Harer J, et al. Imaging and analysis platform for automatic phenotyping and trait ranking of plant root systems. Plant Physiol. 2010;152:1148–57.

45. Passioura JB. The perils of pot experiments. Funct Plant Biol. 2006;33:1075–9.

46. Passioura JB. Scaling up: the essence of effective agricultural research. Funct Plant Biol. 2010;37:585–91.

47. Poorter H, Bühler J, van Dusschoten D, Climent J, Postma JA. Pot size matters: a meta-analysis of the effects of rooting volume on plant growth. Funct Plant Biol. 2012;39:839–50.

48. Gregory PJ, Bengough AG, Grinev D, Schmidt S, Thomas WTB, Wojciechowski T, et al. Root phenomics of crops: opportunities and challenges. Funct Plant Biol. 2009;36:922–9.

49. Chenu K, Cooper M, Hammer GL, Mathews KL, Dreccer MF, Chapman SC. Environment characterization as an aid to wheat improvement: interpreting genotype–environment interactions by modelling water-deficit patterns in North-Eastern Australia. J Exp Bot. 2011;62:1743–55.

50. Chenu K, Deihimfard R, Chapman SC. Large-scale characterization of drought pattern: a continent-wide modelling approach applied to the Australian wheatbelt – spatial and temporal trends. New Phytol. 2013;198:801–20.

51. Chenu K. Chapter 13 - Characterizing the crop environment – nature, significance and applications. In: Calderini VOSF, editor. Crop Physiology (Second Edition). San Diego: Academic; 2014. p. 321–48.

52. Slafer GA. Genetic basis of yield as viewed from a crop physiologist's perspective. Ann Appl Biol. 2003;142:117–28.

53. Gomez-Macpherson H, Richards R. Effect of sowing time on yield and agronomic characteristics of wheat in south-eastern Australia. Aust J Agric Res. 1995;46:1381–99.

54. Rebetzke GJ, Chenu K, Biddulph B, Moeller C, Deery DM, Rattey AR, et al. A multisite managed environment facility for targeted trait and germplasm phenotyping. Funct Plant Biol. 2013;40:1–13.

55. Blum A, Shpiler L, Golan G, Mayer J. Yield stability and canopy temperature of wheat genotypes under drought-stress. Field Crop Res. 1989;22:289–96.

56. Rebetzke GJ, Rattey AR, Farquhar GD, Richards RA, Condon A, Tony G. Genomic regions for canopy temperature and their genetic association with stomatal conductance and grain yield in wheat. Funct Plant Biol. 2012;40:14–33.

57. Mitchell JH, Chapman SC, Rebetzke GJ, Bonnett DG, Fukai S. Evaluation of a reduced-tillering (tin) gene in wheat lines grown across different production environments. Crop Pasture Sci. 2012;63:128–41.

58. Rich SM, Watt M. Soil conditions and cereal root system architecture: review and considerations for linking Darwin and Weaver. J Exp Bot. 2013;64:1193–208.

59. Saxena DC, Sai Prasad SV, Chatrath R, Mishra SC, Watt M, Prashar R, et al. Evaluation of root characteristics, canopy temperature depression and stay green trait in relation to grain yield in wheat under early and late sown conditions. Ind J Plant Physiol. 2014;19:43–7.

60. Chapman S, Cooper M, Podlich D, Hammer G. Evaluating plant breeding strategies by simulating gene action and dryland environment effects. Agron J. 2003;95:99–113.

61. Hammer GL, Chapman S, van Oosterom E, Podlich DW. Trait physiology and crop modelling as a framework to link phenotypic complexity to underlying genetic systems. Aust J Agric Res. 2005;56:947–60.

62. Hickey L, Dieters M, DeLacy I, Kravchuk O, Mares D, Banks P. Grain dormancy in fixed lines of white-grained wheat (Triticum aestivum L.) grown under controlled environmental conditions. Euphytica. 2009;168:303–10.

63. Hickey LT, Wilkinson PM, Knight CR, Godwin ID, Kravchuk OY, Aitken EAB, et al. Rapid phenotyping for adult-plant resistance to stripe rust in wheat. Plant Breed. 2011;131:54–61.

64. Hickey L, Dieters M, DeLacy I, Christopher M, Kravchuk O, Banks P. Screening for grain dormancy in segregating generations of dormant × non-dormant crosses in white-grained wheat (Triticum aestivum L). Euphytica. 2010;172:183–95.

65. Schneider CA, Rasband WS, Eliceiri KW. NIH Image to ImageJ: 25 years of image analysis. Nat Meth. 2012;9:671–5.

66. Patterson HD, Thompson R. Recovery of inter-block information when block sizes are unequal. Biometrika. 1971;58:545–54.

67. Butler D, Cullis B, Gilmour A, Gogel B, Gogel B. {ASReml}-R Reference Manual. 2009.

20

Plant phenotyping: from bean weighing to image analysis

Achim Walter*, Frank Liebisch and Andreas Hund

Abstract

Plant phenotyping refers to a quantitative description of the plant's anatomical, ontogenetical, physiological and biochemical properties. Today, rapid developments are taking place in the field of non-destructive, image-analysis -based phenotyping that allow for a characterization of plant traits in high-throughput. During the last decade, 'the field of image-based phenotyping has broadened its focus from the initial characterization of single-plant traits in controlled conditions towards 'real-life' applications of robust field techniques in plant plots and canopies. An important component of successful phenotyping approaches is the holistic characterization of plant performance that can be achieved with several methodologies, ranging from multispectral image analyses via thermographical analyses to growth measurements, also taking root phenotypes into account.

The conceptual and methodological basis of phenotyping

The terms phenotype and genotype were coined by the Danish plant scientist Wilhelm Johannsen [1,2]. Half a century after Mendel's experiments on the basis of inheritance and in a time of dispute between the Darwinian and Lamarckian view of evolution, he performed experiments on the heritability of seed size in self-fertilizing beans. Johannsen selected large and small beans of a variety and observed significant difference in seed sizes of the progenies. He concluded that there must be a genetic effect influencing seed size. However, when he selected again within individual plants of the progenies, he could not influence seed size anymore. He concluded that he had selected pure lines for which the phenotype was only driven by environmental effects, such as the seed position on the plant. In his own words Johannsen [2] stated:

"All 'types' of organisms, distinguishable by direct inspection or only by finer methods of measuring or description, may be characterized as 'phenotypes'. Certainly phenotypes are real things; the appearing (not only apparent) 'types' or 'sorts'of organisms are again and again the objects for scientific research. All typical phenomena in the organic world are eo ipso

phenotypical, and the description of the myriads of phenotypes as to forms, structures, sizes, colors and other characters of the living organisms has been the chief aim of natural history, –which was ever a science of essentially morphological-descriptive character.... Hence we may adequately define this conception as a 'phenotype-conception' in opposition to the 'genotype-conception'."

Since then, the term phenotype has been used to describe a wide range of traits in plants, microbes, fungi and animals. In plant breeding and quantitative genetics, usually hundreds or even thousands of measurements are performed to select superior individuals or identify regions in the genome controlling a trait. This demands for high-throughput phenotyping, which has been and is still most widely accomplished by quantitative and qualitative assessments (rating) performed by plant breeders. The term 'phenotyping' was beginning to be used in the 1960s. In plants, the increasing capabilities of analytical chemistry allowed to broaden the concept of a quantitative analysis of traits to the description of the variability of proteins [3], of metabolic pathways [4] and of other 'real things' connected to the character of living plants. From Johannsen's description, it is clear that phenotyping – no matter whether in plants, bacteria, fungi or animals – is characterized by an enormous amount of processes, functions, structures, or – most generally spoken – dimensions (Figure 1). In this sense, phenotyping can be considered as far more complex than

* Correspondence: achim.walter@usys.ethz.ch
Institute of Agricultural Sciences, ETH Zürich, Universitätstrasse 2, 8092 Zürich, Switzerland

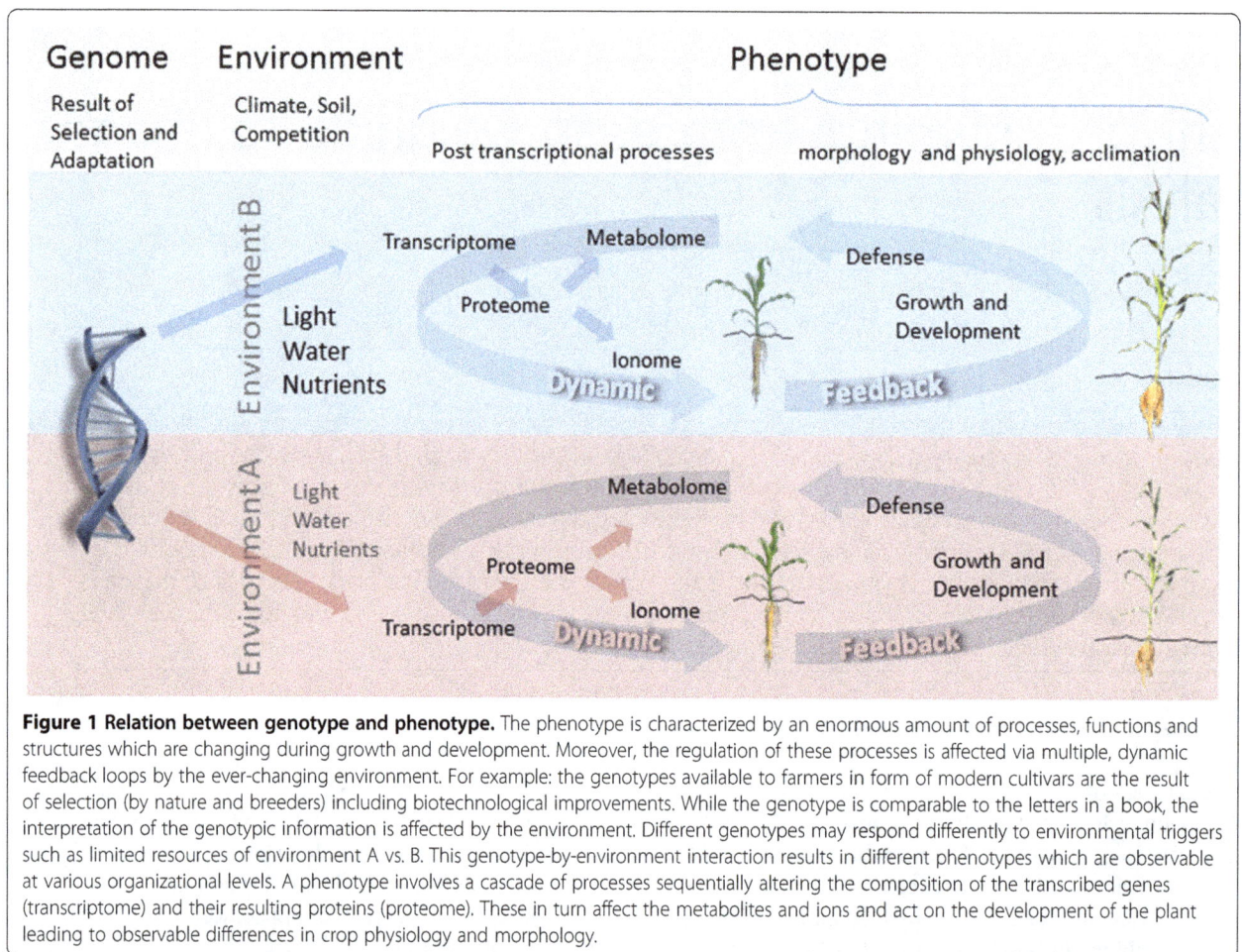

Figure 1 Relation between genotype and phenotype. The phenotype is characterized by an enormous amount of processes, functions and structures which are changing during growth and development. Moreover, the regulation of these processes is affected via multiple, dynamic feedback loops by the ever-changing environment. For example: the genotypes available to farmers in form of modern cultivars are the result of selection (by nature and breeders) including biotechnological improvements. While the genotype is comparable to the letters in a book, the interpretation of the genotypic information is affected by the environment. Different genotypes may respond differently to environmental triggers such as limited resources of environment A vs. B. This genotype-by-environment interaction results in different phenotypes which are observable at various organizational levels. A phenotype involves a cascade of processes sequentially altering the composition of the transcribed genes (transcriptome) and their resulting proteins (proteome). These in turn affect the metabolites and ions and act on the development of the plant leading to observable differences in crop physiology and morphology.

the analysis of the linear arrangement of genes in the genotype [5]. A comprehensive characterization of the phenotype of any plant – no matter whether it is a model plant or a crop – is far out of reach of the research capabilities of our generation. Therefore, a comprehensive, phenotypic model description of a plant – similar to the model description of an engine – will remain a distant aim of future, 'Systems Biology'.

Today, high-throughput plant phenotyping refers to the characterization of the whole cascade of changes happening after DNA is transcribed into RNA (transcriptomics) leading to the formation of proteins (proteomics). This cascade from DNA via RNA to proteins, known as the central dogma of molecular biology, determines other plant phenotypic traits, such as metabolites (metabolomics), ions (ionomics), and, last but not least morphological or architectural parameters [6]. According to Guo and Zhu [7], "the purpose of phenotyping is to produce a description of the plant's anatomical, ontological, physiological, and biochemical properties". Throughout the last decade, the terms phenotyping and phenomics have more and more often been linked to non-destructive optical analyses of plant traits based on images [8,9]. Thereby,

phenotype analysis is turning its focus back towards the object of interest of Johannsen, but instead of counting, weighing or measuring the length of beans, it is now using image analysis to quantitatively determine Johannsens 'real things' [2] of plants in an increasingly holistic and integrative manner. This reorientation began with a study investigating growth of several Arabidopsis genotypes [10]. Utilizing digital imaging to resolve plant rosette area, this pioneering study revealed growth differences between wild-type plants and plants deficient in their photosynthetic capacity within a few days. Today, after 15 years of development within this new scientific field, phenotyping has begun to become a toolbox applicable also to plant breeders to select desirable genotypes for their specific field of interest – be it salt-tolerance in *Triticum* [11], drought-tolerance in barley [12] or maize [13]. Looking back at somewhat more than a decade of non-destructive, image analysis-based plant phenotyping, one can state that the focus of phenotyping has broadened to a certain extent from basic-science oriented analysis of phenotypic differences between a wild-type and a mutant plant from experiments with potted single plants to the analysis of plant plots and canopies in field experiments in the context of plant breeding [14,15] or precision agriculture [16,17]. Of course, a lot of current, image-based phenotyping approaches have originated from the use of non-imaging sensors which have been applied in the field, such as thermography point sensors. Other methods have been introduced from the field of remote sensing, such as the satellite-based calculation of spectral indices.

Based on the pioneering approach of Leister [10], in greenhouses and growth chambers, numerous automated facilities and robots have been set up that allow for the comparison of several hundred plants per day in an automated manner. These setups form the working horse for a lot of scientific investigations in basic research of the public and private sector alike. Moreover, these setups are continuously being refined and their image-processing capabilities form the basis for a next generation of phenotyping platforms that are operating in the field from different carriers such as tractors [18,19], blimps [20,21] or unmanned aerial vehicles [22]. What we are experiencing today is the beginning of a combined use of multiple

imaging (or non-imaging, but remote sensing based) technologies for the quantitative description of the performance of plants during their entire ontogeny in their environment (Figure 1). In this early phase of computer-vision-based plant phenotyping methods, concepts and approaches are proposed, which allow for a characterization of the overall performance of a plant in its given environment. Of course, these developments would not have been possible without decades of pioneering work in photogrammetry and remote sensing, which is the science and technology of obtaining information about physical objects and the environment through the process of recording, measuring and interpreting imagery derived from non – contact sensor systems [23].

To get a more comprehensive overview on the achievements of plant phenotyping, it may be helpful to structure the state of the art into four main classes of methods that are currently being used (Figures 2 and 3). These methods are related a) to the spectral reflectance and absorbance of leaf, plant and canopies, b) to the plant or canopy

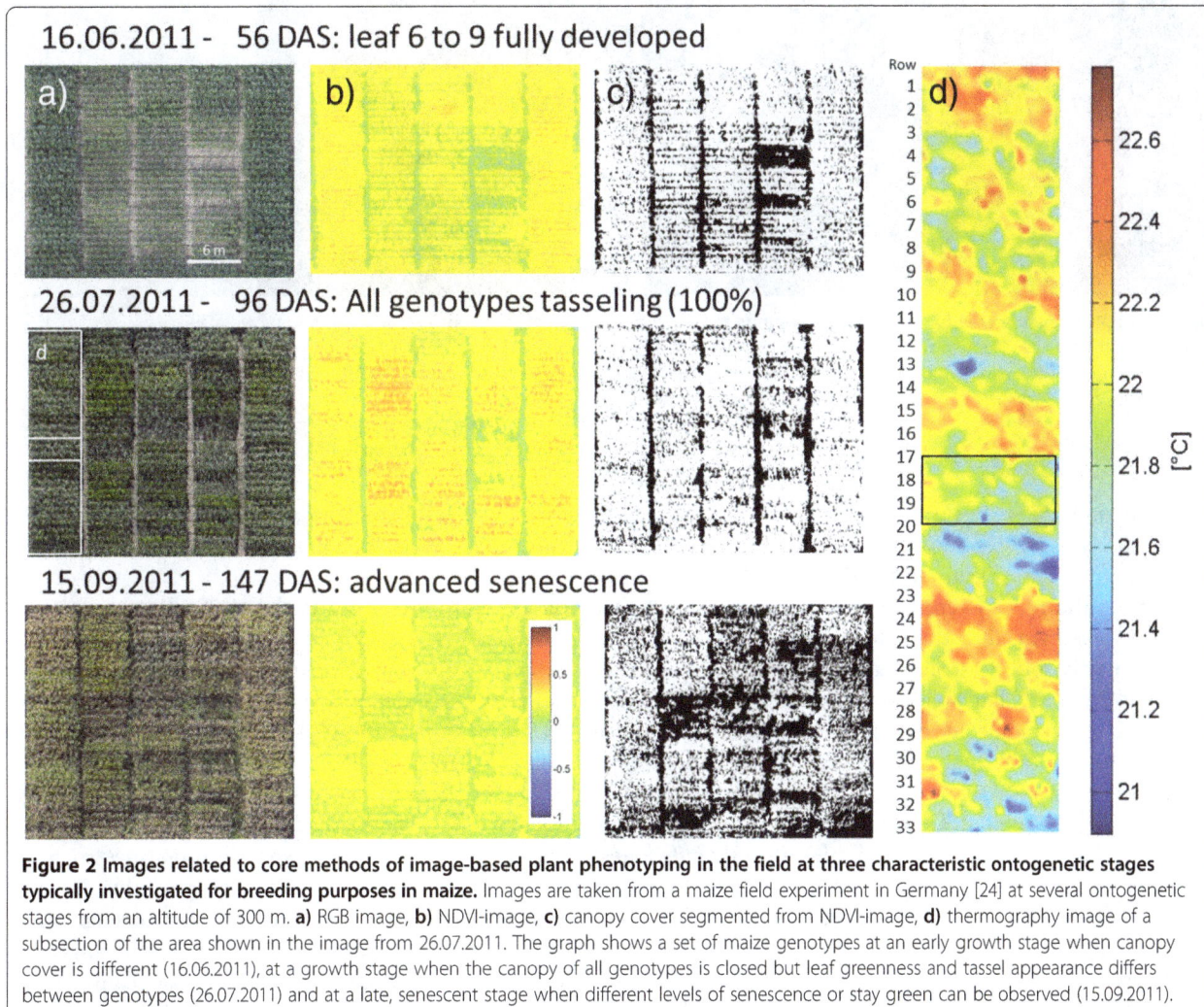

Figure 2 Images related to core methods of image-based plant phenotyping in the field at three characteristic ontogenetic stages typically investigated for breeding purposes in maize. Images are taken from a maize field experiment in Germany [24] at several ontogenetic stages from an altitude of 300 m. **a)** RGB image, **b)** NDVI-image, **c)** canopy cover segmented from NDVI-image, **d)** thermography image of a subsection of the area shown in the image from 26.07.2011. The graph shows a set of maize genotypes at an early growth stage when canopy cover is different (16.06.2011), at a growth stage when the canopy of all genotypes is closed but leaf greenness and tassel appearance differs between genotypes (26.07.2011) and at a late, senescent stage when different levels of senescence or stay green can be observed (15.09.2011).

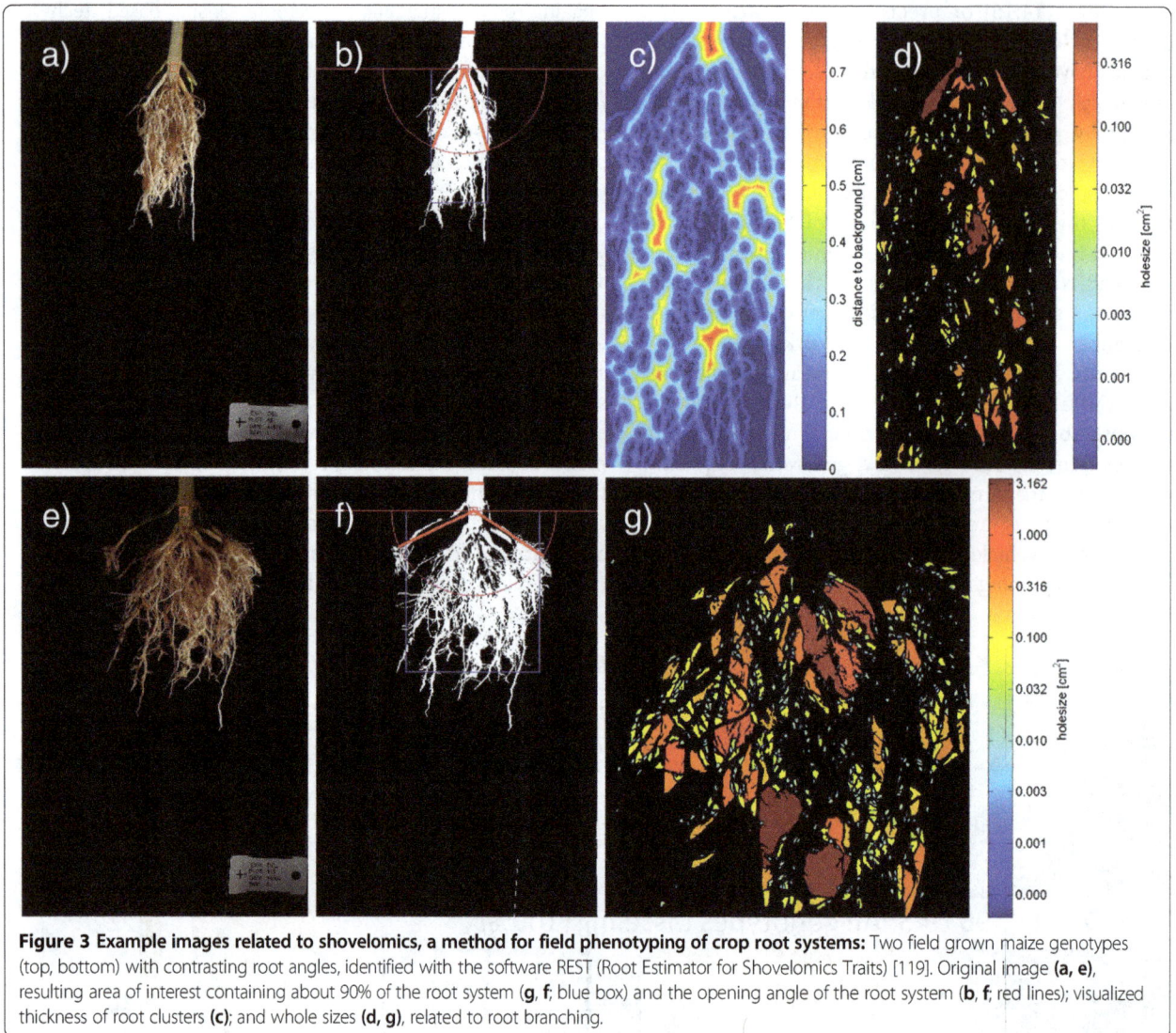

Figure 3 Example images related to shovelomics, a method for field phenotyping of crop root systems: Two field grown maize genotypes (top, bottom) with contrasting root angles, identified with the software REST (Root Estimator for Shovelomics Traits) [119]. Original image **(a, e)**, resulting area of interest containing about 90% of the root system (**g, f**; blue box) and the opening angle of the root system (**b, f**; red lines); visualized thickness of root clusters **(c)**; and whole sizes **(d, g)**, related to root branching.

temperature and derived indicators for transpiration and water status, c) to size, morphology, architecture or growth of plants or their canopies and finally d) to the architecture of the root system analyzed in the lab and in the field. Distinguishing these four methodological classes only partly reflects structural or functional core categories of the plant, but it shows the current activity of the plant phenotyping community driven by available sensor technologies and analysis methods. In all technologies and for all research aspects, it is crucial to attempt a precise positioning of the required sensors and to perform reliable measurements at high-throughput and high precision to advance our capabilities and to arrive at a more holistic characterization of plant or crop performance [14]. Future phenotyping approaches will most probably analyze several aspects of plant performance at the same time, potentially using multiple

sensors, thereby resolving complex traits, such as demonstrated e.g. by Liebisch et al. [24].

Spectral assessment of plant shoots and canopies

Spectral indicators used for plant trait detection and phenotyping range from simple ratios calculated from responses at two wavelengths [13,16], via normalized indices [13,16] (example discussed below) to very complex equations and algorithms [25,26]. A very immediate indicator of plant performance is its leaf color. Our eye is highly sensitive to different shades of green. Leaf greenness is determined by genotype specific properties such as content and development of leaf chlorophyll, by plant health and by leaf morphological characteristics such as thickness and surface structure. Leaf greenness changes according to plant development, is affected by plant nutrition and environmental stresses such as cold, heat and drought stress. The most frequently used indicator for leaf greenness in remote sensing is the normalized difference vegetation index (NDVI) [27,28], which exploits the difference between reflectance in certain regions of the visible light spectrum (VIS), where absorption of chlorophyll is maximal, and in the near-infrared part of the spectrum (NIR), which is not affected by photosynthesis. Most often, reflectance and absorption in the visual range are narrowed down either to the red or to the blue region of the spectrum, where chlorophyll and light harvesting antenna pigments are absorbing maximally.

The exact calculation of NDVI with respect to the wavelengths used depends on the objectives of the study and on the sensor, but in general it is calculated as NDVI = (NIR-VIS)/(NIR + VIS), thereby normalizing the difference between reflection in the visible (VIS) and the near-infrared range (NIR) to the sum of reflected light in both ranges. For the visible range often red or blue bands are used for detection of NDVI [24,27,28]. This allows for a comparison of plants or canopies in different illumination situations, such as in a field during a somewhat cloudy day or in order to compare between measurements taken at different days (Figure 2). With respect to plant phenotyping, NDVI has been applied to study phenology changes in crops [29-31], to study vegetation ecology [32,33], stress [34,35] and nitrogen status [36-38]. Remote measurement of leaf greenness on the canopy scale is affected by the angle of the optics towards the canopy, by illumination conditions and by canopy characteristics such as canopy height and leaf angle distribution. Recent studies in wheat and maize correlate NDVI with important crop properties such as biomass [39,40], chlorophyll content [41,42] and nitrogen status [34,43]. The development of NDVI during a season or an extended period of time can be used to investigate traits important for breeding such as stay green [44,45] and growth rates [46].

It has to be pointed out that NDVI is just one of a huge number of spectral indices that can be utilized for remote characterization of plant performance [16,26,47,48] in the field and in laboratory studies. Other indices, such as the 'modified chlorophyll absorption ratio index' take spectral components of green light into account and provide thereby some information about the density of the canopy, since green light is also reflected to the sensor from deeper layers of the canopy, penetrating the top layers. For some phenotyping applications, the analysis of canopy coloration is performed not from purely reflected sunlight, but the plant canopy is actively illuminated. In agricultural management, sensors such as the 'GreenSeeker[TM]' (NTech Industries, Inc., USA) or 'Crop Circle[TM]'(Holland Scientific Inc., USA) have been introduced to the market years ago: There, the canopy is actively illuminated by hand-held or tractor-mounted devices that also perceive

and interpret the reflected radiation. Based on the calculated NDVI or greenness indicator, fertilization of the investigated crop patch is performed (low greenness – more nitrogen fertilizer required), taking species-specific crop models into account.

The analysis of a certain part of the visible light spectrum following an induction by active illumination is also the foundation for phenotyping analyses based on chlorophyll fluorescence of a canopy [49]. Chlorophyll fluorescence has been used to describe the performance of the photosynthetic apparatus from the analysis of light emitted at longer wavelengths and at later times (between μs and a few minutes) following up on an illumination pulse. It has been noted in the 1930s [50] that due to electron transfer processes within the photosystem, characteristic intensities of photons are emitted that can be used to derive potential photosynthetic yield, quantum efficiency of photosystem II (ϕ_{PSII}) and other parameters. Chlorophyll fluorescence has successfully been used in phenotyping studies in the laboratory [51-53] and in the field [54]. Traditionally, chlorophyll fluorescence has been measured using hand-held devices. It was successfully applied to select maize with greater cold tolerance of the photosynthetic apparatus [55]. In the field, laser-induced chlorophyll florescence was for example used to determine biomass and nitrogen status in oilseed rape [56]. A problem of actively remotely sensed chlorophyll fluorescence is that it needs a true saturating light pulse in order to determine crucial parameters, such as ϕ_{PSII}, which can be achieved by applying a laser from a long distance. Alternatively, very high spectral resolution in ideally sub-nm range is used today for passive estimation of chlorophyll fluorescence from solar reflectance spectra by for example the Fraunhofer line depth technique (FLD) [57-59].

Thermography-based investigations of transpiration in the soil-plant-atmosphere continuum

Another important line of research utilizes sensors that detect canopy temperature from long wavelength infrared radiation according to the relation between body temperature and the light spectrum emitted from that body [60,61]. Since plant tissues are cooled by transpiration of water, canopy temperature can be linked to transpiration rates, if the temperature of the canopy and of the surrounding environment can be analyzed precisely enough. Therefore, thermal imaging offers a large potential for non-destructive measurement of plant water status for irrigation management and for phenotyping [62] in the context of stress tolerance or drought stress avoidance [63]. Yet, it is far from being trivial to interpret plant temperatures correctly, since they depend strongly on the microclimate of the plant stand and they need to be balanced carefully with reference temperatures of non-transpiring and/or fully transpiring canopies in close spatial and temporal vicinity (see [61] for more details). Another constraint of thermal imaging is the high temporal and spatial variability caused by a) environmental conditions changing rapidly in the field e.g. on cloudy days [63-65], and b) different canopy densities of different genotypes that can lead to non-comparable microclimatic conditions in multi-plot field experiments [62]. Plant canopy temperatures may also be strongly affected by differences in development of examined genotypes. Early flowering and a concomitantly earlier start of senescence e.g. affect canopy temperature by reducing transpiration per square meter from an ageing canopy, which carefully needs to be taken into account. Plant density differences might be caused by different germination rate resulting from field variation in soil properties, by genotypic differences or by different sowing density. The effect of background temperatures can be separated by normalization to background temperatures [63], but other climate parameters such as radiation or wind speed affect leaf temperature as well [66] and their quantitative effect is not well understood under field conditions. Nevertheless, it has to be pointed out that thermography is a powerful, integrative tool to differentiate between phenotypes, especially if it is used to test the overall effect of precisely defined physiological aberrances on certain plant genotypes in the field. An example for such an application is the detection of early stress symptoms of plant diseases [67] which affect transpiration. Thereby, thermography facilitates phenotyping for disease-resistant plant genotypes – one of the most important plant breeding aims in all major crops.

Optical analysis of aboveground plant size, organ and canopy growth

The most direct, overall plant performance indicator – at least during the ontogenetic phase of vegetative development – is the growth of plant biomass or plant size. As mentioned above, this is often monitored in a global way by assessing the number of pixels, which an individual plant or the total canopy of an experimental plot is covering within an image of calibrated size [10]. Such methods have been successfully applied in the laboratory to assess the performance of Arabidopsis [68], tobacco [69] or cereal grain crops [11,12]. All global players of the agrobiotech business do have such monitoring platforms, with which they test differences between genotypes [51] or effects of plant protection or plant strengthening substances applied to a crop of interest. Size analysis of the plant is not as straightforward as it may seem since the precision of the measurement depends strongly on the orientation between canopy and sensor, on the precise distinction between object and background and other pitfalls of image analysis in the context of plant phenotyping [70].

In field experiments, plant size is not only estimated from top-view images, but – especially for monocotyledonous crops – by analysis of canopy height from measurements of light barriers mounted on tractors that analyze the top level of a canopy as the tractor pulls the light barriers along the seeded rows of the crop [19,71]. Other field applications comprise the analysis of canopy cover (CC), which simply refers to the fraction of the ground that is covered by the canopy [21,24,72-74]. The CC trait can be used to detect temporal and genotypic differences and it is linked to important plant traits such as early vigor and senescence that have long been used for crop breeding, turning CC into one of the key traits for 'next generation phenotyping' [15]. CC can be calculated from digital images with a red, green and blue channel (RGB) or NDVI images, segmenting the green plant from the non-green background or even from images that assess chlorophyll fluorescence. In principle, it is also possible to determine plant shape, number of leaves, and structure of the canopy or leaf area index from such images – especially when they are used to reconstruct the 3D-shape of the canopy either from multiple images or from scanning the canopy. 3D-reconstruction already works in the lab [75-77], but is challenging to be reliably performed in the field [78]. Yet, with the increase of computing power and with modern imaging capacities of unmanned aerial vehicles [16,22] and other devices that are capable of generating plant images in the field from multiple perspectives, it should be possible to advance enormously in this area in the near future. Then, automatic counting of tiller numbers, ear densities, fractions of damaged leaves and other traits relevant in classical breeding programs can be performed. Also, the dynamic development (when does a plant grow how intensely) and the relation of plant growth and environmental parameters (which genotypes grow best at certain temperatures) will form an important focus of next generation phenotyping. In a proof-of-concept study Grieder et al. [79] investigated wheat genetic variation in growth response to temperature using image based phenotyping in the field.

Root phenotyping at high-throughput in controlled conditions

Root phenotyping is as important as shoot phenotyping, since the performance of any plant strongly depends on its root architecture and function [80-82]. The added value of root phenotyping becomes obvious e.g. in breeding programs, in which it is shown that root traits sometimes have a higher heritability than the aboveground target trait (e.g. grain yield). Good examples of the high importance of root architecture are a) the benefit of shallow rooting in phosphorous-poor soils, which maximizes P uptake from the topsoil [83] and b) the benefit of aluminium tolerance in acidic tropical soils which enhances deep rooting [84].

For methodological reasons, root phenotyping capabilities have been developed in the laboratory first and are now evolving towards field applicability – in a similar manner, but with some temporal delay compared to shoot phenotyping capabilities. Laboratory-based methods to study root growth were recently reviewed by Zhu et al. [82]. Root phenotyping platforms and methodologies usually combine some degree of automation with imaging and image processing. To facilitate the inspection of roots, special care has to be taken how to cultivate plants in a way that allows for normal plant development and for access to the root. The most basic and hence most widely used systems to observe roots are based on soil-free growth media. There, the root either grows in paper rolls [85] on the surface of germination papers [86,87], or gels [88-90], in air regularly sprayed with nutrient solution [91] or in aerated aqueous solutions [92]. Another version of hydroponics, which includes some degree of mechanical resistance, is to cultivate roots in transparent plexiglas nail board sandwiches filled with 1.5 mm glass beads through which a nutrient solution is circulated [93]. In all of these systems, total root length, branching angles and other parameters are determined, using manual measurement, visual rating or imaging. Imaging needs to be performed with high resolution scanners or cameras to be able to resolve lateral roots for image processing. The basic global evaluation of images extracts root length. Often, individual root diameters are used as decision criterion to distinguish between the main roots and their lateral branches [86,94] typically upon usage of the software WinRhizo [95,96]. With the development of suitable software such as SmartRoot [97] that allows for topology analyses, root system architecture can be analyzed in detail for branching angles etc. and growth kinematics of individual roots within the root system [87]. Still substantial manual input is required for such analyses [87]. Thus, there is still the need for significant improvement of image processing, even for soil-free systems in which roots are comparably easy to detect. In case of soil as growth medium, image processing becomes even more challenging. However, more natural systems like soil-filled rhizotrons or growth columns are indispensable to study the interaction of roots with edaphic factors. For example, soil compaction or the effects of drying soil are difficult to establish in soil-free systems.

Rhizotrons or columns, filled with soil or other growth substrates, enable a direct inspection of roots along a transparent wall [98] or within a small soil column by using x-ray based computed tomography [99,100] to visualize the 3D-configuration of roots. The most advanced versions of these systems combine large soil volumes with high-throughput and automation. These are the soil-filled 2D rhizotrons of the GROWSCREEN-Rhizo platform with a rooting depth of 90 cm [101] and 25 cm diameter-by-100 cm growth columns in combination with µCT imaging at the Hounsfield facility of University of Nottingham [102]. Lysimeters in form of tall columns that are placed with a distance to each other in order to simulate a planting density as under field conditions are well suited to get an indirect measure of rooting depth and water uptake by means of regular weighing [103]. Such systems can serve as an excellent bridge between controlled conditions and real field conditions. Other methodologies with the potential to study root system architecture and functioning in the future are nuclear magnetic resonance [104-106], neutron radiography [107] and positron emission tomography [108], which allow segmenting the root from the surrounding substrate.

Root phenotyping in the field

Due to the hidden nature of roots, it is extremely difficult to assess them optically in the field – unless one is digging them out or one approaches them using a tunnel. Therefore, the most widely used traditional methodology to study roots in the field is the so-called 'trench profile' method, in which soil is carefully removed from the side, often using fine brushes, and in which the root system is then sequentially revealed and drawn layer by layer from successive profile walls [109-111]. In other approaches, soil cores are taken in order to sample vertical root length densities or weights, sometimes using semiautomatic extraction methods [112,113]. A far more rapid method to evaluate the maximum rooting depth from soil samples is the core break method, developed by Bohm [114]. In this method, soil cores of up to 2 m length are broken into sections of 10 cm to determine the maximum rooting depth, corresponding to the depth of the last interface at which a root is observed [85].

Another promising 'field-technique' widely practiced is the analysis of excavated upper parts of the main root system [115-117]. This method is termed "Shovelomics" [117] and is performed by excavating a few liters of soil with one crop plant in the center of the surface. Soil is gently washed away from the top part of the root system and the core skeleton of the main root branches is then analyzed for parameters such as root angles and densities (Figure 3). Analysis methods comprise a wide range of different techniques from simple rating and counting [117] to imaging in combination with custom image analysis software [115,118,119], allowing to measure basic root characteristics related to branching, root dimensions and structure (Figure 3). The shovelomics method still has to prove its value for trait-based selection by delivering new yield-related traits that cannot be measured sufficiently above-ground.

Other techniques applicable to field studies are based on so-called mini-rhizotron systems which consist of plexiglas tubes inserted into the soil, in which a small camera or a scanner, is inspecting the surrounding root-soil continuum (see review by Johnson et al. [120]). Limited numbers of genotypes may be monitored using these mini-rhizotrons [121,122]. Several other, indirect methods were proposed and used to analyze root system architecture or overall root performance, such as root pulling resistance [123] or the analysis of leaf abscisic acid content [124]. Total root mass has been proposed as a trait to be measured by electrical capacitance measurements that analyze the response behavior of currents applied to one electrode inserted at the base of the stem and to another electrode in the rooting substrate [125,126]. This method has been used in high-throughput analyses of root mass in the field [127,128], but recent studies indicate that the "root capacitance" may be more related to the cross sectional area (or circumference) of the root at the soil [129] or solution surface [130]. These observations cast some doubt on the reliability of this otherwise promising approach to explore root-soil interactions and root phenotypes based on electrical properties. Clearly, the intensity of water uptake is related to transpiration (and thereby can be assessed with thermography as shown above) and it alters electrical properties of the soil in a way that can be determined by changes of the total electrical resistivity of soil situated between two electrodes [131]. Maybe, a dynamic analysis of ion and water content in the rhizosphere, which can be performed on the basis of electrical analyses, will become an element of our capabilities to characterize an important trait of the multidimensional plant phenotype: water and nutrient uptake. Of course, this does not depend only on the root system architecture, but also on intrinsic hydraulic properties and the uptake and transport efficiency of tissues [132,133]. Therefore, the set of methods to analyze overall indicators of plant performance in plant phenotyping will increase surely in the near future, allowing then to obtain a more and more holistic view of plant performance.

Conclusion

The field of plant phenotyping is still under rapid development at the moment. Image-based plant phenotyping is beginning to prove its value not only in basic science, but also in crop breeding and precision agriculture, providing a quantitative basis of the description of plant-environment-interactions. Key to the success is the ease and applicability of modern image analysis approaches that are applied at multiple points in time throughout crop development, thereby allowing for cost-efficient high-throughput phenotyping at appropriate ontogenetical stages. Since the potential of image analysis in the context of plant phenotyping is far from being adequately exploited, the scientific field of plant phenotyping can be expected to continue prospering throughout the coming years.

Competing interests

The authors declare that they have no competing interests.

Authors' contributions

All authors contributed to manuscript writing. All authors read and approved the final manuscript.

References

1. Johannsen W. Erblichkeit in Populationen und reinen Linien. Jena: Gustav Fischer Verlag; 1903.
2. Johannsen W. The genotype conception of heredity. Am Nat. 1911;45:129–59.
3. Schulze WX, Usadel B. Quantitation in Mass-Spectrometry-Based Proteomics. Annu Rev Plant Biol. 2010;61:491–516.
4. Schauer N, Fernie AR. Plant metabolomics: towards biological function and mechanism. Trends Plant Sci. 2006;11:508–16.
5. Houle D, Govindaraju DR, Omholt S. Phenomics: the next challenge. Nat Rev Genet. 2010;11:855–66.
6. Normanly J. High-Throughput Phenotyping in Plants - Methods and Protocols. Methods in Molecular Biology, vol. 918. Springer New York Heidelberg Dordrecht London: Humana Press; 2012. p. 365.
7. Guo Q, Zhu Z. Phenotyping of plants. Encyclopedia of Analytical Chemistry. 2014, published online: http://onlinelibrary.wiley.com/doi/10.1002/9780470027318.a9934/full.
8. Furbank RT, Tester M. Phenomics - technologies relieve the phenotyping bottleneck. Trends Plant Sci. 2011;16:635–44.
9. Fiorani F, Schurr U. Future scenarios for plant phenotyping. Annu Rev Plant Biol. 2013;64:267–91.
10. Leister D, Varotto C, Pesaresi P, Niwergall A, Salamini F. Large-scale evaluation of plant growth in Arabidopsis thaliana by non-invasive image analysis. Plant Physiol Biochem. 1999;37:671–8.
11. Rajendran K, Tester M, Roy SJ. Quantifying the three main components of salinity tolerance in cereals. Plant Cell Environ. 2009;32:237–49.
12. Hartmann A, Czauderna T, Hoffmann R, Stein N, Schreiber F. HTPheno: An image analysis pipeline for high-throughput plant phenotyping. BMC Bioinf. 2011;12:148.
13. Winterhalter L, Mistele B, Jampatong S, Schmidhalter U. High throughput phenotyping of canopy water mass and canopy temperature in well-watered and drought stressed tropical maize hybrids in the vegetative stage. Eur J Agron. 2011;35:22–32.
14. Araus JL, Cairns JE. Field high-throughput phenotyping: the new crop breeding frontier. Trends Plant Sci. 2014;19:52–61.
15. Cobb J, DeClerck G, Greenberg A, Clark R, McCouch S. Next-generation phenotyping: requirements and strategies for enhancing our understanding of genotype–phenotype relationships and its relevance to crop improvement. Theor Appl Genet. 2013;126:867–87.
16. Mulla DJ. Twenty five years of remote sensing in precision agriculture: key advances and remaining knowledge gaps. Biosyst Eng. 2013;114:358–71.
17. Mistele B, Schmidhalter U. Estimating the nitrogen nutrition index using spectral canopy reflectance measurements. Eur J Agron. 2008;29:184–90.
18. Deery D, Jimenez-Berni J, Jones H, Sirault X, Furbank R. Proximal remote sensing buggies and potential applications for field-based phenotyping. Agronomy. 2014;4:349–79.
19. Montes JM, Technow F, Dhillon BS, Mauch F, Melchinger AE. High-throughput non-destructive biomass determination during early plant development in maize under field conditions. Field Crop Res. 2011;121:268–73.
20. The High Resolution Plant Phenomics Centre [http://www.csiro.au/Outcomes/Food-and-Agriculture/HRPPC.aspx], 26.11.2014.
21. Gerard B, Buerkert A. Aerial photography to determine fertiliser effects on pearl millet and Guiera senegalensis growth. Plant and Soil. 1999;210:167–77.
22. Zhang C, Kovacs JM. The application of small unmanned aerial systems for precision agriculture: a review. Precis Agric. 2012;13:693–712.

23. Deren L. From photogrammetry to inconic informatics - on the historical development of photogrammetry and remote sensing, vol. XXIX. Washington: ISPRS Archives; 1992.

24. Liebisch F, Kirchgessner N, Schneider D, Walter A, Hund A. Remote, aerial phenotyping of maize traits with a mobile multi-sensor approach Plant Methods, this issue, Editorially accepted.

25. Malenovský Z, Homolová L, Zurita-Milla R, Lukeš P, Kaplan V, Hanuš J, et al. Retrieval of spruce leaf chlorophyll content from airborne image data using continuum removal and radiative transfer. Remote Sens Environ. 2013;131:85–102.

26. Haboudane D, Miller JR, Pattey E, Zarco-Tejada PJ, Strachan IB. Hyperspectral vegetation indices and novel algorithms for predicting green LAI of crop canopies: Modeling and validation in the context of precision agriculture. Remote Sens Environ. 2004;90:337–52.

27. Pettorelli N. The Normalized Difference Vegetation Index. Oxford: OUP; 2013.

28. Tucker CJ. Red and photographic infrared linear combinations for monitoring vegetation. Remote Sens Environ. 1979;8:127–50.

29. Kipp S, Mistele B, Schmidhalter U. Identification of stay-green and early senescence phenotypes in high-yielding winter wheat, and their relationship to grain yield and grain protein concentration using high-throughput phenotyping techniques. Funct Plant Biol. 2013;41:227–35.

30. Cairns JE, Sanchez C, Vargas M, Ordonez R, Araus JL. Dissecting maize productivity: ideotypes associated with grain yield under drought stress and well-watered conditions. J Integr Plant Biol. 2012;54:1007–20.

31. Jansen M, Pinto F, Nagel KA, van Dusschoten D, Fiorani F, Rascher U, et al. Non-invasive phenotyping methodologies enable the accurate characterization of growth and performance of shoots and roots. In: Tuberosa R, editor. Genomics of Plant Genetic Resources. Dordrecht: Springer Science+Business Media; 2014.

32. Nijland W, de Jong R, de Jong SM, Wulder MA, Bater CW, Coops NC. Monitoring plant condition and phenology using infrared sensitive consumer grade digital cameras. Agr Forest Meteorol. 2014;184:98–106.

33. Soudani K, Hmimina G, Delpierre N, Pontailler JY, Aubinet M, Bonal D, et al. Ground-based network of NDVI measurements for tracking temporal dynamics of canopy structure and vegetation phenology in different biomes. Remote Sens Environ. 2012;123:234–45.

34. Rambo L, Ma B-L, Xiong Y, Regis Ferreira da Silvia P. Leaf and canopy optical characteristics as crop-N-status indicators for field nitrogen management in corn. J Plant Nutr Soil Sci. 2010;173:434–43.

35. Behmann J, Steinrücken J, Plümer L. Detection of early plant stress responses in hyperspectral images. ISPRS J Photogramm Remote Sensing. 2014;93:98–111.

36. Arnall DB, Tubaña BS, Holtz SL, Girma K, Raun WR. Relationship between nitrogen use efficiency and response index in winter wheat. J Plant Nutr. 2009;32:502–15.

37. Wang Y, Wang D, Zhang G, Wang J. Estimating nitrogen status of rice using the image segmentation of G-R thresholding method. Field Crop Res. 2013;149:33–9.

38. Gerard B, Buerkert A, Hiernaux P, Marschner H. Non-destructive measurement of plant growth and nitrogen status of pearl millet with low-altitude aerial photography (reprinted from plant nutrition for sustainable food production and environment, 1997). Soil Sci Plant Nutr. 1997;43:993–8.

39. Moriondo M, Maselli F, Bindi M. A simple model of regional wheat yield based on NDVI data. Eur J Agron. 2007;26:266–74.

40. Winterhalter L, Mistele B, Jampatong S, Schmidhalter U. High-throughput sensing of aerial biomass and above-ground nitrogen uptake in the vegetative stage of well-watered and drought stressed tropical maize hybrids. Crop Sci. 2011;51:479–89.

41. Eitel JUH, Long DS, Gessler PE, Hunt ER, Brown DJ. Sensitivity of ground-based remote sensing estimates of wheat chlorophyll content to variation in soil reflectance. Soil Sci Soc Am J. 2009;73:1715–23.

42. Hunt ER, Doraiswamy PC, McMurtrey JE, Daughtry CST, Perry EM, Akhmedov B. A visible band index for remote sensing leaf chlorophyll content at the canopy scale. Int J Appl Earth Obs Geoinf. 2013;21:103–12.

43. Erdle K, Mistele B, Schmidhalter U. Comparison of active and passive spectral sensors in discriminating biomass parameters and nitrogen status in wheat cultivars. Field Crop Res. 2011;124:74–84.

44. Lopes MS, Reynolds MP. Stay-green in spring wheat can be determined by spectral reflectance measurements (normalized difference vegetation index) independently from phenology. J Exp Bot. 2012;63:3789–98.

45. Lopes MS, Araus JL, van Heerden PDR, Foyer CH. Enhancing drought tolerance in C4 crops. J Exp Bot. 2011;62:3135–53.

46. Hill MJ, Donald GE, Hyder MW, Smith RCG. Estimation of pasture growth rate in the south west of Western Australia from AVHRR NDVI and climate data. Remote Sens Environ. 2004;93:528–45.

47. Thenkabail PS, Lyon JG, Huete A. Advances in hyperspectral remote sensing of vegetation and agricultural croplands. In: Hyperspectral remote sensing of vegetation. Boca Raton: CRC Press, Taylor & Francis Group; 2012. p. 3–36.

48. Liebisch F, Küng G, Damm A, Walter A. Characterization of crop vitality and resource use efficiency by means of combining imaging spectroscopy based plant traits, Workshop on Hyperspectral Image and Signal Processing: Evolution in Remote Sensing, vol. 6. 24–27 June. Lausanne, Switzerland: IEEE International; 2014.

49. Baker NR. Chlorophyll fluorescence: a probe of photosynthesis in vivo. Annu Rev Plant Biol. 2008;59:89–113.

50. Kautsky H, Hirsch A. Chlorophyll-fluorescence and carboxylic acid assimilation. I. Announcement: The fluorescence performance of green plants. Biochem Z. 1934;274:423–34.

51. Jansen M, Gilmer F, Biskup B, Nagel KA, Rascher U, Fischbach A, et al. Simultaneous phenotyping of leaf growth and chlorophyll fluorescence via GROWSCREEN FLUORO allows detection of stress tolerance in Arabidopsis thaliana and other rosette plants. Funct Plant Biol. 2009;36:902–14.

52. Lootens P, Devacht S, Baert J, Van Waes J, Van Bockstaele E, Roldán-Ruiz I. Evaluation of cold stress of young industrial chicory (Cichorium intybus L.) by chlorophyll a fluorescence imaging. II. Dark relaxation kinetics. Photosynthetica. 2011;49:185–94.

53. van der Heijden G, Song Y, Horgan G, Polder G, Dieleman A, Bink M, et al. SPICY: towards automated phenotyping of large pepper plants in the greenhouse. Funct Plant Biol. 2012;39:870–7.

54. Pieruschka R, Albrecht H, Muller O, Berry JA, Klimov D, Kolber ZS, et al. Daily and seasonal dynamics of remotely sensed photosynthetic efficiency in tree canopies. Tree Physiol. 2014;34:671–3.

55. Fracheboud Y, Haldimann P, Leipner J, Stamp P. Chlorophyll fluorescence as a selection tool for cold tolerance of photosynthesis in maize (Zea mays L.). J Exp Bot. 1999;50:1533–40.

56. Thoren D, Schmidhalter U. Nitrogen status and biomass determination of oilseed rape by laser-induced chlorophyll fluorescence. Eur J Agron. 2009;30:238–42.

57. Rascher U, Damm A, van der Linden S, Okujeni A, Pieruschka R, Schickling A, et al. Sensing of photosynthetic activity of crops. In Precision Crop Protection - the Challenge and Use of Heterogeneity. Edited by Oerke E-C, Gerhards R, Menz G, Sikora RA: Springer Science+Business Media B.V.; 2010: 87–99.

58. Damm A, Elbers JAN, Erler A, Gioli B, Hamdi K, Hutjes R, et al. Remote sensing of sun-induced fluorescence to improve modeling of diurnal courses of gross primary production (GPP). Glob Chang Biol. 2010;16:171–86.

59. Meroni M, Rossini M, Guanter L, Alonso L, Rascher U, Colombo R, et al. Remote sensing of solar-induced chlorophyll fluorescence: Review of methods and applications. Remote Sens Environ. 2009;113:2037–51.

60. Berger B, Parent B, Tester M. High-throughput shoot imaging to study drought responses. J Exp Bot. 2010;61:3519–28.

61. Maes WH, Steppe K. Estimating evapotranspiration and drought stress with ground-based thermal remote sensing in agriculture: a review. J Exp Bot. 2012;63:4671–712.

62. White JW, Andrade-Sanchez P, Gore MA, Bronson KF, Coffelt TA, Conley MM, et al. Field-based phenomics for plant genetics research. Field Crop Res. 2012;133:101–12.

63. Jones HG, Serraj R, Loveys BR, Xiong L, Wheaton A, Price AH. Thermal infrared imaging of crop canopies for the remote diagnosis and quantification of plant responses to water stress in the field. Funct Plant Biol. 2009;36:978–89.

64. Jones HG. Application of thermal imaging and infrared sensing in plant physiology and ecophysiology. In Advances in Botanical Research. Edited by Callow JA: Academic Press; 2004, 41:107–163.

65. Costa JM, Grant OM, Chaves MM. Thermography to explore plant–environment interactions. J Exp Bot. 2013;64:3937–49.

66. Schymanski SJ, Or D, Zwieniecki MA. Stomatal control and leaf thermal and hydraulic capacitances under rapid environmental fluctuations. PLoS One. 2013;8:e54231.

67. Mahlein A-K, Oerke E-C, Steiner U, Dehne H-W. Recent advances in sensing plant diseases for precision crop protection. Eur J Plant Pathol. 2012;133:197–209.

68. Granier C, Aguirrezabal L, Chenu K, Cookson SJ, Dauzat M, Hamard P, et al. PHENOPSIS, an automated platform for reproducible phenotyping of plant

responses to soil water deficit in Arabidopsis thaliana permitted the identification of an accession with low sensitivity to soil water deficit. New Phytol. 2006;169:623–35.

69. Walter A, Scharr H, Gilmer F, Zierer R, Nagel KA, Ernst M, et al. Dynamics of seedling growth acclimation towards altered light conditions can be quantified via GROWSCREEN: a setup and procedure designed for rapid optical phenotyping of different plant species. New Phytol. 2007;174:447–55.

70. Walter A, Studer B, Kölliker R. Advanced phenotyping offers opportunities for improved breeding of forage and turf species. Ann Bot. 2012;110:1271–9.

71. Fanourakis D, Briese C, Max J, Kleinen S, Putz A, Fiorani F, et al. Rapid determination of leaf area and plant height by using light curtain arrays in four species with contrasting shoot architecture. Plant Methods. 2014;10:9.

72. Kipp S, Mistele B, Baresel P, Schmidhalter U. High-throughput phenotyping early plant vigour of winter wheat. Eur J Agron. 2014;52, Part B:271–8.

73. Bodner G, Himmelbauer M, Loiskandl W, Kaul H-P. Improved evaluation of cover crop species by growth and root factors. Agron Sustain Dev. 2010;30:455–64.

74. Gebhard C-A, Büchi L, Liebisch F, Sinaj S, Ramseier H, Charles R. Beurteilung von Leguminosen als Gründüngungspflanzen: Stickstoff und Begleitflora. Agrarforschung Schweiz. 2013;4:384–93.

75. Paulus S, Behmann J, Mahlein A-K, Plümer L, Kuhlmann H. Low-cost 3D systems: suitable tools for plant phenotyping. Sensors. 2014;14:3001–18.

76. Paulus S, Dupuis J, Mahlein A-K, Kuhlmann H. Surface feature based classification of plant organs from 3D laserscanned point clouds for plant phenotyping. BMC Bioinf. 2013;14:238.

77. Aksoy EE, Abramov A, Wörgötter F, Scharr H, Fischbach A, Dellen B. Modeling leaf growth of rosette plants using infrared stereo image sequences. Comput Electron Agric. 2015;110:78–90.

78. Biskup B, Scharr H, Schurr U, Rascher U. A stereo imaging system for measuring structural parameters of plant canopies. Plant Cell Environ. 2007;30:1299–308.

79. Grieder C, Hund A, Walter A. Image based phenotyping during winter: a powerful tool to assess wheat genetic variation in growth response to temperature. Funct Plant Biol. 2015, published online: http://dx.doi.org/10.1071/FP14226.

80. Lynch J. Root architecture and plant productivity. Plant Physiol. 1995;109:7–13.

81. de Dorlodot S, Forster B, Pagès L, Price A, Tuberosa R, Draye X. Root system architecture: opportunities and constraints for genetic improvement of crops. Trends Plant Sci. 2007;12:474–81.

82. Zhu J, Ingram PA, Benfey PN, Elich T. From lab to field, new approaches to phenotyping root system architecture. Curr Opin Plant Biol. 2011;14:310–7.

83. Bonser AM, Lynch J, Snapp S. Effect of phosphorus deficiency on growth angle of basal roots in Phaseolus vulgaris. New Phytol. 1996;132:281–8.

84. Kochian LV, Pineros MA, Hoekenga OA. The physiology, genetics and molecular biology of plant aluminum resistance and toxicity. Plant and Soil. 2005;274:175–95.

85. Watt M, Moosavi S, Cunningham SC, Kirkegaard JA, Rebetzke GJ, Richards RA. A rapid, controlled-environment seedling root screen for wheat correlates well with rooting depths at vegetative, but not reproductive, stages at two field sites. Ann Bot. 2013;112:447–55.

86. Hund A, Trachsel S, Stamp P. Growth of axile and lateral roots of maize: I development of a phenotyping platform. Plant and Soil. 2009;325:335–49.

87. Le Marié CA, Kirchgessner N, Marschall D, Walter A, Hund A. Rhizoslides: paper-based growth system for non-destructive, high throughput phenotyping of root development by means of image analysis. Plant Methods. 2014;10:13–29.

88. Bengough AG, Gordon DC, Al-Menaie H, Ellis RP, Allan D, Keith R, et al. Gel observation chamber for rapid screening of root traits in cereal seedlings. Plant and Soil. 2004;262:63–70.

89. Nagel KA, Kastenholz B, Jahnke S, Van Dusschoten D, Aach T, Muehlich M, et al. Temperature responses of roots: impact on growth, root system architecture and implications for phenotyping. Funct Plant Biol. 2009;36:947–59.

90. Downie H, Holden N, Otten W, Spiers AJ, Valentine TA, Dupuy LX. Transparent soil for imaging the rhizosphere. PloS One. 2012;7:e44276.

91. de Dorlodot S, Bertin P, Baret P, Draye X. Scaling up quantitative phenotyping of root system architecture using a combination of aeroponics and image analysis. Asp Appl Biol. 2005;73:41–54.

92. Tuberosa R, Sanguineti MC, Landi P, Giuliani MM, Salvi S, Conti S. Identification of QTLs for root characteristics in maize grown in hydroponics and analysis of their overlap with QTLs for grain yield in the field at two water regimes. Plant Mol Biol. 2002;48:697–712.

93. Courtois B, Audebert A, Dardou A, Roques S, Ghneim-Herrera T, Droc G, et al. Genome-wide association mapping of root traits in a japonica rice panel. Plos One. 2013;8(11):e78037.

94. Hund A, Frachboud Y, Soldati A, Frascaroli E, Salvi S, Stamp P. QTL controlling root and shoot traits of maize seedlings under cold stress. Theor Appl Genet. 2004;109:618–29.

95. Zhu JM, Kaeppler SM, Lynch JP. Mapping of QTLs for lateral root branching and length in maize (Zea mays L.) under differential phosphorus supply. Theor Appl Genet. 2005;111:688–95.

96. Trachsel S, Messmer R, Stamp P, Hund A. Mapping of QTLs for lateral and axile root growth of tropical maize. Theor Appl Genet. 2009;119:1413–24.

97. Lobet G, Pages L, Draye X. A novel image-analysis toolbox enabling quantitative analysis of root system architecture. Plant Physiol. 2011;157:29–39.

98. Pfeifer J, Faget M, Walter A, Blossfeld S, Fiorani F, Schurr U, et al. Spring barley shows dynamic compensatory root and shoot growth responses when exposed to localised soil compaction and fertilisation. Funct Plant Biol. 2014;41:581–97.

99. Tracy SR, Roberts JA, Black CR, McNeill A, Davidson R, Mooney SJ. The X-factor: visualizing undisturbed root architecture in soils using X-ray computed tomography. J Exp Bot. 2010;61:311–3.

100. Pfeifer J, Kirchgessner N, Walter A. Artificial pores attract barley roots and can reduce artifacts of pot experiments. J Plant Nutr Soil Sci. 2014;177:903–13.

101. Nagel KA, Putz A, Gilmer F, Heinz K, Fischbach A, Pfeifer J, et al. GROWSCREEN-Rhizo is a novel phenotyping robot enabling simultaneous measurements of root and shoot growth for plants grown in soil-filled rhizotrons. Funct Plant Biol. 2012;39:891–904.

102. The Centre for Plant Integrative Biology [https://www.cpib.ac.uk/], University of Nottingham, 26.11.2014.

103. Zaman-Allah M, Jenkinson DM, Vadez V. A conservative pattern of water use, rather than deep or profuse rooting, is critical for the terminal drought tolerance of chickpea. J Exp Bot. 2011;62:4239–52.

104. Stingaciu L, Schulz H, Pohlmeier A, Behnke S, Zilken H, Javaux M, et al. In situ root system architecture extraction from magnetic resonance imaging for water uptake modeling. Vadose Zone J. 2013;12:1–9.

105. Metzner R, van Dusschoten D, Bueler J, Schurr U, Jahnke S. Belowg round plant development measured with magnetic resonance imaging (MRI): exploiting the potential for non-invasive trait quantification using sugar beet as a proxy. Front Plant Sci. 2014;5:469.

106. Liu Z, Qian J, Liu B, Wang Q, Ni X, Dong Y, et al. Effects of the magnetic resonance imaging contrast agent Gd-DTPA on plant growth and root imaging in rice. Plos One. 2014;9(6):e100246.

107. Leitner D, Felderer B, Vontobel P, Schnepf A. Recovering root system traits using image analysis exemplified by two-dimensional neutron radiography images of lupine. Plant Physiol. 2014;164:24–35.

108. Jahnke S, Menzel MI, Van Dusschoten D, Roeb GW, Bühler J, Minwuyelet S, et al. Combined MRI–PET dissects dynamic changes in plant structures and functions. Plant J. 2009;59:634–44.

109. Weaver JE. Root development of field crops. London: McGraw-Hill Book Company, INC. New York State Museum Memoir; 1926.

110. Kutschera L, Lichtenegger E. Wurzelatlas mitteleuropäischer Ackerunkräuter und Kulturpflanzen. Frankfurt am Main: DLG-Verlag; 1960.

111. Perkons U, Kautz T, Uteau D, Peth S, Geier V, Thomas K, et al. Root-length densities of various annual crops following crops with contrasting root systems. Soil Tillage Res. 2014;137:50–7.

112. Smucker AJM, McBurney SL, Srivastava AK. Quantitative separation of roots from compacted soil profiles by the hydropneumatic elutriation system1. Agron J. 1982;74:500–3.

113. Benjamin JG, Nielsen DC. A method to separate plant roots from soil and analyze root surface area. Plant and Soil. 2004;267:225–34.

114. Bohm W. Methods of studying root systems. Berlin: Springer; 1979.

115. Grift TE, Novais J, Bohn M. High-throughput phenotyping technology for maize roots. Biosyst Eng. 2011;110:40–8.

116. Bohn M, Novais J, Fonseca R, Tuberosa R, Grift TE. Genetic evaluation of root complexity in maize. Acta Agronomica Hungarica. 2006;54:291–303.

117. Trachsel S, Kaeppler SM, Brown KM, Lynch J. Shovelomics: High throughput phenotyping of maize (Zea mays L.) root architecture in the field. Plant and Soil. 2011;341:75–87.

118. Bucksch A, Burridge J, York LM, Das A, Nord E, Weitz JS, et al. Image-based high-throughput field phenotyping of crop roots. Plant Physiol. 2014;166:470–86.

119. Colombi T, Kirchgessner N, Le Marié CA, York L, Lynch J, Hund A. Next generation shovelomics: set up a tent and REST. Plant and Soil. 2015, published online: http://link.springer.com/article/10.1007%2Fs11104-015-2379-7.

120. Johnson MG, Tingey DT, Phillips DL, Storm MJ. Advancing fine root research with minirhizotrons. Environ Exp Bot. 2001;45:263–89.

121. Thorup-Kristensen K. Root growth of green pea (*Pisum sativum* L.) genotypes. Crop Sci. 1998;38:1445–51.

122. Herrera JM, Stamp P, Liedgens M. Dynamics of root development of spring wheat genotypes varying in nitrogen use efficiency. In: Wheat Production in Stressed Environments. Springer, 2007, 197–201.

123. Landi P, Sanguineti MC, Darrah LL, Giuliani MM, Salvi S, Conti S, et al. Detection of QTLs for vertical root pulling resistance in maize and overlap with QTLs for root traits in hydroponics and for grain yield under different water regimes. Maydica. 2002;47:233–43.

124. Giuliani S, Sanguineti MC, Tuberosa R, Bellotti M, Salvi S, Landi P. *Root-ABA1*, a major constitutive QTL, affects maize root architecture and leaf ABA concentration at different water regimes. J Exp Bot. 2005;56:3061–70.

125. Chloupek O. Evaluation of size of a plants-root system using its electrical capacitance. Plant and Soil. 1977;48:525–32.

126. Chloupek O. Relationship between electric capacitance and some other parameters of plant roots. Biol Plant. 1972;14(3):227–30.

127. Messmer R, Fracheboud Y, Baenziger M, Stamp P, Ribaut J-M. Drought stress and tropical maize: QTLs for leaf greenness, plant senescence, and root capacitance. Field Crop Res. 2011;124:93–103.

128. Chloupek O, Forster BP, Thomas WTB. The effect of semi-dwarf genes on root system size in field-grown barley. Theor Appl Genet. 2006;112:779–86.

129. Dietrich RC, Bengough AG, Jones HG, White PJ. Can root electrical capacitance be used to predict root mass in soil? Ann Bot. 2013;112:457–64.

130. Dietrich RC, Bengough AG, Jones HG, White PJ. A new physical interpretation of plant root capacitance. J Exp Bot. 2012;63:6149–59.

131. Srayeddin I, Doussan C. Estimation of the spatial variability of root water uptake of maize and sorghum at the field scale by electrical resistivity tomography. Plant and Soil. 2009;319:185–207.

132. Vadez V. Root hydraulics: the forgotten side of roots in drought adaptation. Field Crop Res. 2014;165:15–24.

133. Saengwilai P, Tian X, Lynch JP. Low crown root number enhances nitrogen acquisition from low-nitrogen soils in maize. Plant Physiol. 2014;166:581–9.

21

Non-invasive assessment of leaf water status using a dual-mode microwave resonator

Said Dadshani[1], Andriy Kurakin[2], Shukhrat Amanov[1], Benedikt Hein[1], Heinz Rongen[2], Steve Cranstone[2], Ulrich Blievernicht[2], Elmar Menzel[4], Jens Léon[1], Norbert Klein[2,3] and Agim Ballvora[1*]

Abstract

The water status in plant leaves is a good indicator for the water status in the whole plant revealing stress if the water supply is reduced. The analysis of dynamic aspects of water availability in plant tissues provides useful information for the understanding of the mechanistic basis of drought stress tolerance, which may lead to improved plant breeding and management practices. The determination of the water content in plant tissues during plant development has been a challenge and is currently feasible based on destructive analysis only. We present here the application of a non-invasive quantitative method to determine the volumetric water content of leaves and the ionic conductivity of the leaf juice from non-invasive microwave measurements at two different frequencies by one sensor device. A semi-open microwave cavity loaded with a ceramic dielectric resonator and a metallic lumped-element capacitor- and inductor structure was employed for non-invasive microwave measurements at 150 MHz and 2.4 Gigahertz on potato, maize, canola and wheat leaves. Three leaves detached from each plant were chosen, representing three developmental stages being representative for tissue of various age. Clear correlations between the leaf- induced resonance frequency shifts and changes of the inverse resonator quality factor at 2.4 GHz to the gravimetrically determined drying status of the leaves were found. Moreover, the ionic conductivity of Maize leaves, as determined from the ratio of the inverse quality factor and frequency shift at 150 MHz by use of cavity perturbation theory, was found to be in good agreement with direct measurements on plant juice. In conjunction with a compact battery- powered circuit board- microwave electronic module and a user-friendly software interface, this method enables rapid in-vivo water amount assessment of plants by a handheld device for potential use in the field.

Keywords: Water content, Microwave resonator, Non-invasive measurements

Background

Drought and salinity stress are undoubtedly important constraints limiting agricultural productivity which can even result in total yield loss [1,2]. To equilibrate the decrease of the uptake of the available water in soils, plants preserve the osmotic potential by reducing stomata conductance. This leads to a reduction of photosynthetic rate and finally reducing plant growth and yield [3,4]. Around 26% of arable land worldwide is suffering from water shortage constituting the most important abiotic stress [5]. In perspective to climate changes in the future an increase of drought stress and consequently problems with plant production [4,6-8] are expected. Understanding

the mechanism of drought stress tolerance is in the focus of current plant research, in order to help breeders developing new cultivars that perform well, even under water scarcity.

The definition of the water status in plant tissue is of importance for the plant researcher to better understand the physiological processes and molecular mechanisms leading to tolerance with respect to water lack stress on the one hand. On the other hand it may help the producers to control the watering procedures. Systematic phenotyping of plants needs standardized and non-invasive methods to define and assess physiological parameters like water status in order to analyze the reactions of single plants or group of plants to environmental.

The water content in vegetative tissues is a parameter of high importance for the photosynthetic performance and an indicator of the plant's health. Currently it is

---
* Correspondence: ballvora@uni-bonn.de
[1]INRES-Plant Breeding, University of Bonn, Katzenburgweg 5, 53115 Bonn, Germany
Full list of author information is available at the end of the article

measured by destructive methods such as comparing the fresh and dry weight of plant tissues [9]. Nevertheless, destructive methods do not allow the instantaneous and continuous monitoring of the water content in living tissue. Therefore, non-destructive techniques that require very weak interaction with the plant tissue in order to avoid altering its physiological activities are highly desired.

Non-destructive analysis by radiation in the microwave to terahertz range is most promising for the development of non-invasive methods to determine the water content because of the strong water absorption in this frequency range [10-12]. The selection of frequency is determined by the size of the assessed objects in comparison to the wavelength, if standard absorption or reflection methods are being used. In the case of plant leaves of centimeter dimension, frequencies above about 30 GHz (wavelength $\lambda = 1$ cm) are advantageous, in particular the THz range with λ below one millimeter.

Recently, THz measurements have been used to measure the water content in leaves [9,13]. However, THz technology is still quite expensive in comparison to the microwave bands below 20 GHz. Our work represents the first systematic study on individual plant leaves by a dielectric resonator based method, similar to the one described by Menzel, et al. [10], which was developed with direct involvement of one of the authors. Other than in the method described by Menzel, et al. [10], the additional use of a low frequency mode being excited in the same cavity at 150 MHz enables independent and simultaneous non-invasive determination of the ionic conductivity [14]. Different to microwave moisture sensors based on planar microwave transmission lines like the one reported by Rezaei, et al. [15] and planar antennae approaches by Sancho-Knapik, et al. [16] our method allows the determination of the real and imaginary components of the complex dielectric permittivity at two well separated frequencies. Moreover, our evanescent field approach overcomes the wavelength limitation and enables the use of much lower frequencies at 150 MHz and 2.4 GHz, with the advantage of cheap electronic components as being used in wireless communication. The potential commercial availability of an evanescent field dual mode microwave sensor system at moderate cost enables the implementation of non-invasive water and conductivity assessment in biological research laboratories.

Microwave properties of plant tissue

The microwave properties of plant tissue strongly correlate to the amount of stored water. The typical water content in healthy plant leaves is around 90% [17].

The interaction of microwaves with water, which is determined by a broad absorption peak due to Debye-type molecular relaxation, centered at around 20 GHz at room temperature, can be described by a strongly frequency dependent complex-valued dielectric permittivity,

$$\varepsilon * (\omega) = \varepsilon'(\omega) + j\varepsilon''(\omega)$$
$$= \varepsilon_\infty + \frac{\varepsilon_s - \varepsilon_\infty}{1 + \omega^2 \tau^2} + j\left[\frac{(\varepsilon_s - \varepsilon_\infty)\omega\tau}{1 + \omega^2 \tau^2} + \frac{\sigma}{\omega\varepsilon_0}\right], \quad (1)$$

with ε_s representing the static dielectric permittivity, ε_∞ the permittivity at $f \to \infty$, τ the dipole relaxation time of the water molecules and σ the ionic conductivity due to dissolved salts or other ions and metabolites [18]. In Eq. 1, the frequency f is expressed by the angular frequency $\omega = 2\pi f$, $\varepsilon_0 = 8.85 \cdot 10^{-12}$ F/m is the vacuum permittivity.

The dielectric properties of liquid water can be well described by Eq. 1 up to about 60 GHz, using temperature dependent values of ε_s, τ and σ [18,19]. At room temperature ($T = 22°C$), experimental data for distilled water can be well fitted using $\varepsilon_s = 78.36$, $\tau = 8.27$ ps, $\varepsilon_\infty = 5.16$ and $\sigma = 0$ [20]. At 2.4 GHz and 150 MHz, where the experiments are conducted, $\varepsilon^*(2.4 \text{ GHz}) = 77 + j\ 9.0$ and $\varepsilon^*(150 \text{ MHz.2}) = 78 + j\ 0.57$, respectively. In particular at 150 MHz, a large contribution of to the conductivity term (3rd term in Eq. 1) by dissolved ions to the imaginary part of ε^* can be expected: broadband microwave dielectric measurement on fluids extracted from wheat leaves revealed equivalent NaCl concentrations of around 1% [21], which results in a conductivity of about 17,600 μS/cm, the corresponding imaginary part of ε^* at 2.4 GHz and 150 MHz are 13 and 211, respectively (3rd term in Eq. 1). Hence, the ratio Im ($\varepsilon^*_{\text{ions}}$)/Im ($\varepsilon^*_{\text{dipole}}$), which describes the ratio of ionic to dipole losses, comes out to be 1.47 at 2.4 GHz and 370 at 150 MHz for the given conductivity. Therefore, the mode at 150 MHz is ideally suited for non-invasive and contact-free conductivity measurements.

It is worth to note that the Debye relaxation parameters and the ionic conductivity are strongly temperature dependent, therefore it is important that the measurements are performed within well-defined temperature intervall. The dielectric response of the leaf can be understood as an effective medium composed of water with ions and of dry bulk material. In contrast to water, the bulk material has a relatively low permittivity $\varepsilon' \leq 10$, and the imaginary part is negligible, as demonstrated by measurements on totally dried leaves (see section about results and discussion) . Therefore, as long as the absolute water content is more than about 10% the contribution of the bulk plant material to the real part of the, dielectric permittivity can be neglected as well. However, as discussed in Ulaby, et al. [21], the calculation of complex permittivity of a representative effective medium would require detailed information about the water distribution within the veins and as inter- and intracellular liquid, because of unequal amounts of water in different tissue compartments. Nevertheless, by assessing dielectric properties of two materials as reported by Sancho-Knapik, et al. [16], a very good correlation between RWC (relative water content) and reflectance at a frequency of 1730 MHz was found both for filter paper and leaves. Therefore, the integral complex permittivity, as determined by microwave dielectric measurement, represents a reasonable experimental quantity which is representative for the water content (or conductivity in case of the imaginary component at 150 MHz) of a leaf under investigation.

According to a comprehensive study within the framework of effective medium theories as described in Ulaby, et al. [21] the static permittivity for fresh wheat leaves is about 35, corresponding to a volumetric moisture of about 60%. This correlation depends on the density of the fresh leaf material, which may vary for different species, but was not analyzed within this study.

Results and discussion
The dual mode cavity as leaf sensor
The patented dual mode cavity sensor, which is discussed in detail in Klein, et al. [14], enables simultaneous dielectric measurements at two distinct and far separated frequencies: For the sensor which was employed in this study, one resonant frequency is at 150 MHz (Mode 0), the second one 2.4 GHz (Mode 1). For the study of the correlation between drying status and permittivity we employed Mode 1 only because of large signal-to-noise ratio, i.e. larger frequency shifts in comparison the resonant halfwidth. In spite of poor signal-to-noise ratio, preliminary data by Mode 0 on fresh wheat leaves are discussed. It is worth to note that Mode 0 is ideally suited for contact-free assessment of the ionic conductivity of bulky plant tissues such as potatoes and sugar beets, where the sample volume and hence the signal-to-noise ratio is much larger.

Mode 1 corresponds to the $TE_{01\delta}$-mode [22] of the cylindrically shaped dielectric resonator, embedded in the dual-mode cavity. The evanescent electric field is presented by concentric circles, the field magnitude increases from zero in the center of the aperture towards its maximum at about 2/3 of the radius of the dielectric resonator (light circle in Figure 1,C), and gradually decreases to zero towards the aperture. From the aperture plane (leaf measurement position), the evanescent field decreases exponentially in axial direction and reaches 50% of its value at the top edge of the aperture at a distance of about 20 mm above the aperture. The evanescent field of the lumped element mode (Mode 0) is strongly concentrated in close vicinity of the radial metallic rod, in particular near the center of the cavity [14].

As it will be discussed along with the experimental data, for Mode 1 the magnitude of the leaf induced alteration of the resonant properties depends on the degree of coverage of the aperture by the leaf under test. In case of a partial coverage, as indicated by the wheat leaf shown in Figure 1(II), a strict protocol how to arrange the leaf on the sensor surface is required for each given type of leaf. A smaller aperture would be tempting for the assessment of smaller leaves, but would cause a strong reduction of the electric field amplitude at the leaf position, which leads to a significant reduction of sensitivity.

During the assessment of a leaf under test, the change of the inverse quality factor Q and the resonant frequency, f_r with respect to the empty resonator is recorded. Both Q and f_r are determined from a fit of a Lorentzian to the measured transmission curve using.

$$U(f) = \frac{U(f_r)}{\sqrt{1 + 4Q^2\left(\frac{f}{f_r} - 1\right)^2}} \tag{2}$$

In Eq. 2 $U(f)$ represents the frequency dependent detector voltage, which is proportional to the power transmitted through the resonator (square law detection) upon sweeping the generator frequency around the resonance frequency f_r. Both modes are excited by a different pair of coaxial probes for each, the signals are generated and recorded by two independent electronic modules. Each of the two PCB (printed circuit-board) - based integrated electronic modules is composed of a digitally controlled synthesizer- PLL (phase locked loop) controlled microwave VCO (voltage controlled oscillator) and a detector unit.

Prior to each measurement with a leaf in place, f_r and Q are recorded for the empty resonator. For the analysis, the negative relative frequency shift due to the sample,

$$FRS \equiv -\frac{f_{r,sample} - f_{r,empty}}{f_{r,empty}} \tag{3a}$$

and the sample induced change of the losses, i.e. change of the inverse Q factor, IQS,

$$IQS \equiv \frac{1}{Q_{sample}} - \frac{1}{Q_{empty}} \tag{3b}$$

are recorded. Since the frequency shift due to a dielectric object is usually negative, FRS is defined to be a positive number. It is important to note that IQS is independent of coupling losses, because coupling leads to a constant $1/Q$ contribution which does not change due the sample in measurement position.

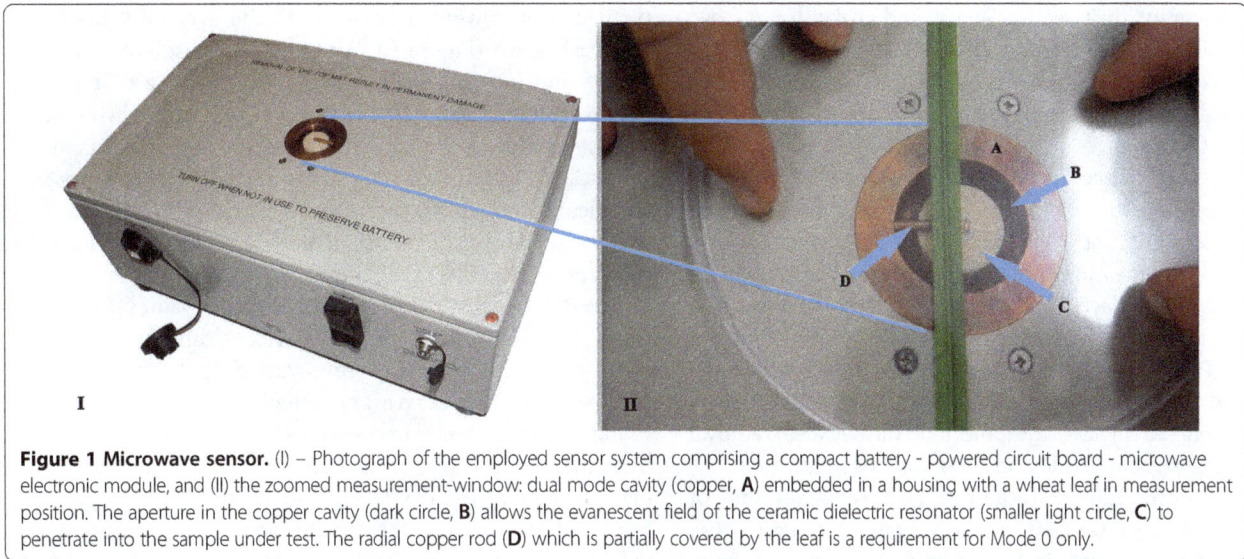

Figure 1 Microwave sensor. (I) – Photograph of the employed sensor system comprising a compact battery - powered circuit board - microwave electronic module, and (II) the zoomed measurement-window: dual mode cavity (copper, **A**) embedded in a housing with a wheat leaf in measurement position. The aperture in the copper cavity (dark circle, **B**) allows the evanescent field of the ceramic dielectric resonator (smaller light circle, **C**) to penetrate into the sample under test. The radial copper rod (**D**) which is partially covered by the leaf is a requirement for Mode 0 only.

For the case, that the field distribution of the evanescent field is not distorted by the sample, FRS and IQS can be directly related to the complex permittivity of the sample by extended cavity perturbation theory [23].

$$FRS = \frac{\kappa}{2}(\varepsilon'-1) \qquad\qquad IQS = \kappa\varepsilon''$$

$$\kappa = \frac{\varepsilon_0 \int\limits_V E_0^2 dV}{2W}$$

(4)

In Eq. 4, the filling factor κ describes the electric resonant field energy within the sample of volume, the integral in the numerator extends over the volume fraction V of the sample which is exposed to the unperturbed resonator field E_0, normalized to the total electric field energy, W, of the cavity.

In order to test the applicability of the perturbation approach, electromagnetic field simulations of the cavity-leaf system have been performed with CST Microwave Studio [24] for a variety of configurations. The results indicate that the alteration of the magnitude of the electric field at the position of the leaf due to leaf itself is less than 10% in the worst case assuming a homogenous water distribution inside the leaf. Therefore, the analysis by Eq. 4 is justified within the experimental errors. However, we cannot rule out that water being concentrated in veins may lead to some level redistribution of the local electromagnetic field, which is subject of an ongoing study.

The accurate calculation of the filling factor κ requires a detailed analysis of the shape of the leaf and its exact measurement position - along with the electric field distribution of the resonant mode. However, relative

measurements of FRS and IQS for a given leaf in a reproducible measurement position allow the monitoring of relative changes of the complex permittivity. It is worth to mention that the ratio of IQS and FRS is independent of κ, and may represent a size and position independent figure of merit for a given leaf. For Mode 1, even in case of a complete coverage of the aperture, the leaf-induced alteration of resonance frequency and Q factor may depend on the exact measurement position of the leaf under test, because the water distribution in the leaves is inhomogeneous. This means, that a maximum of FRS and IQS is usually achieved if water filled veins are located around the position of maximum field. For the sake of a maximum signal-to-noise ratio, the position was optimized for maximum FRS. In case of elongated leaves like wheat the leaf axis was arranged at an offset of about 50–80% of the radius of the dielectric resonator, corresponding to a field maximum of the TE_{01d} mode (Mode 1). The optimization of the position with regards to Mode 0 is subject to a separate analysis and will not be further addressed in this contribution.

However, as indicated in the section about results and discussion, the leaf-induced alterations can be used for a preliminary analysis.

Although the leaf under test is physically attached to the metallic aperture of the cavity in order to ensure a reproducible measurement position, the measurement is contact-less in nature. A thin plastic foil between aperture and sample would not have any significant effect on the results, because the electric field is coupled to the sample inductively, without any need of an electrical contact.

Measurement of water content in leaves of different plants
The four plant species being analyzed, wheat, maize, potato and canola were selected considering the size and

morphology of their leaves. Wheat and maize leaves have similar shape, both are long but wheat leaves are thinner. On the other hand, the potato and canola have compound leaves with oval leaflets, the canola leaves are larger and thicker.

The three leaves detached from each plant were chosen from three developmental stages in order to characterize tissues of various ages. Shortly after removal from the plant, the leave under test was weighted and subsequently measured with the microwave sensor system. The leave was placed on the window such the measured frequency shift is maximized, as shown in Figure 1 for wheat. This first assessment was representative for the fresh leaf and which was considered as reference of 100% (w/w) water content. In fact, the time interval between removal and measurement was less than 30 seconds in any case.

A significant change in the resonant frequency shift of Mode 1 (2.4 GHz) between fresh and dry leaves was demonstrated, which is far beyond the variation from leaf to leaf for a given plant (Figure 2). It is notable that the frequency shift of the dry leaf is zero within the measurement accuracy limits which indicates that water represents the dominant part of the response.

For Mode 0, only wheat leaves have been investigated till date. The measured values of FRS and IQS are of the same order of magnitude as for Mode 1, but the signal-to-noise ratio is nearly ten times lower than for Mode 1. This is due to the smaller resonant halfwidth of the unloaded resonance, usually expressed by the quality factor Q_{empty} without sample, Q_{empty}(Mode 0) = 350, Q_{empty}(Mode 1) = 4200).

All measurements where performed at room temperature without any room temperature control. Test measurements on canola and wheat leaves at 18°C, 22°C and 27°C showed no significant differences of the FRS or IQS values.

For Mode 1, the measured values of FRS and IQS as a function of percentage of fresh weight for four different types of plants were normalized to the average value of the fresh leave (Figure 3). Although the absolute values of FRS and IQS differ from leaf to leaf due to a different filling factor κ (Eq. 4) the normalized FRS and IQS values exhibit a systematic decrease with increasing weight loss, which indicates that water provides the most significant contribution to the dielectric permittivity of leaf tissue. The data points of individual leaves indicate this trend. The averaged values displayed in Figure 3 as horizontal lines are representative mean values for all leave stages. Based on t-test, significant differences between the mean values were revealed (p < 0.05). Measurement of three leaves and calculation of their mean values show stronger correlations to the water content than the single leaves measurement. Assuming that the mass density of the dry leave tissue is small in comparison to that of water, the weight percentage represents the volumetric water concentration – multiplied by factor of about 0.5-0.7 (maximum water volume concentration in a fresh leaf) [21]. According to our data, the normalized FRS values drop to about 0.45-0.55 for wheat and maize, and to slightly higher values of about 0.65-0.75 for potato and canola - as result of weight reduction or water loss from 100% to 50%. It is remarkable, however, that the FRS – weight dependences exhibit a recognizable positive curvature and deviate from linearity, in contrast to the slight negative curvature being observed by broadband dielectric measurements on wheat leaves and stalks [21]. For potato leaves and in particular for canola leaves, where a considerable portion of water is stored in relatively thick veins, this effect is most pronounced. We presume that the drying process by evaporation works slower for large veins. As a result, the measured weight may not be representative for the real water concentration in the largest veins: if the largest veins are located close to a field maximum of the resonant field, the measured FRS values may overestimate the average water concentration in the leaf, which explains the observed curvature qualitatively.

For the fresh weight data, the average ratio IQS/FRS varies only slightly between 2.8 and 3 for the three different plants. Assuming the validity of the perturbation approach according to Eq. 4, IQS/FRS is equal to two times the loss tangent.

$$\frac{IQS}{FRS} = \frac{2\varepsilon''}{\varepsilon'-1} \approx \frac{2\varepsilon''}{\varepsilon'} = 2\tan\delta \qquad (5)$$

As discussed in the section about microwave properties of plant tissues, the loss tangent of distilled water at 2.4 GHz is 0.12 and 0.3 for an assumed ionic conductivity of 14,000 µS/cm. In other words, within the framework of perturbation theory and the assumption that the losses are due to water and ions only, the anlysis yields a

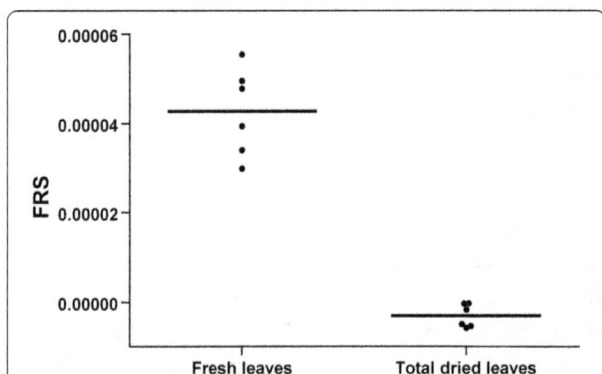

Figure 2 Comparative dielectric conductivity of fresh and totally dried wheat leaves. Six leaves in total were measured fresh and subsequently completely dried. The dots represent the average values of five measurements (technical replicates) and the lines the average values of six leaves.

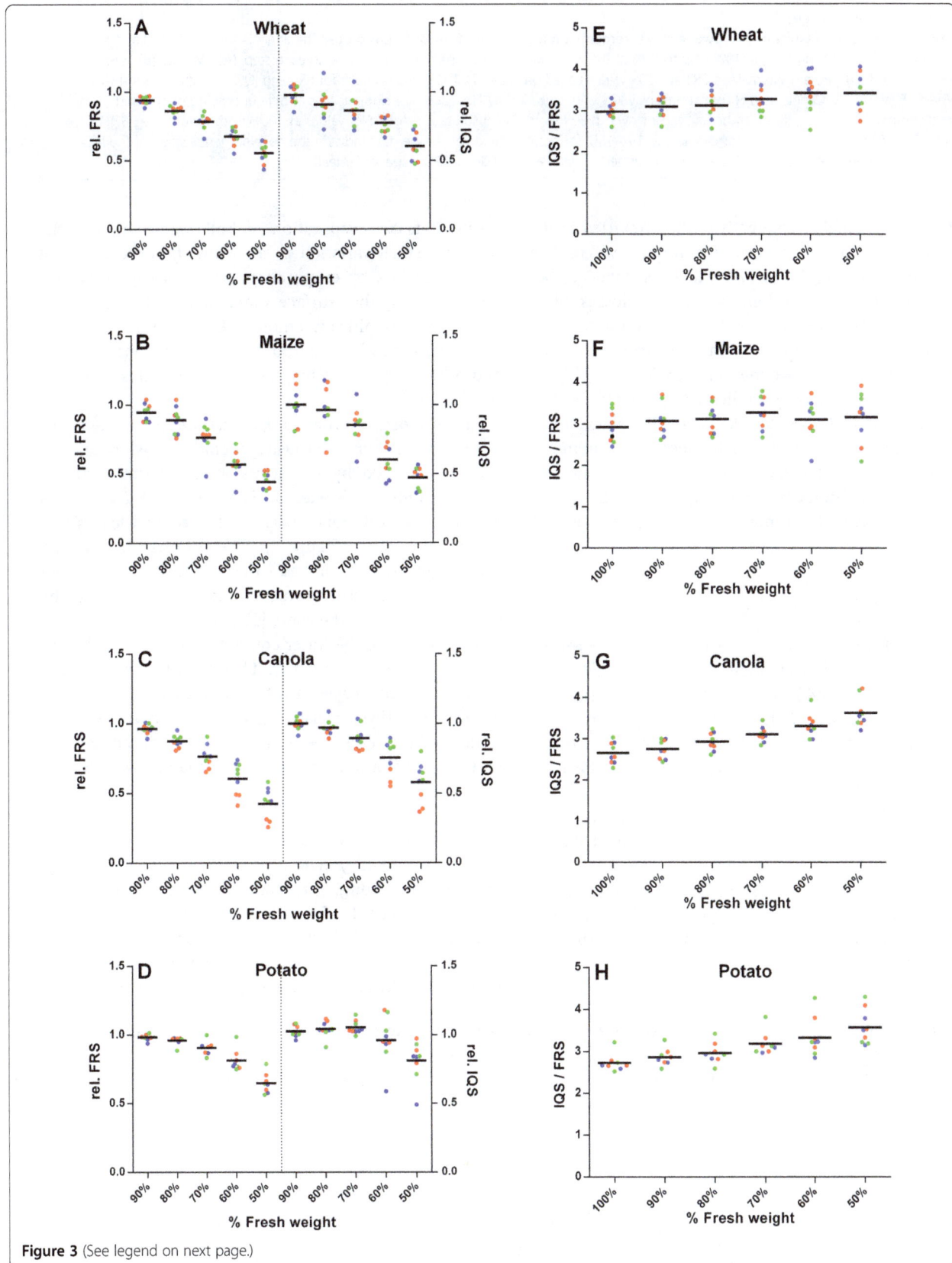

Figure 3 (See legend on next page.)

(See figure on previous page.)
Figure 3 Experimental results of microwave measurements on leaves from four different crops. The analysis was performed at six time-points from initial fresh weight (100%) up to 50% of the initial weight by consecutive 10% drying in each step. **(A)** - Wheat, **(B)**-Maize, **(C)** - Canola, and **(D)** - Potato. Normalized FRS and IQS values based on values at 100% fresh weight for FRS and IQS, respectively are shown. E (Wheat), F (Maize), G (Canola) and H (Potato) display IQS values divided by FRS values for single leaves at different drying stages, and water contents, respectively. Bars indicate arithmetic means over three leaves. The absolute averaged FRS values at 100% fresh weight are $5.5 \cdot 10^{-5}$ for wheat, $1.25 \cdot 10^{-4}$ for maize, $1.41 \cdot 10^{-4}$ for canola and $1.24 \cdot 10^{-4}$ for potato. The colors of the dots indicate the developmental stage of the leaves: blue (first stage, young leaf), green (second stage, intermediate leaf) and red dots (third stage, older leaf).

nearly ten times higher conductivity than reported in the literature. In order to resolve this puzzle, we took a closer look at the *IQS* and *FRS* values of Mode 0, because the separation of ionic conductor losses from water dipole relaxation losses is much more pronounced at this low frequency (Table 1). In order to improve the measurement statistics, we measured *FRS* and *IQS* for 6 fresh leaves from one plant (indicated by the numbers in Table 1). Each of the listed *FRS* and *IQS* values corresponds to the average of five subsequent measurements performed on one leaf, the quoted error represents the standard deviation of these five subsequent measurements.

The conductivity is proportional to *IQS/FRS* (Eqs. 1 and 5)

$$\sigma = \omega \varepsilon_0 \varepsilon_r \tan\delta = \frac{\omega \varepsilon_0 \varepsilon_r}{2} \frac{IQS}{FRS} \qquad (6)$$

with $\varepsilon_r \approx 78$ representing the real part of the permittivity of water at the measurement frequency of 150 MHz. The quoted value (1.46 ± 0.20) µS/cm corresponding to the weighted average of the six leaves is in agreement with literature data [21]. To the best of our knowledge, this is the first non-invasive determination of the conductivity of the fluid inside a plant leaf.

As a possible explanation for the enhanced loss tangent measured at 2.4 GHz, it is likely that higher dielectric relaxation losses than assumed for free water may occur due to a high abundance of surface water, which has a significantly higher loss tangent than bulk water at 2.4 GHz [25,26]. The observed slight increase of *IQS/FRS* at 2.4 GHz with increasing weight loss is likely due to an increase of the ratio of surface to bulk water as

result of faster evaporation of bulk water. In fact, the relatively small variation is far below the expectation of 50% water loss by evaporation, which is supportive for the hypothesis that surface water may contribute to the losses by a significant amount. Comparative measurements with Mode 0 at 150 MHz of sufficient accuracy and other frequencies may help to resolve this puzzle in the future.

Furthermore, in order to demonstrate the practical applicability of our microwave technique, wheat plants being challenged by salt stress were measured (at the moment only by Mode 1). The measured *FRS* values reveal a clear difference between the control leaves and the stressed ones (Figure 4). The decrease in the *FRS* value is likely to be linked to an increase of osmolarity induced by salt stress which is adversely affecting the uptake of water by the roots [27,28].

A reduction of the water content in the plant cell leads to an increase of osmolarity. Therefore, the osmotic potential of canola leaves at 6 time-points was determined. A strong negative correlation (r = − 0.97) between *IQS/FRS* values and the respective osmotic potential of the leaves at different steps of water reduction was found (Figure 5).

Material and methods
Plant material and growth conditions
Four species belonging to different classes of plant kingdom were selected: wheat (*Triticum aestivum* L.) cultivar Zentos, maize (*Zea mays* L.) cultivar Aurelia, potato (*Solanum tuberosum* L.) cultivar Linda and canola (*Brassica napus* L.), cultivar Expert. The plants were grown under

Table 1 Measured FRS and IQS (f = 150 MHz, Mode 0) for 6 different fresh leaves of one wheat plant and calculated ionic conductivity

L. no	FRS	Δ_{FRS}/FRS [%]	IQS	Δ_{IQS}/FRS [%]	σ[µS/cm]	Δ_σ/σ [%]
1	$1.30 \cdot 10^{-5}$	40	$5.25 \cdot 10^{-5}$	27	$1.4 \cdot 10^4$	48
2	$1.47 \cdot 10^{-5}$	26	$6.11 \cdot 10^{-5}$	15	$1.4 \cdot 10^4$	30
3	$1.27 \cdot 10^{-5}$	46	$7.78 \cdot 10^{-5}$	27	$2.0 \cdot 10^4$	53
4	$1.86 \cdot 10^{-5}$	21	$6.84 \cdot 10^{-5}$	52	$1.2 \cdot 10^4$	56
5	$1.94 \cdot 10^{-5}$	7	$8.79 \cdot 10^{-5}$	20	$1.5 \cdot 10^4$	21
6	$1.96 \cdot 10^{-5}$	29	$9.52 \cdot 10^{-5}$	15	$1.6 \cdot 10^4$	14
AVE					$1.46 \cdot 10^4$	14

Each data point corresponds to the average of 5 subsequent measurements and ionic conductivity σ is determinded from *IQS/FRS* by Eq. 6. AVE – weighted average.

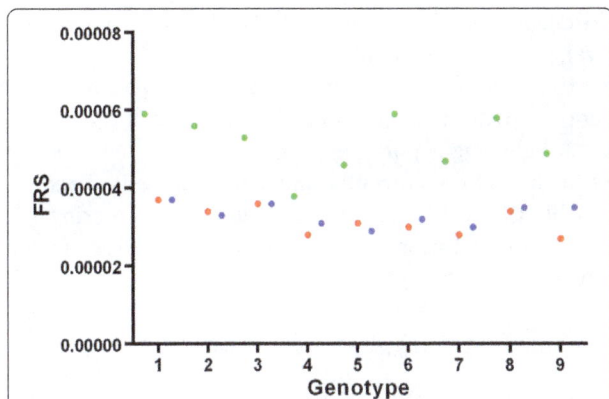

Figure 4 Analysis of wheat leaves from nine genotypes after 15 days of salt stress. Green dots represent control, red and blue dot leaves stressed with 100 mM NaCl and 50 mM Na$_2$SO$_4$, respectively. X-axes represent the genotypes analyzed (numbered 1–9) and the y-axes the corresponding FRS values.

greenhouse conditions in pots filled with soil (clay peat mix) and watered regularly.

For the salt stress experiment, nine wheat genotypes were grown in three replicates in aerated hydroponic system (unpublished data). The tested wheat genotypes were Zentos, Syn086 and 7 progenies of the cross between Zentos and Syn086 [29] which were selected based on their performance under salinity stress, representing salt tolerant and salt sensitive genotypes.

The stress was induced by adding to the nutritional solution either NaCl or Na$_2$SO$_4$, to end-concentration of 100 mM and 50 mM, respectively.

EC at control = 2.5 mS, NaCl = 11.5 mS, Na$_2$SO$_4$ = 9.5 mS), pH was checked every day and adjusted at 6.1

to 6.4. The stress was induced at three leave developmental stage (BBCH 13) and lasted for 15 days.

Measurements of salt stressed plants using the microwave cavity technique

The measurements were performed using a prototype of EMISENS's dual-mode sensor system, which was purpose-designed for this study. The quantities *IQS* and *FRS* were determined from the resonant frequency and *Q* factor, as determined by a fit of a Lorentzian to the measured resonant curves displayed in Figure 6. The leaves were pressed by a transparent plastic cover against the aperture of the dual-mode cavity. In case of small leaves which do not fully cover the aperture the position was optimized for maximum *FRS*, which corresponds to the alignment of an elongated leaf (like the one depicted in in Figure 1) perpendicular to the radial metallic rod at a distance from the center corresponding to about half the radius of the dielectric resonator.

For measurements on different leaves of one plant species care was taken to ensure that nearly identical measurement positions were used. The plants were removed from the hydroponic boxes and one leaf of them was placed on the window of the sensor (Figure 1). Five measurements were performed for each leave without changing the position (technical replicates). Immediately after, the undamaged plants were returned into the hydroponic vessels.

Measurements of water content

In order to follow the kinetics of water content the measurements were performed on detached leaves from the corresponding plants.

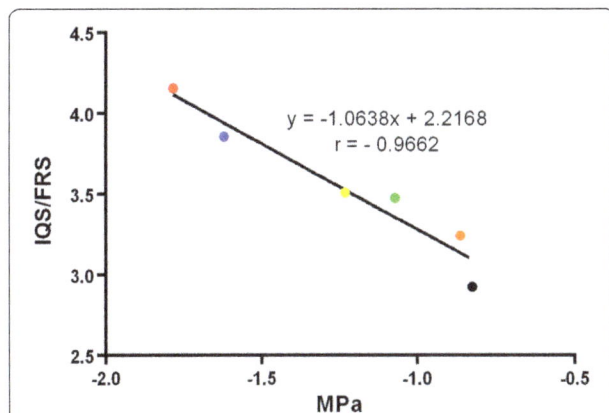

Figure 5 Correlation between osmotic potential and IQS/FRS values for canola leaves. Dot-colors indicate measurement of leaves with water content decreasing stepwise: black – 100% (initial fresh weight); brown – 90%, green – 80%; yellow – 70%; blue – 60% and red – 50% of the initial weight. Each dot represents the mean values from four leaves. For each leave, two measurements were performed to define osmotic potential and five for the *IQS/FRS* value.

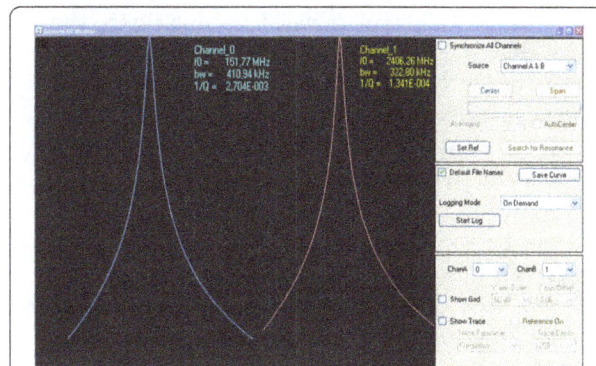

Figure 6 Screenshot of the user interface of the dual-mode sensor system. Blue and red curves display the measured resonance of a measurement of modes 0 and 1, respectively. Shown are the values of resonance frequency (f0), inverse Q factor and 3 dB bandwidth bw = f_0/Q for both modes, as determined by a Lorentz fit (Eq. 2). The axis of these plots (horizontal = frequency), (vertical = detector voltage) are not depicted on the screenshot, the control panel on the right hand side of the screen shot is not relevant for the presented analysis.

The stepwise reduction of water content in leaves was achieved by incubating them at high temperatures. The gravimetric measurement of water loss in the leaves was done by weighting them before and after drying. Shortly, after removal from the plant the leaves were weighted and measured with the microwave sensor system. This first time point was considered as reference for a leaf with 100% (w/w) water. After that, the leaves were placed in an incubator at 45°C until 10% of initial water content was lost and the microwave assessment was performed instantaneously. The drying procedure with 10% loss each step and subsequent microwave measurement was repeated 5 times until reduction to 50% of the initial weight.

Measurement of the osmotic potential

Leaves of canola plants were detached and after the microwave measurements they were analyzed with respect to their osmotic potential. This was repeated for each step of water reduction as described above. The sap of the leaves was extracted by squeezing them using a garlic presser. Fifteen μl sap-solution was employed to define the osmotic potential using an Osmomat (Osmomat 030-D, Gonotec GmbH, Berlin, Germany). The conversion of the osmolality values (osmol/kg) in osmotic potential (MPa) as described by Pariyar, et al. [30].

Conclusions

We have demonstrated non-invasive assessment of the water content by an evanescent field microwave sensor at 2.4 GHz for four different species of plant leaves due to a comparative study with gravimetric data. Our approach was proven to be highly reproducible and applicable for leaves of various size, shape and thickness. The frequency shift versus water content curves are slightly sub linear for the larger leaves, which may result from the inhomogeneous water distribution in the veins. For canola leaves, a strong correlation between the measured ratio of loss and frequency shift data to the osmotic potential was found, which indicates that the microwave method can be used for contact-free assessment of the osmolytes status of a plant. Due to the combination of a microwave ($f = 2.5$ GHz) and a sub-microwave frequency ($f = 150$ MHz) in one sensor device the method has a strong potential for simultaneous non-invasive assessment of water and salt status in a single leaf under test.

For the future, a down-scaled system operated at higher frequencies may be developed in order to achieve a higher reproducibility for the assessment of smaller leaves. The optimization of the design of the dual mode sensor and a further refinement of the electronic modules and the employed algorithm for accurate measurements of small changes of the resonant parameters should enable the simultaneous study of water content and average mineral content.

We expect that our technique may advance to a standard tool for hydration monitoring in plants in the near future. A lightweight portable version for assessment of plants in the field is currently under development. This may enable the realization of knowledge-based watering systems as integral procedure of precision agriculture in the future.

Competing interests

The authors declare that they have no competing interests.

Authors' contributions

SD, NK, JL and AB designed the study, analyzed and interpreted the data and drafted the manuscript. AK, HR, SC, EM, UB and NK developed and constructed the resonator, the electronic module and the software. SD, SA, BH performed the measurements with the plants using the microwave resonator. All authors read and approved the final manuscript.

Acknowledgements

The work has been done in laboratories of EMISENS Company and of INRES-Plant Breeding, University of Bonn. We thank EU for financial support in frame of network "CROP.SENSE.net" (EFRES grant Nr. z1011bc001 and BMZ for Project no.: 09.7860.1-001.00.

Author details

[1]INRES-Plant Breeding, University of Bonn, Katzenburgweg 5, 53115 Bonn, Germany. [2]EMISENS GmbH, Zur Rur 25, 52428 Juelich, Germany. [3]Department of Materials, Imperial College London, South Kensington Campus, London SW7 2AZ, UK. [4]Dr.- Ing. Elmar Menzel Ingenieurbüro, Birkenstr. 18, 63533 Mainhausen, Germany.

References

1. Khan MA, Ashraf MY, Mujtaba SM, Shirazi MU, Khan MA, Shereen A, et al. Evaluation of high yielding canola type brassica genotypes/mutants for drought tolerance using physiological indices as screening tool. Pak J Bot. 2010;42(6):3807–16.
2. Lugojan C, Ciulca S. Evaluation of relative water content in winter wheat. Journal of Horticulture, Forestry and Biotechnology. 2011;15(2):173–7.
3. Lawlor DW, Tezara W. Causes of decreased photosynthetic rate and metabolic capacity in water-deficient leaf cells: a critical evaluation of mechanisms and integration of processes. Ann Bot. 2009;103(4):561–79.
4. Brestic M, Zivcak M. PSII fluorescence techniques for measurement of drought and high temperature stress signal in crop plants: protocols and applications. In: Molecular Stress Physiology of Plants. Springer; 2013: 87–131.
5. Sade B, Soylu S, Soylu E. Drought and oxidative stress. Afr J Biotechnol. 2013;10(54):11102–9.
6. Trnka M, Eitzinger J, Dubrovský M, Semerádová D, Štěpánek P, Hlavinka P, et al. Is rainfed crop production in central Europe at risk? Using a regional climate model to produce high resolution agroclimatic information for decision makers. J Agric Sci. 2010;148(06):639–56.
7. Entrup NL, Berendonk C, Demmel M, Dietzsch H, Dissemond A, Estler M, Haumann G, Herrmann A, Hochberg H, Holtschulte B. Lehrbuch des Pflanzenbaues: Kulturpflanzen/Hrsg.: Norbert Lütke Entrup; Bernhard Carl Schäfer: AgroConcept; 2011.
8. Born N, Behringer D, Liepelt S, Beyer S, Schwerdtfeger M, Ziegenhagen B, et al. Monitoring plant drought stress response using terahertz time-domain spectroscopy. Plant Physiol. 2014;164(4):1571–7.
9. Jordens C, Scheller M, Breitenstein B, Selmar D, Koch M. Evaluation of leaf water status by means of permittivity at terahertz frequencies. J Biol Phys. 2009;35(3):255–64.
10. Menzel MI, Tittmann S, Buehler J, Preis S, Wolters N, Jahnke S, et al. Non-invasive determination of plant biomass with microwave resonators. Plant Cell Environ. 2009;32(4):368–79.

11. Ferrazzoli P, Paloscia S, Pampaloni P, Schiavon G, Solimini D, Coppo P. Sensitivity of microwave measurements to vegetation biomass and soil moisture content: A case study. Geoscience and Remote Sensing, IEEE Transactions on. 1992;30(4):750–6.

12. Castro-Camus E, Palomar M, Covarrubias A. Leaf water dynamics of Arabidopsis thaliana monitored in-vivo using terahertz time-domain spectroscopy. Sci Report. 2012;3:2910–0.

13. Gente R, Born N, Voß N, Sannemann W, Léon J, Koch M, et al. Determination of leaf water content from terahertz time-domain spectroscopic data. Journal of Infrared, Millimeter, and Terahertz Waves. 2013;34(3–4):316–23.

14. Klein N, Vitusevich S, Danylyuk S. Resonator arrangement and method for analyzing a sample using the resonator arrangement. Alexandria VA: U.S. Patent No. 8,410,792. 2; 2013.

15. Rezaei M, Ebrahimi E, Naseh S, Mohajerpour M. A new 1.4-GHz soil moisture sensor. Measurement. 2012;45(7):1723–8.

16. Sancho-Knapik D, Gismero J, Asensio A, Peguero-Pina JJ, Fernández V, Alvarez-Arenas TG, et al. Microwave l-band (1730MHz) accurately estimates the relative water content in poplar leaves. A comparison with a near infrared water index (R 1300 /R 1450). Agr Forest Meteorol. 2011;151(7):827–32.

17. Shry C, Reiley E. Introductory horticulture. New York: Cengage Learning; 2010.

18. Kaatze U. The dielectric properties of water in its different states of interaction. J Solut Chem. 1997;26(11):1049–112.

19. Stogryn A. Equations for calculating the dielectric constant of saline water (correspondence). Microwave Theory and Techniques, IEEE Transactions on. 1971;19(8):733–6.

20. Barthel J, Buchner R. High-frequency permittivity and its use in the investigation of solution properties. Pure Appl Chem. 1991;63(10):1473–82.

21. Ulaby FT, Jedlicka R. Microwave dielectric properties of plant materials. Geoscience and Remote Sensing, IEEE Transactions on. 1984;4(4):406–15.

22. Gillon P, Kajfez D. Dielectric resonators. Atlanta: Noble; 1998.

23. Pozar DM. Microwave engineering, Ch. 8. New York: Wiley; 1998.

24. Studio CM. Computer simulation technology. Darmstadt, Germany: GmbH; 2009.

25. Nandi N, Bhattacharyya K, Bagchi B. Dielectric relaxation and solvation dynamics of water in complex chemical and biological systems. Chem Rev. 2000;100(6):2013–46.

26. Basey-Fisher TH, Guerra N, Triulzi C, Gregory A, Hanham SM, Stevens MM, et al. Microwaving blood as a non-destructive technique for haemoglobin measurements on microlitre samples. Adv Healthcare Mater. 2014;3(4):536–42.

27. Gorham J, Jones RW, McDonnell E. Some mechanisms of salt tolerance in crop plants. In: Biosalinity in Action: Bioproduction with Saline Water. Netherlands: Springer; 1985. p. 15–40.

28. Munns R, Tester M. Mechanisms of salinity tolerance. Annu Rev Plant Biol. 2008;59:651–81.

29. Kunert A, Naz AA, Dedeck O, Pillen K, Léon J. AB-QTL analysis in winter wheat: I. Synthetic hexaploid wheat (T. turgidum ssp. dicoccoides × T. tauschii) as a source of favourable alleles for milling and baking quality traits. Theor Appl Genet. 2007;115(5):683–95.

30. Pariyar S, Eichert T, Goldbach HE, Hunsche M, Burkhardt J. The exclusion of ambient aerosols changes the water relations of sunflower (Helianthus annuus) and bean (Vicia faba) plants. Environ Exp Bot. 2013;88:43–52.

Permissions

All chapters in this book were first published in Plant Methods, by BioMed Central; hereby published with permission under the Creative Commons Attribution License or equivalent. Every chapter published in this book has been scrutinized by our experts. Their significance has been extensively debated. The topics covered herein carry significant findings which will fuel the growth of the discipline. They may even be implemented as practical applications or may be referred to as a beginning point for another development.

The contributors of this book come from diverse backgrounds, making this book a truly international effort. This book will bring forth new frontiers with its revolutionizing research information and detailed analysis of the nascent developments around the world.

We would like to thank all the contributing authors for lending their expertise to make the book truly unique. They have played a crucial role in the development of this book. Without their invaluable contributions this book wouldn't have been possible. They have made vital efforts to compile up to date information on the varied aspects of this subject to make this book a valuable addition to the collection of many professionals and students.

This book was conceptualized with the vision of imparting up-to-date information and advanced data in this field. To ensure the same, a matchless editorial board was set up. Every individual on the board went through rigorous rounds of assessment to prove their worth. After which they invested a large part of their time researching and compiling the most relevant data for our readers.

The editorial board has been involved in producing this book since its inception. They have spent rigorous hours researching and exploring the diverse topics which have resulted in the successful publishing of this book. They have passed on their knowledge of decades through this book. To expedite this challenging task, the publisher supported the team at every step. A small team of assistant editors was also appointed to further simplify the editing procedure and attain best results for the readers.

Apart from the editorial board, the designing team has also invested a significant amount of their time in understanding the subject and creating the most relevant covers. They scrutinized every image to scout for the most suitable representation of the subject and create an appropriate cover for the book.

The publishing team has been an ardent support to the editorial, designing and production team. Their endless efforts to recruit the best for this project, has resulted in the accomplishment of this book. They are a veteran in the field of academics and their pool of knowledge is as vast as their experience in printing. Their expertise and guidance has proved useful at every step. Their uncompromising quality standards have made this book an exceptional effort. Their encouragement from time to time has been an inspiration for everyone.

The publisher and the editorial board hope that this book will prove to be a valuable piece of knowledge for researchers, students, practitioners and scholars across the globe.

List of Contributors

Alessandra Rocchetti
Biological and Medical Sciences, Oxford Brookes University, Oxford OX3 0BP, UK

Chris Hawes
Biological and Medical Sciences, Oxford Brookes University, Oxford OX3 0BP, UK

Verena Kriechbaumer
Biological and Medical Sciences, Oxford Brookes University, Oxford OX3 0BP, UK

Fabio Pasin
Centro Nacional de Biotecnología (CNB-CSIC), Darwin 3, Madrid 28049, Spain

Satish Kulasekaran
Centro Nacional de Biotecnología (CNB-CSIC), Darwin 3, Madrid 28049, Spain

Paolo Natale
Centro Nacional de Biotecnología (CNB-CSIC), Darwin 3, Madrid 28049, Spain

Carmen Simón-Mateo
Centro Nacional de Biotecnología (CNB-CSIC), Darwin 3, Madrid 28049, Spain

Juan Antonio García
Centro Nacional de Biotecnología (CNB-CSIC), Darwin 3, Madrid 28049, Spain

Martin T Jahn
School of Environmental Sciences, University of East Anglia, Norwich Research Park, Norwich NR4 7TJ, UK
Department of Botany II, Julius-Maximilians University Würzburg, Julius-von-Sachs-Platz 3, 97082 Würzburg, Germany

Katrin Schmidt
School of Environmental Sciences, University of East Anglia, Norwich Research Park, Norwich NR4 7TJ, UK

Thomas Mock
School of Environmental Sciences, University of East Anglia, Norwich Research Park, Norwich NR4 7TJ, UK

Kerry O'Donnelly
Institute of Chemical Biology, Department of Chemistry, Imperial College, Flowers Building, South Kensington Campus, Exhibition Road, London SW7 2AZ, UK

Guangyuan Zhao
Department of Chemistry, Imperial College, South Kensington Campus, Exhibition Road, London SW7 2AZ, UK

Priya Patel
Department of Chemistry, Imperial College, South Kensington Campus, Exhibition Road, London SW7 2AZ, UK

M Salman Butt
Institute of Chemical Biology, Department of Chemistry, Imperial College, Flowers Building, South Kensington Campus, Exhibition Road, London SW7 2AZ, UK

Lok Hang Mak
Department of Chemistry, Imperial College, South Kensington Campus, Exhibition Road, London SW7 2AZ, UK

Simon Kretschmer
Department of Chemistry, Imperial College, South Kensington Campus, Exhibition Road, London SW7 2AZ, UK

Rudiger Woscholski
Institute of Chemical Biology, Department of Chemistry, Imperial College, Flowers Building, South Kensington Campus, Exhibition Road, London SW7 2AZ, UK

Laura M C Barter
Institute of Chemical Biology, Department of Chemistry, Imperial College, Flowers Building, South Kensington Campus, Exhibition Road, London SW7 2AZ, UK

Anamika Mishra
Global Change Research Centre, Academy of Sciences of the Czech Republic, v. v. i, Bělidla 986/4a, 603 00, Brno, Czech Republic

Arnd G Heyer
Institute of Biomaterials and Biomolecular Systems, Department of Plant Biotechnology, University of Stuttgart, Stuttgart, Germany

Kumud B Mishra
Global Change Research Centre, Academy of Sciences of the Czech Republic, v. v. i, Bělidla 986/4a, 603 00, Brno, Czech Republic

Christoph-Martin Geilfus
Institute of Plant Nutrition and Soil Science, Christian-Albrechts-Universität zu Kiel, Hermann-Rodewald-Str. 2, 24118 Kiel, Germany

Karl H Mühling
Institute of Plant Nutrition and Soil Science, Christian-Albrechts-Universität zu Kiel, Hermann-Rodewald-Str. 2, 24118 Kiel, Germany

Hartmut Kaiser
Botanisches Institut, Christian-Albrechts-Universität zu Kiel, Am Botanischen Garten 3-9, 24118 Kiel, Germany

Christoph Plieth
Zentrum für Biochemie und Molekularbiologie, Christian-Albrechts-Universität zu Kiel, Am Botanischen Garten 3-9, 24118 Kiel, Germany

Gunnar Huep
Center for Biotechnology & Department of Biology, Bielefeld University, Universitaetsstrasse 25, D-33615 Bielefeld, Germany

Nils Kleinboelting
Center for Biotechnology & Department of Biology, Bielefeld University, Universitaetsstrasse 25, D-33615 Bielefeld, Germany

Bernd Weisshaar
Center for Biotechnology & Department of Biology, Bielefeld University, Universitaetsstrasse 25, D-33615 Bielefeld, Germany

Lachlan James Palmer
School of Biological Science, Flinders University, Bedford Park, South Australia 5042, Australia

Daniel Anthony Dias
Metabolomics Australia, School of Botany, The University of Melbourne, Parkville, Melbourne 3010, Victoria, Australia

Berin Boughton
Metabolomics Australia, School of Botany, The University of Melbourne, Parkville, Melbourne 3010, Victoria, Australia

Ute Roessner
Metabolomics Australia, School of Botany, The University of Melbourne, Parkville, Melbourne 3010, Victoria, Australia

Robin David Graham
School of Biological Science, Flinders University, Bedford Park, South Australia 5042, Australia

James Constantine Roy Stangoulis
School of Biological Science, Flinders University, Bedford Park, South Australia 5042, Australia

Jun Ding
Key Laboratory of Analytical Chemistry for Biology and Medicine (Ministry of Education), Department of Chemistry, Wuhan University, Wuhan 430072, China
Chinese Acad Sci, Key Lab Plant Germplasm Enhancement & Specialty A, Wuhan Bot Garden, Wuhan 430074, China

Jian-Hong Wu
College of Chemical Engineering, Wuhan Textile University, Wuhan 430200, China

Jiu-Feng Liu
Key Laboratory of Analytical Chemistry for Biology and Medicine (Ministry of Education), Department of Chemistry, Wuhan University, Wuhan 430072, China

Bi-Feng Yuan
Key Laboratory of Analytical Chemistry for Biology and Medicine (Ministry of Education), Department of Chemistry, Wuhan University, Wuhan 430072, China

Yu-Qi Feng
Key Laboratory of Analytical Chemistry for Biology and Medicine (Ministry of Education), Department of Chemistry, Wuhan University, Wuhan 430072, China

Patti E Stronghill
Department of Biology, University of Toronto, 1265 Military Trail, Scarborough, Canada

Wajma Azimi
Kingston General Hospital, Kingston, Canada

Clare A Hasenkampf
Department of Biology, University of Toronto, 1265 Military Trail, Scarborough, Canada

Alex P Whan
CSIRO Plant Industry, GPO Box 1600, Canberra ACT 2601, Australia

Alison B Smith
National Institute for Applied Statistics and Research Australia, Univeristy of Wollongong, Wollongong NSW 2522, Australia

Colin R Cavanagh
CSIRO Plant Industry, GPO Box 1600, Canberra ACT 2601, Australia

Jean-Philippe F Ral
CSIRO Plant Industry, GPO Box 1600, Canberra ACT 2601, Australia

Lindsay M Shaw
CSIRO Plant Industry, GPO Box 1600, Canberra ACT 2601, Australia

Crispin A Howitt
CSIRO Plant Industry, GPO Box 1600, Canberra ACT 2601, Australia

Leanne Bischof
CSIRO Computational Informatics, North Ryde NSW 2113, Australia

David Rousseau
Université de Lyon, Laboratoire CREATIS, CNRS, UMR5220, INSERM, U1044, Université Lyon 1, INSA-Lyon, Villeurbanne, France

Yann Chéné
Laboratoire Angevin de Recherche en Ingénierie des Systèmes (LARIS), Université d'Angers, 62 avenue Notre Dame du Lac, 49000 Angers, France

Etienne Belin
Laboratoire Angevin de Recherche en Ingénierie des Systèmes (LARIS), Université d'Angers, 62 avenue Notre Dame du Lac, 49000 Angers, France

Georges Semaan
Laboratoire Angevin de Recherche en Ingénierie des Systèmes (LARIS), Université d'Angers, 62 avenue Notre Dame du Lac, 49000 Angers, France

Ghassen Trigui
GEVES, Station Nationale d'Essais de Semences (SNES), rue Georges Morel, 49071 Beaucouzé, France

Karima Boudehri
GEVES, Station Nationale d'Essais de Semences (SNES), rue Georges Morel, 49071 Beaucouzé, France

Florence Franconi
La Plateforme d'Ingénierie et Analyses Moléculaires (PIAM), Université d'Angers, 49000 Angers, France

François Chapeau-Blondeau
Laboratoire Angevin de Recherche en Ingénierie des Systèmes (LARIS), Université d'Angers, 62 avenue Notre Dame du Lac, 49000 Angers, France

Frank Liebisch
Institute of Agricultural Sciences, ETH Zürich, Universitätstrasse 2, 8092 Zürich, Switzerland

Norbert Kirchgessner
Institute of Agricultural Sciences, ETH Zürich, Universitätstrasse 2, 8092 Zürich, Switzerland

David Schneider
Norddeutsche Pflanzenzucht, Hohenlieth Holtsee D-24363, Germany

Achim Walter
Institute of Agricultural Sciences, ETH Zürich, Universitätstrasse 2, 8092 Zürich, Switzerland

Andreas Hund
Institute of Agricultural Sciences, ETH Zürich, Universitätstrasse 2, 8092 Zürich, Switzerland

Wei Guo
Institute for Sustainable Agro-ecosystem Services, Graduate School of Agricultural and Life Sciences, The University of Tokyo, 1-1-1. Midori-cho, Nishi-Tokyo, Tokyo 188-0002, Japan

Tokihiro Fukatsu
National Agriculture and Food Research Organization, 3-1-1 Kannondai, Tsukuba, Ibaraki 305-8666, Japan

Seishi Ninomiya
Institute for Sustainable Agro-ecosystem Services, Graduate School of Agricultural and Life Sciences, The University of Tokyo, 1-1-1. Midori-cho, Nishi-Tokyo, Tokyo 188-0002, Japan

Fang Wang
College of Science, Hunan Agricultural University, Changsha 410128, China

Deng-wen Liao
Forestry Department of Hunan Province, Quality Testing and Inspection Centre of Forest Products, Changsha 410007, China

Jin-wei Li
Agricultural Information Institute, Hunan Agricultural University, Changsha 410128, China

Gui-ping Liao
Agricultural Information Institute, Hunan Agricultural University, Changsha 410128, China

Monika Chylińska
Institute of Agrophysics, Polish Academy of Sciences, Doswiadczalna 4, 20-290 Lublin, Poland

Monika Szymańska-Chargot
Institute of Agrophysics, Polish Academy of Sciences, Doswiadczalna 4, 20-290 Lublin, Poland

Artur Zdunek
Institute of Agrophysics, Polish Academy of Sciences, Doswiadczalna 4, 20-290 Lublin, Poland

Prateek Gupta
Repository of Tomato Genomics Resources, Department
of Plant Sciences, School of Life Sciences, University of
Hyderabad, Hyderabad 500046, India

Yellamaraju Sreelakshmi
Repository of Tomato Genomics Resources, Department
of Plant Sciences, School of Life Sciences, University of
Hyderabad, Hyderabad 500046, India

Rameshwar Sharma
Repository of Tomato Genomics Resources, Department
of Plant Sciences, School of Life Sciences, University of
Hyderabad, Hyderabad 500046, India

Ralf Gente
Faculty of Physics and Material Sciences Center, Philipps-
Universität Marburg, Renthof 5, 35032 Marburg, Germany

Martin Koch
Faculty of Physics and Material Sciences Center, Philipps-
Universität Marburg, Renthof 5, 35032 Marburg, Germany

Cecile AI Richard
The University of Queensland, QAAFI, St Lucia, QLD
4072, Australia

Lee T Hickey
The University of Queensland, QAAFI, St Lucia, QLD
4072, Australia

Susan Fletcher
Department of Agriculture, Fisheries and Forestry, Leslie
Research Facility, Toowoomba, QLD 4350, Australia

Raeleen Jennings
Department of Agriculture, Fisheries and Forestry, Leslie
Research Facility, Toowoomba, QLD 4350, Australia

Karine Chenu
The University of Queensland, QAAFI, 203 Tor Street,
Toowoomba, QLD 4350, Australia

Jack T Christopher
The University of Queensland, QAAFI, Leslie Research
Facility, Toowoomba, QLD 4350, Australia

Achim Walter
Institute of Agricultural Sciences, ETH Zürich,
Universitätstrasse 2, 8092 Zürich, Switzerland

Frank Liebisch
Institute of Agricultural Sciences, ETH Zürich,
Universitätstrasse 2, 8092 Zürich, Switzerland

Andreas Hund
Institute of Agricultural Sciences, ETH Zürich,
Universitätstrasse 2, 8092 Zürich, Switzerland

Said Dadshani
INRES-Plant Breeding, University of Bonn,
Katzenburgweg 5, 53115 Bonn, Germany

Andriy Kurakin
EMISENS GmbH, Zur Rur 25, 52428 Juelich, Germany

Shukhrat Amanov
INRES-Plant Breeding, University of Bonn,
Katzenburgweg 5, 53115 Bonn, Germany

Benedikt Hein
INRES-Plant Breeding, University of Bonn,
Katzenburgweg 5, 53115 Bonn, Germany

Heinz Rongen
EMISENS GmbH, Zur Rur 25, 52428 Juelich, Germany

Steve Cranstone
EMISENS GmbH, Zur Rur 25, 52428 Juelich, Germany

Ulrich Blievernicht
EMISENS GmbH, Zur Rur 25, 52428 Juelich, Germany

Elmar Menzel
Dr.- Ing. Elmar Menzel Ingenieurbüro, Birkenstr. 18,
63533 Mainhausen, Germany

Jens Léon
INRES-Plant Breeding, University of Bonn,
Katzenburgweg 5, 53115 Bonn, Germany

Norbert Klein
EMISENS GmbH, Zur Rur 25, 52428 Juelich, Germany
Department of Materials, Imperial College London, South
Kensington Campus, London SW7 2AZ, UK

Agim Ballvora
INRES-Plant Breeding, University of Bonn,
Katzenburgweg 5, 53115 Bonn, Germany